Mammal Phylogeny

Mesozoic Differentiation, Multituberculates, Monotremes, Early Therians, and Marsupials

The family tree of mammals, 1932. Courtesy American Museum of Natural History, New York, New York.

Frederick S. Szalay   Michael J. Novacek   Malcolm C. McKenna
Editors

Mammal Phylogeny

Mesozoic Differentiation, Multituberculates, Monotremes, Early Therians, and Marsupials

With 115 Illustrations in 288 Parts

Springer-Verlag
New York   Berlin   Heidelberg   London
Paris   Tokyo   Hong Kong   Barcelona   Budapest

Frederick S. Szalay
Departments of Anthropology and
  Ecology and Evolutionary Biology
Hunter College and
  CUNY Graduate Center
New York, NY 10021, USA

Michael J. Novacek
Department of Vertebrate Paleontology
American Museum of Natural History
New York, NY 10024, USA

Malcolm C. McKenna
Department of Vertebrate Paleontology
American Museum of Natural History
New York, NY 10024, USA

Cover illustration: Marsupial neonate attached to nipple shortly after birth. Drawing by Nancy Hong, adapted from Hill and Hill, *Trans. Zool. Soc. Lond.* Vol. 28, 1955. Illustration courtesy of W. Patrick Luckett.

Library of Congress Cataloging-in-Publication Data
Mammal phylogeny / Frederick S. Szalay, Michael J. Novacek, Malcolm C.
  McKenna, editors.
    p.   cm.
    Includes bibliographical references and index.
    Contents: v. 1. Mesozoic differentiation, multituberculates,
  monotremes, early therians, and marsupials — v. 2. Placentals.
    ISBN 0-387-97854-2 (Springer-Verlag New York Berlin Heidelberg :
  v. 1 : alk, paper). — ISBN 3-540-97854-2 (Springer-Verlag Berlin
  New York Heidelberg : v. 1. : alk. paper). — ISBN 0-387-97853-4
  (Springer-Verlag New York Berlin Heidelberg : v. 2. : alk. paper). —
  ISBN 3-540-97853-4 (Springer-Verlag Berlin New York Heidelberg : v.
  2. : alk. paper)
    1. Mammals—Phylogeny—Congresses.   I. Szalay, Frederick S.
  II. Novacek, Michael J.   III. McKenna, Malcolm C.   IV. NATO.
  QL708.5.M36   1992
  599′.038—dc20   91-29232
                                                                  CIP

Printed on acid-free paper.

© 1993 Springer-Verlag New York, Inc.

All rights reserved. This work may not be translated or copied in whole or in part without the written permission of the publisher (Springer-Verlag New York, Inc., 175 Fifth Avenue, New York, NY 10010 USA), except for brief excerpts in connection with reviews or scholarly analysis. Use in connection with any form of information storage and retrieval, electronic adaptation, computer software, or by similar or dissimilar methodology now known or hereafter developed is forbidden.
The use of general descriptive names, trade names, trademarks, etc., in this publication, even if the former are not especially identified, is not to be taken as a sign that such names, as understood by the Trade Marks and Merchandise Act, may accordingly be used freely by anyone. Materials prepared by individuals as part of their official duties as U.S. government employees are not covered by copyright.

Production managed by Terry Kornak and coordinated by Faye Zucker Editorial Service. Manufacturing supervised by Jacqui Ashri.
Typesetting by Asco Trade Typesetting Ltd., North Point, Hong Kong.
Printed and bound by Edwards Brothers, Ann Arbor, MI.
Printed in the United States of America.

9 8 7 6 5 4 3 2 1

Mammal Phylogeny: Mesozoic Differentiation, Multituberculates, Monotremes, Early Therians, and Marsupials
ISBN 0-387-97854-2 Springer-Verlag New York Berlin Heidelberg
ISBN 3-540-97854-2 Springer-Verlag Berlin Heidelberg New York

Mammal Phylogeny: Placentals
ISBN 0-387-97853-4 Springer-Verlag New York Berlin Heidelberg
ISBN 3-540-97853-4 Springer-Verlag Berlin Heidelberg New York

Mammal Phylogeny, two-volume set
ISBN 0-387-97676-0 Springer-Verlag New York Berlin Heidelberg
ISBN 3-540-97676-0 Springer-Verlag Berlin Heidelberg New York

Preface

The roots of this book and its sister volume, *Mammal Phylogeny: Placentals*, go back to discussions and plans, shelved for a while, between F.S. Szalay and W.P. Luckett during the international and multidisciplinary symposium on rodent evolution sponsored by NATO, July 2–6, 1984, in Paris. That conference, organized by W.P. Luckett and J.-L. Hartenberger, the proceedings of which were published in 1985, proved an inspiring experience to all of the participants, as this was repeatedly expressed both during and after the meetings. In addition to issues relating to rodents, general theoretical topics pertaining to the evolutionary biology and systematics of other groups of mammals regularly surfaced during the presentations and discussions. M.J. Novacek, who was also a participant in the rodent symposium, shared with Luckett and Szalay the enthusiasm acquired there, and he also expressed strong interest for a meeting on mammal evolution with a general focus similar to that of the rodent gathering.

In 1988, Szalay and Luckett, after having planned in detail a program, direction, and core list of participants, were awarded a $30,000 grant by the Alfred P. Sloan Foundation through the Research Foundation of the City University of New York. The grant was contingent upon obtaining additional funds sufficient to assure that the symposium would be held. Raising the remaining funds proved to be a problem. Pat Luckett, for reasons of "conference organization fatigue," stepped back from the project and decisively directed Szalay to approach the American Museum of Natural History for the remaining funds and facilities. M.J. Novacek, Chair of Vertebrate Paleontology of the AMNH at that time, gave his support for the plans and secured matching funds and the excellent and atmospheric facilities of that institution for the meetings. Matching support came primarily from the Childs Frick Laboratory Endowment, which has long supported basic research in the field of vertebrate paleonotology. M.C. McKenna was asked to join the organizing committee, and a final list of participants, plans, and procedures were agreed to by Szalay, Novacek, and McKenna. The Mammal Conference, the basis of these volumes, was eventually held at the American Museum of Natural History, May 28–June 1, 1990.

The running of the conference was especially facilitated by three people who have rendered their talented services. Barbara Werscheck, administrative assistant to M.J. Novacek, was the perfect coordinator between the joint efforts of the "CUNY team" and the host institution. Sharon Saia, a CUNY graduate student at that time, did magnificent work, securing complex airline and hotel reservations for an international group of over fifty participants. Robert K. Costello, PhD candidate at CUNY, has been the indispensable right hand to Szalay, both in the running of the conference and in helping to edit this volume. To all three, the organizers of the Mammal Conference express their deeply felt thanks.

A special debt of gratitude is owed to the Alfred P. Sloan Foundation, the Childs Frick Laboratory Endowment, and the American Museum of Natural History for making the Mammal Conference and this ensuing volume possible.

This volume is dedicated to **W.P. Luckett** for helping to make the early, and therefore causally critical, stages of the Mammal Conference a reality, and for fostering the idea, for two decades, of a multidisciplinary approach to the study of mammalian evolutionary history.

F. S. SZALAY
M. J. NOVACEK
M. C. MCKENNA

Contents

Preface   *v*

Contributors   *ix*

1 Introduction   *1*
FREDERICK S. SZALAY

2 Ontogeny, Genetic Control, and Phylogeny of Female Reproduction in Monotreme and Therian Mammals   *4*
MARILYN B. RENFREE

3 Development and the Phylogenetic Features of the Middle Ear Region   *21*
R. PRESLEY

4 Relationships of the Liassic Mammals *Sinoconodon*, *Morganucodon oehleri*, and *Dinnetherium*   *30*
ALFRED W. CROMPTON AND ZHEXI LUO

5 Basicranial Evidence for Early Mammal Phylogeny   *45*
JOHN R. WIBLE AND JAMES A. HOPSON

6 Cranial Morphology and Multituberculate Relationships   *63*
DESUI MIAO

7 Reconsideration of Monotreme Relationships Based on the Skull and Dentition of the Miocene *Obdurodon dicksoni*   *75*
MICHAEL ARCHER, PETER MURRAY, SUZANNE HAND, AND HENK GODTHELP

8 Ontogenetic Evidence for Cranial Homologies in Monotremes and Therians, with Special Reference to *Ornithorhynchus*   *95*
ULRICH ZELLER

9 Pedal Evolution of Mammals in the Mesozoic: Tests for Taxic Relationships   *108*
FREDERICK S. SZALAY

10 Phylogenetic Systematics and the Early History of Mammals   *129*
TIMOTHY ROWE

11 Phylogeny of Multituberculata   *146*
NANCY B. SIMMONS

12 Cranial Morphology of the Therian Common Ancestor, as Suggested by the Adaptations of Neonate Marsupials   *165*
    WOLFGANG MAIER

13 An Ontogenetic Assessment of Dental Homologies in Therian Mammals   *182*
    W. PATRICK LUCKETT

14 Theria of Metatherian–Eutherian Grade and the Origin of Marsupials   *205*
    RICHARD L. CIFELLI

15 Metatherian Taxon Phylogeny: Evidence and Interpretation from the Cranioskeletal System   *216*
    FREDERICK S. SZALAY

Topic Index   *243*

Taxon Index   *247*

Contributors

Michael Archer, School of Biological Science, University of New South Wales, Kensington, New South Wales, Australia 2033.

Richard L. Cifelli, Oklahoma Museum of Natural History and Department of Zoology, University of Oklahoma, Norman, OK 73019, USA.

Alfred W. Crompton, Museum of Comparative Zoology, Harvard University, Cambridge, MA 02138, USA.

Henk Godthelp, School of Biological Science, University of New South Wales, Kensington, New South Wales, Australia 2033.

Suzanne Hand, School of Biological Science, University of New South Wales, Kensington, New South Wales, Australia 2033.

James A. Hopson, Department of Organismal Biology and Anatomy, University of Chicago, Chicago, IL 60637, USA.

W. Patrick Luckett, Department of Anatomy, University of Puerto Rico, Medical Science Campus, San Juan, Puerto Rico 00936-5067, USA.

Zhexi Luo, Department of Biology, College of Charleston, Charleston, SC 29424, USA.

Wolfgang Maier, Department of Systematic Zoology, Eberhard Karls-University, D-7400 Tübingen, Germany.

Malcolm C. McKenna, Department of Vertebrate Paleontology, American Museum of Natural History, New York, NY 10024, USA.

Desui Miao, Museum of Natural History, Department of Systematics and Ecology, The University of Kansas, Lawrence, KS 66045-2454, USA.

Peter Murray, Northern Territory Museum and Art Gallery, Alice Springs, Northern Territory, Australia 5750.

Michael J. Novacek, Department of Vertebrate Paleontology, American Museum of Natural History, New York, NY 10024, USA.

R. Presley, Department of Anatomy, University of Wales, College of Cardiff, Cardiff CF1 3YF, United Kingdom.

Marilyn B. Renfree, Department of Zoology, University of Melbourne, Parkville, Victoria 3150, Australia.

Timothy Rowe, Department of Geological Sciences, and Vertebrate Paleontology Laboratory, The University of Texas at Austin, Austin, TX 78712, USA.

Nancy B. Simmons, Department of Mammalogy, American Museum of Natural History, New York, NY 10024, USA.

Frederick S. Szalay, Departments of Anthropology and Ecology and Evolutionary Biology, Hunter College and CUNY Graduate Center, New York, NY 10021, USA.

John R. Wible, Department of Anatomical Sciences and Neurobiology, School of Medicine, University of Louisville, Louisville, KY 40292, USA.

Ulrich Zeller, Institute of Anatomy, University of Göttingen, 3400 Göttingen, Germany.

CHAPTER 1

Introduction

FREDERICK S. SZALAY

This book (along with its sister volume *Mammal Phylogeny: Placentals*) presents a series of authoritative and synthetic contributions to the understanding of all the major groups of the Mammalia, including the extinct taxa. The chapters in these two volumes present a state of phylogenetic affairs of the most critical features studied in mammalian phylogenetics, of the higher groupings themselves, as well as the critical issues of both conceptual and empirical methodologies. These latter are at the heart of not only phylogenetics but also of all science.

The multiple aims of presenting the most tested hypotheses of relationships of mammalian features and of the taxa, and openly airing the serious and unresolved issues relating to both the empirical aspects of phylogeny and the methodological disagreements, cannot be neatly accomplished. The process of evolution (the result of which is phylogeny) includes four distinct components, even if these are separable only for the necessary heuristics to carry out research. The phyletic (anagenetic, or patristic), cladistic (branching, or splitting) relationships, and the phenetic and temporal distances of the stages of lineages or of distinct lineages all require their specific testing procedures, each tied by invisible threads to the other three. All of this occured of course in a biogeographic context tied to the history of the earth itself.

Such conceptual foundations are not always considered in analyses of phylogenies. Added is the problem that the issue of causality and testing, the combined central goal of all science, has had a curious relationship to the theoretical structure of phylogenetic studies. If hypotheses do not contain a causality-directed structure in their testing procedures, then a number of challenging questions arise concerning the conceptual framework of the approach within which these are undertaken. Yet these issues, as hoped and anticipated, have arisen, even if indirectly sometimes and have taken on a reality during the conference and preparation of this volume, and are evident in many of the chapters.

In reading this volume, it should be recognized that the changing views of mammal relationships during the past decades are inseparable from a number of general issues bearing on the process of generating the new understanding. These may be grouped as: (a) new information from the fossil record, (b) new information from the study of many different kinds of evidence available from living mammals (c) connections made between evidence sampled from living animals to the (causally plausible) inferences from the extinct forms, (d) conceptual methodology, which strongly influences the way the objective evidence is ordered or the manner in which hypotheses are tested against it, and, finally, but perhaps most importantly, (e) whether the aim of the investigator is, primarily, to produce either a classification or a historical narrative explanation of specific monophyletic segments, or of entire clades. These issues enter into understanding phylogeny (i.e., evolutionary history) and were the general areas of contention, and therefore of potential progress, during the conference and the prepartion of this volume.

There are a number of specific areas of important developments which have had both stimulating and productive effects on the study of mammals and other organisms. These include the beginnings of an explicit enunciation of a conceptual approach, first formalized by Hennig (phylogenetic systematics, later to become cladistics); the increased clarification of the relationship of taxonomic theory to evolutionary theory and its potential to harmonize the somewhat variant approaches to phylogenetics; the increasing importance of molecular and genetic approaches to phylogeny; and a new recognition of the importance of evolutionary morphology (developmental, functional, and ecological) for linking the fossil record with the living. In testing hypotheses of phylogeny, the combined taxonomic, functional, and ecomorphological approaches to morphological information of both the fossils and of the living have proved themselves to be as powerful and "modern" as any of the other newer areas of inquiry.

Statements about the unreliability of morphology should be examined and comparatively evaluated in the marketplace of phylogenetic contributions. The enduring significance of the rigorous and (tested) theory-laden study of morphology in evolutionary biology is reflected in many of the chapters in this volume.

It became clear that while some called their conceptual methodology "more," or "less," or "the" phylogenetic, cladistic, or evolutionary one, the conceptual separation, upon closer analysis of the contributions, makes for a creative spectrum of approaches. All of the contributors to this volume are phylogeneticists. Cladistics is not the original Hennigian phylogenetics, and evolutionary systematics, as summarized in the past by Simpson or Mayr, has also been thoroughly influenced and changed since those prescriptions. Both general approaches are decidedly "phylogenetic," albeit with important fundamental disagreements between them. Products of these different perspectives on the evolutionary process, as those in all areas of science, will continue to be tested by others against new objective information—i.e., data that all can agree on as real. In spite of statements of "pure operational methodology" attached to cladistics, real issues of methodological contention always come down to fundamental disagreements between hidden evolutionary assumptions, usually having to do with evolutionary theory and its application in the formal treatment of characters of organisms, and with the taxa themselves.

This introduction is not the place to fully explore this general topic related to the historical analysis of organisms. Nevertheless the differences in conceptual methodologies are an ever present, and importantly explicated, issue throughout the chapters in this volume. The issues of a priori vs. a posteriori weighting of characters, and the reliance on probability-based judgments steeped in functional-adaptive biology vs. the rigorous dictates of parsimony-based algorithms for understanding character and whole animal group evolution cannot be "resolved" here. Perhaps they never can be. But while the drive for consensus and compromise is the proper aim of politics, it is doubtful that it has any conceptually or theoretically legitimate place in the frontiers of science.

There is a decidedly bimodal distribution in the focus of the chapters. While a number of contributors concentrated on better understanding a few selected areas of characters to obtain better tested hypotheses of their evolution against which taxon phylogenies could be tested, many others preferred to use larger numbers of less well understood characters to order taxa cladistically. All in all, a decided change in direction toward more holistic analysis of characters in this volume suggests some reemphasis in priorities in mammalian phylogenetics. It is not so much the novel trees that tend to capture interest, but rather the nature of the evidence, what is known about the characters and the plausibility of character transformations, against which these trees must be tested. Taxon phylogenies are often not resolved in a biologically (or even algorithmically) satisfactory manner. Nevertheless there can be no question that the contributors have not only added much new information to the communal storehouse, but that existing information has been more sharply defined and its relevance more fully explicated.

The nature of reproductive biology and its ontogeny are the sources of some of the oldest known and used traits in taxonomy, reaching back to Aristotle, which required description, functional explanation, and eventually Darwinian phylogenetic understanding in the history of the study of mammals and other vertebrates. All thoroughly tested phylogenetic hypotheses that deal with mammalian evolution (and eventually their classificatory expression) will have to consider the connections of reproductive strategies to other taxonomic properties. In chapter 2, Renfree summarizes the reproductive processes of the three major groups of living mammals, the monotremes, marsupials, and placentals. In the process she goes beyond this task and also reviews the most recent information on genetic control of these attributes, and discusses the phylogenetically and developmentally important features.

Pinpointing the specific origin (ancestry) and elucidating the transformation of the stages of evolution of character complexes of any group will continue to be the most challenging aspect of all future studies of mammalian phylogenetics, in addition to an understanding of cladistic relationships of taxa. While these interrelated aims of mammalian phylogenetics may be often elusive, at least the general ordering of traits and taxa of the most mammal-like cynodonts, and of the major branches of early mammals were given important focus by both the character-oriented and the taxon-oriented contributions.

Review, synthesis, and incorporation of relevant new material from the cranium of early mammals is the aim of chapters 3–8, the papers by Presley; Crompton and Luo; Wible and Hopson; Miao; Archer, Murray, Hand, and Godthelp; and Zeller. This variety of different cranial and dental evidence from advanced cynodonts to multituberculates, monotremes, the early (dentally) atribosphenic groups, and the ancestral (tribosphenic) therians has always been considered some of the most important for understanding early mammalian phylogeny. These studies exemplify the range of approaches in phylogenetic analysis, directed primarily at the evolution of characters, and they place the study of the mammalian cranium into a powerful paleontological, developmental, and phylogenetic perspective.

In chapter 9, Szalay departs from the cranial and dental analyses to examine the probabilities for specific character transformations of the known Mesozoic tarsal evidence, and tests a number of phylogenetic hypotheses of groups against the taxonomic properties generated in that analysis. Both Presley and Szalay in their respective chapters argue for the importance of character analysis (beyond distribution analysis) in order to attain understanding and confidence in the use of attributes as taxonomic properties, and suggest that causal research (whether character or taxon oriented) makes phylogenetics properly scientific.

In chapter 10, Rowe presents on overall cladistic view of some synapsid and early mammal relationships, and in chapter 11, Simmons focuses on multituberculate cladistics, with particular emphasis on algorithm-based analysis of characters.

In chapter 12, it is the form-function and biological role of marsupial neonates that is the basis for a stimulating phylogenetic hypothesis persuasively presented by Maier. He offers evidence on the shared similarities of early placental cranial development with those of the adaptively well understood pump-sucking marsupial neonates. This suggests to him a derived loss of constraints from the ontogenetic cranial differentiation of the earliest reproductively eutherian mammals once the demands of feeding in these were postponed to a later stage in ontogeny. This hypothesis is a powerful one for understanding the evolutionary transformation of the protoplacental cranial development from an ancestral condition represented by the early ontogeny of living marsupials.

In chapter 13, Luckett probes one of the oldest and most critical areas of paleomammalogy: the serial homology of teeth in metatherians and eutherians. His emphasis is on both the nature of ontogenetic information and its application to fossil and living mammals.

Marsupial phylogenetics are dealt with in chapters 14 and 15. Cifelli takes a synthetic view of the difficult dental groups of the Cretaceous therian mammals in order to establish which taxa are cladistically more recently related to marsupials than to other therians. This is a newly emerging and fascinating area of mammalian phylogenetics where the rapidly accumulating new discoveries of American, Mongolian, Polish, and Russian paleontologists will result in a complete revision of our ideas not only of early marsupial phylogenetics but also of the differentiation of the placentals and various other groups of therians. In the last chapter, Szalay briefly reviews the various areas of hard morphology of marsupials employed in phylogenetics and presents a synoptic phylogeny and an outline classification of the higher taxa. His aim is to diagnose (rather than to define) each monophyletic group by the derived traits of the postulated last common ancestor of the members. He makes a case for the inevitability of using both paraphyletic and holophyletic monophyla if taxonomists want to reflect both the cladistic and phyletic components of phylogeny in classifications.

CORRESPONDENCE ADDRESS. Frederick S. Szalay, Departments of Anthropology and Ecology and Evolutionary Biology, Hunter College, 695 Park Avenue, New York, NY 10021, USA.

CHAPTER 2

Ontogeny, Genetic Control, and Phylogeny of Female Reproduction in Monotreme and Therian Mammals

MARILYN B. RENFREE

Overview

In all living groups of mammals, prolonged protection of the developing young is a characteristic strategy of reproduction. Lactation, however, is the only uniquely mammalian characteristic, and all three groups have characteristic adaptations for achieving the transfer of energy from mother to young first via the uterus and placenta, and subsequently via the milk.

In mammals the simple oviduct of lower vertebrates has differentiated into the uterus and vagina. There is a tendency toward fusion of the posterior end of the paired oviducts. The vagina remains paired in monotremes and marsupials, but is a single fused tube in all eutherians; the uterus may be paired, partly fused, or completely fused as in humans. This fusion occurs independently in unrelated mammalian families as a correlate of reduction in litter size. The relative position of the ureters and the genital ducts is the character that most clearly separates the major groups of living mammals. The ureters of monotremes open into the urogenital sinus opposite the urethral openings of the bladder. In therians, the ureters migrate from a dorsal position adjacent to the Wolffian duct to a direct connection to the bladder, but the ureters pass medially between the genital duct in marsupials, not laterally as in eutherians. The conclusion must be that the original selection for the migration of the urinary ducts to the bladder in therian mammals was for an excretory and not a reproductive function, though the subsequent consequences for the development of the reproductive tract were profound. This is a true dichotomy, and allows for no intermediate stage.

When the reproductive processes of the three major groups of mammals are compared, many features emerge as synapomorphies. Examples are the Graafian follicles, functional corpora lutea, bilaminar blastocysts, uterine secretion, yolk sac placentae, mammary glands, and lactation. Although monotremes retain some plesiomorphic characters (e.g, egg laying), which show an early divergence from the stock leading to therians, comparisons between marsupials and eutherians suggest a dichotomy in development from a common ancestral group not a derivation of one from the other. Most of the apomorphic therian characters of marsupial reproduction are associated with the greater emphasis on lactation rather than gestation.

There appears to be a major dichotomy in the control of sexual differentiation in therian mammals. In eutherians, the sexual development of the male and female reproductive tracts has been assumed to be solely the result of the presence or absence of hormones secreted by the fetal testis, but in marsupials, several sexually dimorphic characters (the scrotum, mammary anlagen, gubernaculum, and processus vaginalis) are under direct genetic control. It now appears that certain sexual dimorphisms are also under direct genetic control in eutherians.

Tyndale-Biscoe and Renfree (1987) suggested that the basic mode of mammalian reproduction evolved simultaneously with the origin of mammals in the Triassic, but remained almost unchanged until the late Mesozoic because it was the most appropriate mode for small nocturnal insectivorous mammals. Tyndale-Biscoe and Renfree (1987) have further

Contents

Introduction, 5
Mammalian Reproductive Characters, 6
Anatomy And Ontogeny of the Female Reproductive Tract:
    Capture of the Nephric System by the Reproductive System, 8
The Role of Genes and Hormones in the Development of the Reproductive Tract, 10
Scrotum versus Pouch, 12
Delivery of the Young, 13
The Mammary Gland and Its Evolution, 15
Lactation versus Placentation: Energy Transfer from Mother to Young, 16
Conclusions, 17
Acknowledgments, 17
References, 17

suggested that the present-day differences appear to have evolved during the adaptive radiation of the therian mammals in response to the metabolic requirement of increasing body size and the constraints imposed as these mammals moved into new ecological niches. Recent comparative studies detailed here on the anatomy and physiology of marsupials support these ideas.

Introduction

In all living groups of mammals, prolonged protection of the developing young is a characteristic strategy of reproduction. Lactation has become the central control of reproduction, and mammals have highly specialized adaptations for the transfer of energy from mother to young, first via the uterus and placenta, and subsequently via the milk.

When Haeckel wrote in 1898 that marsupials were direct descendants of the monotremes, and that the eutherian mammals were simply a result of the development of a placenta and the loss of the pouch and the epipubic bones, he did not know how long this mistaken paradigm would persist. The affinities between the living mammals are clear, but as new knowledge has become available on the remarkable therian specializations within each group of mammals, the general consensus has emerged that marsupials and eutherians arose from a common and more primitive stock that, in many of its characters, resembled the monotremes. This is not to say that living monotremes have no apomorphies of their own. One example (though not a reproductive character) is the recently discovered sophisticated from of electrosensory system used by both the platypus and echidna to locate food, which depends on a totally different type of electroreceptors from the known vertebrate electroreceptors such as those found in amphibians, the electric eel, and sharks (Scheich et al., 1986; Gregory et al., 1987; Proske, 1990).

Marsupials differ principally from eutherian mammals in their mode of reproduction, but reproductive parts do not make good fossils, and so the new information on the reproductive physiology of marsupials is only slowly being incorporated into modern evolutionary arguments (e.g., Tyndale-Biscoe and Renfree, 1987; Blackburn et al., 1989). In terms of reproduction, there is, in fact, only one hard-and-fast distinction, and this one is of the utmost importance. In the marsupials the ureters pass between the oviducts; in the eutherians the oviducts pass between the ureters (Fig. 2.1). This difference is one that is established very early in embryonic life, at the time when the kidneys are developing (see Fig. 2.2), and it is the essential factor that determines the reproductive peculiarities of these two great subdivisions of the Mammalia. Since the ureters pass

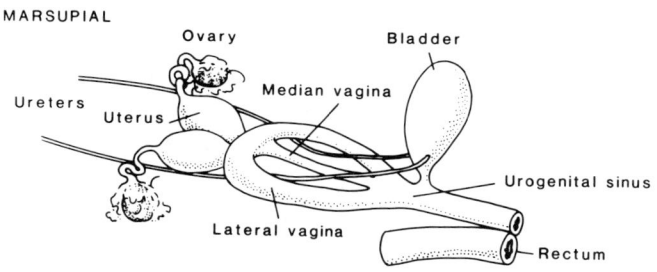
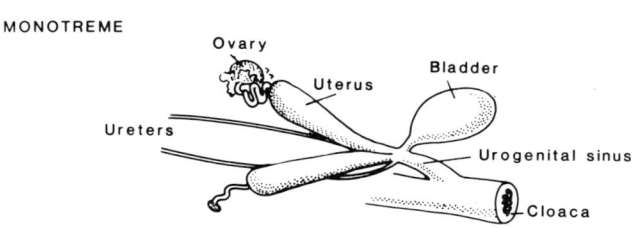
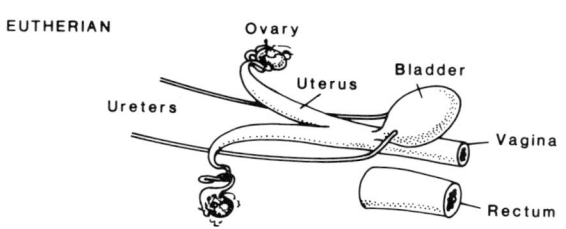

FIGURE 2.1. The relationships of the excretory and genital ducts in marsupials (top), monotremes (center) and eutherians (bottom). The monotreme condition is a plesiomorphic amniotic condition. Note that in this example the right ovary is nonfunctional, as in the platypus. The ureters open into the urogenital sinus directly opposite the opening of the bladder, whereas in the synapomorphic therian condition the ureters, while taking either the medial (marsupial) or lateral (eutherian) route, open directly into the bladder.

between, and hence separate, the oviducts of the marsupials, these ducts cannot meet in the midline to form a large single median uterus with a wide median vagina opening to the exterior. This may be one of the reasons that marsupial offspring are so remarkably small at birth, when compared with the size of the parent (Wood-Jones, 1923).

From the condition of the oviducts and ureters in the monotremes, it is easy to derive either the marsupial or eutherian pattern, just as either may be derived from the condition present in the early embryo. The evolutionary divergence of marsupials and eutherians from the common therian stock resulted in the Marsupialia developing along a line characterized by, among other things, the degeneration of the allantoic placenta, the curtailment of intrauterine life, the production of immature offspring, and the development of a marsupium or pouch in which to shelter them. The Eutheria, on the other hand, capitalized on the allantoic placenta, a pro-

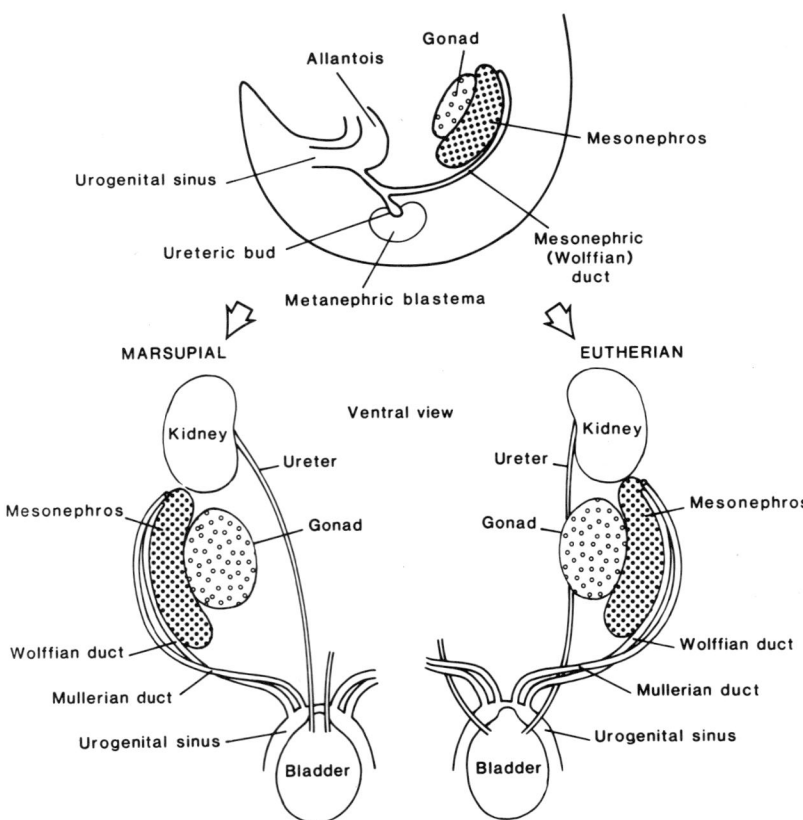

FIGURE 2.2. The ontogeny of the genitourinary tracts. In the indifferent embryo of both marsupials and eutherians (lateral view shown), the ureter buds off from the dorsal wall of the mesonephric (or Wolffian) duct. As the base of the ureter and Wolffian ducts become incorporated into the urogenital sinus, the ureter becomes either medial or lateral to the Wolffian (and later the Mullerian (or paramesonephric) ducts). As the allantoic stalk becomes the bladder, the ureters eventually enter it directly on the dorsal surface. Subsequent medial fusion of the Mullerian duct occurs to form the vagina or the anterior vaginal culs-de-sac in eutherians and marsupials, respectively.

longation of intrauterine life, and the birth of somewhat more mature offspring.

This chapter is an elaboration of selected arguments given in Tyndale-Biscoe and Renfree (1987), where we suggested that the basic mode of mammalian reproduction evolved simultaneously with the origin of mammals in the Triassic and remained almost unchanged until the late Mesozoic, because it was the most appropriate mode for small nocturnal insectivorous mammals. We further suggested that the major differences seen in living mammals today (Table 2.1) evolved during the great adaptive radiations of the marsupials and eutherians in response to the metabolic requirements of increasing body size and the ecological constraints imposed by the radiation into new ecological niches and modes of life.

As an Antipodean biologist used to looking at things upside down, I intend to concentrate on recent discoveries concerning the marsupial and monotreme modes of reproduction, noting only where eutherians differ from this normal Antipodean pattern.

Mammalian Reproductive Characters

As was already evident to Wood-Jones in 1923, knowledge of the physiology of reproduction meant that it was no longer possible to derive marsupials from existing monotremes, and primitive eutherians from primitive marsupials. Yet even today the number of characters shared by the latter two groups are often overlooked, as are some of the unique reproductive specializations of the monotremes. Some of the characters detailed earlier (Tyndale-Biscoe and Renfree, 1987) and listed again in Table 2.1 will be highlighted here in the light of new information that has become available since 1987.

As an overall generalization, it is clear that there are many synapomorphic characters of the three groups of mammals (Tyndale-Biscoe and Renfree, 1987). These may be divided into ovarian (Graafian follicles and functional corpora lutea), uterine (endometrial secretion), embryonic (trophoblast, bilaminar blastocysts, yolk sac placentae), and lacteal (mammary glands and lactation functions). We are often more concerned with the rare exceptions than the gross similarities.

The important apomorphic characters that are unique to marsupials are (1) the fusion of the vaginae anterior to the ureters, (2) the development of a median pseudo-vaginal canal, (3) a pouch, and (4) special endocrine controls of early lactation and milk secretion. Most of these are causally associated with greater emphasis on lactation than gestation (Renfree, 1983). The important apomorphic characters unique to the Eutheria are (1) the lack of a shell membrane around the fertilized egg, (2) early differentiation of the inner cell mass in most, but not all, species, (3) the universal precocious de-

Chapter 2. Ontogeny, Genetic Control, and Phylogeny of Reproduction

TABLE 2.1. A comparison of reproductive characters among the three groups of living mammals

Character	Monotremes	Marsupials	Eutherians
1 Karyotype	>50	<30	>40
2 Dosage compensation	Incomplete, tissue-specific X inactivation	Paternal X inactivation	Random X inactivation
3 Bulbo-urethral glands	Present	Present	Present
4 Prostate gland	Present, disseminate	Present, disseminate	Present
5 Seminal vesicles	Absent	Absent	Present
6 Glans penis	Bifid	Bifid or single	Single
7 Scrotum	Absent	Pre-penial	Pre- or post-penial
8 Testes	Abdominal	Inguinal or scrotal	Abdominal, inguinal, or scrotal
9 Testicular blood supply	Simple	Rete mirabile	Pampiniform plexus
10 Sperm head	Long, fusiform	Short	Short
11 Ureter's entry	Urogenital sinus, dorsal	Bladder, ventromedial	Bladder, ventrolateral
12 Endometrium	Secretory	Secretory	Secretory
13 Ovarian follicles	No antrum, liquor folliculi	Antral, liquour folliculi	Antral, liquour folliculi (no antrum in some species)
14 Corpora lutea	Secretory, autonomous	Secretory, autonomous	Secretory, pituitary or placenta dependent
15 Ovum	Large, yolk-filled	Small, yolk extrusion	Very small, no yolk
16 Cleavage	Meroblastic to blastocyst	Holoblastic to blastocyst	Holoblastic to morula (or blastocyst)
17 Bilaminar blastocyst with trophoblast	Present	Present	Present
18 Mucoid coat	Present	Present	Present in few, absent in most
19 Shell membrane	Present	Present	Absent
20 Shell	Present	Absent	Absent
21 Embryo formation	From outer layers	From outer layers	Inner cell mass in most
22 Vascularized yolk sac	Present in all	Present in all	Present in some
23 Vascularized chorioallantois	Present	Absent (expect Peramelidae)	Present in all
24 Invasive villous placenta	Absent	Absent (expect Peramelidae)	Present in all
25 Endocrine function of placenta	—	Slight to unknown	Various
26 Immunoprotection of fetus	—	Uncertain	Present
27 Delivery of young	Altricial from egg	Altricial from uterus	Altricial to precocial from uterus
28 Hair and sweat glands	Present	Present	Present
29 Mammary gland, alveolar, and myoepithelial cells	Present	Present	Present
30 Mammary hairs	Present	Present	Absent
31 Mammary anlagen	Areola patches	Areola patches	Mammary lines
32 Teats	Absent	Present	Present
33 Mammary glands in males	Present	Absent	Present
34 Crural spurs and glands	Present	Absent	Absent
35 Epipubic bones	Present	Present	Absent
36 Pouch	Present/absent	Present/absent	Absent
37 Lactation	Long duration	Long duration	Short duration
38 Milk composition	Major changes through lactation	Major changes through lactation	Minor changes through lactation

Slightly modified from Tyndale-Biscoe and Renfree (1987).

velopment of the chorioallantoic villous placenta (with or without an associated choriovitelline placenta), and (4) the tendency to develop a variety of luteotrophic and luteolytic controls of the corpora lutea in several orders. All of these features are causally associated with the relatively longer retention of the embryo in the uterus and its delivery at a more advanced stage of development than in any marsupial or monotreme.

There are differences in early embryonic development between marsupials and eutherians, such that the marsupial blastocyst forms as a hollow ball of cells with no inner cell mass, but both have a true a blastocyst cavity. It is incorrect (Lillegraven, 1975; Lillegraven et al., 1987) to refer to the blastocyst cavity as the blastocoel. The blastocyst cavity is a special feature of eutherian and marsupial embryos, which are considered secondarily alecithal, and this is emphatically not homologous with the blastocoel found in holoblastic embryos (Denker, 1983). The early development of both marsupials and eutherians is characterised by precocious differentiation of an extra-embryonic tissue, the trophoblast, which is specialized for nutrition of the embryo, for metabolic exchange, for hormone production, and for establishing contact with the maternal tissues at implantation. These functions are found in both marsupial and eutherian embryos. It is again incorrect, on both morphological and functional grounds, to classify the trophoblast as a neomorph that is unique to Eutheria (Lillegraven, 1975, 1985; Lillegraven et al., 1987). Taylor and Padykula (1978) clearly refuted Lillegraven's arguments, and there is a growing body of experimental evidence to show the fallacy underlying his hypothesis (Kirsch, 1977; Tyndale-Biscoe and Renfree, 1987; Blackburn et al., 1988). Even among eutherians, the spatial distribution of early embryonic cells does not always accurately reflect cell lineage. Like marsupials, some eutherians have no visible inner cell mass (reviewed in Wimsatt, 1975; Tyndale-Biscoe and Renfree, 1987), and Denker (1981, 1983) believes that "inner cell mass" is not an appropriate term for all eutherian species, because in some (such as the rabbit), not all the embryoblast cells take up an internal position before blastocyst formation. They may also lose this position again (such as in carnivores and ungulates with a central type of implantation), so the segregation and early determination of embryonic versus extra-embryonic cells (sensu Lillegraven et al., 1987) are not clear-cut. Trophoblast therefore cannot be a neomorphic tissue.

The monotremes retain some primitive amniote characters such as (1) a relatively large egg containing yolk, (2) development of an egg tooth and caruncle in the fetus, (3) separate Mullerian ducts, (4) dorsal entry of the urethra to the urogenital sinus, (5) laying of the egg when the embryo is at the somite stage, (6) incubation of the egg outside the body, and (7) lack of teats, all of which indicate a divergence from the common therian stock.

Monotremes are ovoviviparous. They do not have sufficient yolk to sustain them until hatching, and they must obtain substantial nutrients from the uterus before the shell (but after the shell membrane) is deposited (Griffiths, 1978, 1988). Viviparity is said to be achieved when eggs are retained in the uterus to term, as in reptiles (see Weekes, 1935). Therians are therefore viviparous (Amoroso et al., 1980; Tyndale-Biscoe and Renfree, 1987; Blackburn and Evans, 1986; Blackburn et al., 1988, 1989). Lillegraven et al. (1987) considers monotremes oviparous and erroneously refers to marsupials as ovoviviparous in contradistinction to eutherians because they possess a keratinous shell membrane, which is hardly more substantial than the zona pellucida of eutherians, and from which, like the zona "hatching" of Eutheria, the marsupial fetus has to "hatch" after about two-thirds of pregnancy (Renfree, 1973; Hughes, 1974). This "hatching" a result of the action of proteolytic enzymes (Denker and Tyndale-Biscoe, 1986). This is quite unlike the hatching of a reptile, monotreme, or bird from its egg shell after birth. Lillegraven's use of the terms *ovoviviparous* and *hatching* to characterize marsupials is therefore incorrect and implies unsubstantiated homologies.

Anatomy and Ontogeny of the Female Reproductive Tract: Capture of the Nephric System by the Reproductive System

Although adult metanephric kidneys and gonads have little in common, the testes capture the embryonic mesonephric ducts and use them to carry spermatozoa to the exterior. In contrast to these seminal ducts, however, the oviducts do not connect directly with the gonads, and eggs are shed from the ovaries into the peritoneal cavity, where they are drawn by ciliary action into the ostium of the oviduct, which leads to the exterior.

As stated earlier, the relative position of the ureters and the genital ducts is the character that most clearly diagnoses the three groups of living mammals. In monotremes, as in sauropsids, the ureters open into the dorsal wall of the urogenital sinus opposite the urethral opening of the bladder (Fig. 2.3), but in both marsupials and eutherians, the ureters migrate from a dorsal position where they bud off from the Wolffian duct, to make a direct ventral connection with the bladder. There are only two possible routes for this; in marsupials the ureters take a medial route so that they keep the paired female genital ducts apart, whereas in all eutherians they take the lateral route, so that the paired genital ducts can fuse (Fig. 2.2).

The exact steps in this developmental process have not always been clearly explained. The ureteric bud arises from the lower end of the mesonephric or Wolffian duct on its dorsomedial aspect: At this stage of development, the paramesonephric or Mullerian duct has not developed caudally and is not involved in the early stages of ureteric development. In marsupials, the ureteric bud migrates ventrally until it eventually arises from the medial side of the Wolffian duct, whereas in Eutheria it migrates to the lateral side (Buchanan and Fraser, 1918). This occurs around the end of the fifth week of gestation in humans, the second week of gestation in rats and rabbits, and by the third week of gestation in the tammar wallaby. With the gradual incorporation of the posterior ends of the Wolffian ducts into the urogenital sinus, the ureters are also drawn down into the urogenital sinus. As the bladder develops, the ureters open independently into the base of the bladder primordium, which is itself derived from the intra-abdominal allantoic stalk. As the metanephric kidneys migrate cranially in the dorsal coelom, the testes move caudally in the ventral coelom. When testicular descent into the scrotum eventually occurs, the Wolffian ducts (now vasa deferentia) loop over the ureters (Eutheria) or run laterally beside them (Marsupialia) (Fig. 2.4).

The early workers (e.g., Buchanan and Fraser, 1918; Baxter, 1935) found that this was the most profound distinction between marsupials and eutherians because it was an exclusive one (reviewed in Tyndale-Biscoe, 1973). The alternative view, advocated more recently

Chapter 2. Ontogeny, Genetic Control, and Phylogeny of Reproduction

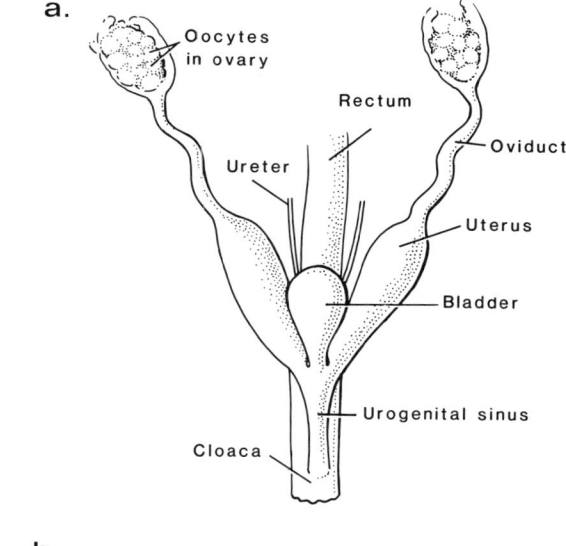

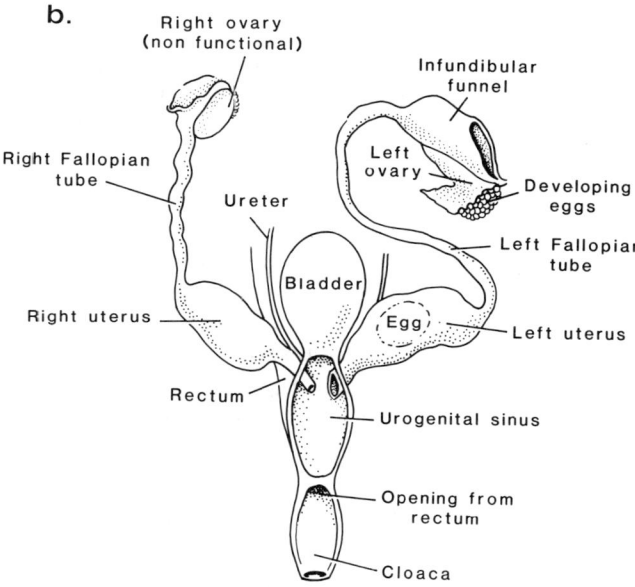

FIGURE 2.3. The anatomy of the reproductive tract of monotremes, as seen in the echidna, *Tachyglossus aculeatus* (a), and the platypus, *Ornithorhyncus anatinus* (b). Note that the ureters enter into the urogenital sinus, and in platypus, as in many Chiroptera, the right ovary is nonfunctional. (a): redrawn from Griffiths (1984); b: redrawn from Griffiths (1988).

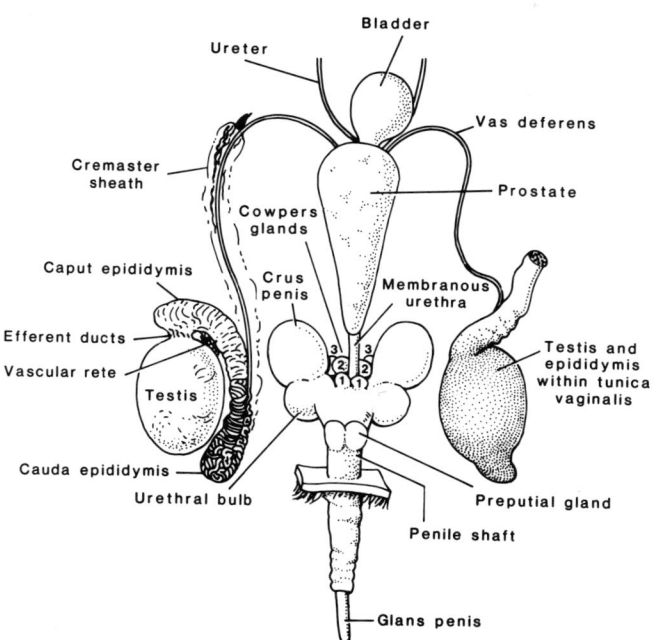

FIGURE 2.4. The urogenital system of the male tammer wallaby, *Macropus eugenii*. Note that the ureters do not loop over the vas deferens as they do in eutherian males.

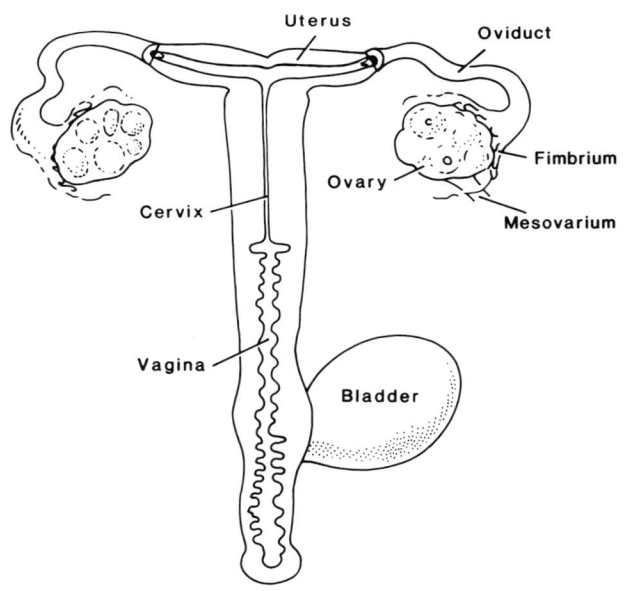

FIGURE 2.5. The genital system of the white-toothed shrew, *Crocidura russula*. The bursa that surrounds the ovary is not shown. The uterus is represented by the top part of the "T," with a very long cervix entering the vagina. Litter size varies from three to ten young, which weigh 1 to 1.2 g at birth. Redrawn from Kress and Millian (1987).

by Lillegraven (1969, 1975, 1985), Sharman (1970), and Luckett (1977), is that the marsupial pattern is essentially the same as that of monotremes and sauropsids. This, however, denies the common ontogeny of the ureteric buds from Wolffian ducts. These authors further contend that the midline fusion of the Mullerian ducts in the Eutheria to form a single vagina and uterus (e.g., Fig. 2.5) was a necessary prerequisite for prolonged fetal retention in the uterus, and for growth and eventual delivery of a large neonate, a point to which I will return later. Based on this argument, with which we disagree (Tyndale-Biscoe and Renfree, 1987), the marsupial condition is seen by some as an ineffectual attempt at midline fusion, thus denying the possibility for the development of larger fetuses.

Against this interpretation is the fact that in many eutherian species, such as rabbits, rats, cows, pigs, and bats, the uteri are either completely or extensively sepa-

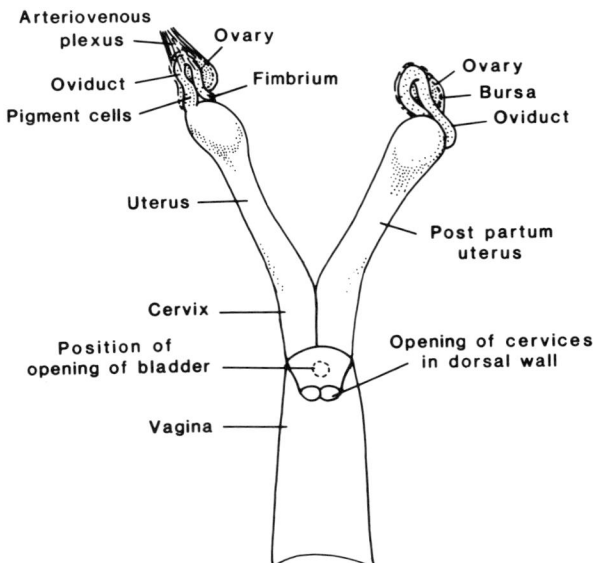

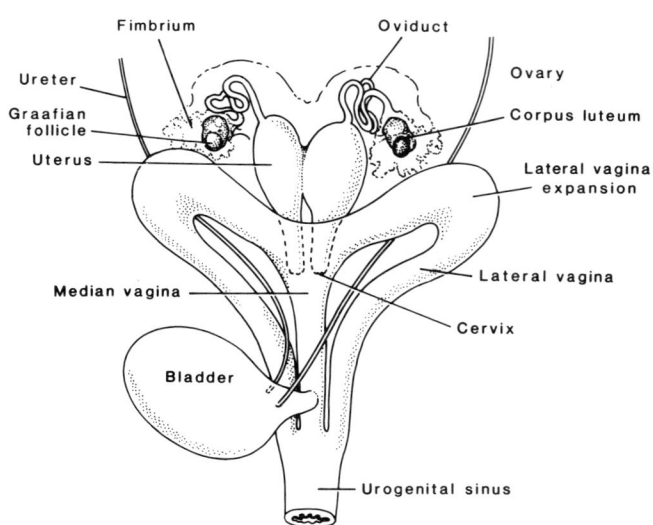

FIGURE 2.6. The genital system of the fruit bat, *Pteropus scopulatus* (body weight about 300 g). Birth weight of the young is about 50 g. The ovary is always heavily encapsulated by the bursa with the follicles and corpora lutea interior. There are pigment cells on the oviduct. There is a rich arterio-venous plexus, and the ovarian artery is within the venous drainage, allowing for countercurrent exchange between the two supplies. In this species both ovaries are functional, and the vagina, which is hormone unresponsive, is always cornified. Drawing based on photographs provided by Dr. L. Martin, University of Queensland.

FIGURE 2.7. The urogenital system (ventral view) of the female tammar wallaby, *Macropus eugenii*. Note that postpartum left uterus is slightly larger and is on the ipsilateral side to the corpus luteum of pregnancy. The single follicle is on the contralateral ovary. In the macropodids the median vagina remains open after the first birth.

rate and each is capable of accommodating one or more fetuses (e.g., Figs. 2.5 and 2.6). In addition, there obviously remains much potential plasticity in uterine morphology. For example, in human females, a wide range of abnormalities of the uterus exist, from a complete double uterus with double vagina ("uterus didelphys") to a condition in which there is a single unilateral horn ("uterus unicollis"), all of which can support a pregnancy. Conversely, in no marsupials do the uteri fuse in the midline even though the adjacent portions of the Mullerian ducts invariably do so to form the anterior vaginal cul de sac (Fig. 2.7). Because the combined volume of the fetus and assoicated membranes of marsupials is comparable to the fully formed monotreme egg, which is delivered through a single oviduct, there again seems to be no reason why the absence of midline fusion should have posed a particular constraint on fetal size. On the other hand, an essential adaptation for retention and nurture of the embryo was the development of a glandular endometrium capable of producing a copious secretion under the stimulus of progesterone. This is a feature shared by all three groups of mammals.

Indeed, based on present knowledge there seems to be no well-substantiated adaptive biological role for the midline fusion of the oviducts anterior to the ureters in all marsupials. Tyndale-Biscoe and Renfree (1987) conclude that the original selection pressures shaping the pattern of urinary and genital ducts in both eutherians and marsupials were not for a reproductive function but for an excretory function. For the ureters to enter the bladder directly, instead of via the urogenital sinus as they do in monotremes, the genital ducts had to be circumvented on one side or the other, and it did not matter which, although the subsequent consequences for the development of the reproductive tract were profound. This is a true dichotomy between marsupials and eutherians, which allows for no intermediate stage, and one pattern cannot be derived from the other (Tyndale-Biscoe and Renfree, 1987).

A point sometimes forgotten is that although this dichotomy results in the bizarre (that is, different from the accustomed eutherian pattern) vaginal anatomy of the female marsupial (Fig. 2.7), it also results in the convoluted arrangement of the vasa deferentia of the male eutherian after testicular descent. Thus, the male marsupial pattern is simpler (Fig. 2.4), as is the female eutherian one, again supporting the argument that the selection was initially for an excretory rather than a reproductive role.

The Role of Genes and Hormones in the Development of the Reproductive Tract

Apart from the obvious morphological differences in the anatomy of the reproductive tract, marsupials differ in other less well-known but perhaps more fundamental ways from eutherians and monotremes. One of these is the way in which the genes regulate sex determination and the differentiation of the reproductive tract.

Sex in all mammals is determined genetically: females usually have an XX sex chromosome constitution, and males an XY. The eutherian X is very large (about 5% of genome with 2,000 to 3,000 genes) but the Y is smaller (2% of genome) (Graves, 1990a). The Y is largely heterochromatic, with highly repeated sequences, and is almost devoid of genes recognizable by inheritance of gene markers. The X and Y chromosomes of eutherians during meiosis pair only at a tiny region at the tip of the short arms of both chromosomes—the pseudoautosomal region. Marsupials, like eutherians, have an XX-female, XY-male chromosomal constitution, but the sex chromosomes are proportionately smaller than those of eutherians, apparently as a result of loss of the pseudoautosomal pairing region. The Y chromosome of many marsupials is exceedingly small, suggesting that the marsupial X has captured more of the ancestral Y chromosome by Y-X translocation than is the case in eutherians (Shaw et al., 1990; Graves, 1990b); this would also account for the loss of the homologous pairing segment between the X and Y during meiosis in all marsupials (Sharp, 1982), since no synaptonemal complex is ever formed. Furthermore, in the few marsupial species for which we have information, X-chromosome inactivation is almost invariably confined to the paternal X (reviewed in Graves, 1987), whereas in eutherians it is random except in trophoblast cells in which it is also paternal. Despite these differences, many sex-linked genes on the mammalian X have been conserved in common between marsupials and eutherians (reviewed in Graves, 1987, 1990b).

In monotremes, the echidnas have an $X_1X_1X_2X_2/X_1X_2Y$ sex determination mechanism resulting from a Y-autosome translocation (Murtagh, 1977), while the platypus has the more usual mammalian system of XX/XY (Murtagh, 1977; Wrigley and Graves, 1988). Monotremes have an incomplete, tissue-specific X inactivation, and it is confined to the unpaired segment of the X. The monotreme X chromosome is largely homologous to the Y chromosome, and Wrigley and Graves (1988) suggest that monotremes represent an early stage of X-Y chromosome differentiation in mammals.

Another interesting discovery in eutherians relates to the exciting recent work of Surani and colleagues (1984, 1987) on the disproportionate roles that the maternal and paternal genomes play in the formation of the embryo and the fetal membranes (reviewed by Monk, 1987). Using chimaeric mice, they found that although both parental genomes are essential for development of the conceptus to term, the early expression of the maternal genome is confined to the embryo, while the trophoblast expresses the paternal genome. The inner

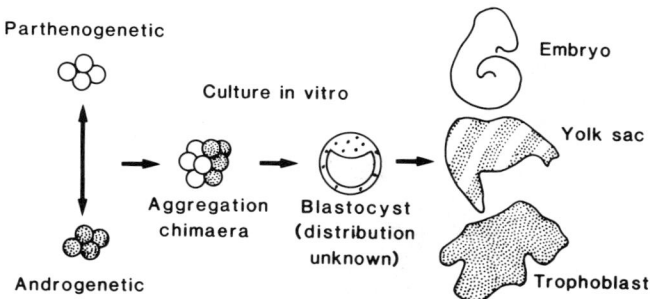

FIGURE 2.8. The preparation of chimaeras from embryonic cells containing only the paternal genome (androgenetic) and maternal genome (parthenogenetic) and the results of reaggregation of the cells. Note that of the embryos formed, only maternal genes were expressed, while the trophoblast expressed only the paternal genome. Yolk sac membranes had both types. Redrawn from Surani et al. (1987).

layer of the yolk sac mesoderm is derived from embryonic ectoderm, and the outer layer from the extra-embryonic endoderm, so that the yolk sac contains cells derived from both the trophoblast and the inner cell mass (Fig. 2.8). This specific imprinting of the parental genome occurs during gametogenesis. Since the yolk sac is such a prominent feature of marsupial embryos and fetuses, it would be interesting to know how the parental genomes are expressed in unilaminar, bilaminar, and trilaminar marsupial blastocysts.

Sex determination in mammals is thought to result from the action of a gene or genes on the Y chromosome that transforms the indifferent gonad into a testis. Subsequent sexual differentiation of the male and female reproductive tracts has hitherto been attributed solely to the action of hormones secreted by the testis in embryonic and fetal life (Jost, 1970; Wilson et al., 1981). Although this model is attractive because of its simplicity, new evidence derived in the main from our studies of marsupials suggests that there may be a more widespread genetic control of sexual differentiation, since some structures have already become sexually dimorphic prior to gonadal differentiation. The evidence to date, although fragmentary, suggests that monotremes and marsupials, like eutherians, normally require the presence of a Y chromosome for testicular formation. Thus XX and XO marsupials fail to develop testes, while XY and XXY individuals have well-developed testes (Sharman et al., 1970, 1990; Renfree and Short, 1988).

Our research has revealed a major new marsupial/eutherian dichotomy in the mechanisms involved in sexual differentiation. Some sexually dimorphic characters in marsupials such as the scrotum, mammary anlagen, gubernaculum, and processus vaginalis appear to be under direct genetic rather than secondary hormonal

 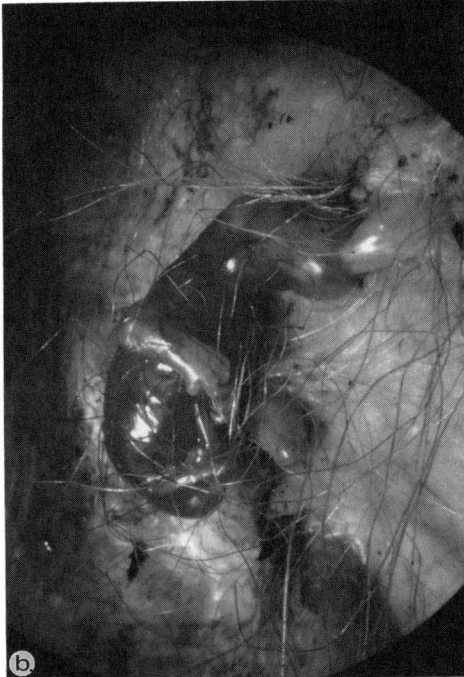

FIGURE 2.9. A litter of young attached to the teats in the pouchless South American marsupial, *Marmosa mitis* (**A**) compared to the single young of the tammar *Macropus eugenii* (**B**) within the deep and well-formed pouch characteristic of macropodids. Photographs kindly provided by Dr. C.H. Tyndale-Biscoe, CSIRO Wildlife and Ecology (**A**) and Dr. D.D. Parer, A.B.C. Natural History Unit (**B**).

control (O et al., 1988; Renfree and Short, 1988; Shaw et al., 1988, 1990; Hutson et al., 1988; Short et al., 1988). Scrotal development occurs when only a single X chromosome is present (XY, XO), while two X chromosomes are necessary for pouch formation (XX, XXY) (Fig. 2.9). There is also some evidence (Hartman and League, 1925; Sharman et al., 1970; Tyndale-Biscoe and Hinds, 1989) to suggest that genetic sex may also control the sexual dimorphism in body size of marsupials (Renfree and Short, 1988). It is not clear at present whether genetic or hormonal control is the derived state. Thus a study of sexual differentiation in monotremes would be most informative.

Scrotum versus Pouch

Although marsupial scrotal development is genetically determined, whereas eutherian scrotal development is under hormonal control (Shaw et al., 1988, 1990), it seems unlikely that the marsupial scrotum has evolved independently from the manner of the eutherian pattern. Adult marsupials have a single urogenital opening in both males and females, so there are no female external genitalia, and hence no female genital homologue of the scrotum. However, in eutherians the labia majora of the vulva are the homologue of the scrotum and will develop in genetic males in the absence of androgen. Thus it is perhaps not surprising that differing scrotal control mechanisms have evolved in marsupials and eutherians. Again, at this stage of knowledge it is not easy to discern any particular evolutionary reason for, or advantage of, these alternative control mechanisms.

No male marsupials possess a pouch (Tyndale-Biscoe and Renfree, 1987), and it has been suggested that the pouch has become the marsupial homologue of the scrotum (McCrady, 1938; Bolliger, 1944) or that both are derived from different parts of the same anlagen (Tyndale-Biscoe and Renfree, 1987). McCrady (1938) considered that the pouch and the scrotum of the Virginia opossum were derived from a common primordium early in pouch life. In the tammar wallaby, we have shown that the scrotum differentiates several days before birth, whereas the pouch first forms several days after birth, either from a different anlagen or from differing regions of the same anlagen. The two structures never coexist, except in the rare case of bilateral gynandromorphs. Since a pouch is found in XX and XXY individuals, but not in normal XY individuals, the pouch may be under the control of an X-linked gene that is functional only when two X chromosomes are present (Renfree and Short, 1988; Shaw et al., 1990; Sharman et al., 1970, 1990).

Monotremes are an enigma in terms of the pouch and scrotum. Male and female echidnas have pouches, whereas the platypus never does. Neither echidnas nor platypus has a scrotum, as the testes are intraabdominal (Griffiths, 1968, 1978); only two marsupials have nonscrotal testes, and these are inguinal and not abdominal (see Tyndale-Biscoe and Renfree, 1987). Some female marsupials do not have pouches, e.g., *Antechinus stuartii* and *Monodelphis domestica* (Fig.

2.9), and it is believed that the ancestral marsupials were pouchless (Tyndale-Biscoe and Renfree, 1987). If the pouch is a more recently evolved structure, or has evolved independently several times, (discussed later), this would also suggest that it may not be derived from a common anlagen with the scrotum; the two structures certainly appear to be under completely different genetic control.

What evidence is there for a genetic control of sexual dimorphism in eutherian mammals? In rats on day twelve of gestation, male fetuses are heavier and contain more protein than females, although gonadal differentiation cannot be detected until after this stage (Scott and Holsen, 1977). Tsunoda et al. (1985) cultured eight-cell mouse embryos and separated them on the basis of time taken to form the blastocyst stage. They found that fast-growing embryos give rise to 71% male young, while the slow-growing embryos produced 80% female young. At a later stage of development, somite number is greater in male than female mouse embryos (Seller and Perkins-Cole, 1987). Since these differences are sex-specific and obviously precede gonadal differentiation, a sex-linked genetic influence on early embryonic growth must be suspected.

The hormonal control of sexual dimorphism may be a relatively recent evolutionary development. In invertebrates there is abundant evidence for a cell-autonomous genetic control of sexual phenotype. The best understood example is *Drosophila melanogaster*; in this species, the X-chromosome–autosome ratio controls the expression of the three genes *transformer, transformer 2,* and *intersex*. These in turn regulate other genes that control either male or female development of the cell (Baker and Belote, 1983; Boggs et al., 1987). One of these genes, *sex-lethal*, also appears to control dosage compensation of the X chromosomes. *Sex-lethal* and *daughterless* also play a role in determining sexual expression in the germ line, but not necessarily by the same mechanisms used to determine somatic sex (Cline, 1985; Cronmiller and Cline, 1987). In mammals too, germ cell sex appears to be genetically determined in a cell-specific manner (Short, 1972; Burgoyne, 1987).

Thus studies in marsupials have directed attention to the fact that the hormonal control of sexual differentiation overlies a more fundamental genetic control system (Fig. 2.10). In marsupials genetic and hormonal controls obviously operate in concert, and the same may also be true to a more limited extent in eutherian mammals. The evolution of mammalian X chromosome inactivation seems to have proceeded in a stepwise fashion from incomplete, tissue-specific inactivation, which is characteristic of monotremes, through the incomplete paternal inactivation of marsupials and the fetal membranes of the eutherians, to random, complete inactivation of eutherian embryos (Graves, 1987; Wrigley and Graves, 1988). We might therefore find a

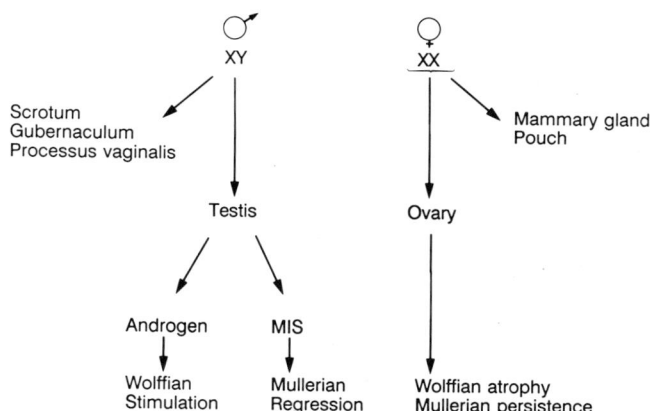

FIGURE 2.10. Hypothesis proposed by Shaw et al. (1990) to account for the control of sexual differentiation in marsupials.

similar genetic to hormonal progression in the regulation of sexual differentiation.

Delivery of the Young

Much has been written about the precocial nature of monotreme and marsupial young. What has not been realized, however, is the variation in the size and state of development of the neonatal marsupial. Although not as dramatic as the eutherian variation from the highly altricial shrews (see Fig. 2.11) and even bears to the precocious ungulates—there is no marsupial guinea pig, for example, that is so precocious that it can almost survive without mother's milk or a marsupial giraffe that can stand within minutes of birth—there is variation in marsupial size and development at birth. Hughes and Hall (1988) have recognized three grades of organization within neonatal marsupials: dasyurid marsupials being the smallest in size and least developed,

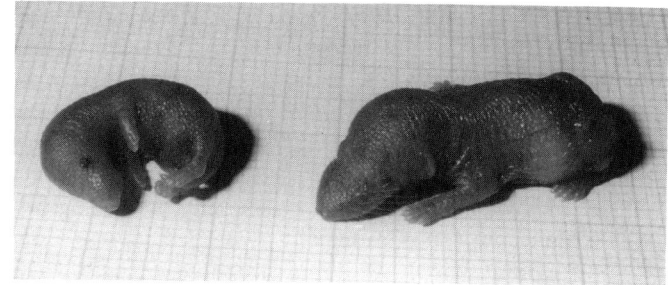

FIGURE 2.11. Newborn young of the shrews *Sorex araneus* (left) and *Crocidura russula* (right), showing the altricality characteristic of insectivorans. Like macropodid marsupials, *Sorex* (and *Neomys*) have a postpartum oestrus and a period of embryonic diapause, while *Crocidura*, which deliver young at a more advanced stage after a longer gestation, do not undergo diapause (Vogel, 1972, 1981). In this latter group, lactation is shorter than gestation. Photograph courtesy of Dr. P. Vogel, University of Lausanne.

perameloids being intermediate, and macropodoids being the most advanced. Lillegraven et al. (1987) have argued that early birth of a highly altricial young is plesiomorphic in the Mammalia, but Cockburn (1989) provides a convincing reassessment of the arguments and concludes that small is not only beautiful, but for marsupials it is probably apomorphic. In noting that small dasyurids have the special adaptation of a sternal swelling to uncurl the neonate and permit its mouth to connect with the teat, he follows Hughes and Hall's (1988) suggestion that in marsupials there has been a major shift from a requirement for motor capacity at birth toward direct ejection into the pouch of young with almost no mobility, as occurs in dasyurids (Hughes and Hall, 1988) and in *Monodelphis* (J.L. VandeBerg, personal communication). In at least the smaller species of these two marsupial groups, the opening of the birth canal approaches the opening of the pouch or the mammary area when the parturient mother bends over to lick the urogenital opening. Although such dasyurid and didelphid neonates are passive in the birth process, the macropodid neonate is not, and must climb unaided into the pouch. Recent data have shed new light on this fascinating process.

The physiology of parturition is a complex process in all mammals, but one that also involves appropriate maternal behavior. Successful birth ultimately depends on a precise synchronization with the appropriate maternal and fetal behavior. Maternal behavior at birth is well defined in eutherians, but aside from Sharman and Pilton's (1964) and Sharman and Calaby's (1964) studies, little had been recorded for marsupials. However, the adoption of a characteristic and highly stereotyped behavior must be essential for the birth of the tiny macropodid neonate and its successful journey to the pouch. In macropodid marsupials, females show an increased tendency to lick and clean the pouch about twenty-four hours before birth. (Sharman and Calaby, 1964; Renfree et al., 1989). Females lick the urogenital opening with increasing frequency for a few hours before birth, but within five minutes of birth the female adopts the characteristic birth posture with the tail passed forward between the legs, sitting on her lower back with hips rotated upwards, with the upper body hunched forward. The intensity of licking declines gradually from about ten minutes after birth.

Nonpregnant female wallabies injected with PGF-2α on twenty-three and twenty-six of the cycle show parturient behavior within minutes (Hinds et al., 1990). This behavior continued for over three hours, quite unlike the very brief period seen at normal birth; the duration of birth behavior corresponded with the period that PGFM levels were highest. This provided the first evidence that PGF-2α might be involved in controlling birth behavior. However, the doses of PGF-2α administered were pharmacological, and it was unclear whether the effect was a direct one on the brain or an indirect response to the prostaglandin-induced uterine and vaginal contractions.

In a subsequent study, Shaw (1990) has clearly demonstrated that PGF-2α injection in physiological doses induces parturient behavior not only in adult females, but in nonbreeding nulliparous females and in males, who avidly clean their scrota (Fig. 2.1). R. Rose (personal communication) has confirmed this behavior in four other macropodid and one peramelid marsupial. Shaw (1990) concludes that the action of the prostaglandin is a direct one on the brain. In the only comparable study on a eutherian, prostaglandins have behavioral effects in pigs. Injection of PGF-2α into nonpregnant sows induces the main elements of nest building, a major maternal behavior (Blackshaw and Blackshaw, 1982). These authors believe this was an indirect behavioral effect of prolactin, since PGF-2α induces an immediate rise in plasma prolactin (Taverne et al., 1979), and prolactin induces maternal behaviour patterns in rats (Riddle et al., 1935). In tammars, PGF-2α injection also induces an immediate rise in plasma prolactin. However, this rise in prolactin cannot be responsible for the birth behavior, since injection of prolactin alone has no obvious behavioral sequelae (Hinds et al., 1990).

A remarkable aspect of the behavioral response to PGF-2α in tammars is that it occurs so readily in males and nulliparous females in the nonbreeding season. In other species, most sexual behaviors are dependent on an appropriate priming with appropriate sex steroids, and the expression can also be influenced by prior experience. The responses of males of other species may also differ from those of females. Thus, PGF-2α induces nesting behavior in sows, as mentioned above, but copulatory behavior in boars (Blackshaw and Blackshaw, 1982). This specialized use of the same hormone appears to be synapomorphic, at least between the sow and the tammar. It will be interesting to see just how widespread prostaglandin-induced behavior is among mammals and whether this is another parameter that may shed light on evolutionary relationships.

Until recently (Tyndale-Biscoe et al., 1974; Tyndale-Biscoe, 1979), it was not acknowledged that marsupials had any endocrine recognition of pregnancy (reviewed in Tyndale-Biscoe and Renfree, 1987), as it was assumed that such a small conceptus could not redirect maternal physiology. It is now clear that marsupials do have a maternal recognition of pregnancy, the blastocyst (or trophoblast) influences the secretory activity of the uterus (Renfree, 1972, 1973), the fetus influences the duration of the luteal phase (Merchant, 1979), and the fetus directs the time of parturition (Kirsch and Poole, 1972; Poole, 1973, 1975) by stimulating (directly or indirectly) the release of prolactin and the cascade of events that results in a sharp rise in prostaglandin and regression of the corpus luteum (Tyndale-Biscoe et al., 1983; Lewis et al., 1986; Fletcher et al., 1990; Hinds et al., 1990). Marsupials are therefore now known to have a similar and complex control of parturition comparable to eutherians, and the wallaby pattern bears many similarities to that of the sheep. These are presumably synapomorphies of the two groups.

The Mammary Gland and Its Evolution

Bresslau (1912) recognized the similarity of monotreme mammary glands to those of marsupials and eutherians both in development and adult structure, and this has since been amply confirmed by Griffiths et al. (1972, 1973) with the light and electron microscope. Monotremes and marsupials share a common pattern of mammary development (areolar patches, mammary hairs) different from Eutheria (mammary line); marsupials and eutherians share the character of having the galactophores gathered together in clusters opening at the ends of teats, which are lacking in monotremes; monotremes and eutherians share the character that adult males of most species possess potentially functional mammary glands (Haacke, 1885), which, under certain hormonal conditions, can secrete milk (Griffiths, 1978; Cowie, 1984). It is not clear which of these shared characters is primitive and which is derived unless it is determined on other criteria that Eutheria are the "most advanced group" (Tyndale-Biscoe and Renfree, 1987). The absence of all vestiges of mammary glands in adult male marsupials is understandable because, during the first period of lactation, the young must remain permanently attached to one teat (see Fig. 2.9) and the opportunity for the male to share lactation is therefore ruled out. Nonetheless, it is of interest that neonatal and juvenile didelphid males do have a reduced number of mammary anlagen, and most disappear by eight to nine weeks postpartum (Renfree et al., 1990; Robinson et al., 1991). Among Eutheria, rats show a similar loss or suppression of mammary anlagen during late fetal life. It is less clear why no males of any species of Monotremata or Eutheria normally lactate, though the phenomenon of the lactating billy goat is well known. In mammals with exclusively monogamous mating patterns where the male spends much of his time carrying the young, as in marmoset monkeys, or in canids where the feeding of the young is shared by several members of the group, it is hard to understand why males do not lactate, despite several suggestions by Daly (1979).

Lactation is undeniably central to mammalian reproduction, and in all extant mammals is an exceedingly complex process. This could not have arisen *de novo*, but until recently there have been no convincing hypotheses to account for the development of a milk-secreting gland. Recently Blackburn et al. (1989) have presented an appealing new hypothesis for the origins of lactation. Previous theories postulated that the primordial mammary gland arose from a highly vascularized region of skin that served to warm the eggs, that secretions of maternal cutaneous glands enhanced the survival of eggs or hatchlings, and that there was a concomitant evolution of sucking behavior. Blackburn et al. (1989) argue that these theories have little predictive value and do not explain the selective pressures leading to postpartum nutrient provision and the production of altricial young. Their alternative hypothesis is based on the fact that the degree of altriciality exhibited by present-day monotremes, marsupials, and many eutherians could not have preceded the evolution of milk. Since mammary glands are apparently derived from epitrichial (sebaceous and apocrine) glands, rather than eccrine glands, it is possible that the mammary precursor could synthesize carbohydrates, proteins, and lipids as do modern epitrichial glands. Hayssen and Blackburn (1985) noted structural and functional links between the components of milk and the antimicrobial secretions of other integument-derived glands, and Blackburn et al. (1989) point out that the antimicrobial properties of milk are often overlooked. If the secretory products of the mammary gland precursor enhanced survival of the eggs or young by virtue of their antimicrobial properties, and assuming the prior development of a vascularized abdominal incubation patch, the following sequence of events might have occurred:

1. Protolacteal secretions controlled or destroyed bacteria, fungi, or other potential pathogens on or near the surface of the egg shell.
2. Protolacteal secretions, ingested in small quantities by the hatchlings, enhanced offspring survival by controlling microflora of the pharynx and digestive tract. These secretions may also have provided immunity to the young (a function perhaps reflected in the important role that 1gA plays in present-day mammalian milks, where it acts at the mucosal surface of the infant's gut).
3. Hypertrophy of the cutaneous glands of the incubation patch with more copious secretion.
4. Gradual shift to maternal secretion of a more nutritious milk, and an increase in efficiency of suckling.

The hypertrophy and specialization of cutaneous glands accompanied by the controlled production of a copious, nutritious secretion would allow the eventual evolution of milk as a primary source of energy for development and growth of the offspring.

Blackburn et al. (1989) point out that one of the potential advantages of lactation is that it spreads out the maternal energy investment in reproduction. Females could extend their reproductive investment beyond gestation. Lactation made the production of altricial young possible, again suggesting that the extremely altricial young and long lactation periods of marsupials may be highly derived characteristics. An important corollary of this is that by making extreme altriciality possible, lactation made pouches highly adaptive. These presumably evolved long after the development of lactation (Blackburn et al., 1989), and apparently have evolved on more than one occasion (Tyndale-Biscoe and Renfree, 1987).

Lactation versus Placentation: Energy Transfer from Mother to Young

Lactation in marsupials lasts for relatively longer than in most eutherian species to support their young which are more immature at birth (Renfree, 1983; Tyndale-Biscoe, Stewart, and Hinds, 1984; Findlay and Renfree, 1984). Although the processes of mammary development and secretion are similar in all mammals, the temporal course of these events is different in marsupials (Stewart and Tyndale-Biscoe, 1983). One of the characteristic features of lactation in marsupials is a major change in the composition of the milk as lactation progresses. As in all mammals, the first milk is a colostral-type secretion, a low-fat high-protein fluid with free floating cells (Griffiths et al., 1972) and with immunoglobulin G- and A-like proteins (Hindes and Mizell, 1976; Deane et al., 1990). The data available from the seven or eight species studied suggest that all marsupials have a similar pattern of milk composition: a relatively dilute milk (10% solids) produced to supply the new pouch young, becoming progressively more concentrated (25% to 50% milk solids) with time (Green, 1984; Green et al., 1987; Green and Merchant, 1988). The growth rate of the young is regulated by the rate of milk production and by its changing composition. The main constituents of marsupial milk are similar to those of eutherian mammals, but there are marked changes throughout lactation as the pouch young changes from continuous to intermittent sucking (Griffiths et al., 1972; Green et al., 1980; Green, 1984; Green and Merchant, 1988). Total solids are higher than in bovine or human milk; proteins, carbohydrates, and fats increase quantitatively and change qualitatively during lactation (Messer and Green, 1979).

Although eutherian milks do not show such dramatic qualitative and quantitative changes in milk composition, perhaps because the young do not undergo so many developmental changes during lactation, in some eutherian species early and late milks do differ. For example, increases in the milk fat content occur in the black-tailed deer *Odocoileus hemionus*, and changes in milk composition occur in giraffes (*Giraffa camelopardalis*) (Ben Shaul, 1962). Even human milk differs in composition between mothers giving birth prematurely and those giving birth at term (Atkinson et al., 1978), suggesting that human milk may be uniquely suited to optimizing growth and development in the premature infant—just as in marsupials.

Concurrent, asynchronous lactation occurs in continuously breeding macropodids, such as the red kangaroo *Macropus rufus* and the agile wallaby *M. agilis*, or even in the tammar under experimental conditions. The female gives birth immediately after the pouch has been vacated by the older young, which continues to suckle from outside the pouch. Thus two of the four mammary glands simultaneously secrete milk of entirely different composition. The mechanisms that control changing milk composition in marsupials and that allow the mammary glands to differentiate and regress independently are unclear, but local factors are obviously important.

Although there is as yet little information on crude growth efficiency in marsupials, their energy and growth conversions fall within the range spanned by eutherian mammals (Green, 1984; Rose, 1987; Green and Merchant, 1988; Green et al., 1988; Cork and Dove, 1989; Dove and Cork, 1989), so that the relatively long lactation and comparatively slow rate of growth of the young are not due to inefficient conversion of milk, and must therefore be regulated by maternal milk supply (Green and Merchant, 1988). This idea is supported by several observations. Transfer of young to teats of larger species or to teats of more developed glands results in accelerated growth rates of pouch young, whereas transfer to smaller glands growth (Merchant and Sharman, 1966; Findlay and Renfree, 1984; Green and Merchant, 1988). Only transitory effects on milk composition were produced by such asynchronous transfers, but these effects were not maintained, suggesting that changes in the composition of the milk are an intrinsic characteristic of the mammary epithelial cells and occur as the cells age (Findlay and Renfree, 1984). Such a system allows simultaneous asynchronous lactation to exist independently in two adjacent glands, which can then supply different milks to young at different physiological stages of development with different dietary requirements.

Few attempts have been made to measure the cost of lactation in marsupials (Rose, 1987; Green and Merchant, 1988; Green et al., 1988). One way is to compare the maternal energy investment required to rear a young to weaning. Using the tammar and the sheep as models, it appears that the pattern of milk intake is similar in both species if eutherian lactation is compared only with the post-pouch exit life of marsupials (Dove and Cork, 1989). During pouch life, the marsupial pattern of milk energy intake is similar to that of the energy deposition in the eutherian fetus during pregnancy (Dove and Cork, 1989; Cork and Dove, 1989). These authors conclude that the energy cost of lactation in the tammar can thus be regarded as equivalent to the sum of the energy costs of pregnancy and lactation in

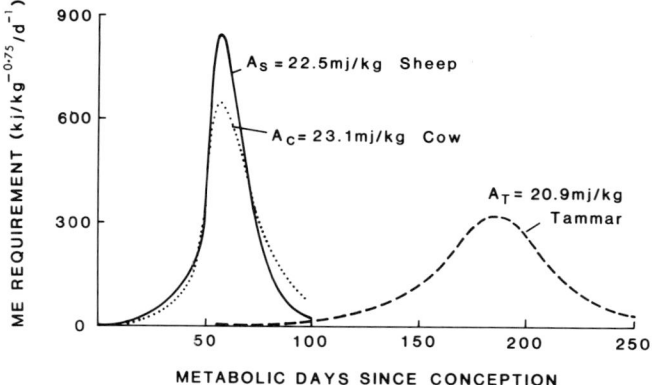

FIGURE 2.12. Estimated requirements for metabolizable energy (ME) kJ.kg$^{-0.75}$.d^{-1}) to support pregnancy/lactation in 500 kg dairy cows (...), 50 kg sheep (——), and 5 kg tammar wallabies (– – – –), in relation to metabolic days (days−3.5/W$^{0.27}$) since conception. Plots for cattle, sheep, and tammars were derived as described in Cork and Dove (1989). Birth in cattle and sheep occurs near 50 metabolic days. Areas under the curves (A_c, A_s, and A_T for cattle, sheep, and tammars, respectively) represent total energetic requirements, and were determined as definite integrals of the original equations (1–281 and 1–147 days for pregnancy in cattle and sheep, respectively; 1–256 and 1–140 days for lactation in cattle and sheep, respectively; and 1–360 days for lactation in tammars). The total energy requirements thus calculated are remarkably similar. Redrawn from Cork and Dove (1989).

the sheep. Furthermore, there are not large differences between these two herbivores in terms of weight-related allocation of energy to reproduction (Cork and Dove, 1989) (Fig. 2.12). Thus the marsupial strategy of spreading the offspring's intake of energy over a long lactation appears to be an important adaptation for a small grazing mammal living in a relatively aseasonal environment that is not faced with brief predictable periods of seasonal food abundance. The female tammar, therefore, unlike the ewe, can afford to have an extensive, as opposed to an intensive, grazing time, giving it a relative ecological advantage.

These important analyses of Cork and Dove show that the energy requirements for reproduction and development in their model marsupial, the tammar, fall within the range of comparable eutherians (in this case, the sheep and cow). Furthermore, they confirm that the transfer of energy in the form of milk is limited by the mother (Dove and Cork, 1989). Another way in which the mother can limit the milk supply is by dramatically constricting her pouch as in the Tasmanian rat-kangaroo (Rose, 1986, 1987); this ejects the young and denies them access to the teats. These data once again contradict the suggestion of Lillegraven et al. (1987) that energy transfer between mother and young during reproduction is less efficient in marsupials than in eutherian mammals.

Conclusions

The data presented here suggest that the morphotypic marsupial reproduction, rather than being a primitive therian feature, is highly derived. The major differences seen in the mode of reproduction in the ancestors of the three groups of living mammals today must have evolved in response to the metabolic requirements of increasing body size and the ecological constraints imposed during the great adaptive radiations of mammals (Tyndale-Biscoe and Renfree, 1987). Much of the evolutionary debate about patterns of mammalian reproduction has suffered from an obsession with only phylogenetic considerations and a paucity of data about monotremes and marsupials, which have by default been relegated to the role of "primitive types." I believe that this Antipodean perspective will help to redress the balance and dispel some of the unwarranted misjudgments that have grown out of past theoretical shackles and empirical ignorance.

ACKNOWLEDGMENTS. I thank Dr. G. Shaw for helpful discussions on the development of the urogenital system. I especially thank my husband, Professor R.V. Short, for a critical appraisal of the whole manuscript. I am grateful to Dr. Len Martin for providing the dissections used for the illustration of the bat reproductive system, and to Drs. D. Parer, C.H. Tyndale-Biscoe, and P. Vogel for the photographs of the marsupial and the shrew young. This chapter is dedicated to Dr. Hugh Tyndale-Biscoe, with whom I have had many an evolutionary debate, and who has contributed so much to marsupial reproductive biology.

CORRESPONDENCE ADDRESS. Marilyn B. Renfree, Department of Zoology, University of Melbourne, Parkville, Victoria 3150, Australia.

References

Amoroso, E.C., Heap, R.B., and Renfree, M.B. 1980. Hormones and the evolution of viviparity. In: *Hormones and evolution vol. 2* (Barrington, J.W., ed.). London: Academic Press, pp. 925–989.

Atkinson, S.A., Bryan, M.H., and Anderson, G.H. 1978. Human milk: Difference in nitrogen concentration in milk from mothers of term and premature infants. *J. Pediat.* 93:67–69.

Baker, B.S., and Belote, J.M. 1983. Sex determination and dosage compensation in *Drosophila melanogaster*. *Ann. Rev. Genet.* 17:345–397.

Baxter, J.S. 1935. Development of the female genital tract in the American opossum. *Carnegie Institute Contributions to Embryology.* 25:15–35.

Ben Shaul, D.M. 1962. The composition of the milk of wild animals. *International Zoo Yearbook* 4:333–342.

Blackburn, D.G., and Evans, HE. 1986. Why are there no viviparous birds? *Amer. Natural.* 128:-165–190.

Blackburn, D.G., Hayssen, V., and Murphy, C.J. 1989. The origins of lactation and the evolution of milk: A review with new hypotheses. *Mammal Rev.* 19:1–26.

Blackburn, D.G., Taylor, J.M., and Padykula, H.A. 1988. Trophoblast concept as applied to therian mammals. *J. Morph.* 196:127–136.

Blackshaw, J.K., and Blackshaw, A.W. 1982. The effects of prostaglandin (PGF-2α) on the behaviour of the domestic non-pregnant sow and boar. *Animal Production in Australia* 14:550–552.

Boggs, R.T., Gregor, P., Idriss, S., Belote, J.M., and McKeown, M. 1987. Regulation of sexual differentiation in *D. melanogaster* via alternative splicing of RNA from the *transformer* gene. *Cell* 50:739–747.

Bolliger, A. 1944. An experiment on the complete transformation of the scrotum into a marsupial pouch in *Trichosurus vulpecula*. *Med. J. Aust.* 2:56–58.

Bresslau, E. 1912. Die Entwickelung des Mammarapparates der Monotremen, Marsupialier und einiger Placentalier. In: *Zoologische Forschungreisen in Australien, vol. 4* Semon, R., ed. Jena: Gustav Fischer, pp. 653–874.

Buchanan, G., and Fraser, E.A. 1918. The development of the urogenital system in the marsupialia, with special reference to *Trichosurus vulpecula*. *J. Anat.* 53:35–95.

Burgoyne, P.S. 1987. The role of the mammalian Y chromosome in spermatogenesis. *Hum. Genet.* 61:85–90.

Cline, T.W. 1985. Primary events in the determination of sex in *Drosophila melanogaster*. In: *Origin and evolution of sex* (Halvorson, H.O., and Monroy, A., eds.). New York: Alan R. Liss Inc., pp. 301–327.

Cockburn, A. 1989. Adaptive patterns in marsupial reproduction. *Trends in Ecol. & Evol.* 4:126–130.

Cork, S.J., and Dove, H. 1989. Lactation in the tammar wallaby (*Macropus eugenii*). II. Intake of milk components and maternal allocation of energy. *J. Zool. London* 219:399–409.

Cowie, A.T. 1984. Lactation. In: *Reproduction in mammals. Book 3: Hormonal control of reproduction* (Austin, C.R., and Short, R.V., eds.). Cambridge Cambridge University Press, pp.195–231.

Cronmiller, C., and Cline, T.W. 1987. The *Drosophila* sex determination gene *daughterless* has different functions in the germ cell line versus the soma. *Cell* 48:479–487.

Daly, M. 1979. Why don't male mammals lactate? *J. Theor. Biol.* 78:325–345.

Deane, E.M., Cooper, D.W., and Renfree, M.B. 1990. Immunoglobulin G levels in fetal and newborn tammar wallabies (*Macropus eugenii*). *Reprod. Fert. Dev.* 2:369–375.

Denker, H.-W. 1981. The determination of trophoblast and embryoblast cells during cleavage in the mammal: New trends in the interpretation of the mechanisms. *Verh. Anat. Ges.* 75:435–448.

Denker, H.-W. 1983. Cell lineage, determination and differentiation in earliest developmental stages in mammals. In: Bibliotheca anatomica, No. 24 (Lierse, W., ed.). Basel, Switzerland: Karger, pp.22–58.

Denker, H.W., and Tyndale-Biscoe, C.H. 1986. Embryo implantation and proteinase activities in a marsupial (*Macropus eugenii*). *Cell and Tissue Research* 246:279–291.

Dove, H., and Cork, S.J. 1989. Lactation in the tammar wallaby (*Macropus eugenii*). I. Milk consumption and the algebraic description of the lactation curve. *J. Zool. London* 219:385–397.

Findlay, L., and Renfree, M.B. 1984. Growth, development and secretion of the mammary gland of macropodid marsupials. *Symp. Zool. Soc. London* 51:403–432.

Fletcher, T.P., Shaw, G., and Renfree, M.B. 1990. Effects of bromocriptine at parturition in the tammar wallaby (*Macropus eugenii*). *Reprod. Fert. Dev.* 2:79–88.

Graves, J.A.M. 1987. The evolution of mammalian sex chromosomes and dosage compensation: Clues from marsupials and monotremes. *Trends Genet.* 3:252–256.

Graves, J.A.M. 1990a. The search for the mammalian testis-determining factor is on again. *Reprod. Fert. Dev.* 2:199–204.

Graves, J.A.M. 1990b. Sex chromosome function in marsupials and monotremes. *Aust. J. Zool.* 37:409–410.

Green, B. 1984. Composition of milk and energetics of growth in marsupials, *Symp. Zool. Soc London.* 51:369–387.

Green, B., and Merchant, J.C. 1988. The composition of marsupial milk. In: The developing marsupial (Tyndale-Biscoe, C.H., and Janssens, P.A., eds.). New York: Springer-Verlag, pp. 41–54.

Green, B., Merchant, J.C., and Newgrain K. 1987. Milk composition in the eastern quoll, *Dasyurus viverrinus* (Marsupialia: Dasyuridae). *Aust. J. Biol. Sci.* 40:379–387.

Green, B., Merchant, J.C., and Newgrain, K. 1988. Milk consumption and energetics of growth in pouch young of the tammar wallaby *Macropus eugenii*. *Aust. J. Zool.* 36:217–227.

Green, B., Newgrain, K., and Merchant, J.C. 1980. Changes in milk composition during lactation in the tammar wallaby (*Macropus eugenii*). *Aust. J. Biol. Sci.* 33:35–42.

Gregory, J.E., Iggo, A., McIntyre, A.K., and Proske, U. 1987. Electroreceptors in the platypus. *Nature* 326:386–387.

Griffiths, M. 1968. *Echidnas*, International Series of Monographs in Pure and Applied Biology, Zoology Division, vol. 38. Oxford: Pergamon Press, pp. 1–254.

Griffiths, M. 1978. *The biology of the monotremes*. New York: Academic Press, pp. 1–341.

Griffiths, M. 1984. Mammals: Monotreme. In: *Marshall's physiology of reproduction* (Lamming, G.E., ed.). London: Churchill Livingstone, pp. 351–385.

Griffiths, M. 1988. The platypus. *Scientific American* 256:84–91.

Griffiths, M., Elliot, M.A., Leckie, R.M.C., and Schoefl, G.I. 1973. Observations of the comparative anatomy and ultrastructure of mammary glands, and on the fatty acids of the triglycerides in platypus and echidna milk fats. *J. Zool. London.* 169:255–279.

Griffiths, M., McIntosh, D.L., and Coles, R.E.A. 1969. The mammary gland of the echidna, *Tachyglossus aculeatus*, with observations on the incubation of the egg and on the newly hatched young. *J. Zool London* 158:371–386.

Griffiths, M., McIntosh, D.L., and Leckie, R.M.C. 1972. The mammary glands of the red kangaroo, with observations on the fatty acid components of the milk triglycerides. *J. Zool. London* 166:265–275.

Haacke, W. 1885. On the marsupial ovum, the mammary pouch, and the male milk glands in *Echidna hystrix*. *Proc. Roy. Soc. London Series B* 38:72–74.

Haeckel, E. 1898. The place of the didelphia in the mammalian phylum. In: *The last link*. Berlin: Georg Reimer, pp. 66–67.

Hartman, C.G., and League, B. 1925. Description of a sex-intergrade opossum, with an analysis of the constituents of its gonads. *Anat. Rec.* 29:283–298.

Hayssen, V., and Blackburn, D.G. 1985. α-lactalbumin and the origins of lactation. *Evolution* 39: 1147–1149.

Hindes, R.D., and Mizell, M. 1976. The origin of immunoglobulins in opossum "embryos." *Dev. Biol.* 53:49–61.

Hinds, L.A., Tyndale-Biscoe, C.H., Shaw, G., Fletcher, T.P., and Renfree, M.B. 1990. Effects of prostaglandin and prolactin on luteolysis and parturient behaviour in the non-pregnant tammar, *Macropus eugenii*. *J. Reprod. Fert.* 88:323–333.

Hughes, R.L. 1974. Morphological studies on implantation in marsupials. *J. Reprod. Fert.* 39:173–186.

Hughes, R.L., and Hall, L.S. 1988. Structural adaptations of the newborn marsupial. In: *The developing marsupial* (Tyndale-Biscoe, C.H., and Janssens, P.A. eds.). New York: Springer-Verlag, pp. 8–27.

Hutson, J.M., Shaw, G., O, W.-S., Short, R.V., and Renfree, M.B. 1988. Mullerian inhibitory substance production and testicular migration and descent in the pouch young of a marsupial. *Development.* 104:549–556.

Jost, A. 1970. Hormonal factors in the sex differentiation of the mammalian foetus. *Phil. Trans. Roy. Soc. London* B259:119–130.

Kirsch, J.A.W. 1977. Biological aspects of the marsupial-placental dichotomy: A reply to Lillegraven. *Evolution* 31:898–900.

Kirsch, J.A.W., and Poole, W.E. 1972. Taxonomy and distribution of the grey kangaroos, *Macropus giganteus* Shaw and *Macropus fuliginosus* (Desmarest), and their subspecies (Marsupialia: Macropodidae). *Aust. J. Zool.* 20:315–339.

Kress, A., and Millian, J. 1987. the female genital tract of the shrew *Crocidura russula*. In: Advances in anatomy, embryology and cell biology (Beck, F. et al., eds.). Berlin, Heidelberg, New York: Springer-Verlag, 101:1–76.

Lewis, P.R., Fletcher, T.P., and Renfree, M.B. 1986. Prostaglandin in the peripheral plasma of tammar wallabies during parturition. *J. Endocr.* 111:103–109.

Lillegraven, J.A. 1969. Latest Cretaceous mammals of upper part of Edmonton formation of Alberta, Canada and review of marsupial-placental dichotomy in mammalian evolution. *Paleontological Contributions from the University of Kansas.* 50:1–122.

Lillegraven, J.A. 1975. Biological considerations of the marsupial-placental dichotomy. *Evolution.* 29:707–722.

Lillegraven, J.A. 1985. Use of the term "trophoblast" for tissues in therian mammals. *J. Morph.* 183:293–299.

Lillegraven, J.A., Thompson, S.D., McNab, B.K., and Patton, J.L. 1987. The origin of eutherian mammals. *Biol. J. Linnean Soc.* 32:281–336.

Luckett, W.P. 1977. Ontogeny of amniote fetal membranes and their application to phylogeny. In: Major patterns in vertebrate evolution Hecht, M.K., Goody, P.C. and Hecht, B.M., eds.). New York: Plenum Press, pp. 439–516.

McCrady, E. 1938. The embryology of the opossum. *Am. Anat. Memoirs* 16:1–233.

Merchant, J.C. 1979. The effect of pregnancy on the interval between one oestrus and the next in the tammar wallaby, *Macropus eugenii*. *J. Reprod. Fert.* 56:459–463.

Merchant, J.C., and Sharman, G.B. 1966. Observations on the attachment of marsupial pouch young to the teats and on the rearing of pouch young by foster-mothers of the same or different species. *Aust. J. Zool.* 14:593–609.

Messer, M., and Green, B. 1979. Milk carbohydrates of marsupials. II. Qualitative and quantitative changes in milk carbohydrates during lactation in the tammar wallaby (*Macropus eugenii*). *Aust. J. Biol. Sci.* 32:519–531.

Monk, M. 1987. Genomic imprinting: Memories of mother and father. *Nature* 328:203–204.

Murtagh, C.E. 1977. A unique cytogenetic system in monotremes. *Chromosoma* 65:37–57.

O, W.-S., Short, R.V., Renfree, M.B., and Shaw, G. 1988. Primary genetic control of somatic sexual differentiation in a mammal. *Nature* 331:716–717.

Poole, W.E. 1973. A study of breeding in grey kangaroos, *Macropus giganteus* Shaw and *M. fuliginosus* (Desmarest), in central New South Wales. *Aust. J. Zool.* 21:183–212.

Poole, W.E. 1975. Reproduction in the two species of grey kangaroos, *Macropus giganteus* Shaw, and *M. fuliginosus* (Desmarest). II. Gestation, parturition and pouch life. *Aust. J. Zool.* 23:333–353.

Proske, U. 1990. The electric monotreme. *Aust. Nat. History* 23:289–295.

Renfree, M.B. 1972. Influence of the embryo on the marsupial uterus. *Nature* 240:475–477.

Renfree, M.B. 1973. The composition of fetal fluids of the marsupial *Macropus eugenii*. *Develop. Biol.* 33:62–79.

Renfree, M.B. 1983. Marsupial reproduction: the choice between placentation and lactation. In: Oxford reviews of reproductive biology 5 (Finn, C.A., ed.). Oxford University Press.

Renfree, M.B., Fletcher, T.P., Blanden, D.R., Lewis, P.R., Shaw, G., Gordon, K., Short, R.V., Parer-Cook, E., and Parer, D. 1989. Physiological and behavioural events around the time of birth in macropodid marsupials. In: *Kangaroos, wallabies and rat kangaroos* (Grigg, G., Jarman, P., and Hume, I.D., eds.). Sydney, Australia: Surrey Beatty & Sons, Pty. Ltd., pp. 323–337.

Renfree, M.B., Robinson, E.S., Short, R.V., and VandeBerg, J.L. 1990. Mammary glands in male marsupials. 1. Primordia in neonatal opossums, *Didelphis virginiana* and *Monodelphis domestica*. *Development* 110:385–390.

Renfree, M.B., and Short, R.V. 1988. Sex determination in marsupials: Evidence for a marsupial-eutherian dichotomy. *Phil. Tans. R. Soc. London* B. 322:41–53.

Riddle, O., Bates, R.W., and Lahr, E.L. 1935. Maternal behaviour induced in virgin rats by prolactin. *Proc. Soc. Exp. Biol. Med.* 32:730–734.

Robinson, E.S., Renfree, M.B., Short, R.V., and VandeBerg, J.L. 1991. Mammary glands in male marsupials. 2. Development and regression of mammary primordia in *Monodelphis domestica* and *Didelphis virginiana*. *Reprod. Fertil & Develop* 3:295–301.

Rose, R.W. 1986. The control of pouch vacation in the Tasmanian bettong, *Bettongia gaimardi*. *Aust. J. Zool.* 34:485–491.

Rose, R.W. 1987. Reproductive energetics of two Tasmanian rat-kangaroos (Potoroinae: Marsupialia). *Symp. Zool. Soc. London* 57:149–165.

Scheich, H., Langner, G., Tidemann, C., Coles, R.B., and Guppy, A. 1986. Electroreception and electrolocation in platypus. *Nature* 319:401–402.

Scott, W.J., and Holsen, J.F. 1977. Weight differences in rat embryos prior to sexual differentiation. *J. Embryol. Exp. Morph.* 40:259–263.

Seller, M.J., and Perkin-Cole, K.J. 1987. Sex difference in mouse embryonic development at neurulation. *J. Reprod. Fert.* 79:159–161.

Sharman, G.B. 1970. Reproductive physiology of marsupials. *Science, New York* 167:1221–1228.

Sharman, G.B., and Calaby, J.H. 1964. Reproductive behaviour in the red kangaroo, *Megaleia rufa*, in captivity. *CSIRO Wild. Res.* 9:58–85.

Sharman, G.B., Hughes, R.L., and Cooper, D.W. 1990. The chromosomal basis of sex differentiation in marsupials. *Aust. J. Zool* 37:451–466.

Sharman, G.B., and Pilton, P.E. 1964. The life history and reproduction of the red kangaroo (*Megaleia rufa*). *Proc. Zool. Soc. London.* 142:29–48.

Sharman, G.B., Robinson, E.S., Walton, S.M., and Berger, P.J. 1970. Sex chromosomes and reproductive anatomy of some intersexual marsupials. *J. Reprod. Fertil.* 21:57–68.

Sharp, P. 1982. Sex chromosome pairing during male meiosis in marsupials. *Chromosoma* 86:27–47.

Shaw, G. 1990. Control of parturient behaviour by prostaglandin F-2α in the tammar wallaby (*Macropus eugenii*). *J. Reprod. Fert.* 88:335–342.

Shaw, G., Renfree, M.B., and Short, R.V. 1990. Primary genetic control of sexual differentiation in marsupials. *Aust. J. Zool.* 37:443–450.

Shaw, G.S., Renfree, M.B., Short, R.V., and O, W.-S. 1988. Experimental manipulation of sexual differentiation in wallaby pouch young with exogenous steroids. *Development* 104:689–701.

Short, R.V., 1972. Germ cell sex. In: *Edinburgh Symposium on the Genetics of the Spermatozoon* (Beatty, R.A., and Glueckson-Waelsch, S., eds.). Copenhasen: Bogtrykkereit Forum, pp.325–345.

Short, R.V., Renfree, M.B., and Shaw, G. 1988. Sexual development in marsupial pouch young. In: The developing marsupial, models for biomedical research (Tyndale-Biscoe, C.H., and Janssens, P.A., eds.). Berlin: Springer-Verlag, pp.200–210.

Stewart, F., and Tyndale-Biscoe, C.H. 1983. Pregnancy and lactation in marsupials. *Curr. Top. Exp. Endoc.* 4:1–33.

Surani, M.A.H., Barton, S.C., and Norris, M.L. 1984. Development of reconstituted mouse eggs suggests imprinting of the genome during gametogenesis. *Nature* 308:548–550.

Surani, M.A.H., Barton, S.C., and Norris, M.L. 1987. Influence of parental chromosomes on spatial specificity in androgenetic-parthenogenetic chimaeras in the mouse. *Nature* 326:395–397.

Taverne, M., Willemse, A.H., Dielman, S.J., and Bevers, M. 1979. Plasma prolactin progesterone and oestradiol-17β concentrations around parturition in the pig. *Anim. Reprod. Sci.* 1:257–263.

Taylor, J.M., and Padykula, H.A. 1978. Marsupial trophoblast and mammalian evolution. *Nature* 271:588.

Tsunoda, Y., Tokunaga, T., and Sugie, T. 1985. Altered sex ratio of live young after transfer of fast-and slow-developing mouse embryos. *Gamete Res.* 12:301–304.

Tyndale-Biscoe, C.H. 1973. *Life of Marsupials*. London: Arnold.

Tyndale-Biscoe, C.H. 1979. Hormonal control of embryonic diapause and reactivation in the tammar wallaby. In: Maternal recognition of pregnancy, Ciba Foundation Symposium 64 (new series). Amsterdam: Excepta Medica, pp. 173–190.

Tyndale-Biscoe, C.H. 1989. The adaptiveness of reproductive processes. In: Kangaroos, wallabies and rat-kangaroos (Grigg, G., Jarman, P., and Hume, I., eds.). New South Wales: Surrey Beatty & Sons Pty Limited, pp. 277–285.

Tyndale-Biscoe, C.H., Hearn, J.P., and Renfree, M.B. 1974. Control of reproduction in macropodid marsupials. *J. Endocr.* 63:589–614.

Tyndale-Biscoe, C.H., Stewart, F., and Hinds, L.A. 1984. Some factors in the initiation and control of lactation in the tammar wallaby. *Symp. Zool. Soc. London* 51:389–399.

Tyndale-Biscoe, C.H., and Hinds, L. 1989. Influence of the immature testis on sexual differentiation in the tammar wallaby, *Macropus eugenii* (Macropodidae: Marsupialia). *Reprod. Fert. Dev.* 1:243–254.

Tyndale-Biscoe, C.H., Hinds, L.A., Horn, C.A., and Jenkin, G. 1983. Hormonal changes at oestrus, parturition and post-partum oestrus in the tammar wallaby (*Macropus eugenii*). *J. Endocr.* 96:155–161.

Tyndale-Biscoe, C.H., and Renfree, M.B. 1987. *Reproductive physiology of marsupials*. Cambridge Cambridge University Press.

Vogel, P. 1972. Vergleichende Untersuchung zum Ontogenesemodus einheimischer Soriciden (*Crocidura russula*, *Sorex araneus* und *Neomys fodiens*). *Revue Suisse de Zoologie* 79:1201–1332.

Vogel, P. 1981 Occurrence and interpretation of delayed implantation in insectivores *J. Reprod. & Fertil.* 29:51–60.

Weekes, H.C. 1935. A review of placentation among reptiles. *Proc. Zool. Soc. London* 2:625–646.

Wilson, J.D., George, F.W., and Griffin, J.E. 1981. The hormonal control of sexual development. *Science* 211: 1278–1284.

Wimsatt, W.A. 1975, Some comparative aspects of implantation. *Biol. Reprod.* 12:1–40.

Wood-Jones, F. 1923–25. *The mammals of south Australia* In: Handbook of the flora and fauna of south Australia. Adelaide, Australia: A.B. James, Government Printer.

Wrigley, J.W., and Graves J.A.M. 1988 Sex chromosome homology and incomplete, tissue-specific X-inactivation suggest that monotremes represent an intermediate stage of mammalian sex chromosome evolution. *J. Heredity* 79:115–118.

CHAPTER 3

Development and the Phylogenetic Features of the Middle Ear Region

R. PRESLEY

Overview

Comparing maturing chondrocranium, early ossification, and maturing ossification stages shows that the facial nerve and ganglion, its palatine and chorda tympani branches, the medial and lateral carotid sympathetics, the internal carotid artery, the tensor tympani muscle and nerve, and the tympanic and auricular branches of IX-X nerves are all fairly consistent landmarks with a definable pattern at each stage in the monotremes, marsupials, and placentals.

Using the above soft tissues, the alicochlear commissure; the prootic incisure; incudal, squamosal, and mastoid components of the crista parotica; and the membranous roof and floor of the cavum supracochleare can be placed in a hypothetical morphotype of early stages of all of the taxa examined.

It is much more problematic to trace this into the pattern of osteological features that are used as diagnostic features in phylogenetic studies. Skeletal development, in both cartilage and bone, is extremely dynamic in later stages. At many sites, different bone or cartilage centers provide essentially similar components that are rationally scored as different because they are parts of different bones. Simple binary logic becomes risky when applied to such components.

The chondrocranial morphotype should provide for an alicochlear commissure, a marked cavum supracochleare, a marked enclosure of the sulcus facialis by the crista parotica, a strong development of the crista supporting incus, a tympanohyal bridge, a chordafortsatz, a prootic venous incisure, and a stapedial blastema pierced by the stapedial artery. The pyriform fenestra may be apomorphic, but which, if any, parts of the roof were from squamosal, petrosal, alisphenoid, etc., and which passages were present or separate can be expected to be variable in closely related forms near to the last common ancestor.

Contents

Introduction, 22
Earliest Stages of Chondrogenesis, 22
   General, 22
   Soft-Tissue Anatomy, 22
   Early Chondrocranial Morphotype, 23
Mature Chondrocranial Stage, 24
   General, 24
   Soft-Tissue Anatomy, 24
   Mature Chondrocranial Morphotype, 24
Osteocranium, 25
   General, 25
   Soft-Tissue Anatomy, 25
   Early Osteocranium: Features of the Common Pattern, 25
   Mammalian Synapomorphies in the Osteocranium, 26
Discussion, 27
Conclusion: A Possible Morphocline, 28
References, 28

Introduction

The phylogeny of petrosal anatomy is increasingly well supported by fossil specimens, but the major basic units are plesiomorphic so that the analysis concentrates on assumed soft-tissue relations associated with the detail of grooves and foramina. The general, but not fully justifiable, tendency has been to assume a morphotype derived from the adult soft-tissue anatomy of living mammals onto a generalized bony framework.

That procedure would assume that muscles, nerves, and vessels can be reliably defined, with decisions on the homologies being clear-cut, and assumed in the terminology. The objective is to say whether, or when, a named structure such as the postglenoid vein or a prootic canal appears in development. Each occurrence is utilized as a component in binary character-state analysis. Often, little attempt is made to ascertain whether the embryonic structure provides a potential for parallelism or forbids the acquisition of certain other characters.

This classical method neglects what has been learned of the interaction of the tissues in development. A number of students of phylogeny are now beginning to tackle this problem: MacPhee (1977, 1979, 1981), discriminating exactly between bullar elements by the study of histogenesis; Maier (1987), emphasizing the great change in proportions of basicranium, brain, and trigeminal ganglion during ontogeny; Presley (1979, 1981), emphasizing the ability of primary embryonic structures to drift with respect to each other before skeletogenesis and also with respect to later-formed boundaries, thus blurring essential homologies; Wible (1986, 1987), defining the widest possible array of anatomies in the developing vascular tree; and Zeller (1985, 1987), showing the need for closely coupled developmental series, have elaborated on the clear challenge to simple assumptions of homology and suggested that mammalian embryos put their own unique demands on the pattern of the skull.

The reproductive and developmental patterns of a species (maturity at birth; relative size at birth; functional demands on the cranial nerves before, at, and after birth) can constrain skull architecture, linking many characters that are often, and inadvisedly, treated as independent in phylogenetic analysis. When such constraints operate before skeletogenesis, structures that by early development and general function should probably be regarded as homologous are rendered apparently nonhomologous. Inclusion of the changing soft-tissue anatomy of the region during development in comparative and phylogenetic analysis could improve the resolution and discriminatory power of such character states. However, even today most analyses involve relatively late stages. The tympanum, for example, should be treated continuously from early stages of development onwards. An anatomical pattern emerges from tissues that are at first mainly a fluid-rich mesenchyme with slight differences in cell density. Firm skeletal tissues such as cartilage and bone appear gradually. The air in the tympanum enters only postnatally (in contrast to reptiles and birds), and therefore the osteological pattern is laid down in essential respects before the tympanum functions mechanically as a drum. The ear ossicles first move in liquid.

In this chapter I seek to recognize both how the elements common to this pattern emerge and those things that are not easily related to precursor anatomy; I also seek features in Recent and fossil groups that may represent mammalian developmental synapomorphies. To provide a concise description of the emergence of the pattern, three developmental stages are used: (1) the earliest indications of major chondrocranial features; (2) the chondrocranium at what was classically referred to as its optimal phase—when considerable detail of structure is present, allowing extensive comparative studies, but when ossification is absent or early; and (3) the established postossification stage, when comparison of features leading to characteristic adult bone structure becomes possible. Non-embryologists should appreciate that these stages are imprecisely defined, and that oviparous, marsupial, and placental reproductive strategies have a considerable effect on relative timing and proportions of the skeleton.

Earliest Stages of Chondrogenesis

GENERAL. Traditional descriptions (de Beer, 1937; Goodrich, 1930) usually show the skeleton at this stage as if it has firm boundaries. In fact, although tracts of denser mesenchyme and zones of early cartilage matrix can be discerned (and with common sense and hindsight can be related to later stages and compared between species), the histological appearance is one of rapid growth that is both interstitial and accretional. There are few areas of sharply differentiated perichondrium, and many regions where boundaries are very ambiguously placed. By contrast, nerves and ganglia, arteries, and the larger veins have clear-cut outlines and are in practice the structures that are used to make identifications at this stage (Fig. 3.1). Utilizing this neurovascular pattern, chondrocranial centers are obvious, though not sharply bounded and showing little characteristic detail. These are the basicranium, the ala temporalis, the pars cochlearis and pars vestibularis of the otic capsule, the occipital arch, and the first (Meckel) and second (Reichert) arch cartilages.

SOFT-TISSUE ANATOMY. A lattice of soft-tissue structures is used to identify future foramina and chondrocranial parts. Its components include the following:

Chapter 3. Development and Phylogenetic Features of Middle Ear

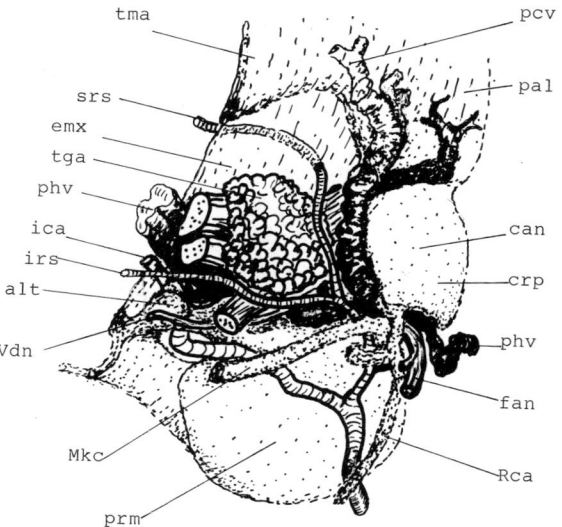

FIGURE 3.1. The suggested morphotypic anatomy of a mammalian embryo at the early chondrocranial stage, showing the main nerves, veins, and arteries related to the early centers of chondrification. See Table 3.1 for abbreviations.

TABLE 3.1. Abbreviations for Figures 3.1–3.3

acc	alicochlear commissure
alt	ala temporalis
aun	auricular branch of X nerve
bcr	basicranium
can	pars canalicularis of otic capsule
crp	crista parotica
emx	ectomeninx
etr	epitympanic recess
fan	facial nerve trunk
foi	fossa incudis
ggn	geniculate (facial) ganglion
gsp	greater superficial petrosal (palatine VII) nerve
ica	internal carotid artery
irs	inferior ramus of stapedial artery
juf	jugular vein in foramen
lcn	lateral carotid sympathetic nerve
lhv	lateral head vein segment of primary head vein
lsp	lesser superficial petrosal (from tympanic IX) nerve
mcn	medial carotid sympathetic nerve
mev	mesencephalic veins
Mkc	Meckel's (first arch) cartilage
ocf	supraoccipito-capsular (mastoid emissary) foramen
otg	otic ganglion
pal	parietal lamina
pcf	parietocapsular (spurious jugular) foramen
pcv	posterior cerebral veins
phv	primary head vein
pos	prootic venous sinus
pov	prootic vein
prm	promontorium of cochlear capsule
prs	posterior ramus of stapedial artery
pss	petrosquamosal venous sinus
ptv	post-trigeminal segment of primary head vein
Rca	Reichert's (second arch) cartilage
rgv	retroglenoid vein
spg	spheno-palatine ganglion
sqa	squamosal bone
sqs	squamosal support of crista parotica
srs	superior ramus of stapedial artery
tes	temporal ramus of stapedial artery
tga	trigeminal (semilunar) ganglion
thy	tympanohyal
tma	taenia marginalis
trs	transverse sinus
tty	tensor tympani muscle and nerve
tyn	tympanic branch of IX nerve (of Jacobson)
Vdn	Vidian (pterygoid canal) nerve

1. *Arteries.* The internal carotid and usually the stapedial as a smallish side branch with its superior and inferior rami.
2. *Veins.* The primary head vein, joined as it runs from medial to the trigeminal nerve ganglion to lateral to the facial nerve ganglion by a plexus of stems from a venous complex draining the posterior aspect of the cerebrum and the mesencephalon. These will later become the prootic sinus or functionally equivalent vein. Posterior to the otocyst the primary head vein turning medially becomes the principal feeder of the internal jugular vein at this stage. A consistent pattern of the external jugular system is usually not yet established.
3. *Nerves.* The trigeminal, facial, tympanic IX, auricular X, medial and lateral carotid sympathetics, palatine facial (greater petrosal), and chorda tympani nerves are usually clear. With care and hindsight, the continuation of tympanic IX to the otic ganglion (lesser petrosal), the auricular branch of the vagus, and the nerve (from mandibular V) to the tensor tympani can be identified, but a variable plexiform network of autonomic side connections also exists. The basic neurovascular pattern is found reasonably consistently in monotremes, marsupials, and placentals, and in most respects it can be regarded as being fairly well conserved throughout amniote embryos at this stage (Tandler, 1899, 1901; van Gelderen, 1924).

EARLY CHONDROCRANIAL MORPHOTYPE. Using this neurovascular network, comparison can be made with squamates, birds, turtles, and crocodiles (Bellairs and Kamal, 1981). The position of the major centers of chondrification listed above are symplesiomorphic, as is the soft-tissue pattern, provided the ala temporalis is homologized with the epipterygoid (Gaupp, 1902). Perhaps the clearest synapomorphy of mammals that affects the skeleton is the passage of the internal carotid artery through the basicranium. Here a dense zone of mesenchyme extends from the pars cochlearis forward to the ala temporalis lateral to the internal carotid artery and medial to the mandibular branch of the trigeminal nerve. This will become alicochlear commissure of therian embryos, but in monotremes the tip of the lagenar housing extends lateral to this site so that the internal carotid appears to pierce the basisphenoid medial to it.

Mature Chondrocranial Stage

GENERAL. Most of the chondrocrania figured in the classical reviews (Parker, 1885; de Beer, 1937; Starck, 1967) are close to this phase. A crisply delineated perichondrium enables very exact reconstructive modeling. The neurovascular pattern, besides retaining its original relations to the major centers, now fits exactly to and defines foramina, fissures, and grooves. These, together with small but exactly delineated outgrowing processes and deepening hollows, support very detailed comparative catalogues of features.

Traditionaly this detail was cross-compared to obtain a morphological history for each feature, emphasizing concepts such as homologies and vestiges. Little attempt was made to look at associated character-suites from a phylogenetic viewpoint or to provide a functional explanation for the architecture relating to the life of the tissues of the embryo at that instant. Such an approach has only recently emerged (Blechschmidt, 1976; Maier, 1987). There now exists the possibility that modern forms can be reduced to some common features, around which detailed differences can be related to differential increases in size, histology, and interconnection. These differences often appear more dramatic in the fully grown skull than is warranted by analysis of embryos.

SOFT-TISSUE ANATOMY. In the tympanic region, some significant changes take place in the soft-tissue framework. The internal carotid artery, in a neutral position in the previous stage, may be displaced medially or laterally by differential growth with respect to the promontorium of the cochlear (or lagenar) capsule (Wible, 1986). The stapedial artery can regress, its territory being supplied by alternative pathways, or it may persist and establish a group-characteristic pattern of branches (Wible, 1987). In therians, but not in monotremes, the portion of the primary head vein lateral to the facial nerve regresses, while in all mammals the post-trigeminal segment of the vein atrophies (Kielan-Jaworowska et al., 1986).

In monotremes, marsupials, and most placentals the posterior cerebral venous tree drains into a channel that communicates with the external jugular system, now much better established, in the region just lateral and anterior to the primary jaw joint. This position corresponds to the retroglenoid vein of adults. However, with respect to the chondrocranium in monotremes and marsupials, the channel emerges through the prootic incisure, mainly continuing in the primary (lateral) head vein in monotremes (though with growth steadily increasing its contribution to the external jugular in marsupials). In placentals, where growth of the cerebrum is more advanced at the time of consolidation of cartilage, the exit vein runs through the more posterior parietocapsular foramen, occipitocapsular foramen, or foramen magnum. In all three cases the tributary veins are homologous by physical and developmental origin; however, because of their later anatomical relations, logic demands that they be distinguished. Care, derived from an understanding of the process, is needed in character-state analysis in such circumstances.

The nerve pattern is essentially similar to that in the previous stage. However, species-consistent departures of the nerve plexus from the classical type, such as the sympathetics coursing separately from the internal carotid artery over the front of the promontorium, become more obvious.

MATURE CHONDROCRANIAL MORPHOTYPE

Plesiomorphic Features. In most cases these are discernible in the earlier stage, but have indistinct detailed relations. Enhancement of the perichondrium makes possible much greater precision in analysis (Fig. 3.2). The main symplesiomorphies with other amniotes include the basicochlear commissure connecting pars cochlearis with the basicranium in a variable number of tracts between carotid foramen and metotic fissure (jugular foramen); the crista parotica, a ridge ventral to the pars canalicularis forming the lateral wall of the sulcus facialis; and the parietal lamina representing the taenia marginalis of the chondrocranium above the pars canalicularis—the prootic sinus runs in the endocranial aspect of a groove at the junction between the two and, when present, the parietocapsular emissary vein discharges to the external aspect of this groove (Padget, 1957). In mammals there is not always a connection of orbitosphenoid with parietal lamina through a continuous taenia marginalis, which would define a prootic fenestra, but in all mammals at least a prootic incisure (notch) exists where the leading edge of the parietal lamina joins the pars canalicularis. The crista parotica projects posteriorly (pars mastoidea) to meet the paracondylar process of the exoccipital.

Apomorphic Features. In most cases these reflect either a different emphasis in mammals on commissural connections between the major chondrocranial units, or a pronounced sculpturing of the lips of nervous or vascular channels. The most significant are (1) the alicochlear commissure, observed in the previous stage; (2) a marked cavum supracochleare where a deepening of the upper rim of the sulcus for palatine VII and a forward extension of the prefacial commissure intervene between trigeminal and facial nerve ganglia; (3) a marked enclosure of the sulcus facialis by the crista parotica, which may reach inward towards the promontorium, thus forming a partial floor to the sulcus; (4) a strong rostral development of the crista supporting in-

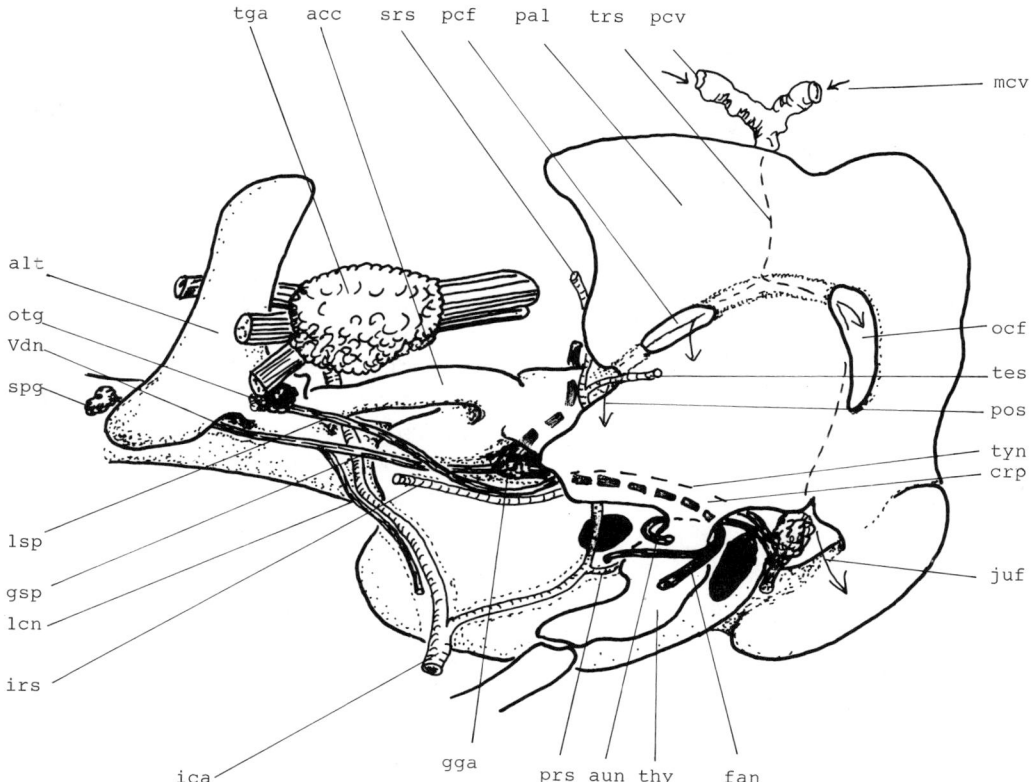

FIGURE 3.2. The chondrocranial morphotype, freely drawn from reconstructions of therian embryos. The vessels are intended to show possible positions explained in the text, based on the hypothesis that variant patterns can be related to the stapedial artery and the various possible exit routes for the blood draining from the transverse sinus. See Table 3.1 for abbreviations.

cus on its lateral face; (5) a tympanohyal, continuous in cartilage with the crista parotica dorsally, but ventrally arching towards the promontorium caudal and ventral to the fenestra ovalis (this arch makes fibrous or cartilaginous connection with the promontorium in monotremes and some marsupials); and (6) a chordafortsatz running forward towards the head of the malleus and body of the incus from the upper part of the tympanohyal, with the chorda tympani lateral and then ventral to it.

Osteocranium

GENERAL. More mature specimens demonstrate that the membrane bone that grows into the definitive osteological pattern follows tissue planes that were predefined at the previous stage as dense tracts of mesenchyme running between the features of the chondrocranium. These planes are reasonably easy to homologize. Which neighboring bones contribute in particular cases is variable; analogues are common. Relative growth has an important effect. In younger stages the connective tissue planes lie parallel to or only slightly obliquely to the nerves and vessels of passage, so that the latter lie in slits or canals. Once the framework is ossified, continued growth with new mechanical tissue properties converts most of the passages into foramina, where descriptively the emphasis comes to be on nerves or vessels piercing a single plane.

SOFT-TISSUE ANATOMY. Although the soft-tissue pattern defining the osteological features is derived from that in the precursor stages, there is an enormous range of variation made possible by additions and deletions, such as replacement of one or all rami of the stapedial artery by another feeder pathway (Tandler, 1899, 1901) or reduction of one or several venous exits (Butler, 1967). The most consistent component appears to be the pattern of nerves (Fig. 3.3), but even here loss of a typical pathway and enhancement of what is in other forms a minor collateral route can occur.

It does not seem possible in these circumstances to suggest a soft-tissue morpotype that differs consistently from that in earlier stages, but it is important to recognize the implication that reappearance of features transiently lost in the adult stages of any group is quite easy to explain. For example, loss of the stapedial artery is not irreversible.

EARLY OSTEOCRANIUM: FEATURES OF THE COMMON PATTERN. In many cases characteristic bony features in one taxon can be homologized with fibrous tracts in many

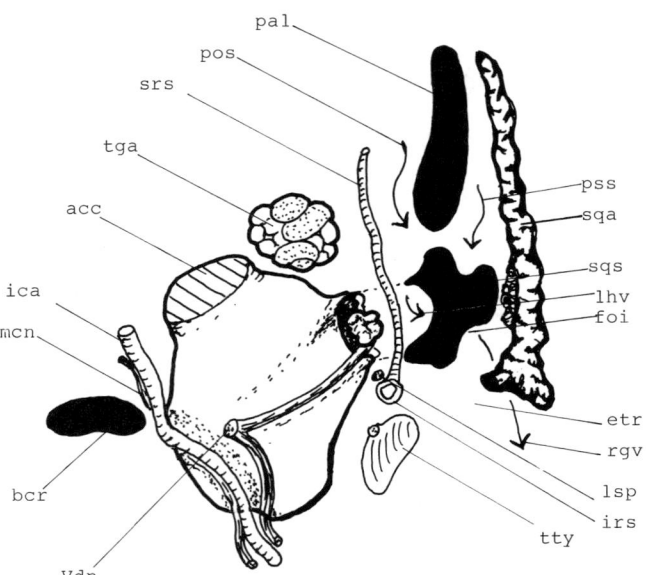

FIGURE 3.3. Generalized relationships of soft tissues in the anterior part of the tympanum, in relation to which the tegmen tympani is ossified. The cavum supracochleare, containing the geniculate ganglion, is roofed and floored by dotted lines, and is flanked by the anterior part of the crista parotica. The roof, floor, or both may be completed by any of the neighboring elements in the osteocranium without strong phylogenetic implications. The formation of the roof of the epitympanic recess, again dotted, similarly shows variation. See Table 3.1 for abbreviations.

other taxa: the potential for arguing plausible cases for polphyly is great. Only vaguely defined, but nevertheless critical, scientific methods such as "common sense" and experience need to be invoked in analysis, and it is clear that using a single tympanic character as a diagnostic feature in binary logic could be unsound. Much weight must therefore be placed on suites of characters—or character complexes, in the parlance of taxonomy.

The pars canalicularis is essentially similar in all mammals, but the roofing structure over the epitympanic recess (tegmen tympani or various equivalents or near equivalents from adjacent bones) projecting forward to close the pyriform fenestra shows great variation. The crista parotica presents an incudal fossa completed laterally by the squamosal; either a prootic venous sinus exists anteriorly or a petrosquamosal sinus, a functionally equivalent external tributary of the jugular system (Butler, 1967; Padget, 1957), lies in the groove between parietal lamina and pars canalicularis, flanked by membrane bone (lamina obturans, squamosal, or parietal); and a specialized zone of the crista supports the squamosal. This zone often projects somewhat laterally from the side of the crista and is opposed by secondary cartilage in the squamosal. Above it lies the passage for the petrosquamosal vessel, anterior to it the incus. This support may represent the cynodont paroccipital process, and the vascular passage the post-temporal canal.

The external auditory meatus contacts the lower border of the squamosal but does not directly contact the crista parotica. A styloid process or tympanohyal projects down from the crista behind the meatus. As in the previous stage, synchondrosis at the junction with the crista is usual. The facial nerve primitively passes behind, and sends the chorda tympani forward lateral to, the styloid. Either a chordafortsatz or a detached cartilage of Spence derived from it directs the chorda toward the medial aspect of the malleus. The styloid and crista form boundaries of a potential foramen stylomastoideum primitivum: This is secondarily carried ventrally to a variable extent by the growth of the squamosal and adjacent elements of the bulla. Behind this, the posterior part of the crista parotica and the squamosal blend in variable proportions to give the mastoid region. The fossa muscularis minor for the stapedius muscle is impressed into the ventral aspect of crista and mastoid.

On the medial wall of the tympanum, that is, the promontorium of the cochlea, the common primitive features are a sulcus for the greater petrosal nerve continuing forward from the sulcus facialis; the fossa muscularis major ventral to this; a promontorial groove for the internal carotid artery; a groove for the root of the stapedial artery; and a bony equivalent of the alicochlear commissure, with dorsomedial to it an impression for the trigeminal ganglion now rendered quasi-intracranial by some form of tegmen tympani.

MAMMALIAN SYNAPOMORPHIES IN THE OSTEOCRANIUM. Comparison with cynodonts indicates major changes in proportion, a considerable reduction of the paroccipital process (possibly by the crista retaining a common embryonic form), and complete loss of a quadratojugal. But the most obvious synapomorphy of mammals is the auditory bulla. It is not possible to deduce the most primitive pattern of the bullar wall from embryology. The only early structures to fulfill the mechanical needs are the ectotympanic and the tympanohyal. The additional elements of the bulla are extensions appearing slightly later in cartilage or membrane bone along planes of fibrous connective tissue attached to petrosal, alisphenoid, squamosal, or basicranium. Alternatively, they are features that appear much later in the fibrous membrane of the tympanic cavity (entotympanics, tubal cartilage). It should be recognized that satellite cartilages, presumably contributing to the growth of neighboring structures by accretion, are frequently found in therian embryos. Such satellites, if they fail to attach to a neighbor, will fall within the general definition of an entotympanic, and this phenomenon could permit ready polyphyletic acquisition of entotympanics with very similar anatomies.

Discussion

The generalizations set out for the three developmental stages have been arrived at by seeking common features in a wide but not complete survey of mammalian embryos. The first and most important caveat is that embryological information is not available for the majority of modern mammalian species.

The firmest generalization to emerge is that the nervous pattern, the primary head vein, and to a lesser extent the carotid vessels afford the most consistent reference framework for anatomy, and this shows most clearly in the early stages. Plesiomorphic and apomorphic features of the skeleton emerge later around this pattern. Because of the traditional approach (when identity is traced back from adult anatomy), the primary head vein in the morphotype has been underemphasized, as has the very common trend for the external jugular vein to capture the deeper veins in the retroglenoid region. The detail of this capture, and the relation to later-appearing cartilages and membrane bones, differ even within species, but the phenomenon itself is synapomorphic for modern mammals. Classical reviews (van Gelderen, 1924) stressing the osteology of exit foramina—for example, putting undue weight on distinguishing between spurious jugular, prootic, and retroglenoid veins—create a false sense of divergence.

The venous system, essentially morphotypic up to the stage of prootic sinus draining into primary head vein, subsequently illustrates how labile (in developmental terms) the subsequent skeletal relationships are. For example, if the parietal lamina develops outside the principal vein, that becomes a prootic sinus. If the lamina lies over and then internal to the principal vein, it is a spurious jugular vein. In marsupials and monotremes, the chondrocranium consolidates early, when the cerebrum is relatively small, and the prootic route is usually established at this stage. In placentals, chondrification is established around a larger cerebrum, and the more posterior venous exits are utilized. This may be an example of the reproductive habit having an effect on cranial osteology.

The early arterial system, shaped by a highly conserved vertebrate developmental pattern (a series of aortic arches joining bilateral dorsal aortae), is constrained to start from an internal carotid artery penetrating the basicranium just caudal to the basal attachment of the maxillo-mandibular visceral arch. The core mesenchyme of the second visceral arch is fed from the dorsal aorta through a dorsal remnant of the second aortic arch. once the third arch has taken over as main feeder to the head—an event common to all gnathostomes (Goodrich, 1930). Thus its distributive side branches such as the stapedial artery are inevitably present at some early developmental stage, even when lost later. It follows from this that the mesenchyme from which the stapes arises will always be penetrated at some developmental stage by a stapedial artery. Thus the potential is there throughout evolution for a group to reacquire a stapedial artery or a stapedial foramen even when it has been lost (Novacek and Wyss, 1986).

Looking at the varied anatomy of the stapedial artery and its analogues in the adults of modern mammals (Wible, 1987), it is not at first obvious why this artery should be accorded special attention, but the early development makes this clearer. Adult vessels such as the external carotid artery, the occipital artery, the arteria diploetica magna, and the posterior auricular artery represent alternatives that tend to arise slightly later than the stapedial and may annex it through anastomosis. It is usually possible to visualize the anatomy of the vascular channels, even when so annexed, in terms of the inferior (parallel to tegmen), superior (meningeal ascending medial to prootic sinus), temporal (post-temporal fossa), and occipital (sulcus facialis) rami of the stapedial, even if the literal truth of the anatomy is only transient and microscopic. It is always likely that companion veins will be present, and they can become the dominant feature in adults (Kielan-Jaworowska et al., 1986; Miao, 1988), but the stapedial tree remains a sufficient developmental explanation of the anatomy.

This statement does not mean that an inviolate bone structure can be predicted in relation to that tree. For example, while in most mammals the inferior ramus runs ventral to the tegmen tympani or its equivalent, in some it lies apparently dorsal to it (Wible, 1987). Preliminary studies suggest that this is because of the relative position of the promontorium, alicochlear commissure, and crista parotica, so that the bulk of the outgrowth closing that part of the pyriform fenestra is from the inferior lip of the groove for the greater petrosal nerve, running back to join the prefacial commissure, whereas in many other mammals it is the superior lip that dominates. Stated in other words, the tegmen is from the floor of the cavum supracochleare in macroscelidids, rather than from its roof. Since the artery runs lateral and parallel to the nerve in this region, it rapidly passes into a plane above the tegmen. Clearly this tegmen has subtly different qualities from the typical mammalian form, but intermediate states in some insectivores and the topological continuity of the relationship to the prefacial commissure forbid the statement that the two forms are completely nonhomologous. It is not clear how one can handle this as a simple binary character statement.

Perhaps the most striking difference between basicrania at the cynodont grade and at the modern mammalian grade is the closure ventrally of the cavum epiptericum. In embryos of modern mammals at an early stage, the main soft-tissue contents here are the trigeminal ganglion and its branches and the primary head vein. The facial nerve ganglion (geniculate) and the trigeminal (semilunar) ganglion have their capsules in contact, and thus the greater superficial petrosal nerve runs forward immediately below the trigeminal ganglion, medial to the mandibular nerve to the point where it acquires a sympathetic component and becomes the nerve of the pterygoid canal (Vidian). Immediately above it the post-trigeminal part of the primary head vein runs laterally to gain the outer surface of the facial ganglion. Later chondrification and ossification of the tegmen tympani in a zone anterior to the prefacial commissure produces a segregated cavum supracochleare, with separation of the ganglia. However, in all cases the cavum trigeminale and the cavum supracochleare communicate, this communication being the hiatus Falopii in mammals for passage of the greater petrosal nerve in modern mammals. In multituberculates this passage is relatively large, and it may be supposed that the primary head vein persisted here. In cynodonts the lateral flange of the petrosal appears to be similar to a partial roof of the cavum supracochleare, and from the size of the vascular apertures it must be supposed that the vein was a dominating feature (Crompton and Sun, 1985). Flooring of the cavum supracochleare seems to have been an advanced feature in the evolution of mammals.

The alicochlear commissure of therian embryos, bounding the carotid canal laterally while placed medial to the mandibular nerve, is a most difficult feature to analyze. It does not correspond easily with any feature of the cynodont skull and is not recognized in traditional accounts of monotremes. In the latter case, a possible solution is to argue that just as the apex of the cochlear capsule in therians is directed toward the commissure, so is the apex of the lagenar capsule in monotremes. In this way it may be supposed that the promonotorium of monotremes has assimilated the alicochlear commissure: Thus the latter is present and lies lateral to the carotid foramen. By this argument, the commissure may have become a tested (but still hypothesized) synapomorphy of modern mammals. Its representation in cynodonts may possibly be seen in the quadrate rami of pterygoid and epipterygoid, but only if it is accepted that the carotid entered the cranial cavity just medial to this, rather than through the carotid foramina in the basisphenoid.

Conclusion: A Possible Morphocline

The foregoing reasoning suggests caution in using individual foramina or neurovascular markings as undoubted synapomorphies or as indices of dichotomy. The potential exists to redevelop many soft-tissue structures even in groups that have apparently lost them in later developmental stages. Nevertheless it is possible to look at the fossil record and relate it in some general features to the Recent forms. The evidence is patchy, and is therefore presented as an indication of possibilities in transformation in the known morphocline rather than as a strictly tested charcter phylogeny.

Cynodont. (Crompton and Sun, 1985). Weak promontorium; lateral flange supporting quadrate ramus possibly representing roof of cavum supracochleare; much vasculature exteriorized; pronounced laterally directed paroccipital process with large post-temporal fossa above it.

Morganucodont. (Kermack et al., 1981; Crompton and Sun, 1985). Some enlargement of promontorium; lateral flange supporting quadrate ramus still cynodont-like; prootic canal between petrosal and anterior lamina (part of lamina obturans) exemplifying some interiorization of vessels; paroccipital process and post-temporal fossa cynodont-like.

Multituberculate. (Kielan-Jaworowska et al., 1986; Miao, 1988). Lagenar promontorium; much of the vasculature interior to anterior lamina, either between it and petrosal or intracranial; prootic sinus and draining both to external jugular and lateral head vein in facial sulcus; incipient to extensive floor of cavum supracochleare, but with large post-trigeminal canal; lateral flange still forming extensive roof; paroccipital process reduced and mastoid region triangular with downward emphasis; post-temporal fossa reduced to small vascular canal.

Monotreme. Much of the vasculature internal to anterior lamina; prootic sinus joining lateral head vein running through facial sulcus; extensive floor to cavum supracochleare but roof incomplete; post-trigeminal canal reduced; complete tympanohyal bridge posterior to epitympanic recess; large post-temporal fossa.

Therian. Tegmen augmented or replced by components of auditory bulla; small hiatus Falopii; much expanded bulla replaces tympanohyal in support of tympanum behind membrane; post-temporal fossa vascular only.

CORRESPONDENCE ADDRESS. R. Presley, Department of Anatomy, University of Wales College of Cardiff, Box 900, Cardiff CF1 3YF, United Kingdom.

References

Bellairs, A.d'A., and Kamal, A.M. 1981. The chondrocranium and the development of the skull in Recent reptiles. In: *Biology of the reptilia, vol. 11* (Gans, C., and Parsons, T.S., eds.). London and New York: Academic Press, pp. 1–263.

Blechschmidt, E. 1976. Principles of biodynamic differentiation. In: *Development of the basicranium* (Bosma, J.F., ed.). Bethesda, MD: U.S. Depertment of Health, NIH pp. 54–81.

Butler, H. 1967. The development of the mammalian dural venous sinuses with special reference to the postglenoid vein. *J. Anat.* 102:33–56.

Crompton, A.W., and Sun, A.L. 1985. Cranial structure and relationships of the liassic mammal *Sinocodon* Zool. J. Linn. Soc. 85:99–119.

de Beer, G.R. 1937. *The development of the vertebrate skull.* London: Oxford University Press.

Gaupp, E. 1902. Uber das ala temporalis des Saugerschadels und die Regio orbitalis einiger anderer Wirbeltierschadel. *Anat. Hefte, 1 Abt* 19:155–230.

Goodrich, E.S. 1930. *Studies on the structure and development of vertebrates.* London: Macmillan.

Kermack, K.A., Mussett, F. and Rigney, H.W. 1981. The skull of Morganucodon. *Zool J. Linn. Soc.* 71:1–158.

Kielan-Jaworowaka, Z., Presley, R., and Poplin, C. 1986. The cranial vascular system in taeniolabidoid multituberculate mammals. *Phil. Trans. Roy. Soc. London* B313:525–606.

MacPhee, R.D.E. 1977. Ontogeny of the ectotympanic-petrosal plate relationship in strepsirhine prosimians. *Folia Primatol.* 27:245–283.

MacPhee, R.D.E. 1979. Entotympanics, ontogeny and primates. *Folia Primatol.* 31:23–47.

MacPhee, R.D.E. 1981. auditory regions of primates and eutherian insectivores. *Contrib. Primatol.* 18:1–282.

Maier, W. 1987. the ontogenetic development of the orbitotemporal region in the skull of *Monodelphis domiestica* (Didelphidae, Marsupialia) and the problem of the mammalian alisphenoid. In: *Morphognesis of the mammalian skull* (Kuhn, H-J., and Zeller, U., eds.). Berlin: Mammalia Depicta, Verlag Paul Parey, pp. 71–90.

Miao, D. 1988. Skull morphology of *Lambdopsalis bulla* (Mammalia, Multituberculata) and its implications to mammalian evolution. *Contr. Geol., Univ. Wyoming, Special Paper* 4:1–104.

Novacek, M., and Wyss, A. 1986. Origin and transformation of the mammalian stapes. *Contr. Geol., Univ. Wyoming, Special Paper* 3:35–53.

Padget, D.H. 1957. The development of the cranial venous system in man, from the viewpoint of comparative anatomy. *Contr. Embryol.* 36:81–151.

Parker, W.K. 1885. The structure and development of the skull in mammalia (Part II, Edentata; Part III, Insectivora). *Phil. Trans. roy. Soc.* B176:1–275.

Presley, R. 1979. The primitive course of the internal carotid artery in mammals. *Acta. Anat.* 103:238–244.

Presley, R. 1981. Alisphenoid equivalents in placentals, marsupials, monotremes and fossils. *Nature* 294:668–670.

Starck, D. 1967. Le crane des mammiferes. In: *Traite de Zoologie XVI(1)* (Grasse, P.P., ed.). Paris: Masson, pp. 405–549.

Tandler, J. 1899. Zur vergleichenden Anatomie der Kopfarterien bei den Mammalia, *Denkschr. Akad. Wiss., Wein* 67:677–784.

Tandler, J. 1901. Zur vergleichenden Anatomie der Kopfarterien bei den Mammalia. *Anat. Hefte.* 18:327–368.

Van Gelderen, C. 1924. Die morphologie der sinus durae matris. *Z. Anat. Entw.* 74:432–508.

Wible, J.R. 1986. Transformations in the extracranial course of the internal carotid artery in mammalian phylogeny. *J. Vert. Paleont.* 6:313–325.

Wible, J.R. 1987. The eutherian stapedial artery: Character analysis and implications for superordinal relationships. *Zool. J. Linn. Soc.* 91:107–135.

Zeller, U. 1985. Die Ontogenese und Morphologie der Fenestra rotunda und des Aquaeductus cochleae von *Tupaia* und anderen Saugern. *Morph. Jb.* 131:179–204.

Zeller, U. 1987. Morphogenesis of the mammalian skull with special reference to *Tupaia.* In: *Morphogenesis of the mammalian skull.* Kuhn, H-J., and Zeller, U., eds.). Berlin: Mammalia Depicta, Verlag Paul Parey, pp. 17–50.

CHAPTER 4

Relationships of the Liassic Mammals *Sinoconodon, Morganucodon oehleri*, and *Dinnetherium*

ALFRED W. CROMPTON AND ZHEXI LUO

Overview

A lack of morphological information makes it difficult to interpret the relationships of advanced cynodonts and early mammals. To address this problem, we studied new skulls of *Sinoconodon* and *Morganucodon oehleri* from the Liassic of Yunnan, China, and *Dinnetherium* from the Kayenta Formation of Arizona.

Sinoconodon possesses a mosaic of primitive and derived cranial features. Many of *Sinoconodon*'s cranial characters are plesiomorphies compared with the successively more distant outgroup cynodonts, *Pachygenelus, Probainognathus, Tritylodon*, and *Thrinaxodon*. *Sinoconodon* retains a large septomaxilla with a transverse shelf and intermediate pterygoid crests on the palate; the prootic vein passes through the cavum epiptericum; the incisors were replaced alternately more than once, and the canines at least four times. The postcanine row is restricted to five multicusped longitudinally ovate teeth that possess only vestigial cingula and do not occlude with one another. The erupting postcanines were successively added to the posterior end of the postcanine row. There is no evidence of replacement of the first four postcanines, but the ultimate postcanine may have been replaced. At least two anterior postcanines were lost in the older specimens.

Sinoconodon shares several derived characters with other mammals. The most notable are the expansion of the brain vault in the parietal region, complete ossification of the medial wall of the orbit, a dentary condyle, and a concave glenoid fossa in the squamosal. These characters suggest that *Sinoconodon* and other mammals form a monophyletic group. *Sinoconodon* lacks a number of diagnostic apomorphies shared by *Morganucodon, Dinnetherium*, and other mammals. In addition, *Sinoconodon* developed some autapomorphic characters: an extraordinarily large canine; a massive dentary condyle; a large occipital condyle; and, relative to skull length, postdentary bones more reduced in size than in the other known Liassic mammals. These characters suggest that *Sinoconodon* is the sistergroup to a taxon that includes all other mammals.

Contents

Introduction, 31
Dentition, 31
Nasal Cavity, 35
Orbital Side Wall, 36
Palate, 37
Petrosal, 38
Discussion and Conclusions, 40
Acknowledgments, 43
References, 43

Chapter 4. Relationships of Early Mammals

Introduction

Several competing hypotheses on the interrelationships of the earliest mammals and their relationship to advanced therapsids have been advanced during last decade (Kemp, 1982, 1983; Crompton and Jenkins, 1979; Crompton and Sun, 1985; Hopson and Barghusen, 1986; Kermack et al., 1973, 1981; Sues, 1985; Rowe, 1988). Both lack of knowledge of the structure of many critical taxa and the uneven emphasis the authors have accorded to various characters have contributed to these conflicting interpretations. Hopson and Barghusen (1986) claim that the trithelodontids (ictidosaurs) are the sistergroup of mammals, whereas the tritylodontids are the sistergroup to the gomphodont (herbivorous) cynodonts. This latter relationship is partly based on a similar pattern of occlusion (bilateral, with a posteriorly directed movement of the lower jaw during the final stages of jaw closure) (Crompton and Ellenburger, 1957; Crompton, 1972). Kemp (1982, 1983), on the other hand, lists several apomorphies shared by mammals and tritylodontids, concluding that they are sistergroups. He (Kemp, 1983: p. 380) dismisses the dental evidence "since convergence of dental structures seems to occur." Sues (1985: p. 216), in a critique of Kemp's character analysis, concludes that "extensive parallel evolution in features of both skull and postcranial skeleton is evident among advanced synapsids, regardless of the preferred hypothesis of tritylodontid relationships." On the basis of his character analysis, Sues agrees with Hopson and Barghusen, that tritylodontids are the sistergroup of gomphodont cynodonts, specifically the *Exaeretodon-Massetognathus* assemblage. Rowe (1988), on the basis of numerous cranial and postcranial characters, argues that the tritylodontids, trithelodontids, and haramiyids are the sistergroup of all the remaining taxa that are generally included within the Mammalia. Several authors (Jenkins and Crompton, 1979; Kermack et al., 1973) suggest a close relationship between the morganucodontids and triconodontids because of dental and petrosal similarities. Rowe (1988), however, concludes that these are not reliable characters for determining phylogenetic relationships of the triconodontids. This controversy underscores the necessity to evaluate more character complexes, carefully interpret the morphology in order to differentiate synapomorphic and homoplasic characters, and perhaps more important, to discover a greater assemblage of late Triassic and early Jurassic synapsids.

Previous studies of *Sinoconodon* (Patterson and Olson, 1961; Crompton and Sun, 1985) are based on isolated skull and postcranial fragments, as well as a complete skull in which the palate and the medial wall of the orbit are damaged. The descriptions of *Morganucodon* are based on an extremely large collection of isolated fragments and a complete skull. This material enabled Kermack et al. (1973, 1981) to give an excellent account of this genus. For the existing account of *Dinnetherium*, only lower and upper jaws and dentitions were discussed (Jenkins et al., 1983; Jenkins, 1984). In this chapter we address the relationship of some Liassic mammals (*Sinoconodon, Morganucodon, Megazostrodon*, and *Dinnetherium*) and the relationships of trithelodontids and tritylodontids to these early mammals.

Thanks to the generosity of Professor Sun Ailin and Mr. Cui Guihai at the Institute of Vertebrate Paleontology and Paleoanthropology (IVPP) at Beijing, we are currently studying a number of nearly complete skulls and partial skulls of the mammals *Sinoconodon* and *Morganucodon oehleri* and a tritylodontid, *Yunnanodon*. All of this material comes from the dark red beds of the Lower Lufeng Formation (Liassic) in the Lufeng Basin of Yunnan Province, China (Sun and Cui, 1986). Other cranial materials available for this study include the extensive but largely undescribed skull materials of *Dinnetherium* and *Pachygenelus*. These studies have revealed a mass of new information.

In assessing the character-state polarity, we primarily rely on outgroup comparison (Wiley, 1979; Eldredge and Cracraft, 1980; Maddison et al., 1984). Trithelodontids, probainognathids, tritylodontids, and galesaurids are taken as successively more distant outgroups to *Sinoconodon* and other early mammals.

Dentition

The dentitions of the new material of *Sinoconodon* can be arranged in series of increasing length (Figs. 4.1 and 4.2). These confirm that *Sinoconodon* did not occlude the postcanine teeth and that no consistent pattern in the positions of upper and lower postcanines can be observed (Crompton and Sun, 1985). Some of the better preserved postcanines possess a faint buccal cingulum on the uppers and a more distinct one on the posterolingual surface of the lowers.

The smallest upper jaw (IVPP 8683) has five upper incisors, all with damaged crowns (Fig. 4.1). Pits for replacing teeth lie internal to I^2 and I^4. The tip of a replacing upper canine is visible in the posterior wall socket for the reception of the lower canine. Three functional postcanines and a fourth in the process of erupting are present. The precise form and size of the erupting tooth have not been determined. With increasing size, the postcanine diastema enlarges. At least two anterior postcanines were lost without replacement. The empty alveoli is filled with spongy bone. In IVPP 4727 the ultimate postcanine is considerably smaller and simpler than other postcanines, whereas in an older specimen (IVPP 8692), it is about the same size as the more anterior postcanines. This suggests replacement of the small ultimate tooth in IVPP 4727.

The size and degree of eruption of the functional

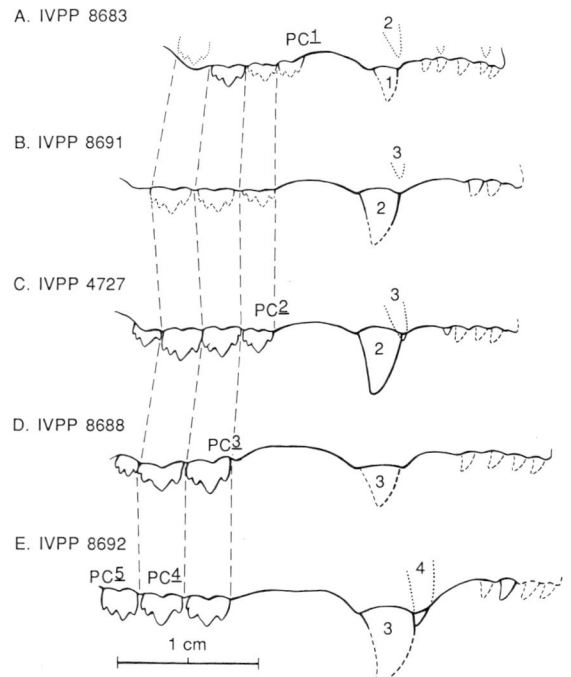

FIGURE 4.1. *Sinoconodon sp.* Possible growth series of upper dentition to illustrate tooth replacement and loss of anterior postcanines.

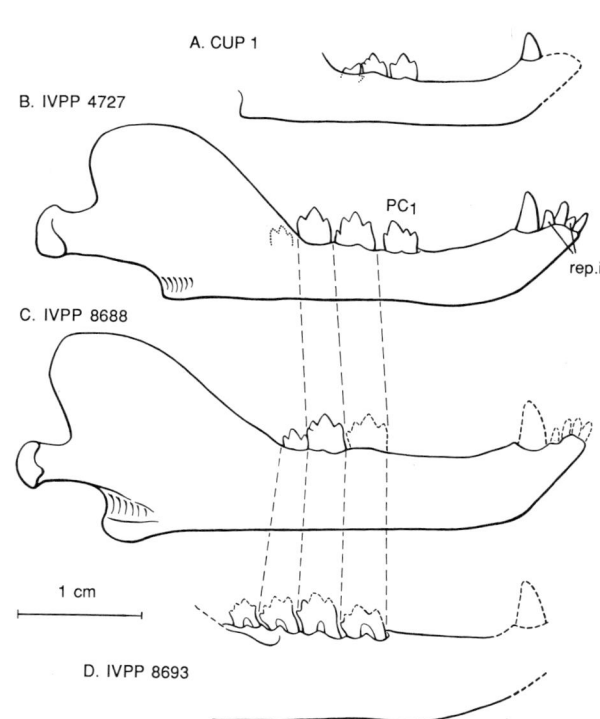

FIGURE 4.2. *Sinoconodon sp.* Possible growth series of lower dentition to illustrate tooth replacement and loss of an anterior postcanine.

canines in different specimens of *Sinoconodon* suggest that canines were replaced at least four times (Figs. 4.1 and 4.2). The number of times the incisors were replaced cannot be determined; however, the lower jaw of IVPP 4727 shows alternate replacement of incisors 2 and 4. As this skull is larger than the smallest (IVPP 8683), several replacements of incisors during the intervening growth period probably occurred. The number of incisors may be reduced to four in older specimens.

Less information is available on the replacement of the lower dentition. The series formed by the three largest *Sinoconodon* jaws (IVPP 4727, 8688, and 8693) suggests that at least one anterior postcanine was lost. A fourth small postcanine is in the process of erupting in IVPP 4727. This tooth appears to have been replaced by a larger postcanine in IVPP 8688, with a new small postcanine added behind. A significant size difference between the postcanines of the smallest lower jaw (CUP Catholic University of Peking 1) and the next larger specimen (IVPP 4727) suggests that the ultimate postcanine of CUP 1 was replaced and a new small postcanine added behind.

Sinoconodon has a pattern of incisor and canine replacement similar to that of some cynodonts. The loss of anterior postcanines and the progressive addition at the end of the postcanine row of a small tooth, which is replaced by a larger tooth, are a pattern seen regularly in gomphodont cynodonts (Hopson, 1971; Crompton, 1972). Loss of anterior postcanines also occurs in carnivorous forms such as *Thrinaxodon* (Crompton, 1963) and *Probainognathus*.

In *Thrinaxodon* (Crompton, 1963) and *Probainognathus*, the dentary has a shallow groove medial to the postcanine alveoli. Replacement teeth developed superficially in pits that originated from this groove. As they matured, they migrated laterally into a space formed by the resorption of the dentary and root of the functional tooth. It is usually assumed that this groove housed the dental lamina. Parrington (1971) describes a similar groove in *Morganucodon*, showing that it communicated with pits housing permanent premolars and the ultimate molar. The groove was retained after the eruption of the permanent dentition. A similar groove without pits, which extends along the length of the postcanine row, was also retained in the lower jaw of the largest specimen of *Sinoconodon* (IVPP 8693).

The new skulls of *Morganucodon oehleri* confirm the observaton by Kermack et al. (1973, 1981) that the postcanine row is divided into three molars and four or five premolars (Fig. 4.3). The only replacement clearly documented in *Morganucodon* is that of the fourth deciduous molar. Parrington's (1971) detailed analyses of numerous *Morganucodon* jaw fragments suppose a mammalian replacement pattern with a single replacement of deciduous incisors, canines, and molars. He

Chapter 4. Relationships of Early Mammals

TABLE 4.1. Abbreviations

	Institutions
BM	British Museum of Natural History
CUP	Catholic University of Peking
IVPP	Institute of Vertebrate Paleontology and Paleoanthropology, Academia Sinica, Beijing
MCZ	Museum of Comparative Zoology, Harvard University

	Morphology
A	articular
a.lam	anterior lamina of petrosal
app	anterior paroccipital process
BO	basioccipital
BS	basisphenoid
bsc	basisphenoid crest
ca.em	canine eminence
ca.ep	cavum epitericum
co.p	coronoid process
d.c.sq	dorsal contact between the squamosal and petrosal
den.c	dentary condyle
eam?	presumed external auditory meatus
EP	epipterygoid
etf	ethmoid-temporal foramen
FO	fenestra ovalis
fo.in	fossa incudis (on petrosal)
for.A	foramen A (through the lateral flange of petrosal)
for.B	foramen B (through the lateral flange of petrosal)
for.oev	foramen for occipital emissary vein
fo.tym.	fossa for posterior extension of cavum tympanum
FR	fenestra rotunda
gl	glenoid fossa for dentary condyle
g.p.f	greater palatine foramen
icf	internal carotid foramen
i.ch.	internal choanae
in.fo	incisive foramen
int.pt.cr	intermediate pterygoid crest
iof	infraorbital foramen
J	jugal
j.f	jugular foramen
L	lacrimal
lac.fo.	lacrimal foramen
lam.cr	lambdoid crest
l.p.f	lesser palatine foramen
lt.fl.	lateral flange of the petrosal
lt.r	lateral ridge on the dentary
M	maxilla
m.cr	median crest of the palate
N	nasal
n.f.	nasal foramen
n.q	notch for quadrate (on squamosal)
oc	occipital condyle
op.f?	optical foramen?
P	parietal
PA	palatine
pang.p	pseudangular ("angular") process
PC	postcanine
p.gl.fl	postglenoid flange
PM	premaxilla
ppf	pterygo-paroccipital foramen
ppp	posterior paroccipital process
pr.	promontorium
pr.cr.	promontorium crest
pro.s.	tympanic opening of the prootic sinus
ptc	post-temporal canal
pt.fl.	pterygoid flange
Q	quadrate
q.r.ep	quadrate ramus of epipterygoid

TABLE 4.1. *Continued*

r.c.	replacing canine
r.c.1–4	replacing canines (first through fourth generation)
r.i.	replacing incisors
r.l	reflected lamina of the angular
r.p.?	replacing postcanine
sc	socket for canine
sg.cr	saggital crest
si	socket for incisor
SM	septomaxilla
smf	septomaxillary foramen
SQ	squamosal
ST.	stapes
st.fo.	stapedial muscle fossa
v.c.sq	ventral facet for the squamosal (on the petrosal)
V_1	foramen for the ophthalmic branch of trigeminal nerve
V_2	foramen for the maxillary branch of trigeminal nerve
V_3	foramen for the mandibular branch of trigeminal nerve
VII	foramen for facial nerve.

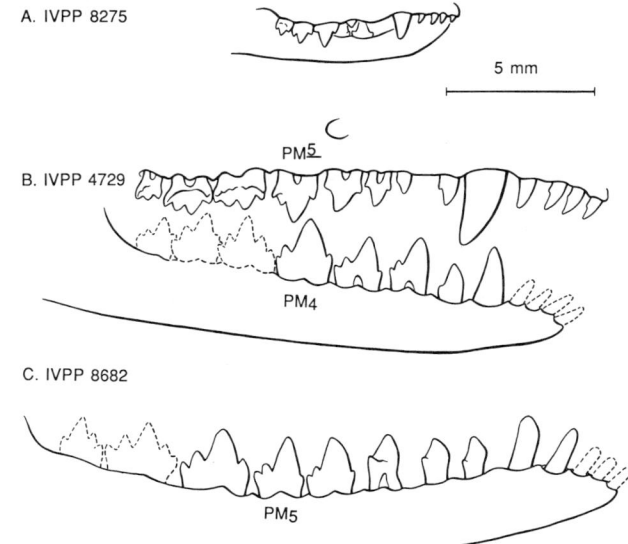

FIGURE 4.3. *Morganucodon oehleri*. **A:** possibly the dentition of a juvenile; **B:** upper and lower dentition of *M. oehleri*; **C:** lower dentition of *M. oehleri* to show variation in the premolar count.

notes the variable size of the ultimate molar in *Morganucodon*. Dismissing the possibility that some of his specimens replaced the smaller ultimate molar with a larger tooth, he instead suggests that the collection of fragments represents a dimorphic or even polymorphic population.

A small skull (IVPP 8275) from the Lufeng may possibly be a juvenile of *Morganucodon oehleri* (Fig. 4.3). Of the four postcanine teeth preserved, the most anterior is an erupting tooth, probably the first premolar; the second deciduous molar is premolariform; the third is more molariform; while the fourth is much smaller than the third. It cannot be determined how often the deciduous dentition was replaced during growth. Gow

(1986) claims that in *Megazostrodon*, which is closely related to *Morganucodon*, the second lower molar was replaced, since the wear facets on this tooth were less developed than those on the adjacent teeth.

A diphyodont dentition suggests the presence of mammary glands and therefore of suckling (Hopson, 1973), because suckling allows substantial cranial growth before tooth eruption. Further, cranial growth after the eruption of the deciduous teeth requires only a single replacement of the deciduous dentition and the eruption of permanent molars. Evidence that teeth of the postcanine row were replaced several times in either *Megazostrodon* or other morganucodontids would undermine the view that lactation, a distinctive characteristic of extant mammals, was present in Liassic mammals.

All the available jaws of *Dinnetherium* appear to be from adult individuals (Figs. 4.4. and 4.5). The postcanine row is differentiated into molars and premolars. In some of the smaller jaws the anterior postcanines were replaced, and in older individuals the molars and premolars are extensively worn. This suggests a diphyodont dentition. In an older specimen, the second premolar is lost and the alveolus is plugged with spongy bone, similar to the pattern seen in *Morganucodon* and *Sinoconodon*.

The structure and occlusion of *Dinnetherium* molars are similar to those of *Morganucodon*. The main cusp a of a lower molar occludes between B and A of an upper molar, as in *Morganucodon* (Fig. 4.5). The difference is that in *Dinnetherium* lower molar cusps b and

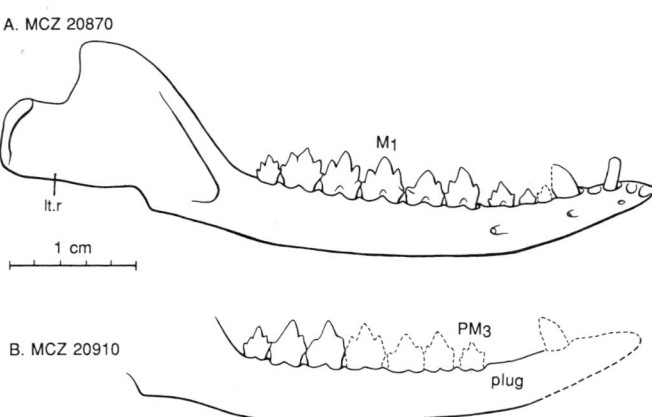

FIGURE 4.4. *Dinnetherium nezorum*. **A:** complete lower jaw with relatively unworn molars; **B:** jaw with extensively worn molars and loss of first two premolars.

c are of equal height in unworn crowns, and are positioned symmetrically on either side of a, whereas in *Morganucodon* cusp b is lower than c and closer to a than c. The wear facets on the lower molars of *Dinnetherium* are close to vertical, while those on the upper are close to horizontal. The lower jaw apparently rotated around its horizontal axis during occlusion. A similar pattern of occlusion is seen in the Jurassic and Cloverly triconodontids (Fig. 4.6), where the lower molars, except for their lack of cingular cusps, resemble those of *Dinnetherium*. Mandibular rotation occurred to a lesser extent in *Morganucodon*. The medial pterygoid

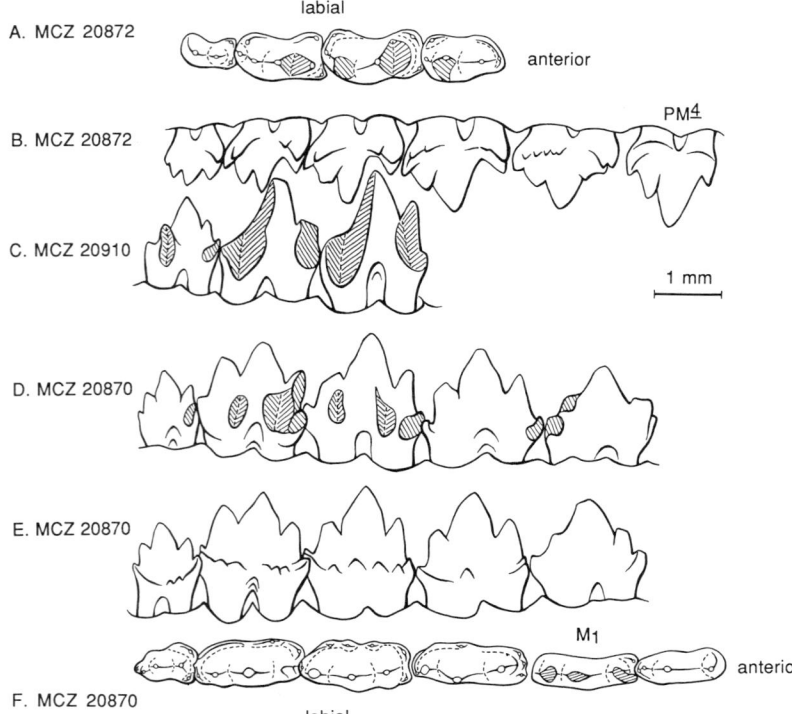

FIGURE 4.5. *Dinnetherium nezorum*. **A:** crown view of last four molars; **B:** external view of upper dentition and last three molars to illustrate how extensive wear modifies crown structure; **C:** external view of lower dentition that has been subjected to only minimal wear; **D, E, and F:** internal and crown view of the same dentition.

Chapter 4. Relationships of Early Mammals

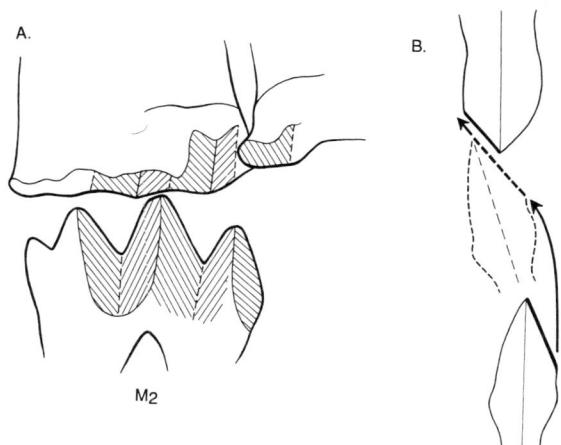

FIGURE 4.6. **A:** Matching wear facets on the inner surface of an upper molar and the outer surface of a lower molar of an undescribed triconodontid from the Cloverly Formation (based on MCZ 19854, 19966). **B:** Different orientation of the wear facets relative to the vertical axes of the upper and lower molars shown in **A**; during the occlusion the lower jaw rotates around its logitudinal axis.

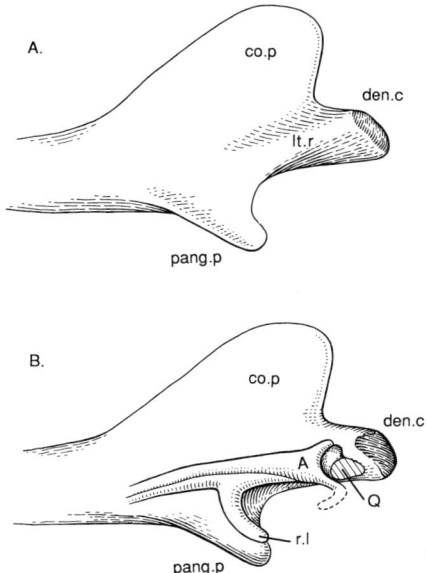

FIGURE 4.7. *Morganucodon oehleri* (IVPP 8682). The posterior region of the lower jaw—**A:** external (lateral) riew; **B:** internal (reversed) view. The reflected lamina is closely applied to a recess on the medial side of the pseudangular process; the quadrate is broken. Note the lateral ridge and the robust dentary condyle. For abbreviations see Table 4.1.

muscle controls mandibular rotation in modern mammals with mobile symphyses. It tends to rotate the mandible counterclockwise (left mandible viewed in front), balancing the masseter and temporal muscles that tend to rotate the mandible clockwise (Oron and Crompton, 1985). In *Morganucodon oehleri*, the reflected lamina of the angular was closely applied to the angular process of the lower jaw (Fig. 4.7). This seems to rule out the insertion of a medial pterygoid on the angle. In contrast to *Morganucodon*, *Dinnetherium* has a greatly expanded lateral ridge in the angular region and near the dentary condyle (Jenkins, 1984). The Cloverly Formation triconodontid has also developed a similar ridge. Muscles (presumably slips of the temporalis and deep masseter) inserting on this ridge would greatly increase clockwise mandibular rotation during occlusion. This may account for the more horizontal upper molar wear facets and more vertical lower molar wear facets in *Dinnetherium* and the Cloverly Formation triconodontid.

In all Liassic mammals (with the exception of *Sinoconodon*) the placement of the lower molars relative to the upper molars remains consistent within each genus. Consequently, teeth occlusion produced consistent wear facets. The relative position of upper and lower molars differs, however, among different taxa. For example, the main cusp of the upper may occlude anterior to lower cusp b, medial to, or behind this cusp. For this reason the term *precise unilateral occlusion* has been used to characterize Liassic mammals other than *Sinoconodon* and the enigmatic haramiyids (Table 4.2). It must be stressed that in none of these forms did upper and lower molars "fit" one another at the time of their eruption. A considerable amount of the crown had to be removed to develop matching wear facets and effective cutting blades. This contrasts sharply, for example, with occlusion in pantotheres, symmetrodonts (with the exception of *Kuehneotherium*), mammals with tribosphenic molars, tritylodontid, and advanced gomphodont cynodonts, where the freshly erupted matching upper and lower teeth "fit" one another and only minimal wear was required to obtain a precise fit.

Nasal Cavity

The new material confirms the suggestion by Wible et al. (1990) that *Sinoconodon* had an internasal process formed by the premaxilla. The septomaxilla is large and forms much of the lateral border of the external nares (Fig. 4.8). A septomaxillary foramen perforates the middle of the bone near the juncture of the premaxillary-maxillary suture. Medially the septomaxilla extends to form a transverse shelf that overlaps the cranial surface of the palatal part of the premaxilla (IVPP 8691). The anterior part of the vomer is Y-shaped in the transverse section and extends almost the entire length of the nasal cavity. The vomer's parallel ridges decrease gradually in height anteriorly. The groove between these ridges received the internasal septum, which either failed to ossify or was dislodged prior to fossilization. A similar Y-shaped vomer is also present in the serial sections of the skull of *Thrinaxodon*

TABLE 4.2. Apomorphic characters of some advanced therapsids and Liassic mammals

A. (Trithelodontids (*Sinoconodon* (all other Liassic mammals)))
   1. Expansion and thickening of the posterior end of the lateral ridge of the dentary.
   2. Unilateral occlusion without consistent positional relationships between the upper and lower postcanines.
   3. Herringbone patterns in enamel prisms.
   4. Greater separation of the fenestra rotunda and the jugular foramen than in *Thrinaxodon*, *Probainognathus*, and *Tritylodon*.
   5. No contact between the distal end of the pterygoid flange and the coronoid bone of the mandible (Hopson and Crompton, in preparation).
   6. Rotation of the dorsal process of the quadrate through 120 degrees relative to the articular trochlea (Hopson and Crompton, in preparation).
   7. Large lateral exposure of the lacimal anterior to the orbit margin.
   8. Slender zygomatic arch (Hopson and Barghusen, 1986).
   9. Dorsomedial lower jaw movement relative to upper jaw during occlusion.

B. Trithelodontids
   1. Reduced number of incisors.
   2. Bulbous upper postcanines.
   3. A large descending process of the frontal.

C. (*Sinoconodon* (*Megazostrodon* ((*Morganucodon*, *Dinnetherium*) *Kuehneotherium*)))
   1. Well-developed squamoso-dentary articulation.
   2. Double-rooted postcanines.
   3. Medial wall of the orbit closed by the orbitosphenoid and ascending process of palatine.
   4. Posterior positioning of the pterygoid flange due to the expansion of the palatine.
   5. Separation of the orbital openings of the greater and lesser palatine foramen (convergence to tritylodontids).
   6. Widening of the posterior part of the palate.
   7. Presence of promontorium.
   8. Prootic sinus (vein) entered the tympanic cavity (convergence to *Probainognathus*).
   9. Posterior border of the incisive foramen formed by the maxilla.
   10. Presence of the postglenoid flange in the squamosal.
   11. Enlargement of the occipital condyles.
   12. Expansion of the brain vault in the parietal region.

D. *Sinoconodon*
   1. Reduction of the cingulum in postcanines.
   2. Enlargment of canines.
   3. Greater reduction of the medial dentary ridges and postdentary bones than in *Morganucodon*.
   4. Extremely large occipital condyle.
   5. Greater separation of the jugular foramen from the round window than in *Morganucodon*, *Dinnetherium*, and triconodontids.

E. (*Megazostrodon*, ((*Morganucodon*, *Dinnetherium*,) *Kuehneotherium*))
   1. Division of the postcanines into premolariform and molariform teeth.
   2. Precise unilateral occlusion and consistent positional relationships between the upper and lower molars.
   3. Greatly reduced mandibular symphysis.
   4. The prootic sinus enclosed in a petrosal canal and the separation of the prootic sinus from the trigeminal ganglion.
   5. Reduction of the promontorial crest on promontorium.

TABLE 4.2. Continued

F. *Megazostrodon*
   1. Occlusion of lower cusp a and upper cusp A between successive teeth.
   2. Strongly divergent anterior and posterior paroccipital processes.

G. ((*Morganucodon*, *Dinnetherium*) *Kuehneotherium*)
   1. Diphyodont dentition.
   2. Tongue and groove interlocking of the adjacent lower molars.
   3. The glenoid formed by the horizontal thickening of the squamosal, with differentiation of the squamosal neck.
   4. Reduced size of the ventral opening of the cavum epiptericum.
   5. Rounded promontorium.
   6. Development of crista parotica on the anterior paroccipital region.
   7. Squamosal withdrawal from the suspensorium for the quadrate.
   8. Presence of a large ventral projection of the posterior paroccipital process.

H. (*Morganucodon*, *Dinnetherium*)
   1. Cusp a occludes between B and A.
   2. Crest on the anterior paroccipital process.

I. *Dinnetherium*
   1. Greatly enlarged lateral mandibular ridge.
   2. Increased mandibular rotation during occlusion.
   3. Enlargement of cusp b in relation to cusp c.
   4. Extremely large posterior paroccipital process.

J. *Morganucodon*
   1. Better developed labial cingula of upper molars than other Liassic mammals.

K. *Kuehneotherium*
   1. Principal cusps of the upper and lower molars in reversed triangle.
   2. Single shear surface between cusp A (paracone) and B (stylacone), and between cusp a (protoconid) and cusp c (metaconid).
   3. Dual shearing surface between A + C, a + b.

L. Tritylodontids
   1. Occlusion of the upper and lower postcanines by eruption.
   2. Multiple-rooted postcanines.
   3. Loss of postcanine tooth replacement (addition of postcanines at the posterior end of the tooth row and loss at the anterior end).
   4. Bilateral occlusion.
   5. Posterior movement of the jaw during occlusion.
   6. Pseudoprismatic enamel arranged in distinct layers.
   7. Incisors reduced to two, with loss of canines.
   8. Loss of the superficial groove for the dental lamina.

(Fourie, 1974) and *Yunnanodon* ("*Yunnania*" of Cui, 1976).

Orbital Side Wall

The complete ossification of the medial wall of the orbit in *Sinoconodon* and *Morganucodon* represents one of

Chapter 4. Relationships of Early Mammals

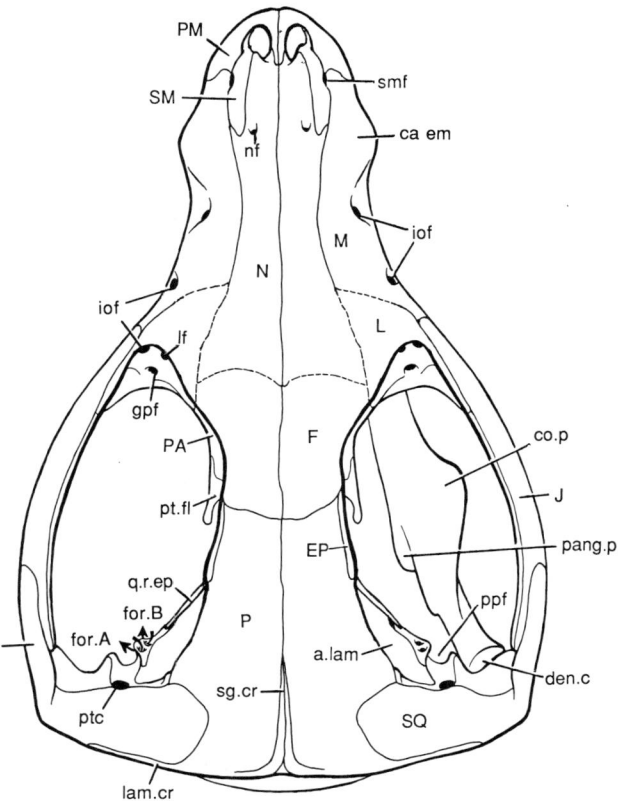

FIGURE 4.8. *Sinoconodon sp.* Dorsal view of the restoration of the skull. Based on IVPP 4727, IVPP 8683, IVPP 8689, IVPP 8692. For abbreviations see Table 4.1.

a list of abbreviations used in the text and the figures). The ethmoid-temporal foramen lies on the suture between the frontal and orbitosphenoid. A large foramen, probably for the optical nerve, perforates the orbitosphenoid. The orbital foramen of the greater palatine canal (spheno-palatine foramen of Kermack et al., 1981) lies on the palatine lacrimal suture, while the lesser palatine foramen is posterior to this suture. A similar arrangement of these two foramina is present in *Morganucodon* (Kermack et al., 1981).

In *Thrinaxodon* (Fourie, 1974), *Probainognathus* (Romer, 1970), and trithelodontids (Crompton, 1958; Hopson and Crompton, in preparation), the medial wall of the orbit is only partially ossified so that a large orbital vacuity is present. In *Thrinaxodon* only a small part of the palatine can be seen in lateral view. The pterygoid and the ectopterygoid form the entire ventral border of the orbit. In *Probainognathus*, Hopson (personal communication) has identified an orbital process of the palatine, but this process is quite small in *Diarthrognathus* (Crompton, 1958) and *Pachygenelus*. In tritylodontids, the orbital vacuity is reduced (more so in large specimens than in small specimens). In large tritylodontids such as *Bienotherium* (Hopson, 1964), *Bienotheroides* (Sun, 1984), and *Kayentatherium* (Sues, 1986), the orbital vacuity is represented by a large sphenopalatine foramen. Compared with these outgroups, the completely ossified orbital wall in *Sinoconodon*, *Morganucodon*, and other mammals is a derived feature.

the most marked differences between the skull structures of advanced cynodonts and those of early mammals (Chapter 12, this volume). The lacrimal and the descending process of the frontal form a part of the medial wall of the orbit in *Sinoconodon* (Fig. 4.9), while the orbitosphenoid partially fills the area anterior to the ascending process of the epipterygoid. There is a significant exposure of the palatine in the lateral view of the orbit. However, a clear suture between the orbital process of the palatine (PA?, Fig. 4.8) and the descending process of the frontal cannot be identified. A large foramen, presumably for V_1, lies on the suture between the orbitosphenoid and the epipterygoid. (See Table 4.1 for

Palate

Sinoconodon (Fig. 4.10) and *Morganucodon oehleri* have retained the five palatal crests characteristic of most therapsids. These consist of a median crest (m.cr.) and two intermediate pterygoid crests (int.pt.cr.), which lie between the median crest and the external transverse processes of the pterygoid flange (pt.fl.). The intermediate crests reach further posteriorly in *Sinoconodon* and *Morganucodon oehleri* than in the outgroup therapsids. The palate at the level of the pterygoid-basisphenoid junction is wider in these mam-

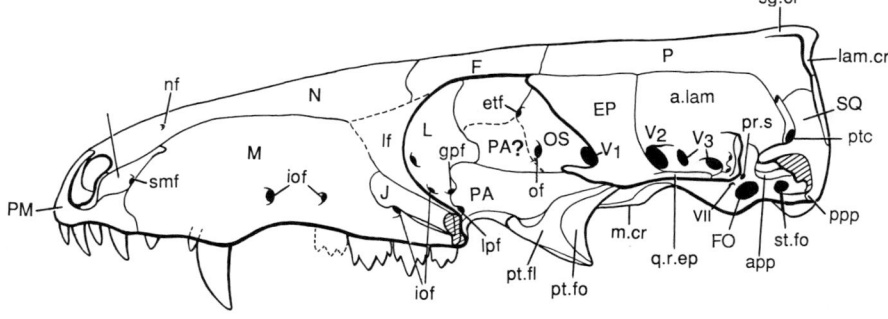

FIGURE 4.9. *Sinoconodon sp.* Lateral view of the restoration of the skull. Based on IVPP 8683, IVPP 4727, IVPP 8692. The zygomatic arch is removed to expose the orbit. For abbreviations see Table 4.1.

38   Alfred W. Crompton and Zhexi Luo

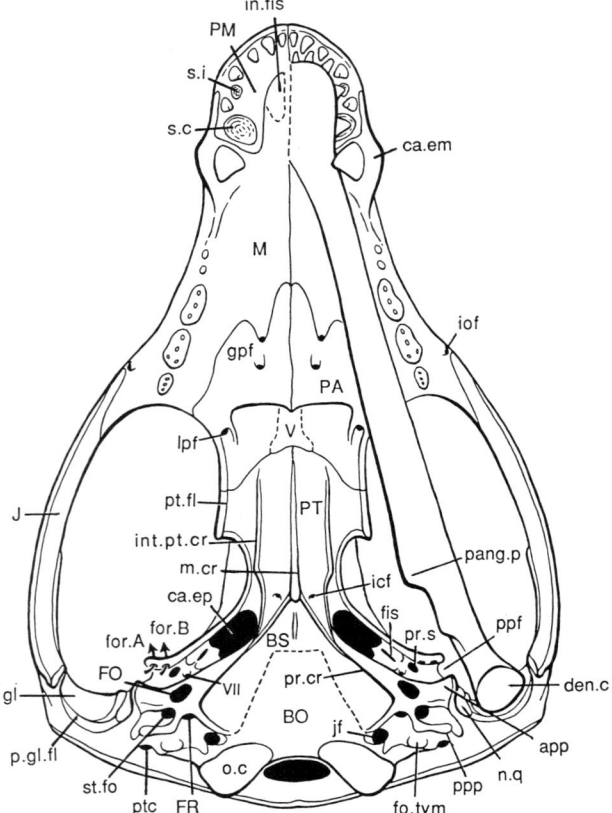

FIGURE 4.10. *Sinoconodon sp.* Ventral view of the restoration of the skull. Based on IVPP 8683, IVPP 4727, IVPP 8692, CUP 5. For abbreviations see Table 4.1.

sive lateral translation of the jaw, but it also precludes the presence of a tensor veli palatini. In the trithelodontid, *Pachygenelus*, the tip of the transverse process of the pterygoid is deflected medially away from the coronoid bone, and in *Sinoconodon*, the contact between this process and the coronoid appears to have been reduced. Consequently *Pachygenelus* and *Sinoconodon* may have possessed the tensor veli palatini. This would allow suckling to take place, but would not prove that suckling occurred.

Petrosal

The petrosal of *Sinoconodon* possesses a broad anterior lamina perforated by three foramina for the branches of the trigeminal nerve (Figs. 4.9 and 4.11). The maxillary branch (V_2) probably exited through the foramen near

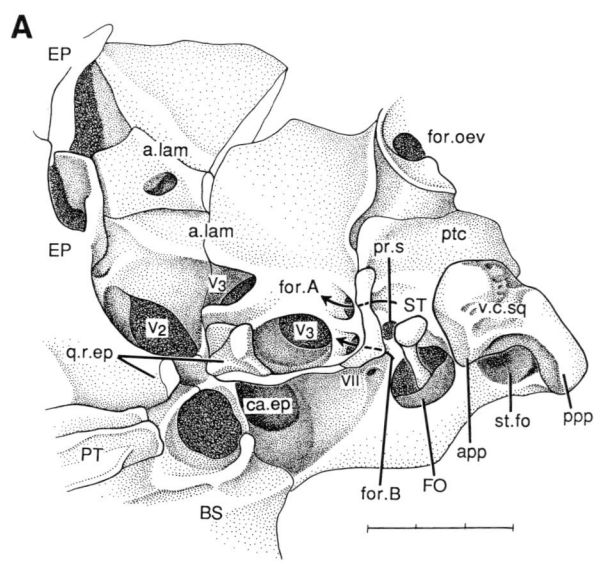

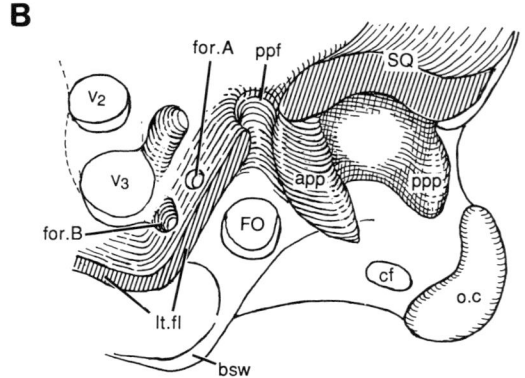

FIGURE 4.11. *Sinoconodon sp.* and tritylodontids. Lateral view of the petrosal structure. **A:** *Sinoconodon sp.* (IVPP 8683). **B:** Generalized petrosal structure of a tritylodontid (based on *Oligokyphus* and *Kayentatherium*).

mals than in *Thrinaxodon* and *Probainognathus*. In all these forms the internal carotid foramina lie between the posterior ends of the median crest and the intermediate pterygoid crests. In *Sinoconodon* the palatal exposure of the palatine is greater than in the outgroup therapsids; consequently the transverse process of the pterygoid flange is situated further posteriorly relative to the postcanine teeth.

Suckling requires a complex oral anatomy. Milk is drawn into the oral cavity and then is either drawn or forced through the fauces to the valleculae and pyriform recesses, where it may be stored prior to swallowing (German and Crompton, 1989). An important component of this mechanism is a seal at the level of the fauces between the oral cavity and the valleculae. This seal is formed by a contraction of palatoglossal muscle and the tensor palatini muscle in synchrony with the elevation of the posterior part of the tongue. In mammals, the muscle fibers of the tensor veli palatini pass over the ventral surface of the pterygoid hamulus to reach the soft palate. In all cynodonts and tritylodontids, the external surface of the transverse flanges of the pterygoid abut the coronoid bone during jaw opening and closing. Not only does this contact prevent exten-

FIGURE 4.12. *Sinoconodon*, triconodontid, and *Dinnetherium*. Ventral view of the petrosal. **A** and **B**: *Sinoconodon sp.* (IVPP 8386); **C**: and unpublished triconodontid from the Cloverly Formation (based on MCZ 19969, 19973); **D**: *Dinnetherium* (based on MCZ 20970, 20971)

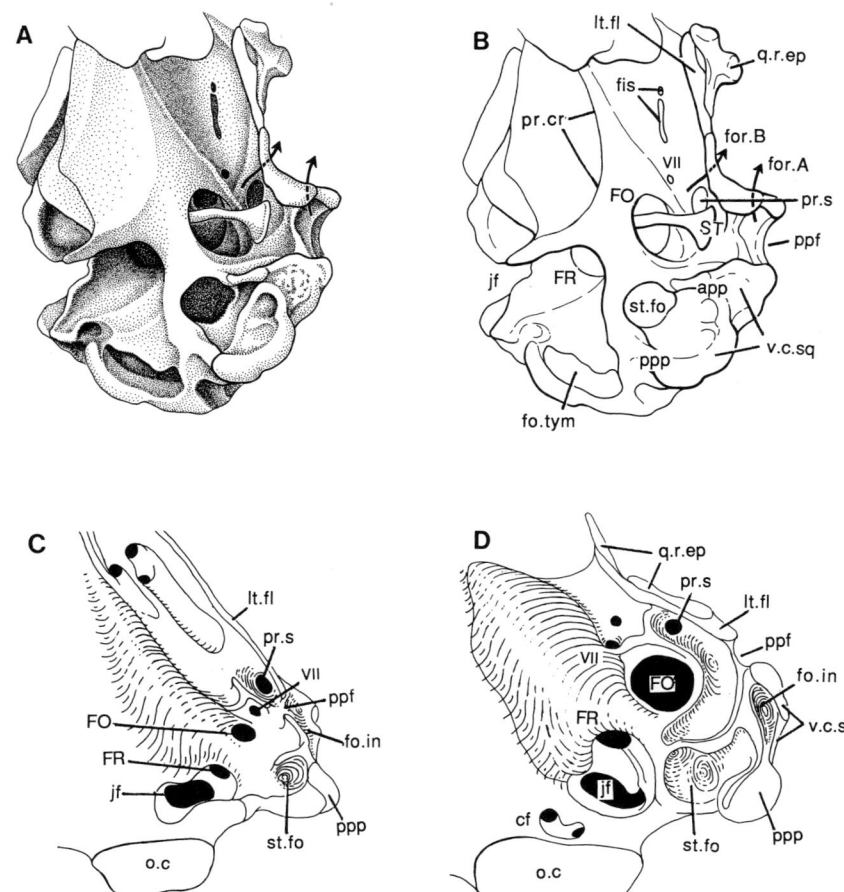

the anterior and lower border of the lamina, and the two branches of the mandibular nerve (V_3) exited through the two remaining foramina. The lateral flange of the petrosal is a broad shelf that is horizontal anterior to and vertical posterior to the foramen for the facial nerve. Therefore, in lateral view the flange appears L-shaped. Two large foramina are present in the lateral flange posterior to the foramen for V_3. In *Morganucodon oehleri* and *Dinnetherium* the lateral flange is less pronounced than in *Sinoconodon* and it is perforated by a single foramen. A lateral flange similar to that of *Sinoconodon* is present in *Megazostrodon* (Gow, 1986). In the Cloverly Formation triconodontids the lateral flange in lateral view is horizontal and lacks the vertical component. Tritylodontids, however, have a very pronounced L-shaped lateral flange perforated by two large foramina (Crompton, 1964; Kemp, 1983; Gow, 1985b) (Fig. 4.11). An L-shaped lateral flange is present in *Luangwa* (Kemp, 1980) and in *Exaeretodon* (MCZ Museum of Comparative Zoology 4781), but the vertical component is absent in *Thrinaxodon*, and tritheledontids.

Sinoconodon has a slightly smaller promontorium of the petrosal than do *Morganucodon oehleri*, *Dinnetherium*, and the triconodontids (Fig. 4.12). In *Sinoconodon* a distinct crest on the ventral surface extends anteriorly to meet the median palatal ridge; and medial to the ridge the surface of the promontorium is broad and flat. Similarly, *Megazostrodon* (Gow, 1986) has a flat ventral promontorium surface with a faint ridge. The promontorium crest of *Sinoconodon* is in the same position as the lateral crests on the parasphenoid-basisphenoid wings of the tritylodontids, trithelodontids, and other therapsids. In *Morganucodon oehleri*, *Dinnetherium*, and triconodontids, the ventral surface of the promontorium is strongly convex and lacks a well-defined crest. The relatively small size of the promontorium and the presence of the promontorium crest suggest that for these features, *Sinoconodon* is more primitive than *Morganucodon oehleri*, *Dinnetherium*, or the triconodontids.

Sinoconodon has a wider separation between the round window and the jugular foramen than do *Morganucodon oehleri*, *Dinnetherium*, or the Cloverly Formation triconodontid. Compared with outgroup therapsids in which the round window and the jugular foramen open into a single foramen, *Sinoconodon* is more derived than in those Liassic mammals where this region is preserved.

The paroccipital region is preserved in several specimens of *Sinoconodon* (IVPP 8683, IVPP 8692, IVPP 4727, CUP 5). They reveal that the paroccipital process terminates laterally in an anterior and posterior process, which are separated by a broad groove that extends medially and joins a deep fossa for the stapedial muscle (st.fo, Fig. 4.12). IVPP 4727 has a bulbous anterior paroccipital process (app), and the lateral surface of the entire paroccipital region (v.c.sq, Figs. 4.11 and 4.12) is in contact with the squamosal. A shallow notch is present on the ventro-medial surface of the squamosal immediately lateral to this bone's contact with the anterior paroccipital process. On the basis of comparison with *Pachygenelus* and other cynodonts, we suggest that this notch supported the dorsal extension of the quadrate (n.q, see Fig. 4.10).

In *Dinnetherium* (MCZ 20971; Fig. 4.12), a thin longitudinal crest is present on the ventral surface of the anterior paroccipital process. A remnant of the quadrate (incus) is preserved medial to this crest. A small depression that the squamosal does not seem to have covered (MCZ 20991 and 20987) lies lateral to this crest. We suggest that this depression is the homologue of the fossa incudis of later mammals (fo. in, Fig. 4.12). The morphology of the anterior paroccipital region in the Cloverly triconodontid (MCZ 19969; Fig. 4.12C) resembles that of *Dinnetherium*. The longitudinal crest is also present on the anterior paroccipital process of two specimens of *Morganucodon oehleri* (IVPP 8682 and 8684). We propose that the trough in the dorsal process of the quadrate was applied to this crest and that the broad dorsal process of the quadrate (incus) reached into the fossa incudis. Many isolated petrosals of *Morganucodon watsoni* (Kermack et al., 1981), however, appear to lack this crest. Although the quadrate has not been found in its natural position in *M. watsoni*, it is assumed that the concave medial surface of this bone articulated with the anterior paroccipital process (Hopson and Barghusen, 1986; Crompton and Sun, 1985). The longitudinal crest on the paroccipital processes of these early mammals and the crista parotica, which in *Ornithorhynchus* supports the incus, appear to be homologous structures (Crompton and Sun, 1985; Zeller, Chapter 8, this volume).

In summary, the anterior paroccipital processes of *Dinnetherium*, *Morganucodon oehleri*, and the Cloverly triconodontid differ from *Sinoconodon* in that they are relatively small, they possess a well-defined crest for the support of the quadrate (incus), and the relative size of the contact with the squamosal is reduced. The characters of *Sinoconodon* are plesiomorphic by outgroup comparison. Among the outgroup therapsids, a large part of the lateral end of the anterior paroccipital process is covered by the squamosal, such as in *Thrinaxodon* (Fourie, 1974), *Probainognathus* (Romer, 1970), and tritheledontids *Diarthrognathus* (Crompton, 1958) and *Pachygenelus* (Crompton and Hylander, 1986; Hopson and Crompton, in preparation). In these forms a flange of the squamosal lies between the dorsal process of the quadrate and the anterior paroccipital process. In tritylodontids, however, the squamosal withdraws dorsally to expose the extremely large and bulbous anterior paroccipital process (Crompton, 1964; Sun, 1984). As a result, the dorsal process of the quadrate has broad contact with the anterior paroccipital process.

Dinnetherium (MCZ 20987), *Morganucodon oehleri* (IVPP 8684), and the Cloverly Formation triconodontids possess a posterior paroccipital process (ppp, Fig. 4.12) that is ventrally directed and does not contact the squamosal. In the plesiomorphic condition in therapsids, the lateral side of the posterior region of the paroccipital process is covered by the squamosal. The posterior paroccipital processes in the *Sinoconodon* skulls available for this study are not preserved as well as those in *Dinnetherium* and *Morganucodon oehleri*.

The tympanic foramen for the prootic vein (sinus) lies in almost the same position relative to the pterygoparoccipital foramen in *Sinoconodon*, *Morganucodon oehleri*, *Dinnetherium*, and the Cloverly triconodontid (pr. s, Fig. 4.12). The tympanic foramen of the prootic sinus in *Sinoconodon* (IVPP 8683) opens into the cavum epitericum, suggesting that the passage of the prootic vein passed through the cavum epitericum. This contrasts with *Megazostrodon* (an unnumbered MCZ specimen), *Morganucodon* (Kermack et al., 1981), *Dinnetherium* (MCZ 20991), *Trioracodon* (BM British Museum of Natural History 47781; Kermack, 1963), and the Cloverly triconodontid (Crompton and Jenkins, 1979), in all of which the prootic vein (sinus) was isolated from the cavum epitericum and passed through a canal within the petrosal. It appears to be a shared derived character of all mammals except *Sinoconodon* (Luo, 1989). In the lizard, the prootic vein passes through the cavum epitericum in an open groove (Oelrich, 1956). This primitive condition was retained by many therapsids, such as *Thrinaxodon* (Crompton and Sun, 1985; Luo, 1989) and tritylodontids (Hopson, 1964; Luo 1989). In *Probainognathus*, the proximal course of the prootic sinus passed through the cavum epitericum, but the distal part of the prootic sinus appears to have been enclosed within the petrosal in a canal (Luo, 1989). This represents an intermediate condition between earlier cynodonts on one hand and all Liassic mammals on the other.

Discussion and Conclusions

Lack of knowledge of early mammalian evolution has made it difficult, if not impossible, to determine the relationships between the Liassic mammals, haramiyids, tritheledontids, and tritylodontids. What light does the new information on cranial and dental structures throw

Chapter 4. Relationships of Early Mammals

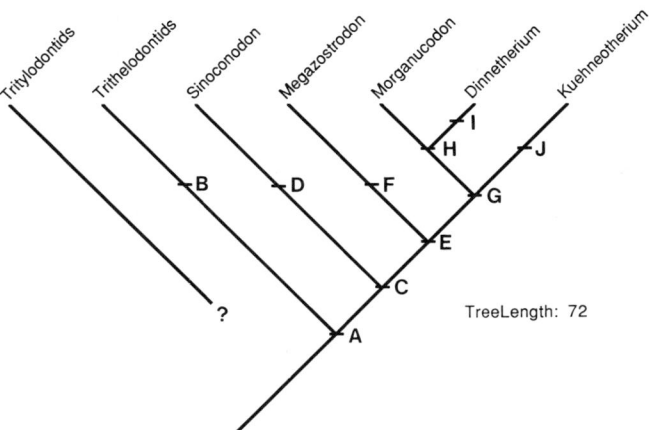

FIGURE 4.13. Relationships of the Liassic mammals. For the list of apomorphies see Table 4.2.

on the interrelationship of the Liassic mammals and their relationship to later mammals as well as advanced therapsids, such as tritylodontids and trithelodontids?

Sinoconodon, Dinnetherium, Morganucodon, and *Megazostrodon* share numerous apomorphies (Table 4.2) with other mammals (haramiyids cannot be included because they are known only from isolated teeth). Such apomorphies include a well-developed squamoso-dentary articulation, a large brain vault in the parietal region of the skull, complete ossification of the medial wall of the orbit, a more posteriorly positioned pterygoid flange, an anterior lamina to the petrosal perforated by at least two branches of the trigeminal nerve, a floor to the cavum epiptericum, and a petrosal promontorium housing the cochlear duct. Such an abundance of shared derived characters supports the view that these Liassic mammals together with all later mammals constitute a monophyletic group (Fig. 4.13).

Sinoconodon lacks many of the apomorphies of other Liassic and later mammals, especially the tooth replacement pattern and a postcanine row that is not divided into premolariform and molariform teeth. Its temporomandibular joint is formed by a large and robust dentary condyle and a broad concave glenoid facing anteriorly and sligtly medially (Crompton and Sun, 1985). This contrasts with the horizontal glenoid that lies below a distinct thickening of the squamosal in both *Dinnetherium* (MCZ 20872) and *Morganucodon watsoni* (Kermack et al., 1981; Crompton and Sun, 1985). *Megazostrodon, Dinnetherium,* and other mammals, besides *Sinoconodon,* form a monophyletic group (Fig. 4.13) that can be diagnosed by this horizontal squamosal glenoid as well as the petrosal canal for the prootic sinus (vein), the geniculate ganglion and division of the facial nerves that lay above the floor of the cavum epiptericum, and a more bulbous promontorium. The dentitions of *Morganucodon, Megazostrodon, Dinnetherium,* and *Kuehneotherium* are characterized by precise unilateral occluion, well-developed cingular cusps, a division of the postcanine row into premolariform and molariform teeth, and the consistent wear facets on the molariform teeth. This probably also holds true for *Helvetiodon, Brachyzostrodon,* and *Woutersia* (Sigogneau-Russell, 1983a,b), which are known only from isolated teeth found in older strata (Rhaetic).

Sinoconodon has several apomorphies shared by neither the outgroup therapsids nor the remaining Liassic mammals. These include greater separation of the fenestra rotunda and the jugular foramen, a large occipital condyle, loss without replacement of the anterior postcanines, greater reduction of postdentary bones and the medial ridge supporting them than in *Morganucodon* or *Dinnetherium,* and enlarged upper canines. Such autapomorphies set *Sinoconodon* apart from the morganucodontids and *Kuehneotherium,* leading us to suggest that *Sinoconodon* is the sistergroup to all other mammals.

The pattern of *Dinnetherium* molar occlusion is similar to that of the later triconodontids. This suggests a close relationship, but other characters must be considered before this can be established. The interlocking device between lower molars in *Morganucodon* and *Megazostrodon* is similar: Lower cusp d fits between lower cusps b and e of the next molar. In *Dinnetherium* and *Kuehneotherium,* the lower cusp d fits between the lower cusps e and f of the next molar. One molariform tooth appears to be replaced in *Megazostrodon* (Gow, 1985a, b), and cusp a of the lower molar occludes anterior to cusp B of the corresponding upper molar.

However, the evidence for differentiation of the postcanine row into molars and premolars in *Kuehneotherium* is not strong (Mills, 1984). Because the upper and lower molariform teeth of *Kuehneotherium* form reversed triangles, this genus is probably a sistergroup to all later mammals that possess embrasure shearing (e.g., symmetrodonts, pantotheres, and mammals with tribosphenic molars).

A major morphological gap separates the dentitions of the Liassic mammals discussed in this chapter from the Liassic therapsids (tritylodontids and trithelodontids). The dentition of trithelodontids is, in most aspects, primitive (Fig. 4.14). Replacement is alternate along the entire length of the dentition; the single long roots of fully erupted postcanines are fused to the alveolar walls; and the postcanine row lacks a division into premolariform and molariform teeth. Although upper and lower postcanines have developed internal cingula on the lowers and external cingula on the uppers (a morganucodontid characteristic), only the lower postcanines retain the primitive longitudinally ovate patterns characteristic of primitive cynodonts and the morganucodontids. Apomorphic features of tritheledontids include bulbous upper postcanines and a reduction and

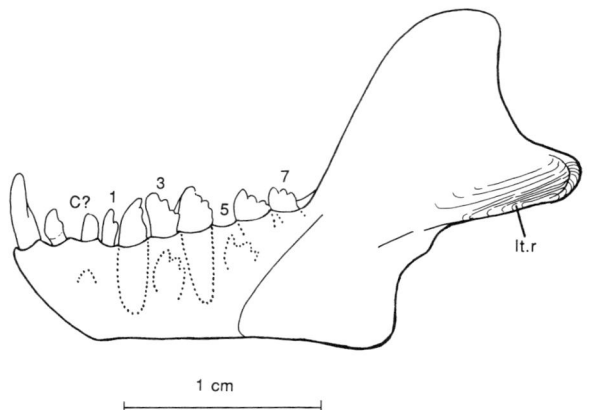

FIGURE 4.14. *Pachygenelus sp.* Lateral view of lower jaw to illustrate tooth replacement pattern and lengths of roots prior to fusion to the alveolar bone.

specialization of the incisors. A number of features relate *Pachygenelus* and *Diarthrognathus* to the Liassic mammals, including thickening on the terminal end of the lateral ridge on the dentary (suggesting an incipient squamoso-dentary contact) (Figs. 4.7 and 4.14) and an enamel structure similar to that of *Megazostrodon*, i.e., pseudoprisms with a herringbone arrangement of the crystalites (Stern, personal communication).

Tritylodon lacks a squamoso-dentary articulation, and its postcanine dentition lacks the primitive features that characterize the trithelodontid dentition. In older specimens of *Tritylodon*, postcanines are lost in front and added to the back of the row; there is no evidence of replacement of the ultimate tooth. The postcanines are more advanced than those of the Liassic mammals, with the possible exception of the haramiyids, in that the occlusal surfaces of freshly erupted upper and lower postcanines fit one another and their matching shearing surfaces do not require extensive wear to produce the multiple complex sickle-shaped shearing crests. Occlusion is bilateral, the postcanines possess more than two roots, and they are transversely widened. The lower jaws of tritylodontids moved posterodorsally during occlusion. Although the enamel of tritylodontids contains pseudoprisms, these (Stern, personal communication) are arranged in distinct layers rather than the symmetrical arrangement of the pseudoprisms in *Pachygenelus* and *Megazostrodon*.

Kemp's (1982, 1983) and Rowe's (1988) views on the relationships of tritylodonts are based upon several cranial and postcranial characters that they recognize as synapomorphies. Rowe (1988), for example, includes the following dental characters in his list of synapomorphies: postcanine teeth with well-developed shearing surfaces and consistent wear patterns, completely divided roots, and division of these roots by a transverse plane. If the relationships proposed by Kemp and Rowe are correct, then features such as transversely widened postcanines, more than two roots, posterior movement of the lower jaw during occlusion, and postcanine teeth that fit one another must be specialized characters of tritylodontids, as comparable features are absent in Liassic mammals, appearing only in later mammals. Hopson (Personal communication) has shown that if only cranial characters are considered, the Liassic mammals plus tritheledonts and chiniquodont cynodonts form the sistergroup to gomphodont cynodonts and tritylodontids. Both the latter groups possessed transversely widened postcanine teeth, posteriorly directed movement of the lower jaw during occlusion, sickle-shaped shearing crests, and postcanine teeth that fit one another when they erupted. However, he also points out that if both cranial and postcranial characters are considered, then tritylodontids form the sistergroup of tritheledontids plus mammals. Such a relationship implies a loss of the posterior directed power stroke and a reversal to primitive dental features in the tritheledontids (alternate tooth replacement, fusion of fully erupted teeth to the jaws, and loss of multirooted teeth). If this relationship is accepted, the interpretation of the dental evolution among these advanced groups would become very complex.

A characteristic of all cynodonts, with the exception of the procynosuchids, is a large masseter muscle and a zygomatic arch that bows away from the external surface of the dentary. With this muscle arrangement, more complex movements are possible. A feature common to gomphodont cynodonts, tritylodontids, tritheledontids, and all mammals is that jaw movements are complex and not limited to the vertical plane, but include movement in a horizontal plane as well-mesiodorsally in primitive mammals and tritheledontids, posterodorsally in gomphodonts and tritylodontids. In modern mammals, the jaw is held in a sling of muscle, and differential contraction of this muscle complex controls jaw movements during occlusion. We suggest that the gomphodonts and tritylodontids evolved a postcanine dentition in conjunction with a posteriorly directed occlusal stroke, whereas the transversely directed occlusal stroke of tritheledontids and mammals arose later from forms with a simple orthal jaw movement during occlusion.

We do not have sufficient information at present to clarify the competing hypotheses on the relationships between tritylodontids, tritheledontids, and mammals. Judging from the diversity of Liassic mammals, especially the highly specialized dentition of the haramiyids, the precursors of Liassic and Rhaetic mammals must have existed in the late Triassic. We need a better fossil record of late Triassic and early Jurassic synapsids and better understanding of the morphology of functional complexes, particularly those involved in feeding, hearing, and locomotion, to determine whether characters upon which phylogenies are based represent shared derived characters or whether they developed in parallel.

ACKNOWLEDGMENTS. Our most sincere thanks go to Professor Sun Ailin and Mr. Cui Guihai of the Institute of Vertebrate Paleontology and Paleoanthropology for providing the early mammal skull specimens that they have patiently collected from Lufeng over last two decades. Their graciousness in permitting us to study the Lufeng Liassic mammal collection and their generous help with this project are greatly appreciated. We are indebted as well to Professor Farish Jenkins, Jr., for allowing us to study the Cloverly Formation triconodontid and *Dinnetherium* skull materials. We are also thankful for the help of Professor James A. Hopson. Mr. William Amaral helped to prepare all the specimens; without his extraordinary skills, this study would not have been possible. We also thank Mr. Laszlo Meszoley for his help with illustrations, and Catherine Musinsky for her editorial assistance.

CORRESPONDENCE ADDRESS. Alfred W. Crompton, Museum of Comparative Zoology, Harvard Unitversity, 26 Oxford Street, Cambridge, MA 02138, USA.

References

Barghusen, H.R. 1968. The lower jaw of the cynodonts (Reptilia, Therapsida) and the evolutionary origin of the mammalian adductor jaw musculature. *Postilla* 116:1–49.

Barghusen, H.R. 1986. On the evolutionary origin of the therian tensor veli palatini and tensor tympani muscles. In: *The ecology and biology of mammal-like reptiles*. Hutton, N., MacLean, P.D., Roth, J.J., and Roth, E.C., eds. Washington and London: Smithsonian Institution Press, pp. 253–262.

Crompton, A.W. 1958. The cranial morphology of a new genus and species of ictidosaurian. *Proceedings of the Zoological Society of London* 140:697–750.

Crompton, A.W. 1963. Tooth replacement in the cynodont *Thrinaxodon liorhinus* Seeley. *Annals of the South African Museum* 46:479–521.

Crompton, A.W. 1964. On the skull of *Oligokyphus*. *Bulletin of British Museum of Natural History, Geology* 9:70–82.

Crompton, A.W. 1972. Postcanine occlusion in cynodonts and tritylodonts. *Bulletin of British Museum (Natural History)* 21:27–71.

Crompton, A.W. 1974. The dentitions and relationships of the Southern African Triassic mammals, *Erythrotherium parringtoni* and *Megazostrodon rudnerae*. *Bulletin of the British Museum of Natural History, Geology* 24(7):399–437.

Crompton, A.W. 1989. The evolution of mammalian mastication. In: *Complex organismal functions: Integration and evolution in vertebrates* (Wake, D.B., and Roth, G., eds.) New York: John Wiley and Sons Ltd., p. 23–40.

Crompton, A.W., and Ellenburger, F. 1957. On a new cynodont from the Molteno Beds and the origin of the tritylodontids. *Annals of the South African Museum* 44:1–13.

Crompton, A.W., and Hylander, W.L. 1986. Changes in mandibular function following the acquisition of a dentary-squamosal jaw articulation. In: *The ecology and biology of mammal-like reptiles*. Hutton, N., MacLean, P.D., Roth, J.J., and Roth, E.C., eds., Washington and London: Smithsonian Institution Press, pp. 263–282.

Crompton, A.W., and Jenkins, F.A. 1979. Origin of mammals. In: *Mesozoic mammals: The first two-thirds of mammalian history* (Lillegraven, J.A., Kielan-Jaworowska, Z., and Clemens, W.A., eds.). Berkeley; University of California Press, pp. 59–72.

Crompton, A.W., and Sun, A.L. 1985. Cranial structure and relationships of the Liassic mammal *Sinoconodon*. *Zoological Journal of Linnean Society* 85:99–119.

Cui, G. 1976. *Yunnania*, a new tritylodontid from Lufeng, Yunnan. *Vertebrata PalAsiatica* 25:1–7.

Edgeworth, F.H. 1935. *Cranial muscles of vertebrates*. Cambridge: Cambridge University Press.

Eldredge, N., and Cracraft, J. 1980. *Phylogenetic patterns and the evolutionary process: Method and theory in comparative biology*. New York: Columbia University Press.

Fourie, S. 1974. the cranial morphology of *Thrinaxodon liorhinus* Seeley. *Annals of the South African Museum* 65:337–400.

German, R.Z., and Crompton, A.W. 1989. Mechanism of suckling in infant pigs. *American Association of Dental Research*, abstract 1100, p. 319.

Gow, C.E. 1980. The dentitions of the Trithelodontidae (Therapsida: Cynodontia). *Proc. R. Soc. London*, 208:461–481.

Gow, C.E. 1985a. Apomorphies of the Mammalia. *South African Journal of Science* 81:558–560.

Gow, C.E. 1985b. The side wall of the braincase in cynodont therapsids and a note on the homology of the mammalian promontorium. *South African Journal of Zoology* 21:136–148.

Gow, C.E. 1986. A new skull of *Megazostrodon* (Mammalia: Triconodonta) from the Elliot Formation (lower Jurassic) of southern Africa. *Palaeontologia Africana* 26:13–23.

Hopson, J.A. 1964. The braincase of the advanced mammal-like reptile *Bienotherium*. *Postilla* 87:1–30.

Hopson, J.A 1971. Postcanine replacement in the gomphodont cynodont *Diademondon*. In: Early mammals (Kermack, D.M., and Kermack K.A., eds.). *Zoological Journal of Linnean Society* 50(Suppl. 1):1–21.

Hopson, J.A. 1973. Endothermy, small size, and the origin of mammalian reproduction. *The American Naturalist* 107:446–452.

Hopson, J.A. In press. Systematics of the non-mammalian Synapsida and implications for patterns of evolution in synapsids. In: *Controversial views on the origins of the higher categories of tetrapods* (Sehultz, H.P., and Trueb, L., eds.). Ithaca, NY: Cornell University Press.

Hopson, J.A., and Barghusen, H.R. 1986. An analysis of therapsid relationships. In: *The ecology and biology of mammal-like reptiles*. (Hutton, N. III, MacLean, P.D.,

Roth, J.J., and Roth, E.C., eds.). Washington, D.C.: Smithsonian Institution Press, pp. 83–106.

Jenkins, F.A. Jr. 1984. A survey of mammalian origins. In: Mammals: Notes for a short course (Gingerich, P.D., and Badgley, C.E., eds.). *University of Tennessee Department of Geological Sciences, Studies in Geology* 8: 32–47.

Jenkins, F.A. Jr., and Crompton, A.W. 1979. Triconodonta. In: *Mesozoic mammals: the first two-thirds of mammalian history* (J.A. Lillegraven, Z. Kielan-Jaworowska and W.A. Clemens, eds.). Berkeley: University of California Press, pp. 74–90.

Jenkins, F.A. Jr., Crompton, A.W., and Downs, W.R. 1983. Mesozoic mammals from Arizona: New evidence on mammalian evolution. *Science* 222:1233–1235.

Kemp, T.S. 1980. Aspects of the structure and functional anatomy of the Middle Triassic cynodont *Luangwa*. *Journal of Zoology* (London) 191:193–239.

Kemp, T.S. 1982. *Mammal-like reptiles and the origin of mammals.* London: Academic Press.

Kemp, T.S. 1983. The interrelationships of mammals. *Zoological Journal of the Linnean Society* 77:353–384.

Kermack, K.A. 1963. The cranial structure of the triconodontids. *Philosophical Transactions of the Royal Society of London* (B)246:83–103.

Kermack, K.A., Mussett, F., and Rigney, H.W. 1973. The lower jaw of *Morganucodon*. *Zoological Journal of the Linnean Society* 53(2):87–175.

Kermack K.A., Mussett F., and Rigney H.W. 1981. The skull of *Morganucodon*. *Zoological Journal of the Linnean Society* 71:1–158.

Luo, Z. 1989. *The petrosal structures of Multituberculata (Mammalia) and the molar morphology of the early arctocyonids (Condylarthra: Mammalia).* Ph.D. Dissertation, Museum of Paleontology, University of California at Berkeley.

Maddison, W.P., Donahue, M.J., and Maddison, D.R. 1984. Outgroup analysis and parsimony. *Systematic Zoology* 33:83–103.

Mills, J.R.E. 1971. The dentition of *Morganucodon*. In: Early mammals (Kermack, D.M., and Kermack, K.A., eds.). *Zoological Journal of the Linnean Society* 50(Suppl. 1):26–63.

Mills, J.R.E. 1984. The molar dentition of a Welsh pantothere. *Zoological Journal of the Linnean Society* 82:189–205.

Oelrich, T.M. 1956. The anatomy of the head of *Ctenosauria pectinata* (Iguanidae). *Miscellaneous Publications, Museum of Zoology, University of Michigan* 94:1–122.

Oron, U., and Crompton, A.W. 1985. A cineradiographic and electromyographic study of mastication in *Tenrec ecaudatus*. *Journal of Morphology* 185:155–182.

Parrington, F.R. 1971. On the Upper Triassic mammals. *Philosophical Transactions of the Royal Society of London* B261:231–272.

Parrington, F.R. 1973. The dentition of the earliest mammals. *Zoological Journal of the Linnean Society* 52:85–95.

Patterson, B., and Olson, E.O. 1961. A triconodontid mammal from the Triassic of Yunnan. In: *International colloquium in the evolution of lower and non-specialized mammals*. Brussels: Koninklijke Vlaamse Academiie voor Wetenschapen, Letteren en Schone Kunsten van Belgie, pp. 129–191.

Romer, A.S. 1970. The Chañares (Argentina) Triassic reptile fauna. VI. A chiniquodontid cynodont with an incipient squamosal-dentary jaw articulation. *Breviora.* 344:1–18.

Rowe, T. 1986. *Osteological diagnosis of Mammalia, Linne. 1758, and its relationship to extinct Synapsida.* Ph.D. Dissertation, Museum of Paleontology, University of California at Berkeley.

Rowe, T. 1988. Definition, diagnosis and origin of Mammalia. *Journal of Vertebrate Paleontology* 8:241–264.

Sigogneau-Russell, D. 1983a. Nouveaux taxon de mammiféres rhétiens. *Acta Paleontologica Polonica* 28(1–2):233–249.

Sigogneau-Russell, D. 1983b. A new therian mammal from the Rhaetic locality of Saint-Nicolas-de-Port (France). *Zoological Journal of the Linnean Society* 78:175–186.

Sues, H.-D. 1985. The relationships of the Tritylodontidae (Synapsida). *Zoological Journal of the Linnean Society* 85:205–217.

Sues, H.-D. 1986. The skull and dentition of two tritylodontoid synapsids from the lower Jurassic of western North America. *Bulletin of the Museum of Comparative Zoology, Harvard University* 151:217–268.

Sun, A.L. 1984. Skull norphology of the tritylodont genus *Bienotheroides* of Sichuan. *Scientia Sinica* (Series B) 27:270–284.

Sun, A.L., and Cui, G. 1986. A brief introduction to the lower Lufeng saurischian fauna (lower Jurassic: Lufeng, Yunnan, People's Republic of China) In: *The beginning of the age of dinosaurs-Faunal change across the Triassic-Jurassic boundary* (Padian, K., ed.). London, New York: Cambridge University Press, pp. 275–278.

Wible, J.R., Miao, D., and Hopson, J.A. 1990. The septomaxilla of fossil and recent synapsids and the problem of the septomaxilla of monotremes and armadillos. *Zoological Journal of the Linnean Society* 98:203–228.

Wiley, E.O. 1979. *Phylogenetics: The theory and practice of phylogenetic systematics.* New York: John Wiley and Sons.

CHAPTER 5

Basicranial Evidence for Early Mammal Phylogeny

JOHN R. WIBLE AND JAMES A. HOPSON

Overview

The distribution of thirty-eight basicranial characters is considered among monotremes, marsupials, placentals, and the following extinct taxa—Tritheledontidae, Tritylodontidae, *Sinoconodon*, Morganucodontidae, *Haldanodon*, Triconodontidae, Multituberculata, and *Vincelestes*. PAUP analysis of the ensuing data matrix supports the following conclusions:

1. Marsupialia and Placentalia form a clade supported by an anterior lamina of the petrosal that is greatly reduced or absent, a cavum epiptericum floored primarily by the alisphenoid, major basicranial drainage via the postglenoid foramen, and a squamosal contributing broadly to the cranial wall.
2. A clade with *Vincelestes* from the Early Cretaceous of Argentina as the sister taxon to Marsupialia plus Placentalia is supported by a caudal tympanic process of the petrosal, a post-promontorial tympanic sinus, a true cochlear aqueduct, and a cochlear duct coiled through at least 270 degrees.
3. A clade comprising the taxa in (2) along with Multituberculata plus Monotremata is supported by loss of support for the ventromedial part of incus on the cranium and a greatly reduced quadrate ramus of the alisphenoid. A clade with Multituberculata and Monotremata is supported by a common tympanic aperture for the prootic canal and pterygoparoccipital foramen.
4. A clade including Triconodontidae and the taxa in (3) is supported by loss of the vascular foramen in the periotic lateral flange and the suspension of the postdentary bones from the cranium.
5. A clade with *Haldanodon* and the taxa in (4) is based on the loss of the quadratojugal notch in the squamosal.
6. A clade including the taxa in (5) plus *Sinoconodon* and Morganucodontidae is supported by an anterior lamina of the petrosal expanded forward dorsal to the exit of the maxillary and mandibular nerves, a cavum epiptericum partially floored by the petrosal, facial ganglion floored by petrosal, stapes length less than 5.5% of skull length, ossified base of the pila antotica absent, major basicranial drainage via a large prootic canal that opens endocranially, a paroccipital process with a distinct ventrally directed projection for muscle attachment, a petrosal promontorium, and a well-developed dentary-squamosal contact.
7. Tritylodontidae shares support of the ventromedial part of the quadrate on the cranium via a convex surface on the crista parotica of the petrosal and a fossa for the stapedius muscle on the petrosal with the taxa in (6).

Only the groupings in (1) and (2) are congruent with those produced from analyses of other anatomical systems; the others are controversial.

Contents

Introduction, 46
Materials and Methods, 46
Character Description and Analysis, 47
PAUP Analysis and Discussion, 57
   Node 1, 57
   Node 2, 57
   Node 3, 58
   Node 4, 58
   Node 5, 58
   Node 6, 59
   Node 7, 59
   Node 8, 60
Conclusions, 60
Acknowledgments, 60
References, 60

Introduction

The mammalian basicranium, in particular the ear region, has long fascinated evolutionary biologists. The transformation of the primitive amniote articular and quadrate to the mammalian malleus and incus, which is clearly traceable in detail in the fossil record and is repeated in the ontogeny of modern mammals, remains one of the best documented and oft-cited examples of morphological evolution (Hopson, 1987). In addition, for the specialist in mammalian systematics, the basicranium has been an important source of characters for determining and testing phylogenetic relationships. This is most evident at lower taxonomic levels, e.g., within Carnivora (Hunt, 1974) and within Marsupialia (Archer, 1976), where allocations based on basicranial characters tend more often than not to corroborate those based on other morphological evidence.

The basicranium has also played a significant role in studies of the interrelationships of monotremes, marsupials, placentals, and various Mesozoic taxa. The consensus view in the 1970s (e.g., Hopson, 1970; McKenna, 1975) that Mammalia is divided into Prototheria (monotremes, triconodonts, docodonts, and multituberculates) and Theria (marsupials, placentals, and Mesozoic relatives) was based largely on a single character of the petrosal bone—presence or absence of an anterior lamina in the side wall of the braincase. This view has been abandoned by most recent authors (but see Luo, 1988, 1989) because the pattern of closure of the side wall of the braincase exhibits developmental variability among extant mammals (Presley, 1981). However, another basicranial character had a profound influence in the 1980s (e.g., Patterson, 1980; Kemp, 1982, 1983): The suspension of the malleus and incus from the cranium has been used to ally monotremes (and multituberculates) with marsupials and placentals.

Despite the importance of the basicranium in mammalian systematics, there have been few attempts to integrate a range of basicranial characters into phylogenetic analyses of monotremes, marsupials, placentals, and their Mesozoic relatives. A notable exception is that of Rowe (1988). Of the 158 characters of the skull and postcranial skeleton analyzed by Rowe (1988), more than 40 are from the basicranium. Yet not all relevant taxa for which basicrania have been described (e.g., tritheledontids, Crompton, 1958; triconodontids, Kermack, 1963; Crompton and Sun, 1985 are treated by Rowe (1988). Additionally, basicrania have been described for several relevant taxa (e.g, paulchoffatiid multituberculates, Hahn, 1988; *Vincelestes*, Bonaparte and Rougier, 1987; Rougier and Bonaparte, 1988; various Late Cretaceous marsupials, Wible, 1990) subsequent to Rowe's (1988) study. The inclusion of these "missing" taxa into an analysis of the basicranium among the higher-level mammalian taxa is the impetus for this chapter. A critique of some of the basicranial characters employed by Rowe (1988) has been published by one of us elsewhere (Wible, 1991). Here we consider the distribution of thirty-eight characters of the basicranium among the following taxa: Tritheledontidae, Tritylodontidae, *Sinoconodon*, Morganucodontidae, *Haldanodon*, Triconodontidae, Multituberculata, *Vincelestes*, Monotremata, Marsupialia, and Placentalia.

Materials and Methods

Two criteria were followed in choosing extinct taxa for our analysis: relevance to the ancestry of Recent mammals and representation by basicrania that have been either described in the literature or studied by us. Of taxa so represented by basicrania, Tritheledontidae, Tritylodontidae, *Sinoconodon*, Morganucodontidae, Triconodontidae, Multituberculata, and *Vincelestes* have been identified in recent analyses as those most relevant to the ancestry of monotremes, marsupials, and placentals (Rowe, 1988; Hopson, 1991; Wible, 1991). The literature sources consulted for descriptions of basicrania are as follows: Tritheledontidae (Crompton, 1958); Tritylodontidae (Kühne, 1956; Ginsburg, 1962; Crompton, 1964; Hopson, 1964, 1965; Sun, 1984; Gow, 1986a; Sues, 1986); *Sinoconodon* (Patterson and Olson, 1961; Crompton and Sun, 1985; Crompton and Luo, this volume); Morganucodontidae (Kermack et al., 1981; Crompton and Sun, 1985; Gow, 1986b; Crompton and Luo, this volume); Triconodontidae (Kermack, 1963; Crompton and Sun, 1985; Crompton and Luo, this volume); Multituberculata (Simpson, 1937; Sloan, 1979; Kielan-Jaworowska et al., 1986; Miao and Lillegraven, 1986; Hahn, 1988; Miao, 1988; Luo, 1989); *Vincelestes* (Bonaparte and Rougier, 1987; Rougier and Bonaparte, 1988); Monotremata (Kuhn, 1971; Zeller, 1989); Marsupialia (Archer, 1976; Wible, 1990); and Placentalia (MacPhee, 1981; Novacek, 1986). Citations for specimens studied by us are in the text. We also include the Upper Jurassic docodontid *Haldanodon exspectatus*. J.A. Lillegraven and G. Krusat have allowed us to include relevant observations from their then unpublished manuscript, subsequently published as Lillegraven and Krusat (1991). Though Rowe (1988) has recently questioned the monophyly of Triconodontidae, we treat it as monophyletic in our analysis based on shared occlusal pattern and molar cusp morphology (Jenkins and Crompton, 1979).

Characters were sampled from all surfaces of the petrosal (prootic and opisthotic) and its neighboring bones in the basicranium. The distribution of the basicranial character states described in the text is recorded in a taxon/character matrix in Table 5.1. This matrix was

Chapter 5. Basicranial Evidence for Early Mammal Phylogeny

TABLE 5.1. Distribution of the 38 basicranial characters among 12 taxa subjected to PAUP analysis

Outgroup	00000	00000	00N00	000NN	N??00	00000	00000	000
Tritheledontidae	00000	00000	0100?	000NN	N0000	000?1	00000	000
Tritylodontidae	00000	01?0?	0110?	000NN	N1100	00011	00001	000
Sinoconodon	01011	?00??	1????	001?0	01?0?	00111	00001	?01
Morganucodontidae	01011	01010	11011	00100	01001	00111	00001	?01
Haldanodon	010?1	0???0	11111	0?100	01001	00111	00101	?01
Triconodontidae	01?21	?1???	1111?	10100	00011	00111	10101	?01
Multituberculata	01121	02211	11001	11101	00001	00111	11101	101
Vincelestes	01021	?21??	1???3	0?1?0	0??1?	11111	11111	?01
Monotremata	01111	02211	11102	01101	00001	00111	11100	101
Marsupialia	1N031	12111	11113	02110	10010	11111	11111	111
Placentalia	1N031	12111	11113	020NN	10010	11111	11111	111

Characters appear in the order that they are described in the text. Characters scored "0" represent the ancestral state, except where the outgroup condition is "?" (nonpreservation) or "N" (nonapplicable); characters scored "1", "2", or "3" represent derived states.

subjected to the branch-and-bound algorithm of the computer program PAUP (Swofford, 1989). All multistate characters were entered as unordered data, so that any state in the transformation series can change to any other state without adding extra steps to the tree length. Two sorts of missing data are recognized: "?" (nonpreservation) and "N" (not applicable). Following Maddison et al. (1984), the ancestral conditions for all character transformations were determined using two outgroups. Unfortunately, there is no consensus as to what are the immediate outgroups to the taxa considered here (cf. Rowe, 1988; Hopson, 1991; Wible, 1991). Nevertheless, all of the potential outgroups identified by the above authors (i.e., *Exaeretodon*, *Diademodon*, *Probainognathus*, Chiniquodontidae, and *Cynognathus*) exhibit the same character states for all but two of the basicranial characters considered here. Therefore, these taxa are treated collectively as the outgroup in polarizing the majority of characters, and the two exceptions are entered with a "?" as the ancestral state.

Our exclusive use of basicranial characters in this analysis is not a reflection of a belief in the basicranium as a taxonomic touchstone. We do not advocate phylogenies produced from a single anatomical (or biochemical) system. Our purpose here is to identify characters to be included in analyses of the entire anatomy and to detail how the basicranium evolves relative to other systems.

Mammalia is used in the traditional sense here, to include those taxa with a well-developed dentary condyle contacting a squamosal glenoid (Kermack and Mussett, 1958; Crompton and Sun, 1985; Hopson, 1991). Therefore, among the extinct taxa considered here, *Sinoconodon*, morganucodontids, *Haldanodon*, triconodontids, multituberculates, and *Vincelestes* are mammals, while tritylodontids, tritheledontids, and the outgroups are non-mammalian cynodonts.

Character Description and Analysis

1. Anterior lamina of petrosal (prootic)—present (0) or greatly reduced or absent (1).
2. Contribution of anterior lamina of petrosal (prootic) to cranial wall—extends forward approximately to posterior border of foramen for maxillary and mandibular nerves (0) or expanded forward anterior and dorsal to exit of maxillary and mandibular nerves (1). In the extinct taxa considered here, an ossification continuous with the petrosal (prootic), referred to as the anterior lamina, contributes in varying degrees to the lateral wall of the cavum epiptericum, the extradural space within which the trigeminal ganglion lies. We use the exit of the maxillary and mandibular nerves to characterize the extent of the anterior lamina's contribution to the braincase. In non-mammalian cynodonts, the anterior lamina forms the posterior border of the foramen for the maxillary and mandibular nerves and does not extend much further anteriorly than the level of this foramen (Hopson and Crompton, 1969; Fig.5.1A). This foramen is a single opening in most taxa, including the tritheledontid *Pachygenelus*, though the tritylodontids *Bienotheroides* (Sun, 1984), *Kayentatherium* (Sues, 1986), and *Tritylodon* (Gow, 1986a) possess separate foramina for the maxillary and mandibular nerves. In the last taxon, the mandibular nerve foramen in some specimens is enclosed entirely within the anterior lamina (Gow, 1986a). In contrast to the pattern in non-mammalian cynodonts, the anterior lamina in the extinct mammalian taxa is expanded forward dorsal to the exit of the maxillary and mandibular nerves (Fig.5.1B–D). A single trigeminal opening between the anterior lamina and alisphenoid is found in the morganucodontid *Megazostrodon* (Gow, 1986b) and the docodontid *Haldanodon* (Lillegraven and

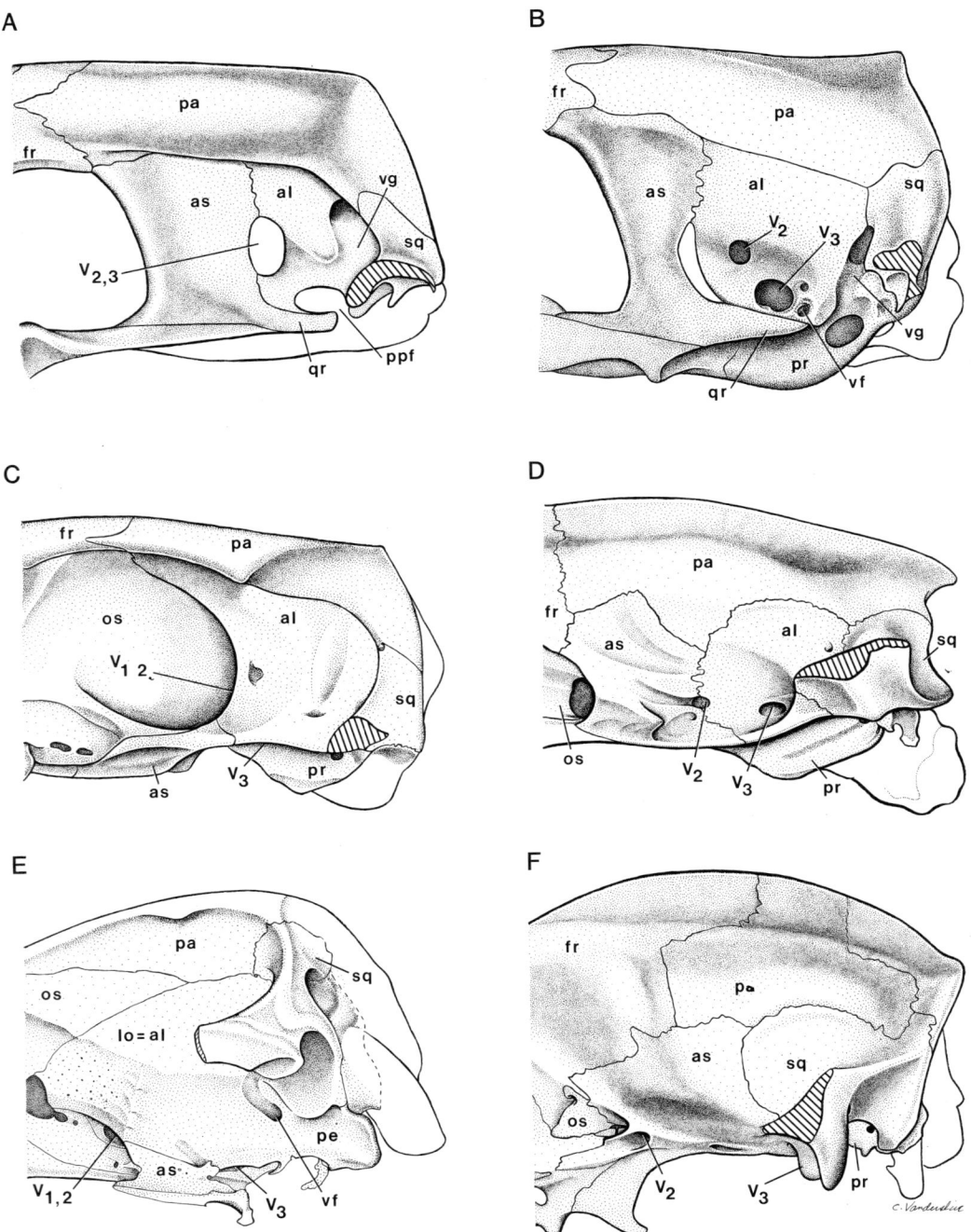

FIGURE 5.1. Braincases in lateral view with zygoma removed. **A**: the tritheledontid *Pachygenelus monus*, based on SAM K1329 and K1350. **B**: the morganucodontid *Morganucodon oehleri*, based on CUP 2320 and Kermack et al. (1981). **C**: the taeniolabidoid multituberculate *Chulsanbataar vulgaris*, ZPAL MgM-I/168 (see Kielan-Jaworowska et al, 1986). **D**: *Vincelestes neu-* *quenianus*, based on MACN-NO5 (Hopson and Rougier, 1993). **E**: the monotreme *Ornithorhynchus anatinus*, sutures reconstructed on adult skull (Hopson, personal collection), based on Vandebroek (1961. fig. 181). **F**: the marsupial *Didelphis marsupialis*, redrawn from Jollie (1962. fig. 3–5B). For abbreviations, see Table 5.2.

Krusat, 1991). The mandibular nerve foramen lies entirely within the anterior lamina in the remaining taxa, *Sinoconodon* (Crompton and Sun, 1985); *Morganucodon* (Kermack et al., 1981; Crompton and Sun, 1985); triconodontids (Kermack, 1963; Crompton and Sun, 1985; Fig. 5.3B), multitubercu-

lates (Sloan, 1979; Kielan-Jaworowska et al., 1986; Hahn, 1988; though not in *Lambdopsalis*, Miao, 1988); and *Vincelestes* (Rougier and Bonaparte, 1988). In addition, the maxillary nerve is also surrounded by the anterior lamina in *Sinoconodon* (Crompton and Luo, this volume), *Morganucodon*

Chapter 5. Basicranial Evidence for Early Mammal Phylogeny

TABLE 5.2. Abbreviations

	Institutions
AMNH	American Museum of Natural History, New York
BMNH	The Museum of Natural History, London
CUP	Catholic University of Peking (housed in the Field Museum of Natural History, Chicago)
MACN	Museo Argentino de Ciencias Naturales, Buenos Aires
MCZ	Museum of Comparative Zoology, Harvard University
SAM	South African Museum, Cape Town
SGP	Museo des Serviços Geológicos, Lisbon
YPM-PU	Yale Peabody Museum, Princeton University Collection
ZPAL	Institute of Paleobiology, Polish Academy of Sciences, Warsaw

	Morphology
ac	aqueductus cochleae
al	anterior lamina of petrosal
as	alisphenoid (epipterygoid)
bo	basioccipital
bs	basisphenoid
ce	cavum epitericum
ctpp	caudal tympanic process of petrosal
e	ectotympanic
ep	ectopterygoid ("Echidna pterygoid")
er	epitympanic recess
fc	fenestra cochleae
fi	fossa incudis
fr	frontal
fv	fenestra vestibuli
gf	glenoid fossa
hF	hiatus Fallopii
i	incus
jf	jugular foramen
ju	jugal
lf	lateral flange of petrosal
lo	lamina obturans
lt	lateral trough
m	malleus
os	orbitosphenoid
pa	parietal
pc	prootic canal
pe	petrosal
pf	perilymphatic foramen
pff	primary facial foramen
pgf	postglenoid foramen
pi	pila antotica
pmp	postmeatal process
pp	paroccipital process
ppf	pterygoparoccipital foramen
pr	promontorium
ptf	post-temporal foramen
q	quadrate
qj	quadratojugal
qr	quadrate ramus of epipterygoid
s	stapes
sf	stapedius fossa
sff	secondary facial foramen
spd	sulcus for perilymphatic duct
sq	squamosal
th	tympanohyal
tp	tympanic process of alisphenoid
ttf	tensor tympani fossa
vf	vascular foramen
vg	vascular groove
V_1, V_2, V_3	foramina for ophthalmic, maxillary, and mandibular divisions of the trigeminal nerve
V_3	foramen ovale

(Kermack et al., 1981; Crompton and Sun, 1985; Fig. 5.1B), and *Vincelestes* (Hopson and Rougier, 1993; Fig. 5.1D).

In marsupials and placentals, the petrosal does not contribute to the lateral wall of the trigeminal fossa and the side wall of the braincase is completed by the alisphenoid and squamosal (Fig. 5.1F). Therefore, an anterior lamina such as that described above is absent. "Greatly reduced" is included in the derived state of character 1 because a possible remnant of the anterior lamina may enclose the prootic canal (see character 18) that occurs in some marsupials (Wible, 1990). Because an anterior lamina of the petrosal is essentially absent in marsupials and placentals, these taxa are scored "N" for character 2.

Scoring monotremes for characters 1 and 2 is problematic. In *Ornithorhynchus*, the lateral wall of the cavum epitericum is formed by an independent intramembranous ossification, the lamina obturans, which fuses in the nestling with the petrosal and in subsequent stages with other adjacent elements (Zeller, 1989; Wible, unpublished observations) (Fig. 5.1E). In *Tachyglossus*, the lateral wall of the cavum epitericum is formed by the lamina obturans, squamosal, "ectopterygoid" ("Echidna pterygoid"), and ossified appositional cartilage and appositional bone from the petrosal (Kuhn and Zeller, 1987). Fusion of the petrosal and lamina obturans, which appears as several ossification centers, may be later than with adjacent elements (Griffiths, 1978). Several views exist on the homologies of the monotreme lamina obturans. This element may be a neomorph of monotremes (Kuhn and Zeller, 1987) or it may be equivalent to the anterior lamina of the petrosal described above for the extinct taxa and/or the intramembranous part of the alisphenoid of marsupials and placentals (Patterson, 1980; Kemp, 1982, 1983). We currently favor the view that the monotreme lamina obturans is equivalent to the anterior lamina of the petrosal and score the primitive state of character 1 for Monotremata. Positionally, the lamina obturans, in particular that of the platypus, and the anterior lamina of the petrosal are virtually identical; for example, the prootic canal is bounded rostrally by the anterior lamina in *Morganucodon* (Kermack et al., 1981), triconodontids (Kermack, 1963), and multituberculates (Kielan-Jaworowska et al., 1986; Hahn, 1988) and passes between the lamina obturans and petrosal in *Ornithorhynchus* (Zeller, 1989). Additionally, Lillegraven and Krusat (1991) report a suture between the anterior lamina and lateral flange of the petrosal for *Haldanodon*, which to us suggests that the docodontid anterior lamina forms by a developmental process resembling that of the monotreme lamina obturans. Given these resemblances, we see no reason for invoking a neomorphic origin for the lamina obturans. Moreover, given that both a well-developed anterior lamina of the petrosal and alisphenoid occur in *Morganucodon* (Kermack et al., 1981) (Fig. 5.1B) and in *Vincelestes*, the oldest therian for which the braincase is known (Rougier and Bonaparte, 1988; Fig. 5.1D), it is unlikely that the monotreme lamina obturans is equivalent to both the anterior lamina and the intramembranous part of the alisphenoid. Monotremata is scored "1" for character 2 because the lamina obturans is expanded far forward dorsal to the maxillary and mandibular nerve foramina in both the echidna and platypus (Fig. 5.1E).

3. Contribution of alisphenoid (epipterygoid) to cranial wall—tall element in side wall of braincase, contacting frontal (0) or reduced and not contacting frontal (1). In most of the taxa considered here, the alisphenoid (epipterygoid) is a substantial element in the orbitotemporal region, reaching dorsally from the basicranium to contact the frontal in the skull roof (Fig. 5.1A, B, D, F). In contrast, the alisphenoid is considerably reduced and does not contact the frontal in monotremes (Fig. 5.1E) and in Late Cretaceous taeniolabidoid multituberculates from Mongolia (i.e., *Kamptobaatar*, Kielan-Jaworowska, 1971; *Chulsanbaatar*, *Nemegtbaatar*, Kielan-Jaworowska et al., 1986; Fig. 5.1C). *Lambdopsalis* from the Paleocene of China, the only other taeniolabidoid preserving sutures in the orbitotemporal region, has a very large alisphenoid that contacts the frontal (Miao, 1988), but we consider this a unique feature of this highly specialized taxon. The only other multituberculate for which the extent of the alisphenoid has been described is the ptilodontoid *Ectypodus* from the early Eocene of Wyoming. Sloan (1979) tentatively reconstructed a well-developed alisphenoid for this taxon, but we are unable to confirm sutures delimiting the parietal, orbitosphenoid, and alisphenoid in the only preserved skull (YPM-PU 14724). Until the extent of the alisphenoid is known for paulchoffatiids and ptilodontoids, we score the derived state for Multituberculata based on the condition in Late Cretaceous taeniolabidoids.

4. Cavum epiptericum floor—open ventrally (0), partially floored by petrosal (1), floored primarily by petrosal (2), or floored primarily by alisphenoid (3). Most non-mammalian cynodonts have no osseous floor beneath the trigeminal ganglion (Fig. 5.2A). The exceptions are *Exaeretodon* (Bonaparte, 1966) and tritylodontids (Kühne, 1956; Crompton, 1964; Hopson, 1964; Gow, 1986a), in which the caudalmost aspect of the trigeminal ganglion lies in a depression in the petrosal. However, because the resulting floor for the trigeminal ganglion is very small, a separate character state is not recognized here and the ancestral state is scored for tritylodontids. In contrast, a well-developed partial or complete osseous floor beneath the trigeminal ganglion is found in the remaining taxa. A partial floor beneath the caudal part of the trigeminal ganglion is formed by the part of the petrosal that roofs the lateral trough in *Sinoconodon* (Crompton and Sun, 1985) and *Morganucodon* (Kermack et al., 1981; Crompton and Sun, 1985), but rostral to it there is a wide opening in the skull base into the cavum epiptericum (Fig. 5.2B). A complete floor composed of the petrosal (and possibly the alisphenoid) is present in triconodontids (Crompton and Sun, 1985), multituberculates (Sloan, 1979; Kielan-Jaworowska et al., 1986; Hahn, 1988), and *Vincelestes* (Bonaparte and Rougier, 1987). Marsupials and placentals also have a complete floor, but it is formed largely by the alisphenoid (Wible, 1990; Fig. 5.2D). Among monotremes, the platypus has a partial floor formed by the lamina obturans and appositional bone from the petrosal (Wible, unpublished observations; Fig. 5.3D), while the echidna has a complete floor formed by the lamina obturans, appositional bone from the petrosal, ectopterygoid, and palatine (Kuhn and Zeller, 1987). A partial floor ("1") is scored for Monotremata because we interpret the complex, complete floor of the echidna as a highly derived condition. Though *Haldanodon* has a floor beneath the cavum epiptericum, the extent is uncertain because of specimen preservation (Lillegraven and Krusat, 1991).

5. Facial ganglion floor—open ventrally (0) or floored by petrosal (1). In non-mammalian cynodonts, the geniculate or facial nerve ganglion appears to have lain below the primary facial foramen in the back of the cavum epiptericum with no osseous floor (Fig. 5.2A). The only exception is the tritylodontid *Oligokyphus*, in which a narrow bridge of petrosal bone below the ganglion separates the palatine and hyomandibular branches of the facial nerve (Kühne, 1956), which then exit the petrosal via the hiatus Fallopii and secondary facial foramen, respectively. Because this floor is absent in *Tritylodon*, the only other tritylodontid for which this region is described (Gow, 1986a), "O" is scored for Tritylodontidae. In contrast, there is a floor formed by the petrosal below the facial ganglion and separate openings for the hyomandibular and palatine branches in mammals (Crompton and Sun, 1985; Wible, 1990, 1991; Lillegraven and Krusat, 1991; Fig. 5.3).

6. Cavum supracochleare—absent (O) or present (1). The facial ganglion lies in the posterior part of the cavum epiptericum immediately behind the tri-

Chapter 5. Basicranial Evidence for Early Mammal Phylogeny

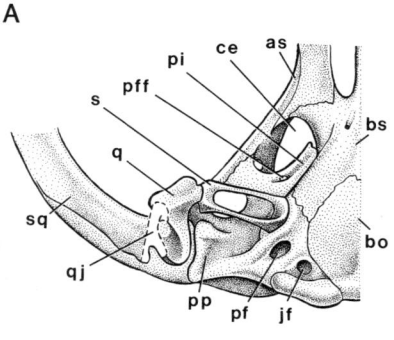
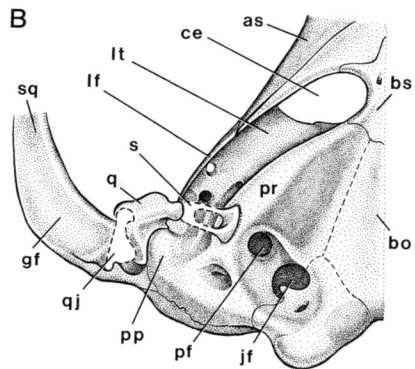
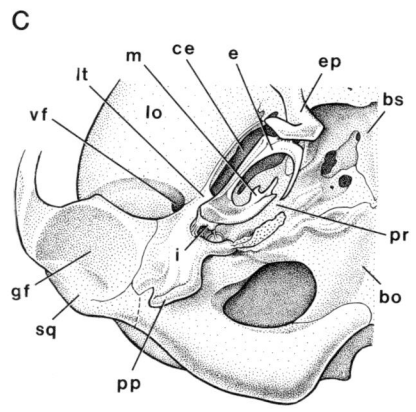
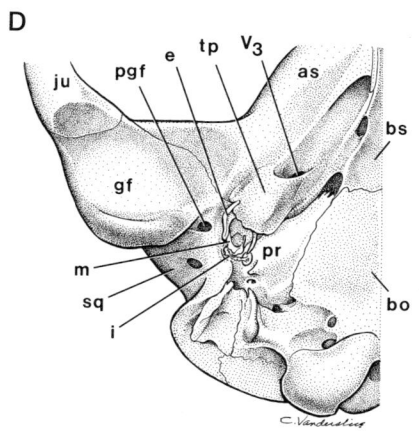

FIGURE 5.2. Basicrania in ventral view. **A**: the tritheledontid *Pachygenelus monus*, based on SAM K1329 and K1350. **B**: the morganucodontid *Morganucodon watsoni*, based on Kermack et al (1981). **C**: the monotreme *Ornithorhynchus anatinus* (Hopson, personal collection). **D**: the marsupial *Didelphis marsupialis* (Hopson, personal collection). Quadratojugal is reconstructed in **A** and **B** based on facets on the quadrate and squamosal. Floor of cavum epiptericum is hidden by ectotympanic in **C** and **D**. For abbreviations, see Table 5.2.

geminal ganglion in most taxa considered. However, in *Tachyglossus*, marsupials, and placentals, an osseous wall encloses the facial ganglion within the petrosal bone and separates a cavum supracochleare (for VII) from the cavum epiptericum (Wible, 1990). Apparently, the enclosed facial ganglion has been derived independently in the echidna (Kuhn, 1971; Wible, 1990), and Monotremata is scored "O" here. The condition in *Sinoconodon*, triconodontids, and *Vincelestes* is unknown.

7. Support of ventromedial part of incus (quadrate) on cranium—via convex surface on squamosal (0), via convex surface on crista parotica of petrosal (1), or no contact (2).
8. Support of dorsal process (crus breve) of incus (quadrate) on cranium—via notch on squamosal (0), via concave surface on petrosal walled laterally by squamosal (1), or via concave surface on petrosal with no substantial lateral wall (2). We identify two multistate characters for the quadrate support on the cranium, reflecting the complexity of the morphologies among the taxa considered. In most non-mammalian cynodonts, the quadrate does not contact the petrosal (prootic and opisthotic); the ventromedial part of the quadrate articulation is with a convex surface on the squamosal and the dorsal process of the quadrate extends into a notch in that bone (Crompton, 1972; Fig. 5.2A). Tritylodontids are the exception; their quadrate contacts the massive, bulbous anterior paroccipital process of the petrosal (Ginsburg, 1962; Crompton, 1964; Hopson, 1964; Sun, 1984; Sues, 1986), which is thought to be equivalent to the crista parotica of Recent mammals (Hopson, 1964). Tritylodontidae is scored "?" for character 8, because it is not certain whether the dorsal process contacts the squamosal in *Oligokyphus* or *Bienotherium*. The quadrate articulation in *Sinoconodon*, as reconstructed by Crompton and Luo (this volume), resembles that in most non-mammalian cynodonts; the major contact of the ventromedial part is with the squamosal and there is a quadrate notch in the

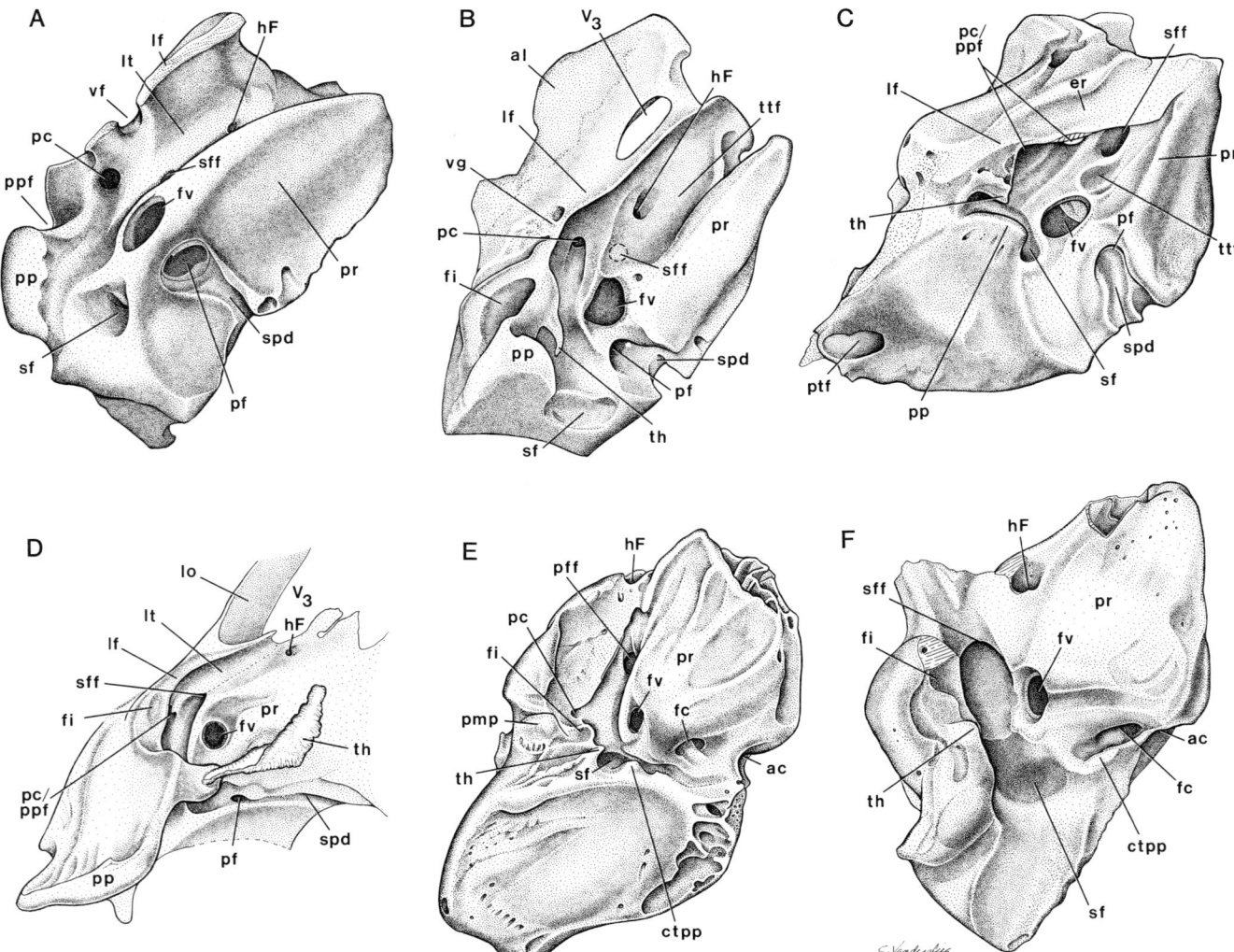

FIGURE 5.3. Right petrosal bones in ventral view. **A**: the morganucodontid *Morganucodon watsoni*, based on Kermack et al. (1981). **B**: the triconodontid *Trioracodon ferox*, BMNH 47781 (see Kermack, 1963). **C**: the taeniolabidoid multituberculate *Catopsalis joyneri*, based on AMNH 119445 and MCZ 19176 (see Kielan-Jaworowska et al., 1986). **D**: the monotreme *Ornithorhynchus anatinus* (Hopson, personal collection). **E**: a Bug Creek marsupial, FMNH PM53907 (see Wible, 1990). **F**: a Bug Creek placental, AMNH 118643 (see Wible, 1990). In **B**, the area of the secondary facial foramen is damaged and the position of this opening is based on the Cloverly triconodontid figured by Crompton and Sun (1985). Dashed line on the roof of the lateral trough in **D** depicts contact between the lamina obturans and periosteal outgrowth from the petrosal. In **E**, the floor of the cavum supracochleare is damaged, exposing the primary facial foramen. For abbreviations, see Table 5.2.

squamosal for the dorsal process. In contrast, in *Morganucodon* the major contact of the ventromedial part of the quadrate is with the anterior paroccipital process (Crompton and Sun, 1985; Fig. 5.2B). Apparently, there are differences among species of *Morganucodon* regarding the housing of the dorsal process; in *M. watsoni* there is a quadrate notch in the squamosal (Kermack et al., 1981; Fig. 5.2B), whereas this depression seems to be in the petrosal (i.e., a true fossa incudis) in *M. oehleri* (Crompton and Luo, this volume). Until we review the recently recovered specimens of *M. oehleri*, we score Morganucodontidae based on *M. watsoni*. A quadrate has not been found for Triconodontidae. However, because the triconodontid paroccipital process resembles that in *M. oehleri* (Crompton and Luo, this volume), the ventromedial part of the quadrate likely contacted there. A true fossa incudis on the petrosal is present in triconodontids (Fig. 5.3B), but the state of the lateral wall, if any, is uncertain. The squamosal is not preserved in the specimens described by Kermack (1963) and is displaced in the specimen figured by Crompton and Sun (1985; MCZ 19973). *Haldanodon* is scored "?"

for both characters, because the exact nature of the quadrate articulation is uncertain (Lillegraven and Krusat, 1991).

In the remaining mammalian taxa, the quadrate (incus) has a more restricted contact with the petrosal, with the dorsal process in a ltrue fossa incudis only. In Multituberculata and *Ornithorhynchus*, the fossa incudis lies on the ventral surface of the crista parotica (or lateral flange) and is not bounded by a lateral wall (Fig. 5.3C, D). In contrast, in *Tachyglossus, Vincelestes*, Marsupialia, and Placentalia, the fossa incudis has a prominent lateral wall formed by the squamosal (Fig. 5.2D). The exceptions among Marsupialia are isolated petrosals from the Late Cretaceous of North America in which the lateral wall of the fossa incudis is formed by a postmeatal process of the petrosal (Wible, 1990; Fig. 5.3E). Monotremata is scored "O" here; the prominent lateral wall present in the adult echidna is lacking in post-hatchling stages and is associated with the highly derived, inflated condition of the squamosal, which walls the temporal fossa laterally.

9. Incus (quadrate) crus longum—absent (0) or present (1). A process on the quadrate that contacts the stapes is known for tritylodontids (Kühne, 1956; Hopson, 1964; Sues, 1986), *Morganucodon* (Kermack et al., 1981), the multituberculate *Lambdopsalis* (Miao and Lillegraven, 1986), and Recent mammals. However, the homologies of the tritylodontid stapedial process with that of the remaining taxa are controversial. Sues (1985) points out that the tritylodontid process exhibits significant differences in its orientation and in the form of its proximal articular facet. We confirm these differences here and add that there are significant functional differences between the broad tritylodontid stapedial process, which is directed at the stapes, and the thin crus longum of the remaining taxa, which meets the stapes at (or near) a right angle. Rowe (1988) and Wible (1991) considered the tritylodontid process to be a crus longum, but based on the above differences we score "O" for Tritylodontidae here. No quadrates have been recovered from *Sinoconodon, Haldanodon*, Triconodontidae, and *Vincelestes*.

10. Stapedial (obturator) foramen of stapes—present (0) or absent (1). Taxa are scored following Wible (1991). Marsupialia and Placentalia are the only taxa in which both character states occur. We follow Novacek and Wyss (1986) in scoring the derived state for these taxa, but add that the polarity of this character in Marsupialia and Placentalia deserves reevaluation. *Haldanodon*, which was not included in Wible (1991), has a stapedial foramen (Lillegraven and Krusat, 1991).

11. Stapes length—greater than 7.5% (0) or less than 5.5% (1) of skull length. Taxa are scored following Wible (1991). Complete stapes are known for only a few of the extinct taxa considered (e.g., the tritheledontid *Pachygenelus monus* SAM K1350 and the multituberculate *Lambdopsalis*, Miao and Lillegraven, 1986). Stapes length is approximated for the remaining taxa by the distance between the fenestra vestibuli and the area on the quadrate for contact of the stapes (or for those taxa with no preserved quadrate, the medial edge of the quadrate articulation on the skull).

12. Separate foramina for cochlear and vestibular nerves (i.e., an internal auditory meatus)—absent (0) or present (1).

13. Common depression housing foramina for facial and vestibulocochlear nerves—absent (0) or present (1). In most non-mammalian cynodonts, a separate foramen for the facial nerve is found on the endocranial surface of the prootic and the vestibulocochlear nerves reach the inner ear via a large gap between the prootic, opisthotic, and supraoccipital. In contrast, a true walled internal auditory meatus with separate foramina for the facial nerve and the cochlear and vestibular branches of the VIII cranial nerve is present in tritheledontids, tritylodontids, and all mammals for which this region is known (Wible, 1991), including *Haldanodon* (Lillegraven and Krusat, 1991). For these taxa, two patterns are found for the internal auditory meatus. In tritylodontids (Kühne, 1956; Crompton, 1964; Hopson, 1964; Gow, 1986a), *Haldanodon* (Lillegraven and Krusat, 1991), triconodontids (Kermack, 1963), and Recent mammals, there is a depression on the endocranial surface of the petrosal at the bottom of which are the foramina for the facial and vestibulocochlear nerves. Such a depression is lacking in tritheledontids (SAM K1329) and *Morganucodon* (Kermack et al., 1981), and the foramen for the facial nerve is well separated from those for the vestibulocochlear nerves. Both conditions occur in multituberculates (Simpson, 1937; Kielan-Jaworowska et al., 1986; Hahn, 1988; Miao, 1988). However, because a common depression is lacking in the Upper Jurassic paulchoffatiids (Hahn, 1988), "0" is scored for Multituberculata.

14. Pila antotica—ossified base present (0) or absent (1). The pila antotica is the part of the primary cranial wall that forms the medial wall of the cavum epiptericum (de Beer, 1937; Fig. 5.2A). Among Recent mammals, a pila antotica is found only in the chondrocranium of monotremes, though most of it is resorbed by the adult and represented by its ossified base, the middle clinoid process of the

basisphenoid (Kühn and Zeller, 1987). Among the extinct taxa considered here, a well-developed ossified base of the pila antotica occurs in non-mammalian cynodonts, including tritheledontids (Crompton, 1958; Fig. 5.2A) and tritylodontids (Kühne, 1956; Crompton, 1964; Hopson, 1964; Gow, 1986a), and in multituberculates (Kielan-Jaworowska et al., 1986; Hahn, 1988; Miao, 1988). In fact, a complete ossified pila antotica is found in taeniolabidoid multituberculates (Kielan-Jaworowska et al., 1986; Miao, 1988). Both Rowe (1988) and Wible (1990, 1991) scored a pila antotica as present in Morganucodontidae. We modify this to absent, because though a pila antotica was tentatively described for *Morganucodon* by Hopson (1964), this very weak process is not on the same scale as in the other extinct taxa. A pila antotica is also absent in *Haldanodon* (Lillegraven and Krusat, 1991) and the triconodontid *Trioracodon ferox* (BMNH 47781).

15. Cochlear duct shape and size—maximum breadth greater than length, oval in cross section (0); elongate (maximum length more than twice breadth), straight or slightly curved, round in cross section (1); elongate, straight with a sharp bend at the distal end, round in cross section (2); or elongate, tightly coiled through at least 270 degrees, round in cross section (3). We score this character only for those taxa for which the details of the shape of the cochlear duct are known and do not rely on the form of the cochlear housing as an indication of the internal morphology. The ancestral state present in the outgroup taxa *Probainognathus* (Allin, 1986) and the chiniquodontid *Probelesodon* (Quiroga, 1979) is a short cochlear duct that is very broad and oval in cross section. In the remaining taxa considered here the cochlear duct is elongate, round in cross section, and ranges from being essentially straight to fully coiled. We recognize three character states based on the degree of curvature. The cochlear duct is straight or only slight curved in *Morganucodon* (Graybeal et al., 1989), *Haldanodon* (Lillegraven and Krusat, 1991), and multituberculates (Sloan, 1979; Kielan-Jaworowska et al., 1986; Miao, 1988; Luo and Ketten, 1989, 1991); straight but with a very sharp bend at its distal end in monotremes (Zeller, 1989); or tightly coiled through a minimum of 270 degrees in *Vincelestes* (G. Rougier, personal communication), marsupials, and placentals.

16. Fossa for tensor tympani muscle on petrosal (prootic)—absent (0) or well-demarcated, immediately in front of fenestra vestibuli (1). Only two taxa considered here have a distinct fossa with prominent walls on the petrosal that accommodates the tensor tympani muscle: Triconodontidae and Multituberculata (Kielan-Jaworowska et al., 1986; Hahn, 1988; Miao, 1988) (Fig. 5.3B, C). In both, the fossa lies immediately in front of the fenestra vestibuli.

17. Course of "sinus canal" vessel dorsal to pterygoparoccipital foramen—extracranial (0), enclosed within the petrosal (1), or endocranial (2). Non-mammalian cynodonts have a foramen or deep notch in the lateral border of the petrosal (prootic and opisthotic) that is closed (or nearly closed) laterally by the lateral flange of the prootic (Parrington, 1946) or by the quadrate ramus of the pterygoid or the alisphenoid (epipterygoid) to form a pterygoparoccipital foramen (Fig. 5.1A). A pterygoparoccipital foramen also occurs in *Sinoconodon* (Crompton and Sun, 1985; Crompton and Luo, this volume) and morganucodontids (Kermack et al., 1981; Gow, 1986b) (Fig. 5.2B), but the quadrate ramus of the alisphenoid appears not to contribute to closure of the lateral border. The vessel (or vessels) that passes through the pterygoparoccipital foramen in the above taxa then runs in a laterally open channel on the lateral face of the petrosal to join the sinus canal system and post-temporal vessels (Fig. 5.1A, B). The ramus superior of the stapedial artery is interpreted to be the homologous vessel in Recent mammals (see Wible, 1989). However, it has different relations to adjacent bones: an intramural course between the lamina obturans (the "anterior lamina"; see character 1) and petrosal in *Ornithorhynchus* and lying entirely within the cranial cavity in marsupials and placentals (Wible, 1987; Zeller, 1989). This vessel is lacking in *Tachyglossus*. The ramus superior reconstructed for multituberculates by Miao (1988) resembles that of the platypus in that it is enclosed within a canal bordered laterally by the anterior lamina of the petrosal. Though the presence of a pterygoparoccipital foramen in triconodontids is uncertain, "0" is scored here because grooves resembling those in *Morganucodon* are found on the lateral surface of the petrosal in *Trioracodon* (Kermack, 1963: Fig. 5.3B). The course of this vessel is not known in the other extinct mammals considered here.

18. Prootic canal—absent (0) or present (1).

19. Position of dorsal opening of prootic canal—within cranial cavity (0) or intramural (between petrosal and squamosal) (1).

20. Position of tympanic opening of prootic canal—separate from other foramina (0) or in common opening with pterygoparoccipital foramen (1).

21. Extracranial exit for major basicranial drainage—prootic canal (0) or postglenoid foramen in squamosal (1). The lateral head vein is the major venous drainage of the cranium in early ontogenetic

stages of all Recent mammals, but is retained in only adult *Ornithorhynchus, Tachyglossus*, didelphids, caenolestids, and some dasyurids (Wible, 1990). In these taxa, the lateral head vein passes through an osseous canal in the skull base, the prootic canal of Gaupp (1908; Fig. 5.3D, E). The prootic canal is large and opens endocranially between the lamina obturans and petrosal in the monotremes, but is reduced in length and width and opens intramurally between the petrosal and squamosal in the marsupials. Absent in non-mammalian cynodonts, a prootic canal resembling that of monotremes has also been identified in *Sinoconodon* (Crompton and Luo, this volume), *Morganucodon* (Kermack et al., 1981; Fig 5.3A), *Haldanodon* (Lillegraven and Krusat, 1991), triconodontids (Kermack, 1963; Crompton and Sun, 1985; Fig. 5.3B), and multituberculates (Kielan-Jaworowska et al., 1986; Hahn, 1988; Miao, 1988; Fig. 5.3C). A prootic cannal also occurs in *Vincelestes* (Bonaparte and Rougier, 1987), but the position of the internal aperture is unknown. Crompton and Luo (this volume) note that the dorsal opening of the prootic canal in *Sinoconodon* is within the cavum epiptericum, whereas in other Mesozoic mammals it is isolated from that space. For now, we do not recognize a separate state, as the dorsal opening of the prootic canal is also in the posterior part of the cavum epiptericum in the triconodontid *Trioracodon ferox* (Kermack, 1963, fig. 1). An additional character (20) identifies a unique arrangement of the prootic canal in *Ornithorhynchus* and ptilodontoid and most taeniolabidoid multituberculates (Wible and Hopson, unpublished observations), in which there is a common tympanic opening that leads to both the prootic canal and pterygoparoccipital foramen (Fig. 5.3C, D). A pterygoparoccipital foramen is lacking in *Tachyglossus*, and the condition in paulchoffatiids is uncertain. The exception among taeniolabidoids is *Meniscoessus* (Luo, 1989), which resembles the remaining mammalian taxa in that the tympanic aperture of the prootic canal is separated from the surrounding basicranial foramina. Character 21 identifies differences in the patterns of basicranial venous drainage among extinct and extant mammals. The prootic canal is well-developed and is the major route for venous blood in all taxa except Marsupialia and Placentalia (Wible, 1990). In the marsupials that retain a prootic canal, this channel is greatly reduced, and the major route for venous blood is via the postglenoid foramen in the squamosal bone (Fig 5.2D). The postglenoid foramen is also the major venous conduit in the remaining marsupials and in placentals, except where the postglenoid vein is secondarily lost. Wible (1990), following Gelderen (1924), has shown that the vein passing through the postglenoid foramen in marsupials has a different ontogenetic history than the vein in the postglenoid foramen in placentals and has raised doubts about the equivalence of these vessels for phylogenetic analysis. Nevertheless, we recognize that marsupials and placentals have reorganized their basicranial venous patterns along similar lines and tentatively treat their postglenoid foramina as homologous structures.

22. Vascular foramen in periotic (prootic) lateral flange—absent (0) or present (1). Taxa are scored following Wible (1991), with the addition that a foramen is present in the lateral flange of *Haldanodon* (Lillegraven and Krusat, 1991). The ancestral state is scored "?" here because the outgroup taxa exhibit both states. A vascular foramen in the lateral flange (Fig. 5.1B, 5.3A) is present in *Exaeretodon*, but absent in *Diademodon*, *Probainognathus*, Chiniquodontidae, and *Cynognathus*.

23. Carotid foramen position—within basisphenoid (0) or within opening into cavum epiptericum (1). Taxa are scored following Wible (1991). The ancestral state is scored "?" here because the outgroup taxa exhibit both states. Carotid foramina in the basisphenoid are present in *Probainognathus* and chiniquodontids, but lacking in *Exaeretodon*, *Diademodon*, and *Cynognathus*. *Haldanodon* has carotid foramina within the basisphenoid (Lillegraven and Krusat, 1991).

24. Craniomandibular joint position relative to fenestra vestibuli—level with (0) or anterior to (1). Taxa are scored following Wible (1991) with the exception of Multituberculata. We score "0" here because the craniomandibular joint is level with the fenestra vestibuli in the Upper Jurassic *Paulchoffatia delgadoi* (SGP 110-115a). *Haldanodon* exhibits the ancestral state (Lillegraven and Krusat, 1991).

25. Paroccipital process with a distinct, ventrally directed projection for muscle attachment—absent (0) or present (1). Differences in the orientation of the paroccipital process (posterior part) have been employed in recent phylogenetic analyses by Rowe (1988) and Wible (1991). In reviewing the morphology of the paroccipital process, we have found it difficult to distinguish in all cases the character states identified by Wible (1991), i.e., ventrally and laterally directed paroccipital processes. We follow Luo (1988) here in recognizing a character state for those taxa in which the paroccipital process has a distinct ventrally directed projection for muscle attachment. Included are *Morganucodon* (Crompton and Sun, 1985; Figs. 5.2B, 5.3A), *Haldanodon* (Lillegraven and Krusat, 1991), triconodontids (Kermack, 1963; Fig. 5.3B), multituberculates (Kielan-Jaworowska et al., 1986; Hahn, 1988; Fig.

26. Caudal tympanic process of petrosal—absent (0) or present (1). Taxa are scored following Wible (1990). *Vincelestes*, marsupials, and placentals have a process arising from the posteroventral surface of the petrosal behind the fenestra cochleae and stapedius fossa (Fig. 5.3E). This process is incomplete in some placentals and only walls the stapedius fossa medially (Fig. 5.3F). A caudal tympanic process of the petrosal is lacking in the remaining taxa, including *Haldanodon* (Lillegraven and Krusat, 1991). *Tachyglossus* has a process arising from the posterosurface of the petrosal, but unlike a caudal tympanic process it lies anterior to the perilymphatic duct.

27. Post-promontorial tympanic sinus—absent (0) or present (1). Taxa are scored following Wible (1990). In most of the taxa considered here, including *Haldanodon* (Lillegraven and Krusat, 1991), the bar of bone dividing the fenestra vestibuli and perilymphatic foramen is extended posteriorly, separating the cochlear fossula from the main middle-ear space (Fig. 5.3A–C). However, this bony bar (between the fenestra vestibuli and cochleae) is not prolonged posteriorly in *Vincelestes*, marsupials (Fig. 5.3E), and placentals (Fig. 5.3F), and a post-promontorial tympanic sinus is formed.

28. Petrosal promontorium—absent (0) or present (1). Following Rowe (1988) and Wible (1991), a bulbous promontorium is present in the mammalian taxa considered here (Figs. 5.2B–D, 5.3), including *Haldanodon* (Lillegraven and Krusat, 1991), but is absent in non-mammalian cynodonts (Fig. 5.2A).

29. Sutural relation of prootic and opisthotic—unfused (0) or fused to form petrosal (1). Taxa are scored following Rowe (1988) and Wible (1991). The prootic and opisthotic are fused in tritylodontids and the mammalian taxa (Figs. 5.2B–D, 5.3), including *Haldanodon* (Lillegraven and Krusat, 1991), but remain separate in the outgroups. Sutures are clearly visible in the juvenile tritheledontid *Pachygenelus monus* (SAM K1350), but the condition in the adult is uncertain.

30. Wall separating perilymphatic foramen from jugular foramen—absent (0) or present (1). Taxa are scored following Wible (1991). With the exception of the outgroups, the perilymphatic foramen (or fenestra cochleae) and jugular foramen are separate in all taxa considered (Fig. 5.2A, B), including *Haldanodon* (Lillegraven and Krusat, 1991).

31. Attachment of postdentary bones—to mandible (0) or suspended from cranium (1). Taxa are scored following Wible (1991). Whereas multituberculates (Miao and Lillegraven, 1986) and Recent mammals are the only taxa for which the post-dentary bones are known to be suspended from the skull, there is evidence that triconodontids and *Vincelestes* exhibit the same state. In addition to having a true fossa incudis (see character 8), triconodontids (see Simpson, 1928, 1929) and *Vincelestes* lack a postdentary trough on the mandible, in which the postdentary bones lie in non-mammalian cynodonts, *Sinoconodon*, morganucodontids (Crompton and Sun, 1985), and *Haldanodon* (Krusat, 1980).

32. Quadrate ramus of alisphenoid (epipterygoid)—elongate (0) or greatly reduced or absent (1). Following Wible (1991), the quadrate ramus of the alisphenoid is greatly reduced or absent in Multituberculata (except *Lambdopsalis*; Miao, 1988), *Vincelestes*, and Recent mammals (Fig. 5.1C–F). An elongate quadrate ramus (Fig. 5.1A, B) appears in the remaining taxa, including *Sinoconodon* (Crompton and Luo, this volume), triconodontids (Crompton and Sun, 1985), and *Haldanodon* (Lillegraven and Krusat, 1991).

33. Quadratojugal notch in squamosal—present (0) or absent (1). Taxa are scored following Rowe (1988) and Wible (1991). With the exception of *Sinoconodon* and Morganucodontidae, a quadratojugal notch in the squamosal is absent in the mammalian taxa considered, including *Haldanodon* (Lillegraven and Krusat, 1991) and triconodontids (Crompton and Sun, 1985).

34. Channel for perilymphatic duct—open (0) or enclosed in osseous canal (cochlear aqueduct) by petrosal (1). Taxa are scored following Wible (1990). A true cochlear aqueduct (and fenestra cochleae) occurs in *Vincelestes*, marsupials (Fig. 5.3E), and placentals (Fig. 5.3F), whereas an open channel for the perilymphatic duct leads to a perilymphatic foramen in the remaining taxa (Fig. 5.3A–D), including *Haldanodon* (Lillegraven and Krusat, 1991).

35. Fossa for stapedius muscle on the petrosal—absent (0) or present (1). Taxa are scored following Wible (1991). With the exception of the outgroups, tritheledontids (Fig. 5.2A), and monotremes, a stapedius fossa is present in all taxa considered (Fig. 5.3A–C, E, F), including *Haldanodon* (ventrolateral petrosal pit in Lillegraven and Krusat, 1991).

36. Manubrium mallei (retroarticular process of articular)—short (0) or elongate (1). Following Wible (1991), the only taxa with a true manubrium of the malleus are Multituberculata and Recent mammals. The condition in *Sinoconodon*, Morganucodontidae, *Haldanodon*, Triconodontidae, and *Vincelestes* is unknown.

37. Squamosal contribution to cranial wall—excluded by petrosal (0) or contributing broadly (1). Taxa are scored following Wible (1991). Only Marsupialia and Placentalia exhibit the derived state.

38. Dentary-squamosal contact—absent (0) or well developed (1). Following Rowe (1988) and Wible (1991), a well-developed dentary-squamosal contact occurs in the mammalian taxa, including *Haldanodon* (Lillegraven and Krusat, 1991).

PAUP Analysis and Discussion

PAUP analysis with the branch-and-bound algorithm of the taxon/character matrix in Table 5.1 resulted in three equally parsimonious trees with a branch length of fifty-six and a consistency index of 0.804. The resulting strict consensus tree is reproduced in Figure 5.4. The three trees differed in the positions of *Sinoconodon* and Morganucodontidae. The clades identified in our analysis of basicranial characters are compared with those produced from other data sets in the following discussion. The numbered nodes correspond to those in Figure 5.4.

NODE 1. A clade including Tritheledontidae and the taxa at Node 2 is supported by two unequivocal synapomorphies of the basicranium—characters 12, internal auditory meatus walled with separate foramina for vestibular and cochlear nerves; and 30, perilymphatic foramen separate from jugular foramen.

NODE 2. A clade composed of Tritylodontidae and the taxa at Node 3 is supported by two unequivocal synapomorphies of the basicranium—characters 7, support of ventromedial part of quadrate on cranium via convex surface on crista parotica of petrosal; and 35, fossa for stapedius muscle on petrosal present. In addition, characters 13 (common depression for facial and vestibulocochlear nerves), 22 (vascular foramen in periotic lateral flange present), and 29 (prootic and opisthotic fused to form petrosal) may be synapomorphies at Node 2, but the condition in the outgroup, Tritheledontidae, and/or *Sinoconodon* is unknown.

There is considerable debate whether the sister group of Mammalia (i.e., Node 3) is Tritylodontidae or Tritheledontidae (Kemp, 1982, 1983; Sues, 1985; Hopson and Barghusen, 1986; Rowe, 1988; Hopson, 1991; Wible, 1991). The basicranial evidence reviewed here supports the Tritylodontidae as the first outgroup to Mammalia (see below) as do analyses of the postcranial skeleton (Kemp, 1982, 1983; Rowe, 1988). Unfortunately, the cranial and postcranial anatomy of tritheledontids to date has not been as well described as that of tritylodontids. Moreover, as has been pointed out by us separately (Hopson, 1991; Wible, 1991), the interrelationships of tritylodontids, tritheledontids, and mammals must be analyzed along with those among advanced non-mammalian cynodonts. If the advanced (eucynodont) non-mammalian cynodonts, both carnivorous and herbivorous species, form a simple paraphyletic series of taxa arranged between *Thrinaxodon* on the one hand and Mammalia on the other, as has been suggested by Rowe (1988), then tritylodontids—based on what is currently known—share more derived features with and are the sister group of mammals. However, if there are in fact two clades of advanced cynodonts, one with transversely widened, multicusped "gomphodont" teeth to which tritylodontids belong, and another with persistently carnivorous morphology to which tritheledontids belong, then the basicranial features that tritylodontids share with mammals have been convergently acquired and tritheledontids are the sister group of mammals (Sues, 1985; Hopson, 1991).

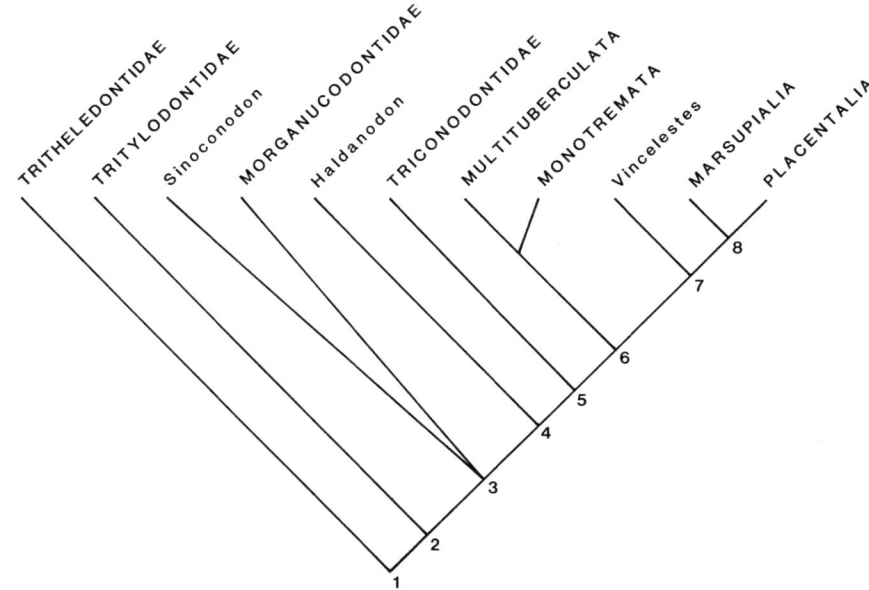

FIGURE 5.4. Strict consensus tree of the three equally parsimonious trees (branch length of fifty-six, consistency index of 0.804) resulting from PAUP analysis with the branch-and-bound algorithm of the data matrix in Table 5.1. Numbers correspond to the nodes discussed in the text.

Only continued discovery and description of extinct taxa in or near the ancestry of mammals will shed light on this debate.

NODE 3. A clade including *Sinoconodon*, Morganucodontidae, and the taxa at Node 4 is supported by twelve unequivocal synapomorphies of the basicranium—2, anterior lamina of petrosal expanded forward dorsal to exit of maxillary and mandibular nerves; 4, cavum epiptericum partially floored by petrosal; 5, facial ganglion floored by petrosal; 11, stapes length less than 5.5% of skull length; 14, ossified base of pila antotica absent; 18, prootic canal present; 19, prootic canal opens endocranially; 20, tympanic opening of prootic canal separate from other foramina; 21, major basicranial drainage via prootic canal; 25, paroccipital process with a distinct ventrally directed projection for muscle attachment; 28, petrosal promontorium present; and 38, well-developed dentary-squamosal contact. The following characters may also be synapomorphies at Node 3, but the condition in Tritheledontidae, Tritylodontidae, *Sinoconodon*, *Haldanodon*, and/or Triconodontidae is unknown—9, quadrate with crus longum, and 15, cochlear duct round in cross section with maximum length more than twice breadth.

Three alternative positions for *Sinoconodon* and Morganucodontidae were identified in the PAUP analysis: as sister taxa; Morganucodontidae as the outgroup to *Sinoconodon* plus the taxa at Node 4; or *Sinoconodon* as the outgroup to Morganucodontidae plus the taxa at Node 4. The latter supports Crompton and Sun's (1985) view that *Sinoconodon* is the most primitive mammal because it lacks several dental characters present in morganucodontids and most other mammals, including unilateral postcanine occlusion and mandibular movement during mastication predominantly with a medial component (see also Crompton and Luo, this volume). This view is not accepted by all authors (see discussion with Node 4). In our analysis, *Sinoconodon* could only be scored for 71% of the characters; further resolution awaits the full description of recently discovered specimens of this taxon.

In an analysis of the petrosal bone among Recent mammals and Mesozoic relatives, Luo (1988) identifies six synapomorphies of a clade comprising morganucodontids, multituberculates, triconodontids, and monotremes. Four of these characters are also included in our analysis. Two (19 and 25) are identified as synapomorphies at Node 3 that are further transformed at a subsequent node. Another character, greater petrosal nerve (palatine ramus of VII) enclosed within a canal, is included in our character 5, because this canal results from the formation of a floor below the facial ganglion. Contra Luo (1988), we identify this canal in the mammals considered here (see Fig. 5.3). The fourth character concerns the quadrate articulation: Quadrate contact with the squamosal that is reduced or lost and an epitympanic recess without a squamosal wall is identified as a synapomorphy of morganucodontids, multituberculates, triconodontids, and monotremes by Luo (1988). This character also occurs in tritylodontids (see character 7). To distinguish among these taxa more fully, we identify two multistate characters for the quadrate articulation here (7 and 8). Two other basicranial characters cited by Luo (1988) concern the position of an artery and a nerve (and ganglion) relative to the cavum epiptericum. Luo (1988) states that the ramus inferior of the stapedial artery runs through at least part of the cavum epiptericum and that the lesser petrosal nerve (and otic ganglion) lies medially within the cavum epiptericum. Whereas the course of the ramus inferior in the extinct taxa is subject to speculation, this vessel clearly runs exclusively ventral to the floor of the cavum epiptericum in *Ornithorhynchus* (Zeller, 1989; Wible, unpublished observations). A complete ramus inferior is absent in *Tachyglossus*, but what is interpreted as a remnant of this vessel by Wible (1984) also lies beneath the floor of the cavum epiptericum. As Luo (1988) noted, inclusion of the lesser petrosal nerve and otic ganglion within the cavum epiptericum distinguishes monotremes from marsupials and placentals, where both structures are usually extracranial (Kuhn and Zeller, 1987; Wible, unpublished observations). Yet these soft structures leave no bony impression in monotremes, and we are uncertain how their location within the cavum epiptericum in fossils is deduced.

NODE 4. A clade including the docodontid *Haldanodon* and the taxa at Node 5 is supported by one unequivocal synapomorphy of the basicranium—33, quadratojugal notch in squamosal absent.

Docodontids have been thought by most recent authors (e.g., Crompton and Sun, 1985) to be closely related to morganucodontids largely because of dental resemblances. Lillegraven and Krusat (1991), however, identify *Haldanodon* as the sister taxon to all other mammals on the basis of the entirety of the available cranial anatomy. Our analysis of the basicranium alone does not support their conclusions, but resolution of this controversy awaits consideration of all available anatomical characters.

NODE 5. A clade including Triconodontidae and the taxa at Node 6 is supported by two unequivocal synapomorphies of the basicranium—22, vascular foramen in periotic lateral flange absent; and 31, postdentary bones suspended from cranium. In addition, there are two equivocal synapomorphies at Node 5—character 4 (cavum epiptericum floored primarily by petrosal), for which the condition in *Haldanodon* is unknown, and character 24 (craniomandibular joint anterior to fenestra vestibuli), for which an equally parsimonious explanation is available.

NODE 6. A clade including Multituberculata, Monotremata, and the taxa at Node 7 is supported by two unequivocal synapomorphies of the basicranium—7, contact of ventromedial part of incus with cranium absent, and 32, quadrate ramus of alisphenoid greatly reduced or absent. Characters 10 (stapes imperforate) and 36 (manubrium mallei) may also be synapomorphies at Node 6, but the condition in *Sinoconodon*, Morganucodontidae, *Haldanodon*, and/or Triconodontidae is unknown. A clade with Multituberculata and Monotremata at Node 6 is supported by one unequivocal synapomorphy—20, a common tympanic aperture for prootic canal and pterygoparoccipital foramen. Equivocal synapomorphies of a Multituberculata/Monotremata clade include characters 3 (alisphenoid reduced and not contacting frontal), 8 (fossa incudis with no substantial lateral wall), 14 (ossified base of pila antotica present), 17 ("sinus canal" vessel dorsal to pterygoparoccipital foramen enclosed in petrosal), and 24 (craniomandibular joint level with fenestra vestibuli). The condition in *Haldanodon*, Triconodontidae, and/or *Vincelestes* is unknown for the first four characters, and an equally parsimonious explanation exists for the last one.

There is no consensus concerning the relationships of the taxa identified at Nodes 5 and 6 in our analysis. As mentioned in the introduction to this chapter, triconodontids, multituberculates, and monotremes at one time were included in the Prototheria along with morganucodontids and docodontids (Hopson, 1970; McKenna, 1975). Recent analyses have allied these three groups more closely with marsupials and placentals. Rowe's (1988) PAUP analysis of 158 characters of the skull and postcranial skeleton among Recent mammals and Mesozoic relatives identifies Multituberculata and Monotremata as the sequential outgroups to Marsupialia and Placentalia. Triconodontidae was not included in his PAUP analysis, but Rowe's (1988) manual estimation of the positions of the triconodontids *Triconodon* and *Trioracodon* has them as paraphyletic taxa at a multichotomous node between Multituberculata and Monotremata. The ninety-one craniodental characters employed by Rowe (1988) were reanalyzed by Wible (1991), and in his PAUP analysis the positions of monotremes and multituberculates were reversed, i.e., Monotremata was identified as the first outgroup to *Vincelestes* and Marsupialia plus Placentalia. Triconodontidae was not included in Wible's (1991) analysis.

Sorting out the relationships of the taxa at Nodes 5 and 6 is complicated by the fact that different anatomical systems support different phylogenies. Postcranially, triconodontids and multituberculates are more like marsupials and placentals than are monotremes (Jenkins and Crompton, 1979; Rowe, 1988). Dentally, monotremes are more like marsupials and placentals, accepting that *Steropodon galmani* from the Early Cretaceous of Australia is a monotreme with a reversed triangle occlusal pattern (Archer et al., 1985; Kielan-Jaworowska et al., 1987). No matter how these taxa are arranged, convergences are unavoidable. Moreover, adding more taxa to the analysis only augments the confusion. For example, *Kuehneotherium* from the Welsh Rhaeto-Liassic (Kermack et al., 1968) shares a reversed triangle occlusal pattern with Monotremata, *Vincelestes*, Marsupialia, and Placentalia, but still had postdentary bones attached to the mandible. Freeing of the postdentary bones from the lower jaw is a synapomorphy at Node 5 here. Either molar cusps with a reversed triangle pattern or the suspension of the postdentary bones from the cranium have evolved more than once among mammals. The reality of our Nodes 5 and 6 will remain a question until more is known about the anatomy of all relevant taxa.

A close relationship between multituberculates and monotremes has been proposed previously (e.g., Kielan-Jaworowska, 1971) but not supported in recent analyses (e.g., Miao, 1988; Rowe, 1988; Wible, 1991). In addition to the derived features identified in our study, another shared by monotremes and multituberculates is "the replacement of the jugal in the zygomatic arch with an elongate maxillary process extending nearly to the squamosal glenoid" (Hopson et al., 1989, p.206). The condition of the zygoma in triconodontids is not known. Though a clade with multituberculates and monotremes seems unlikely based on the molar morphology of *Steropodon galmani* discussed above, we believe this hypothesis deserves further consideration.

NODE 7. A clade allying *Vincelestes* from the Early Cretaceous of Argentina with Node 8 is supported by three unequivocal synapomorphies of the basicranium—26, caudal tympanic process of petrosal present; 27, post-promontorial tympanic sinus present; and 34, perilymphatic duct enclosed in osseous canal by petrosal. Character 16 (a cochlea coiled through at least 270 degrees) also supports this grouping, because though the form of the cochlear duct in triconodontids is uncertain, it is clearly not coiled. Character 8 (fossa incudis walled laterally by squamosal) may also be a synapomorphy at Node 7, but the condition in Triconodontidae is unknown.

A clade with *Vincelestes* as the sister group of Marsupialia and Placentalia is consistent with the evidence from molar morphology presented by Bonaparte and Rougier (1987).

NODE 8. A clade consisting of Marsupialia and Placentalia is supported by four unequivocal synapomorphies of the ear region and braincase—1, anterior lamina of petrosal greatly reduced or absent; 4, cavum epiptericum floored primarily by alisphenoid; 21, major basicranial drainage via postglenoid foramen; and 37, squamosal contributing broadly to cranial wall. Characters 6 (cavum supracochleare present), 17 (endocranial course for "sinus canal" vessel dorsal to pterygoparoccipital foramen), and 25 (paroccipital process flat) may also be synapomorphies at Node 8, but the state in *Vincelestes* is unknown.

Our basicranial data are consistent with recent anatomical and biochemical analyses (e.g., McKenna, 1987; Lillegraven et al., 1987; Novacek et al., 1988) in which a marsupial/placental clade is strongly supported. In the version of PAUP employed here (3.0d), the multiple states present in variable terminal taxa can be interpreted as polymorphisms. Because our data set includes several variable terminal taxa, a second PAUP run was executed with this option. The following taxa and characters were scored for polymorphisms (see discussion above with specific characters): Outgroup (22 and 23), Tritylodontidae (5), Multituberculata (3, 13, 24, and 32), Monotremata (4, 6, 8, and 25), Marsupialia (10 and 18), and Placentalia (10). Analysis of this data matrix produced the same three most parsimonious trees identified in the first PAUP run, with steps added only within the variable terminal taxa. Our polarity assignments for the variable terminal taxa in the first PAUP analysis did not make any difference for overall tree topology.

Conclusions

The basicranium underwent a suite of changes at the boundary between mammalian and non-mammalian cynodonts. These changes were affected, in part, by increasing brain volume, elaborations in the inner ear, alterations to the vascular pattern, and the formation of a new principal jaw articulation. Several more derived levels of organization of basicranial features are apparent within mammals. The marsupial/placental level with its altered pattern of venous drainage and transformed side wall of the braincase is surely a result of shared ancestry, as it is congruent with so many features from diverse anatomical and biochemical systems. Though congruent with the dental evidence, the *Vincelestes*/marsupial/placental level of basicranial organization with its tightly coiled cochlea and enlarged middle-ear space has yet to be tested with characters derived from other parts of the anatomy. Between this and the most primitive mammals are three successive levels of organization, each of which is associated with refinements to the middle-ear ossicular apparatus (e.g., loss of contact between the ventromedial part of the quadrate and petrosal, the suspension of the postdentary bones from the cranium, and loss of the quadratojugal notch). To date, these three levels of organization are associated with few basicranial innovations and are not congruent with other anatomical features, particularly from the dentition and postcranial skeleton. Therefore it is unclear whether a middle-ear ossicular system has evolved only once or multiple times among mammals.

ACKNOWLEDGMENTS. We thank the following individuals and institutions for access to specimens: J.F. Bonaparte and G. Rougier, Museo Argentino de Ciencias Naturales, Buenos Aires; M.A. Cluver and J.A. van den Heever, South African Museum, Cape Town; A.W. Crompton and Z. Luo, Museum of Comparative Zoology, Cambridge; G. Hahn, Institut für Geologie und Paläontologie, Marburg; Z. Kielan-Jaworowska, Paleontologisk Museum, Oslo; H.-J. Kuhn and U. Zeller, Anatomisches Institut, Göttingen; W. Maier, Zoologishes Institut, Tübingen; M.C. McKenna and A.R. Bleefeld, American Museum of Natural History, New York; A. Milner and J.J. Hooker, The Natural History Museum, London; J.H. Ostrom, Peabody Museum, New Haven; B.D. Patterson, L.R. Heaney, and W.D. Turnbull, Field Museum of Natural History, Chicago; and R. Presley, Department of Anatomy, University of Wales College of Cardiff. We are also grateful to J.A. Lillegraven and G. Krusat for permission to cite extensively from their unpublished manuscript on *Haldanodon*. This manuscript benefited from discussions with E.F. Allin, A.W. Crompton, J.A. Lillegraven, Z. Luo, R. Presley, G. Rougier, P.C. Sereno, and R.T. Zanon. Figures 5.1–5.3 were drawn by C. Vanderslice. The research reported here was supported by NSF Grant BSR 87-23162 and 89-96278 to J. R. Wible and NSF Grant BSR 9-06619 to J.A. Hopson.

CORRESPONDENCE ADDRESS. John R. Wible, Department of Anatomical Sciences and Neurobiology, School of Medicine, University of Louisville. Louisville, KY 40292, USA.

References

Allin, E.F. 1986. The auditory apparatus of advanced mammal-like reptiles and early mammals. In: *The ecology and biology of mammal-like reptiles* (Hotton, N. III, MacLean, P.D., Roth, J.J., and Roth, E.C., eds.). Washington, D.C.: Smithsonian Institution, pp. 283–294.

Archer, M. 1976. The basicranial region of marsupicarnivores (Marsupialia), interrelationships of carnivorous marsupials, and affinities of the insectivorous peramelids. *Zool. J. Linn. Soc.* 59:217–322.

Archer, M.A., Flannery, T.F., Ritchie, A., and Molnar, R.E. 1985. First Mesozoic mammal from Australia—an early Cretaceous monotreme. *Nature* 318:363–366.

Bonaparte, J. F. 1966. Sobre las cavidades cerebral, nasal y otras estructuras del craneo de *Exaeretodon* sp. (Cynodontia-Traversodontidae). *Acta Geol. Lilloana* 8:5–31.

Bonaparte, J.F., and Rougier, G. 1987. Mamíferos del Cretácico inferior de Patagonia. *Actas IV Congr. Latinoamer. Paleont., Bolivia* 1:343–359.

Crompton, A.W. 1958. The cranial morphology of a new genus and species of ictidosaurian. *Proc. Zool. Soc. London* 130:183–216.

Crompton, A.W. 1964. On the skull of *Oligokyphus*. *Bull. Brit. Mus. (Nat. Hist.)* 9:69–82.

Crompton, A.W. 1972. The evolution of the jaw articulation in cynodonts. In: *Studies in vertebrate evolution* (Joysey, K.A., and Kemp, T.S., eds.). Edinburgh: Oliver and Boyd, pp. 231–251.

Crompton, A.W., and Sun, A. 1985. Cranial structure and relationships of the Liassic mammal *Sinoconodon*. *Zool. J. Linn. Soc.* 85:99–119.

de Beer, G.R. 1937. *The development of the vertebrate skull.* Oxford: Clarendon Press.

Gaupp, E. 1908. Zur Entwicklungsgeschichte und vergleichenden Morphologie des Schädels von *Echidna aculeata* var. *typica*. *Semon Zool. Forschungsreisen in Australien* 6:539–788.

Gelderen, C.v. 1924. Die Morphologie der Sinus durae matris. Zweiter Teil. Die vergleichenden Ontogenie der neurokraniellen Venen der Vogel und Säugetiere. *Z. Anat. Entwickl.-Gesch.* 74:432–508.

Ginsburg, L. 1962. *Likhoelia ellenbergeri*, tritylodonte du Trias supérieur du Basutoland (Afrique du Sud). *Ann. Paléont.* 48:179–194.

Gow, C.E. 1986a. The side wall of the braincase in cynodont therapsids and a note on the homology of the mammalian promontorium. *S. Afr. J. Zool.* 21:136–148.

Gow, C.E. 1986b. A new skull of *Megazostrodon* (Mammalia, Triconodonta) from the Elliot Formation (Lower Jurassic) of southern Africa. *Palaeont. Afr.* 26:13–23.

Graybeal, A., Rosowski, J.J., Ketten, D.R., and Crompton, A.W. 1989. Inner-ear structure in *Morganucodon*, an early Jurassic mammal. *Zool. J. Linn. Soc.* 96:107–117.

Griffiths, M. 1978. *The biology of the monotremes*. New York: Academic Press.

Hahn, G. 1988. Die Ohr-region der Paulchoffatiidae (Multituberculata, Ober-Jura), *Palaeovert. Montpellier* 18:155–185.

Hopson, J.A. 1964. The braincase of the advanced mammal-like reptile *Bienotherium*. *Postilla* 87:1–30.

Hopson, J.A. 1965. *Tritylodontid therapsids from Yunnan and the cranial morphology of* Bienotherium. Ph.D. Dissertation, University of Chicago, Chicago.

Hopson, J.A. 1970. The classification of non-therian mammals. *J. Mammal.* 51:1–9.

Hopson, J.A. 1987. The mammal-like reptiles: A study of transitional fossils. *Amer. Biol. Teacher* 49:16–26.

Hopson, J.A. 1991. Systematics of the non-mammalian Synapsida and implications for patterns of evolution in synapsids. In: *Origins of the higher groups of tetrapods: Controversy and consensus* (Schultze, H.-P., and Trueb, L., eds.). Ithaca, NY: Cornell University Press, pp. 635–693.

Hopson, J.A., and Barghusen, H.R. 1986. An analysis of therapsid relationships. In: *The ecology and biology of mammal-like reptiles* (Hotton, N. III, MacLean, P.D., Roth, J.J., and Roth, E.C., eds.). Washington, D.C.: Smithsonian Institution Press, pp. 83–106.

Hopson, J.A., and Crompton, A.W. 1969. Origin of mammals. *Evol. Biol.* 3:15–72.

Hopson, J.A., Kielan-Jaworowska, Z., and Allin, E.F. 1989. The cryptic jugal of multituberculates. *J. Vert. Paleont.* 9:201–209.

Hopson, J.A., and Rougier, G.W. 1993. Braincase structure in the oldest known skull of a therian mammal: Implications for mammalian systematics and cranial evolution. *Amer. J. Sci.* 292-A.

Hunt, R.M. Jr. 1974. The auditory bulla in Carnivora: An anatomical basis for reappraisal of carnivore evolution. *J. Morph.* 143:95–115.

Jenkins, F.A. Jr., and Crompton, A.W. 1979. Triconodonta. In: *Mesozoic mammals, the first two-thirds of mammalian history* (Lillegraven, J.A., Kielan-Jaworowska, Z., and Clemens, W.A., eds.). Berkeley: University of California Press, pp. 74–90.

Jollie, M. 1962. *Chordate morphology*. New York: Van Nostrand Reinhold.

Kemp, T.S. 1982. *Mammal-like reptiles and the origin of mammals*. London: Academic Press.

Kemp, T.S. 1983. The interrelationships of mammals. *Zool. J. Linn. Soc.* 77:353–384.

Kermack, D.M., Kermack, K.A., and Mussett, F. 1968. The Welsh pantothere *Kuehneotherium praecursoris*. *Zool. J. Linn. Soc.* 47:407–423.

Kermack, K.A. 1963. The cranial structure of triconodonts. *Phil. Trans. R. Soc. London* B246:83–102.

Kermack, K.A., and Mussett, F. 1958. The jaw articulation of the Docodonta and the classification of Mesozoic mammals. *Proc. R. Soc. London* B148:204–215.

Kermack, K.A., Mussett, F., and Rigney, H.W. 1981. The skull of *Morganucodon*. *Zool. J. Linn. Soc.* 71:1–158.

Kielan-Jaworowska, Z. 1971. Skull structure and affinities of the Multituberculata. *Palaeont. Polonica* 25:5–41.

Kielan-Jaworowska, Z., Crompton, A.W., and Jenkins, F.A. Jr. 1987. The origin of egg-laying mammals. *Nature* 326:871–873.

Kielan-Jaworowska, Z., Presley, R., and Poplin, C. 1986. The cranial vascular system in taeniolabidoid multituberculate mammals. *Phil. Trans. R. Soc. London* B313:525–602.

Krusat, G. 1980. Contribuição pora o conhecimento de Fauna do Kimeridgiano da Mina de Lignito Guimarota (Leiria, Portugal). IV Parte—*Haldanodon exspectatus* Kühne and Krusat 1972 (Mammalia, Docodonta). *Memórias dos Serviços Geológicos de Portugal* 27:1–79.

Kuhn, H.-J. 1971. Die Entwicklung und Morphologie des Schädels von *Tachyglossus aculeatus*. *Abh. Senckenberg. Naturforsch. Ges.* 528:1–224.

Kuhn, H.-J., and Zeller, U. 1987. The cavum epiptericum in monotremes and therian mammals. In: *Morphogenesis of the mammalian skull* (Kuhn, H.-J., and Zeller, U., eds.) Mammalia depicta, 13:51–70. Hamburg, Berlin: Paul Parey.

Kühne, W.G. 1956. *The Liassic therapsid Oligokyphus.* London: Trustees of the British Museum.

Lillegraven, J.A., and Krusat, G. 1991. Cranio-mandibular anatomy of *Haldanodon exspectatus* (Docodonta; Mammalia) from the Late Jurassic of Portugal. *Contr. Geology, Univ. Wyoming, Spec. Paper* 28:39–138.

Lillegraven, J.A., Thompson, S.K., McNab, B.K., and Patton, J.L. 1987. The origin of eutherian mammals. *Biol. J. Linn. Soc.* 32:281–336.

Luo, Z. 1988. Two distinct patterns of apomorphous petrosal characters among major mammalian groups and their phylogenetic implications. *J. Vert. Paleont.* Suppl. 8:20A.

Luo, Z. 1989. *The petrosal structures of Multituberculata (Mammalia) and the molar morphology of the early arctocyonids (Condylarthra: Mammalia)*. Ph.D. Dissertation, University of California, Berkeley.

Luo, Z., and Ketten, D.R. 1989. Observation of the acousto-vestibular structures of multituberculate mammals with computerized tomography. *J. Vert. Paleont.* Suppl. 9:30A.

Luo, Z., and Ketten, D.R. 1991. CT scanning and computerized reconstructions of the inner ear of multituberculate mammals. *J. Vert. Paleont.* 11:220–228.

MacPhee, R.D.E. 1981. Auditory regions of primates and eutherian insectivores: Morphology, ontogeny and character analysis. *Contr. Primatol.* 18:1–282.

Maddison, W.P., Donoghue, M.J., and Maddison, D.R. 1984. Outgroup analysis and parsimony. *Syst. Zool.* 33:83–103.

McKenna, M.C. 1975. Towards a phylogenetic classification of the Mammalia. In: *Phylogeny of the primates: A multidisciplinary approach* (Luckett, W.P., and Szalay, F.S., eds.). New York: Plenum Press, pp. 21–46.

McKenna, M.C. 1987. Molecular and morphological analysis of higher-level mammalian interrelationships. In: *Molecules and morphology in evolution: Conflict or compromise?* (Patterson, C., ed.). Cambridge: Cambridge University Press, pp. 55–93.

Miao, D. 1988. Skull morphology of *Lambdopsalis bulla* (Mammalia, Multituberculata) and its implications to mammalian evolution. *Contr. Geology, Univ. Wyoming, Spec. Paper* 4:1–104.

Miao, D., and Lillegraven, J.A. 1986. Discovery of three ear ossicles in a multituberculate mammal. *National Geograph. Res.* 2:500–507.

Novacek, M.J. 1986. The skull of leptictid insectivorans and the higher-level classification of eutherian mammals. *Bull. Amer. Mus. Nat. Hist.* 183:1–112.

Novacek, M.J., and Wyss, A. 1986. Origin and transformation of the mammalian stapes. In: *Vertebrates, phylogeny, and philosophy* (Flanagan, K.M., and Lillegraven, J.A., eds.). *Contributions to Geology, University of Wyoming, Special Paper 3*, pp. 35–53.

Novacek, M.J., Wyss, A.R., and McKenna, M.C. 1988. The major groups of eutherian mammals. In: *The phylogeny and classification of Tetrapods* (Benton, M.J., ed.). Oxford: Clarendon Press, pp. 31–71.

Parrington, F.R., 1946. On the cranial anatomy of cynodonts. *Proc. Zool. Soc. London* 116:181–197.

Patterson, B., and Olson, E.C. 1961. A triconodontid mammal from the Triassic of Yunnan. *Intern. Colloq. on the Evolution of Lower and Specialized Mammals. Kon. Vl. Acad. Wetensch., Lett. Sch. Kunsten Belgie, Brussels* 1:129–191.

Patterson, C. 1980. Methods of paleobiogeography. In: *Vicariance biogeography: A critique* (Nelson, G., and Rosen, D. E., eds.). New York: Columbia University Press, pp. 446–489.

Presley, R. 1981. Alisphenoid equivalents in placentals, marsupials, monotremes and fossils. *Nature* 294:668–670.

Quiroga, J.C. 1979. The brain of two mammal-like reptiles (Cynodontia-Therapsida). *J. Hirnforsch.* 20:341–350.

Rougier, G., and Bonaparte, J.F. 1988. La pared lateral del cráneo de *Vincelestes neuquenianus* (Mammalia, Eupantotheria) y su importancia en el estudio de los mamíferos mesozoicos. *IV Jornados Argentina Paleont.* 3.

Rowe, T. 1988. Definition, diagnosis and origin of Mammalia. *J. Vert. Paleont.* 8:241–264.

Simpson, G.G. 1928. *A catalogue of the Mesozoic Mammalia in the geological department of the British Museum.* London: British Museum (Natural History).

Simpson, G.G. 1929. American Mesozoic Mammalia. *Memoirs, Peabody Mus., Yale Univ.* 3:1–235.

Simpson, G.G. 1937. Skull structure of the Multituberculata. *Bull. Amer. Mus. Nat. Hist.* 73:727–763.

Sloan, R.E. 1979. Multituberculata. In: *The encyclopedia of paleontology* (Fairbridge, R.W., and Jablonski, D., eds.). Stroudsberg, PA: Dowden, Hutchinson & Ross, pp. 492–498.

Sues, H.-D. 1985. The relationships of the Tritylodontidae (Synapsida). *Zool. J. Linn. Soc.* 85:205–217.

Sues, H.-D. 1986. The skull and dentition of two tritylodontid synapsids from the Lower Jurassic of western North America. *Bull. Mus. Comp. Zool.* 151:217–268.

Sun, A.-L. 1984. Skull morphology of the tritylodont genus *Bienotheroides* of Sichuan. *Scientia Sinica* B27:970–984.

Swofford, D.L. 1989. *PAUP: Phylogenetic analysis using parsimony, version 3.0d.* Champaign: Illinois Natural History Museum.

Vandebroek, G. 1961. *Evolution des Vertébrés: de leur Origine à l'Homme.* Paris: Masson & Cie.

Wible, J.R. 1984. *The ontogeny and phylogeny of the mammalian cranial arterial pattern.* Ph.D. Dissertation, Duke University, Durham, NC.

Wible, J.R. 1987. The eutherian stapedial artery: Character analysis and implications for superordinal relationships. *Zool. J. Linn. Soc.* 91:107–135.

Wible, J.R. 1989. Vessels on the side wall of the braincase in cynodonts and primitive mammals. In: *Trends in vertebrate morphology* (Splechtna H. and Hilgers, H., eds.). *Fortschritte der Zoologie* 35:406–408.

Wible, J.R. 1990. Petrosals of late Cretaceous marsupials from North America and a cladistic analysis of the petrosal in therian mammals. *J. Vert. Paleont.* 10:183–205.

Wible, J.R. 1991. Origin of Mammalia: The craniodental evidence reexamined. *J. Vert. Paleont.* 11:1–28.

Zeller, U. 1989. Die Entwicklung und Morphologie des Schädels von *Ornithorhynchus anatinus* (Mammalia: Prototheria: Monotremata). *Abh. Senckenberg. Naturforsch. Ges.* 545:1–188.

CHAPTER 6

Cranial Morphology and Multituberculate Relationships

DESUI MIAO

Overview

The comparative description of multituberculate crania has provided a solid morphological data base for the character analysis, which reveals that a majority of multituberculate cranial features are merely mammalian plesiomorphies. A few specializations do exist and appear to be either peculiar to all multituberculates or evolved within the group. These include inflated vestibular apparatus in taeniolabidoids, reduced postorbital process and jugal, large premaxilla, nasal and jugular foramen, and exclusion of palatine from orbit. Previously proposed synapomorphies to relate multituberculates as a sister taxon to either monotremes or Recent therians may simply be homoplasies, that is, independent acquisition of three-boned middle ear and independent loss of septomaxilla. Therefore, multituberculates appear to have been a separate lineage in early mammalian radiation either prior to emergence of the latest common ancestor of Recent mammals or before any other mammals even evolved.

"The result of this increased knowledge is to strengthen the growing conviction toward which all the recent accretions of data on the multituberculates have contributed, that these animals were not the ancestors of or closely related to monotremes, marsupials, or placentals, that any phyletic connection between them and the later three groups must have been far back toward the origin of the Mammalia, possibly even before that artificially delimited event, and that taxonomically this means that the Multituberculata form a separate subclass, Allotheria." (Simpson, 1937, p. 761)

Contents

Introduction,   64
Description and Discussion,   64
   Snout,   64
   Skull Roof and Orbit,   66
   Palatal Complex,   66
   Sphenoid Complex,   66
   Ear Region,   67
   Occipital Region,   68
Character Analysis,   68
   Mammalian Plesiomorphies,   68
   Multituberculate Apomorphies,   69
   Mammalian or Therian Synapomorphies Versus Homoplasies,   69
Phylogeny,   69
   Sister Taxon to Monotremata,   70
   Sister Taxon to Marsupialia plus Eutheria,   70
   Sister Taxon to Combined Clades of Remaining Mammals,   72
Conclusions,   72
Acknowledgments,   72
References,   73

Introduction

Multituberculata is the longest lived mammalian order (ranging from the late Triassic to early Oligocene), dominant throughout the Mesozoic, well represented in the fossil record of the Northern Hemisphere, and has figured importantly in both previous and current discussions of phylogenetic relationships of early mammals (e.g., Simpson, 1937; Kermack, 1967; Hopson and Crompton, 1969; Kielan-Jaworowska, 1971; Kemp, 1983; McKenna, 1987; Miao, 1988; Rowe, 1988). Yet the phylogenetic position of multituberculates relative to other mammals remains problematic and controversial. Cranial materials of multituberculates are available, some well preserved, and many already thoroughly studied (e.g., Simpson, 1937; Hahn, 1969, 1981, 1987, 1988; Kielan-Jaworowska, 1971, 1974; Kielan-Jaworowska et al., 1986; Miao, 1988). Thus it is possible to conduct a comparative survey that may shed light on multituberculate relationships.

Despite their scarcity, the described cranial materials of multituberculates are rather diverse in terms of their chronological, ecological, geographic, and taxonomic representations. They render possible the application of ingroup as well as outgroup comparisons to the character analysis in determining the character transformation, functional character complex, and shared derived characters (synapomorphies) (see also Simmons, this volume). In this chapter, therefore, I choose to examine the cranial features of Multituberculata as they bear on the affinities of this order.

Description and Discussion

In an excellent summary account of Multituberculata, Clemens and Kielan-Jaworowska (1979) thoroughly reviewed then available cranial materials of multituberculates, based mainly upon the Mongolian late Cretaceous collection. Since then, an important advance in understanding cranial anatomy of multituberculates (Figs. 6.1–6.3) has been made through the studies of well-preserved skulls of *Lambdopsalis bulla* Chow and Qi, 1978, from the Chinese late Paleocene (Miao and Lillegraven, 1986; Miao, 1988) and reinterpretations of the previously studied materials (Hahn, 1981, 1985, 1987, 1988; Kielan-Jaworowska et al., 1986; Miao, 1988; Hopson et al., 1989). The following review has benefited tremendously from the results of nearly a century-long inquiry into cranial morphology of multituberculates by all investigators, led by Gidley (1909).

SNOUT. The skull of multituberculates is low and broad with a short or modestly long snout and invariably short and wide postglenoid region. The premaxilla is extensive, and bears two to three upper incisors with a usually enlarged I^2. A good correlation seems to exist between a large, procumbent central incisor and an extensive premaxilla in many, often distantly related, mammalian groups in addition to multituberculates. The palatal process of the premaxilla is perforated by an incisive foramen, and at least in *Lambdopsalis bulla* a prenasal process is present and constitutes an internarial bar (Miao, 1988). No septomaxilla has been reported in multituberculates (Wible et al., 1990). The maxilla is an elongated and extensive bone, forming most of the lateral and ventral sides of the snout, the anterior rim and floor of the orbit, and the anterior part of the zygomatic arch. The maxilla usually bears a single large infraorbital foramen, with exception of some paulchoffatiids, arginbaatariids, *Ctenacodon*, and *Meniscoessus*, in which two infraorbital foramina have been described (Clemens, 1963; Archibald, 1982; Hahn, 1985; Kielan-Jaworowska et al., 1987b). However, it has been reported that on at least one specimen of *Lambdopsalis bulla*, one infraorbital foramen is present on the left side and two on the right side (Miao, 1988). This individual variability in the number of infraorbital foramina casts doubt upon its effectiveness in phylogenetic analysis (but see Hahn, 1985; Rowe, 1986).

The nasal is greatly developed, and is correlated with enlargement of the nasal cavity. Inside the nasal cavity,

TABLE 6.1. Abbreviations

acf	ascending canal foramen
al	alisphenoid
bo	basioccipital
bs	basisphenoid
cf	carotid foramen
er	epitympanic recess
f	frontal
fc	fenestra cochleae
fm	foramen masticatorium
fmm	fossa muscularis major
foi	foramen ovale inferium
fv	fenestra vestibuli
gf	glenoid fossa
if	incisive foramen
jf	jugular foramen
l	lacrimal
m	maxilla
n	nasal
oc	occipital
ocl	occipital condyle
os	orbitosphenoid
p	parietal
pe	petrosal
pl	palatine
pm	premaxilla
pop	postorbital process
pp	paroccipital process
pr	promontorium
pt	pterygoid
ptf	post-temporal fossa
sq	squamosal

Chapter 6. Cranial Morphology and Multituberculate Relationships

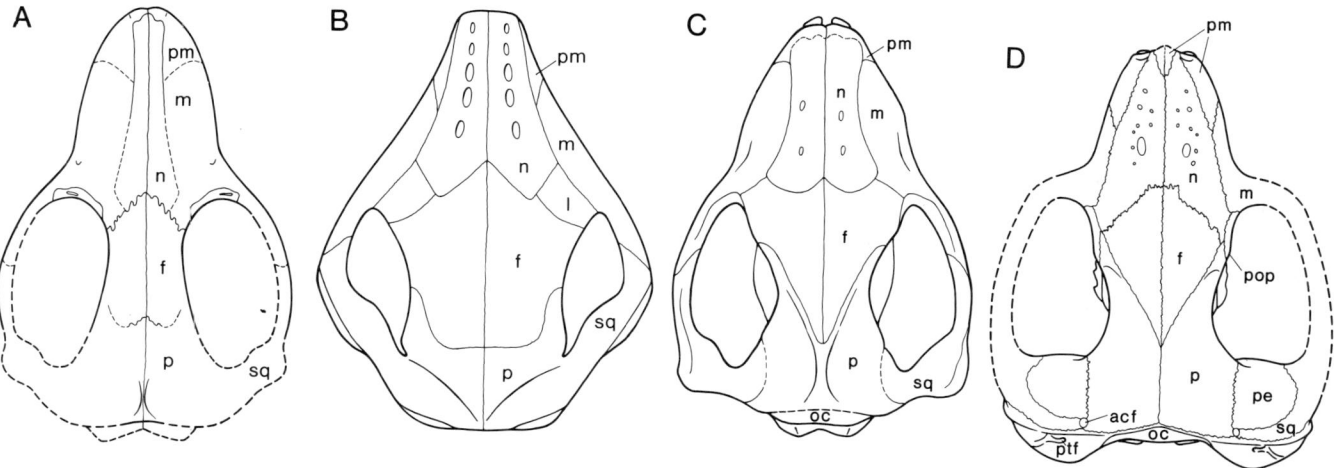

Figure 6.1. Dorsal view of reconstructions of multituberculate crania. **A**: *Paulchoffatia delgadoi* from the Late Jurassic, Portugal (modified from Hahn, 1969); **B**: *Nemegtbaatar gobiensis* from the Late Cretaceous, Mongolia (modified from Kielan-Jaworowska et al., 1986); **C**: *Ptilodus montanus* from middle Paleocene, the United States (modified from Simpson, 1937); **D**: *Lambdopsalis bulla* from late Paleocene, China (modified from Miao, 1988). For abbreviations, see Table 6.1.

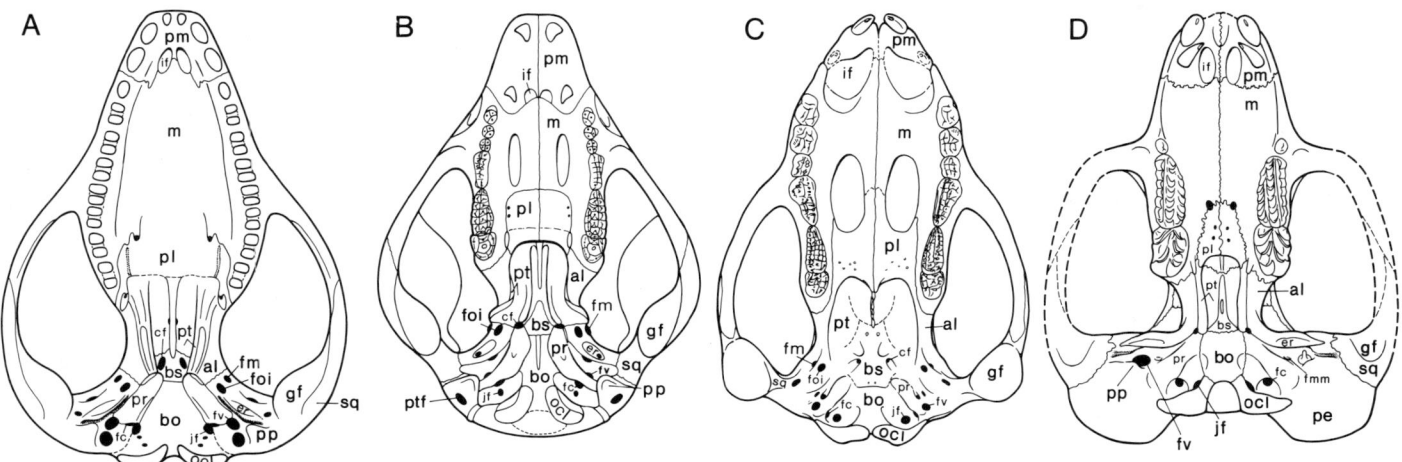

Figure 6.2. Ventral view of reconstructions of multituberculate crania. **A**: Paulchoffatiinae, gen. et sp. indet., from the Late Jurassic, Portugal (modified from Hahn, 1988); **B**: *Nemegtbaatar gobiensis* (modified from Kielan-Jaworowska et al., 1986); **C**: *Ptilodus montanus* (modified from Simpson, 1937); **D**: *Lambdopsalis bulla* (modified from Miao, 1988).

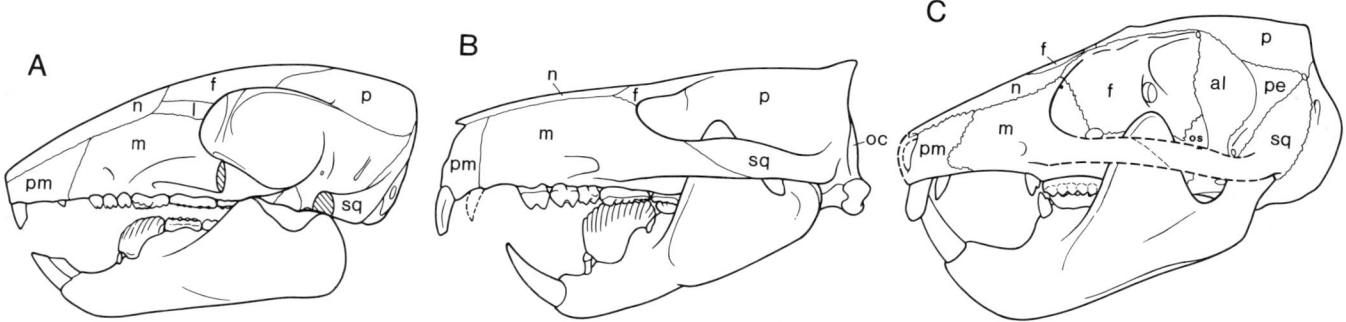

Figure 6.3. Lateral view of reconstructions of multituberculate skulls. **A**: *Nemegtbaatar gobiensis* (modified from Kielan-Jaworowska et al., 1986); **B**: *Ptilodus montanus* (modified from Simpson, 1937; Krause, 1982; Hopson et al., 1989); **C**: *Lambdopsalis bulla* (modified from Miao, 1988).

a complex ridge system is developed on the ventral surface of the nasal, maxilla, and frontal, suggesting probable presence of turbinals. The enlargement of the nasal and development of the complex ridge system perhaps are " . . . related to the increased respiratory requirement associated with the emergence of a mammalian type of metabolism" (Moore, 1981, p. 241). However, a cribriform plate has not been found in multituberculates, contra Rowe (1988). With possible exception of *Paulchoffatia*, all reported multituberculate skulls in which the nasal bone is preserved show presence of nasal foramina. But function of the foramina is unclear. There is also a bony internasal septum.

SKULL ROOF AND ORBIT. The frontal in multituberculates is a large bone that forms much of the cranial roof and the intraorbital wall. In certain taxa, the external exposure of the frontal may be reduced as a result of being overlapped laterally by the neighboring bones (e.g., parietal, lacrimal). However, the orbital process of the frontal is always a dominant component of the orbital mosaic in multituberculates. A lacrimal is present in paulchoffatiids and some of the Mongolian Late Cretaceous taeniolabidiids, but is reduced or lost in other, especially the Tertiary, taxa. Similarly, the palatine has either a minor or no orbital exposure in multituberculates. The postorbital process is not a striking feature and thus renders the reconstruction of multituberculate orbit difficult and uncertain. However, multituberculates generally appear to have a moderate-sized, unfloored orbit. It should be noted that in the taxa with a blade-like P_4, a roofed and floored pocket-like structure is formed in the anterior corner of the orbit, immediately posterior to the infraorbital canal. In such cases, the glenoid fossa tends to be nearly flat rather than concave. Thus a functional correlation seems to be warranted, especially in the light of Krause's study (1982), which elegantly shows the pattern of the jaw movement as a result of the two cycles of mastication in the multituberculates with a blade-like P_4. Multituberculates possess a stout zygomatic arch but, until recently, they are regarded as lacking a jugal bone. Hopson et al. (1989) demonstrate unequivocal presence of a jugal internal to the zygomatic process of the maxilla in several multituberculate genera.

The parietal is extensive and covers the cranial roof in the temporal region. However, the parietal barely extends onto the lateral wall of the braincase. Associated with the parietal are three structures—the postorbital foramen, foramen of ascending canal, and internal parietal groove—that provide osteological evidence for vascular reconstructions (Kielan-Jaworowska et al., 1986; Miao, 1988). Kielan-Jaworowska et al. (1986) concluded that multituberculates retained a continuous orbitotemporal vascular system, a feature consistently occurring in cynodonts and morganucodonts but not in Recent mammals.

Like the other groups of primitive mammals, multituberculates have a small squamosal bone, which shields the auditory capsule and contributes little to the formation of the sidewall of the braincase, in contrast with a large squamosal in therian mammals that usually covers the posteroventral part of the sidewall of the braincase.

PALATAL COMPLEX. The palatine forms the posterior part of the secondary palate and the dorsal, ventral, and lateral walls of the choana. Palatal vacuities may or may not develop, but presence of the minor palatine foramina seems to be a constant feature of the palatine bones in multituberculates. The choanae are separated by a vertical median plate, the vomer. Each side of the choanae is further divided by a pterygopalatine ridge into two longitudinal channels in many Mesozoic multituberculates. Both Barghausen (1986) and R. Presley (personal communication) interpreted the osseous pterygopalatine ridge as a partition between the auditory tube and the soft palate in a shallow choana. As the more derived Mesozoic and Tertiary multituberculates develop a relatively deep choana, the pterygopalatine ridge is lost (Simpson, 1937; Miao, 1988). Ectopterygoid bones are thought to be present at least in certain multituberculates (Clemens and Kielan-Jaworowska, 1979; Kielan-Jaworowska et al., 1986). It should be noted, however, that what has been identified as the ectopterygoid may be part of the alisphenoid. Thus, it is suggested that the ectopterygoid bones are already incorporated into the pterygoid as a ventral element in multituberculates (Miao, 1988).

SPHENOID COMPLEX. Due to lack of clear sutural delineation, the recognition of presphenoid and parasphenoid is difficult. The basisphenoid forms the cranial floor in the middle region anterior to the basioccipital. A pair of the internal carotid foramina perforate the posterior part of the basisphenoid in the late Jurassic paulchoffatiid *Pseudobolodon* (see Hahn, 1988) and the Paleogene *Ectypodus* (see Sloan, 1979) and *Ptilodus* (see Simpson, 1937). In the late Cretaceous Mongolian taeniolabidoid multituberculates, the paired internal carotid foramina are located within the sutural zone between the basisphenoid and petrosals (Kielan-Jaworowska et al., 1986). However, the internal carotid foramina are lost in a more derived taxon, *Lambdopsalis* (see Miao, 1988).

The orbitosphenoid varies in both relative and absolute sizes among different multituberculate taxa. As in advanced cynodonts and other early mammalian groups, the orbitosphenoid in multituberculates lies considerably medial to the alisphenoid (or the epipterygoid in cynodonts), in contrast to extant mammals in which the two bones lie in an edge-to-edge fashion. In other words, the orbitosphenoid in multituberculates is continuous posteriorly with the taenia clino-orbitalis (or ossified pila antotica) in forming the medial wall of cavum epitericum-the primary sidewall of the braincase (Kielan-Jaworowska et al., 1986). A separate optic foramen undoubtedly occurs in *Lambdopsalis* (see Miao, 1988). This suggests that like the reptilian ancestor of mammals, multituberculates retain pila metoptica (as well as pila antotica). Eutherians also retain pila metoptica, but monotremes lost it (Kuhn and Zeller, 1987).

Reconstructions of the alisphenoid in various multituberculate taxa are entangled with varied degrees of uncertainty due to either poor preservation or lack of recognizable demarcation. Early workers (Gidley, 1909; Broom, 1914; Simpson, 1937) avoided speculations on the extent of the alisphenoid in *Ptilodus* and *Taeniolabis*. The alisphenoid was considered a minor element in forming the sidewall of the braincase in both the Jurassic Portuguese plagiaulacoids (Hahn, 1981, 1988) and the late Cretaceous Mongolian taeniolabidoids (Kielan-Jaworowska, 1971, 1974; Kielan-Jaworowska et al., 1986). The mandibular branches of the trigeminal nerve in these taxa were believed to have perforated the anterior lamina of the petrosal instead of the alisphenoid. However, Miao (1988) demonstrated that the alisphenoid in *Lambdopsalis* is well defined by the sutural delineation, and is an extensive element in the orbitotemporal region. As in *Ptilodus* (see Simpson, 1937) and certain rodents (Hill, 1935), the alisphenoid in *Lambdopsalis* is perforated by the two foramina (foramen masticatorium and foramen ovale inferium), which serve for passage of the mandibular branches of the trigeminal nerve.

The cavum epitericum in multituberculates is an extracerebral space between the primary sidewall of the braincase, formed by the ossified taenia clino-orbitalis and orbitosphenoid, and the secondary sidewall that consists of the alisphenoid and anterior lamina of the petrosal (Kielan-Jaworowska et al., 1986; Miao, 1988). The cavum epitericum is completely floored in all described cranial materials of multituberculates (Kielan-Jaworowska et al., 1986; Hahn, 1988; Miao, 1988). Posterior to, and confluent with, the cavum epitericum is the cavum supracochleare, in which the geniculate ganglion lies. Insofar as multituberculates retain a much more extensively ossified medial wall of the cavum epitericum than do monotremes, Miao (1988) suggested that monotremes seem to be more derived.

EAR REGION. It is fair to state that the ear region in multituberculates has been thoroughly studied and is now adequately understood. The excellent descriptions of the multituberculate ear region by Simpson (1937) and Kielan-Jaworowska et al. (1986) have provided a sound basis for the studies of Hahn (1988) and Miao (1988).

Multituberculates have acquired an essentially characteristic mammalian middle ear with a triossicular system (Miao and Lillegraven, 1986). The imperforate and columelliform stapes of multituberculates seems to represent a primitive condition that is similarly seen in monotremes and many marsupials (Novacek and Wyss, 1986). Other primitive characters include the lack of the bony bulla and bony external auditory meatus, round outline of the footplate of the stapes, and evidently loosely attached ectotympanic bone.

The promontorium of the multituberculates is usually narrow, straight, and directed anteromedially. It is moderately convex and houses a nearly straight, rod-like cochlea. The fenestra vestibuli and fenestra cochleae are two subequally sized fenestrae flanking the posterior end of the promontorium. A prominent round process, the paroccipital process, is located posterolateral to the fenestra vestibuli. Anterolateral to the paroccipital process is the stylomastoid notch or foramen. Medial to the fenestra vestibuli lies a distinct fossa, which is called the fossa muscularis major and is the site for the attachment of the tensor tympani muscle. There are two grooves in the vicinity of the fenestra vestibuli: (1) the stapedial artery groove, lying medial to the fenestra vestibuli and along the lateral side of the promontorium; and (2) the facial sulcus, lying lateral to the fenestra vestibuli and leading anteromedially to the foramen facialis secundarium and posteroventrally to the stylomastoid notch or foramen.

The lateral flange in multituberculates is usually a ventral ridge projected from the petrosal (Kielan-Jaworowska et al., 1986; Hahn, 1988). However, in *Lambdopsalis*, it is formed by part of the alisphenoid (Miao, 1988). In both cases, the lateral flange bears on its ventromedial surface distinct concavity, the epitympanic recess, where the fossa incudis must locate.

The subarcuate fossa is generally deep in the multituberculates except *Lambdopsalis*, in which it is rather shallow, probably due to the expansion of the vestibular apparatus (Miao, 1988). A slit-like aquaeductus vestibuli is located posteroventral to the subarcuate fossa, and the prootic canal runs along the anterior edge of the subarcuate fossa.

In post-Jurassic multituberculates, the foramina for the vestibular nerve (VIII) and the facial nerve (VII) are within a generally well-marked depression ventral to the subarcuate fossa on the anteromedial aspect of the petrosal. The depression is termed the internal auditory meatus. However, in more primitive paulchoffatiids, "... a real porus acusticus internus has yet to evolve" (Hahn, 1988, p. 180). It should be noted that even in more derived forms such as *Lambdopsalis*, the internal auditory meatus is still shallow, though distinct, as compared with the condition seen in extant mammals. In this aspect, multituberculates are also more similar to morganucodontids than to Recent mammals.

A striking feature of the inner ear in the known cranial materials of taeniolabidoid multituberculates is the various degree of expansion of the vestibular apparatus. The extraordinary inflation of the vestibule in *Lambdopsalis* is especially well documented (Miao, 1988). Closely related to the expansion of the vestibule, the endolymphatic duct and sac also become enlarged, and the anterior lamina of the petrosal becomes inclined toward anteromedial direction. The specializations in the inner ear of these forms are interpreted as adaptations to the increased sensitivity to low-frequency reception (Miao, 1988). Despite the proposal of the vestibular expansion as a synapomorphy of the multituberculates as a whole (Luo, 1989), it remains to be demonstrated that a similar expansion also occurs in paulchoffatiids and *Ptilodus* among other non-taeniolabidoid multituberculates.

OCCIPITAL REGION. Although there is no sutural demarcation to delimit the individual bones in the occipital region of multituberculates, the previously described tabular bone (Kielan-Jaworowska, 1971; Hahn, 1978; Clemens and Kielan-Jaworowska, 1979) appears absent (Kielan-Jaworowska et al., 1986; Miao, 1988). A distinct feature in the basioccipital region of taeniolabidoids is the large size of the jugular foramen, located in the well-developed jugular fossa. The hypoglossal foramen is double in *Paulchoffatia* (see Hahn, 1969) and is single in *Ptilodus* (see Simpson, 1937), *Ectypodus* (see Sloan, 1979), and *Lambdopsalis* (see Miao, 1988). It is reportedly absent or indiscernible in the Mongolian taeniolabidoids (Kielan-Jaworowska, 1971, 1974; Kielan-Jaworowska et al., 1986).

Character Analysis

Undoubtedly, the morphological information contained in the preceding section can potentially provide numerous characters to keep the computer running for hours in generating countless cladograms. However, the computer is only as sophisticated as its users. Many indiscriminately chosen characters are likely to contain enormous homoplastic features that tend to swamp the relatively few more meaningful characters in a PAUP analysis. The necessity and importance, therefore, of the character scrutiny cannot be overemphasized (see Szalay, this volume, chapter 9).

Instead of attempting an analysis involving all early mammalian groups (see Crompton and Luo, this volume; Maier, this volume; Presley, this volume; Wible and Hopson, this volume; Zeller, this volume), I focus on the issue of whether and what the cranial evidence can bear on the phylogenetic position of multituberculates as related to extant mammals. Likewise, in the followng character analysis, I merely categorize most of the selected characters and discuss in detail only those that are pertinent to the issue.

MAMMALIAN PLESIOMORPHIES. The majority of the above-described cranial features of multituberculates falls under this category.

1. The prenasal process of the premaxilla and the bony internarial bar are present in some multituberculates (Miao, 1988) as well as several other early mammalian taxa such as *Sinoconodon* (see Patterson and Olson, 1961; Crompton and Sun, 1985) and *Haldanodon* (see Miao, 1988; J.A. Lillegraven, personal communication). The prenasal process of the premaxilla and the internarial bar are subsequently lost in monotremes and therian mammals.
2. Presence of complex intranasal ridge system, suggestive of development of turbinals (Miao, 1988).
3. Presence of bony internasal septum (Kielan-Jaworowska et al., 1986; Miao, 1988).
4. The lacrimal is large in many early forms but is reduced or lost in more derived taxa (Miao, 1988).
5. As in advanced cynodonts and other early mammals, the ectopterygoid in multituberculates is already incorporated into the pterygoid as a ventral element (Miao, 1988). The ectopterygoid in monotremes is more reasonably interpreted as retention of a primitive feature instead of a homoplastic reacquisition (Presley and Steel, 1978).
6. Retention of the pila metoptica and presence of the optic foramen (Miao, 1988).
7. As in other early mammalian groups, the orbitosphenoid in multituberculates lies considerably medial to the alisphenoid. These two bones contact each other edge to edge in monotremes and therians (Kielan-Jaworowska et al., 1986; Miao, 1988).
8. The squamosal lacks participation in forming the lateral wall of the cranium (Miao, 1988).
9. Retention of a continuous orbitotemporal vasculature (Kielan-Jaworowska et al., 1986).
10. The taenia clino-orbitalis is extensive in multituberculates but is greatly reduced in monotremes and completely lost in therians (Miao, 1988).
11. Presence of a bony floor to the cavum epiptericum (Kielan-Jaworowska et al., 1986).
12. The cavum epiptericum is confluent with the cavum supracochleare (Miao, 1988).
13. Retention of the post-temporal canal (Kielan-Jaworowska et al., 1986; Hahn, 1988; Miao, 1988).
14. Presence of the prootic canal (Kielan-Jaworowska et al., 1986; Hahn, 1988; Miao, 1988).

15. The promontorium in multituberculates is narrow, elongated, and less eminent as compared with that of monotremes (Simpson, 1937; Griffiths, 1978).
16. Like their reptilian ancestors and other early mammalian groups, multituberculates have a straight and rod-like cochlea (Kielan-Joworowska et al., 1986; Miao, 1988). Monotremes, however, have "...a cochlea with incipient coiling intermediate between the straight cochlea of the Mesozoic mammal *Triconodon* (Simpson, 1928) and the spiral cochlea of the metatherian and eutherian ear..." (Griffiths, 1978, pp. 168–169).
17. The footplate of the stapes has a more or less round outline (Miao and Lillegraven, 1986).
18. The fossa incudis is on the ventromedial surface of the lateral flange (Kielan-Jaworowska et al., 1986; Miao, 1988).
19. The internal auditory meatus in multituberculates is either yet to develop or ill-defined and shallow (Hahn, 1988; Miao, 1988). Only in monotremes and therians does the internal auditory meatus develop into a well-defined and often deep depression (Simpson, 1938).

MULTITUBERCULATE APOMORPHIES. Under this category are included several crainal characters that are either peculiar to multituberculates or developed within various subgroups of multituberculates for developmental, functional, or adaptive reasons.

1. Development of an extensive premaxilla (Simpson, 1937; Kielan-Jaworowska et al., 1986; Miao, 1988).
2. Presence of a large nasal with the nasal foramina (Simpson, 1937; Kielan-Jaworowska et al., 1986; Miao, 1988).
3. The postorbital process is absent or reduced (Kielan-Jaworowska et al., 1986; Miao, 1988).
4. Exclusion of the palatine from the orbital mosaic (Simpson, 1937; Miao, 1988).
5. A reduced jugal lies internal to the zygomatic process of the maxilla (Hopson et al., 1989).
6. Large jugular foramen (Simpson, 1937; Kielan-Jaworowska et al., 1986; Miao, 1988).
7. Expansion of the vestibular apparatus in the taeniolabidoid multituberculates (Miao, 1988).
8. Elastic ossification pattern in the orbitotemporal region as reflected by the variability in extent of the alisphenoid and so-called "anterior lamina of petrosal" (see also Presley, 1981; Maier, 1987; Miao, 1988).

MAMMALIAN OR THERIAN SYNAPOMORPHIES VERSUS HOMOPLASIES. Under this category, I include the cranial characters that multituberculates presumably share exclusively either with marsupials and eutherians or with monotremes, marsupials, and eutherians. To my own surprise, these kinds of characters are so few and so ambiguous that it is a matter of considerable debate whether to view them as synapomorphies or homoplasies.

1. Absence of the septomaxilla. This is probably the only character that is presumably shared exclusively by multituberculates and Recent therians, i.e., marsupials and eutherians (Wible et al., 1990). In the case of absence of the septomaxilla in multituberculates and Recent therians, Wible et al. (1990, p. 224) concluded, "...we consider this bone to have been independently lost in multituberculates and Recent therians, because *Vincelestes*, the sister taxon to Recent therians, has a septomaxilla." It should be added that monotremes are considered to have a true septomaxilla homologous to that of non-mammalian therapsids and many early mammals (Wible et al., 1990; Zeller, this volume). Evidently, this character seems to be of limited validity in deciphering the relationship between multituberculates and Recent therians.
2. Presence of a triossicular system in the middle ear cavity. Besides its possible presence in certain advanced triconodonts (Crompton and Sun, 1985), a triossicular auditory system is known to be present only in multituberculates, monotremes, marsupials, and eutherians (Miao and Lillegraven, 1986). Whether the three-boned middle ear evolved once, twice, or many times is a critical point in early mammalian evolution. Although Patterson (1981, 1982), Kemp (1983, 1988), and Novacek (1990) favor the notion of derivation of the mammalian three middle ear ossicles as a single evolutionary event, the arguments for the polyphyletic origin of the triossicular middle ear in mammals seem at least equally convincing (Allin, 1986; R. Presley, personal communication; Miao, 1988).

In summary, neither of these two characters is conclusively demonstrated to be synapomorphies that multituberculates uniquely share with therians or extant mammals.

Phylogeny

Traditionally, multituberculates were considered a subclass of their own, Allotheria, based mainly upon the "otherness" of their dental characters and the mosaic nature of their "prototherian" cranium and metatherian limbs (Simpson, 1937). A close relationship between multituberculates and monotremes was advocated repeatedly (e.g., Cope, 1888; Broom, 1914; Winge, 1923; Kermack and Kielan-Jaworowska, 1971; Kermack and Kermack, 1984), and was rejected nearly as often (e.g., Simpson, 1928, 1938; Granger and Simpson, 1929; Griffiths, 1978; McKenna, 1987; Miao, 1988). Therian

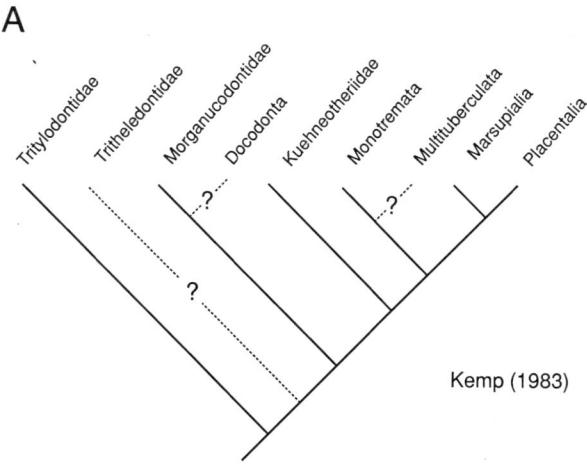

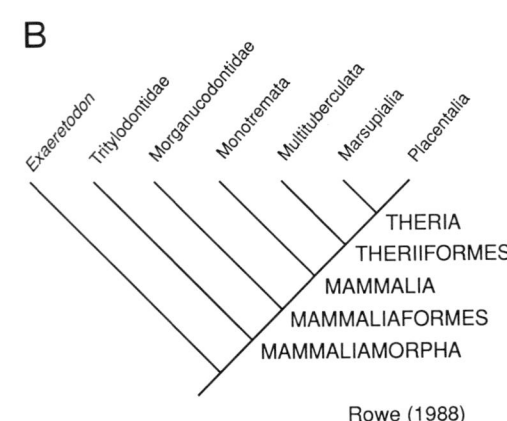

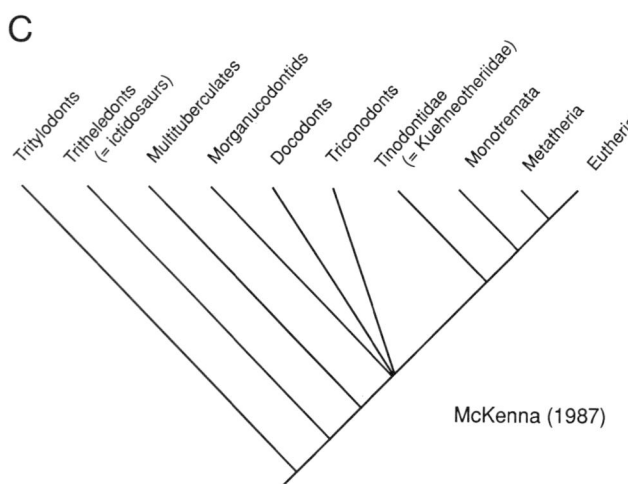

Figure 6.4. Cladograms showing phylogenetic relationships of major mammalian groups. Note different placements of multituberculates in the cladograms. **A and C**: modified from Kemp (1983), McKenna (1987), and Novacek (1990); **B**: modified from Rowe (1988).

affinity of multituberculates has been proposed more recently (Rowe, 1986), and multituberculates are even deemed more closely related to marsupials and eutherians than are monotremes (Rowe and Greenwald, 1987; Rowe, 1988).

I have expressed my own view on multituberculate relationships in considerable detail (Miao and Lillegraven, 1986; Miao, 1988, 1991). Instead of exercising an enthusiastic self-defense, I choose to use the above character analysis to test three current and more influential views (Fig. 6.4) on the phylogenetic ties of multituberculates.

SISTER TAXON TO MONOTREMATA. The cladogram in Figure 6.4A shows a sistergroup relationship between Monotremata and Multituberculata. It should be noted, however, that Kemp (1983, p.378) stated that, "...admittedly rather few similarities that they [i.e., multituberculates] share with monotremes may indicate a relationship between these two groups."

Neither our original analysis (Miao and Lillegraven, 1986) nor the character analysis presented in this chapter shows preference for a close relationship between monotremes and multituberculates. To the contrary, I pointed out (Miao, 1988, p.92) that "...no matter whether the three ear ossicles were assumed to have evolved only once or more than once in mammalian phylogeny, none of the constructed alternative cladograms resulted in close proximity between multituberculates and monotremes (see Miao and Lillegraven, 1986, Fig. 2)." As far as cranial evidence is concerned, not even a single synapomorphy has been unequivocally identified. Therefore, it is extremely unlikely that Multituberculata is a sister taxon to Monotremata.

SISTER TAXON TO MARSUPIALIA PLUS EUTHERIA. The sistergroup relationship between Multituberculata and Theria (Fig. 6.4B) was advocated originally on the basis of one single presumed synapomorphy—number of the infraorbital foramen. Rowe (1986, p. 244) stated that multituberculates "...share one derived character state with Theria, a reduction in the number of exits for the infraorbital canal onto the face (Hahn, 1985)." The invalidity of this character state was discussed at great length elsewhere (Miao, 1988).

Subsequently, based upon a more extensive PAUP analysis, Rowe (1988) reaffirmed the sistergroup relationship between Multituberculata and Theria. Rowe's data matrix consists of ninety-one cranial characters and sixty-seven postcranial characters. The completeness of both Multituberculata (90% complete) and Theria (95% to 96% complete) exceeds that of Monotremata (88% complete). Rowe even operated the separate runs for ninety-one cranial characters and sixty-seven postcranial characters in order to test Hopson's (1987) statement about the discrepancy in pervasiveness of convergent evolution resulting from different sets of data, cranial versus postcranial. Rowe (1988, pp. 246–247) concluded: "Contrary to Hopson's expectation, both data sets yield the same tree as the combined cranial and postcranial data set.... There is consistently a very good fit between each data set (cranial, postcranial, and combined) and the one tree...." Interestingly, Wible (1991) employed essentially Rowe's cranial data set, with certain modifications, and derived a tree with the positions of Monotremata and Multituberculata reversed relative to Theria.

In order to trace the cause of the discrepancy between Rowe's and Wible's analyses, I reexamined the synapomorphies at the node of Theriiformes, a taxon that Rowe (1988) created to include Multituberculata and Theria (Fig. 6.4B). There are twenty-five synapomorphies listed in Rowe (1988, p.263), of which only five are cranial characters and twenty are postcranial ones. Among the five cranial characters (Rowe's 3*, 5, 17*, 70*, and 71), three (Rowe's 3*, 17*, and 70*) are admittedly equivocal and are specially marked with asterisks (*) by Rowe. Two (Rowe's 17* and 70*) of the three equivocal "synapomorphies" marked by Rowe were excluded in Wible's analysis. These are the characters concerning position of the external auditory meatus relative to the zygoma (Rowe's 17*) and presence versus absence of an angular process on the dentary (Rowe's 70*). Rowe's 17* is especially ambiguous because it is used at three consecutive nodes, Mammalia, Theriiformes, and Theria (Rowe, 1988, p. 263). It simply cannot be a synapomorphy at all three taxonomic levels, no matter how good the character is! The third (Rowe's 3*) is absence of the septomaxilla, a character that has been discussed in the preceding section and is considered as an independent loss in Multituberculata and Theria. Apparently, there seems to be no significant disagreement among Rowe, Wible, and myself on exclusion of the third presumed synapomorphy. Therefore, the remaining two synapomorphies (Rowe's 5 and 71) deserve special consideration, despite the fact that both were also excluded in Wible's analysis.

Rowe's synapomorphy 5 is described as the frontals being "...widely expanded to broadly separate orbits..." (Rowe, 1988, p. 260). After reexamining this character, I must concur with Wible (1991). Proportionately, the frontals over forebrain in early multituberculates such as paulchoffatiids are no broader than those in monotremes. As Wible (1991) indicated, this character remains equivocal until a quantitative measure is devised to demonstrate that the subtle difference among these taxa is indeed statistically and phylogenetically significant.

Rowe's synapomorphy 71 is concerned with presence of the pterygoideus shelf on the dentary. I already discussed this character in a functional context at considerable length elsewhere (Miao, 1988). The pterygoideus shelf of dentary is presumably a site for the insertion of the pterygoid muscles, and thus is well developed in mammalian herbivores such as multituberculates. As Adams (1919) demonstrated, the jaw muscles of mammals are remarkably constant throughout the group, except in monotremes and edentates in which the reduced or modified dentition and mandible makes the pterygoid muscles of little functional importance. Therefore, a lack of the pterygoideus shelf in monotremes is likely to be functionally explicable. In general, the phylogenetic significance of highly function-related characters is difficult to assess. Perhaps for this very reason, Wible (1991) eliminated this character from his analysis. At any rate, even if this character were useful, one should in general avoid reliance on a single character for major phylogenetic groupings (Parrington, 1971).

To sum up, as far as crainal characters are concerned, no single unequivocal synapomorphy can be selected from the character analyses conducted by Rowe (1988), Miao (1988, this chapter), and Wible (1991) to denote the node of Theriiformes in Rowe's cladogram.

The cranial character analyses by both Wible (1991) and myself (Miao, 1988, 1991, and this chapter) strongly suggest that monotremes are more derived than multituberculates in possessing a much more reduced taenia clino-orbitalis (ossified pila antotica), an incipient coiling of cochlea, undivided osseous external nares, and pre-tribosphenic molars (see also Archer et al., 1985; Kielan-Jaworowska et al., 1987a). Therefore, monotremes are more closely related to Recent therians than are multituberculates. It seems that the sistergroup relationship between Multituberculata and Theria as suggested by Rowe (1986, 1988) is either invalid or completely subject to the interpretation of the postcranial evidence (but see Szalay, this volume, chapter 9).

It should be noted that the presumed close relationship between Multituberculata and Recent Theria on the basis of postcranial evidence must be put into a broader perspective in which postcranial morphology of monotremes is also scrutinized. Apparently, whether to view the postcranial similarities between multituberculates and Recent therians as synapomorphies or homoplasies is largely dependent on interpretation of the related character states in monotremes. Historically, the seemingly archaic nature of the monotreme postcranial skeleton led to the presumption of the overall primitiveness of monotreme postcranial characters. In his excellent review of monotreme biology, Jenkins (1990) pointed out that modified structural details of the presumed primitive postcranial features in monotremes are highly distinctive and unique. In addition, the evidence drawn from recent studies on monotreme terrestrial locomotion (Pridmore, 1985), cell cycles, body temperature and thermoregulation (Watson et al., 1988), neuroanatomy (Ulinski, 1986), and reproduction (Tyndale-Biscoe and Renfree, 1987) also reveals that the kind of derived "primitiveness" as seen in monotremes bears essential and close similarities to marsupials and eutherians rather than their synapsid ancestors and non-therian cousins.

On the other hand, the postcranial skeletons of early multituberculates are unknown. To use the postcranial characters of the more derived forms in the absence of knowledge of those of the morphotype, one must assume that little or no morphological evolution has occurred in the course of the long evolutionary history of multituberculates. Both common sense and our actual knowledge of the cranial character transformations in multituberculates run afoul of that assumption.

SISTER TAXON TO COMBINED CLADES OF REMAINING MAMMALS. Insofar as multituberculates do not share any synapomorphy with Recent mammals, it is reasonable to suggest that multituberculates diverged from the main lineage leading to Recent mammals prior to emergence of the latest common ancestor of monotremes, marsupials, and eutherians (McKenna, 1987; Miao, 1988, 1991).

Now that Sigogneau-Russell (1989) has convincingly demonstrated the multituberculates' affinity of haramiyids from the Late Triassic of France, the minimum age for the divergence of multituberculates from the other mammalian groups is the Rhaetian. The only pre-Rhaetic mammalian record is reported from the Norian Emborough fissure filling deposits (Fraser et al., 1985; Clemens, 1986; but see Whiteside and Marshall, 1985, for their contention on the age assignment for the deposits). It consists of a single mammalian taxon, *Kuehneotherium*. Therefore, it is not unreasonable to suggest that the divergence of multituberculates and other mammalian lineages may predate the Rhaetian, and that Multituberculata probably represents a sister taxon to all the other mammals, both extant and extinct, combined (Fig. 6.4C). Consequently, the independent acquisition of the three middle ear ossicles in multituberculates and in Recent mammals must be assumed.

Conclusions

With non-mammalian therapsids and other early mammals as general outgroups, the character analysis has shown that (1) an overwhelming majority of multituberculate cranial characters are mammalian or cynodont plesiomorphies; (2) several cranial characters, such as large size of the premaxilla, nasal and jugular foramen, reduction or loss of the postorbital process and jugal, exclusion of the palatine from orbit, and expansion of the vestibule in taeniolabidoids, represent the apomorphies that are either peculiar to multituberculates or developed within certain groups of multituberculates; (3) the presumed synapomorphy between multituberculates and Recent therians, an absence of the septomaxilla, is more likely to be a homoplasy resulting from independent losses (Wible et al., 1990); and (4) presence of a triossicular system in the middle ear cavity of multituberculates and Recent mammals can be viewed as either a synapomorphy (Patterson, 1981, 1982; Kemp, 1983, 1988; Novacek, 1990) or an independent acquisition (Allin, 1986; McKenna, 1987; Miao, 1988).

If morphological information alone is not yet decisive, the time factor may play an interim role (Gingerich, 1988). Now that Sigogneau-Russell (1989) has convincingly demonstrated the multituberculates' affinity of haramiyids, the minimum age of multituberculates' divergence from other mammals may be as early as the Rhaetian. Therefore, assuming that morphology does not fail us, Patterson's (1981, 1982) hypothesis would be acceptable, and multituberculates should be deemed a sister taxon to Recent mammals. Consequently, multituberculates' divergence from the main lineage leading to Recent mammals appears prior to the emergence of the latest common ancestor of monotremes, marsupials, and eutherians (Miao, 1988, 1991). However, if time prevails, multituberculates would represent a sister taxon to all the other mammals combined, both extinct and extant (McKenna, 1987). Neither scenario would support the hypotheses of multituberculates being the sister taxon to monotremes (Kemp, 1983) or to Recent therians (Rowe, 1988), nor does the phylogenetic analysis presented in this chapter. What is really needed is more fossils. Thus, I wish to echo Carroll's (1989, p.236) timely plea: " . . . it's time to put down the straight-edged ruler, and take up the Marsh pick and the pin-vise again."

ACKNOWLEDGMENTS. I wish to thank F.S. Szalay, M.J. Novacek, and M.C. McKenna for inviting me to contribute this chapter. My discussion of the selected characters in this chapter has benefited from the preprint of J.R. Wible (1991) and free exchange of ideas with W. A. Clemens, J.A. Hopson, Z. Kielan-Jaworowska, J.A. Lillegraven, M.C. McKenna, R. Presley, and J.R. Wible. I am especially grateful to W.A. Clemens, J.A. Hopson, J.A. Lillegraven, and H.-P. Schultze for their continued support and constant encouragement. Special thanks are extended to Anne Musser for preparation of the illustrations, Mary Schmalz for typing the manuscript, and J.A. McAllister for creating the computer-generated cladograms. To the trio this contribution is dedicated.

CORRESPONDENCE ADDRESS. Desui Miao, Museum of Natural History, Department of Systematics and Ecology, The University of Kansas, Lawrence, KS 66045-2454, USA.

References

Adams, L.A. 1919. A memoir on the phylogeny of the jaw muscles in recent and fossil vertebrates. *Annals N.Y. Acad. Sci.* 28:51–166.
Allin, E.F. 1986. The auditory apparatus of advanced mammal-like reptiles and early mammals. In: The ecology and biology of mammal-like reptiles (Hotton, N. III, MacLean, P.D. Roth, J.J. and Roth, E.C. eds.). Washington, D.C.: Smithsonian Institution Press, pp. 283–294.
Archer, M., Flannery, T.F., Ritchie, A., and Molnar, R.E. 1985. First Mesozoic mammal from Australia—An early Cretaceous monotreme. *Nature* 18:363–366.
Archibald, J.D. 1982. A study of Mammalia and geology across the Cretaceous-Tertiary boundary in Garfield County, Montana. *Univ. Calif. Publ. Geol. Sci.* 122:1–286.
Barghausen, H.R. 1986. On the evolutionary origin of the therian tensor veli palatini and tensor tympani muscles. In: The ecology and biology of mammal-like reptiles (Hotton, N. III, MacLean, P.D., Roth, J.J., and Roth, E.C., eds.). Washington, D.C.: Smithsonian Institution Press, pp. 253–262.
Broom, R. 1914. On the structure and affinities of the Multituberculata. *Bull. Amer. Mus. Nat. Hist.* 33:115–134.
Carroll, R.L. 1989. Review of: *The phylogeny and classification of tetrapods, vol. 1: Amphibians, reptiles, birds* (Benton, M.J., ed.), Oxford: Oxford University Press, 1988. *J. Vert. Paleontol.* 9:235–236.
Chow, M., and Qi, T. 1978. Paleocene mammalian fossils from Nomogen Formation of Inner Mongolia. *Vert. PalAsia* 16:77–85.
Clemens, W.A. Jr. 1963. Fossil mammals of the type Lance Formation, Wyoming, Part I. Introduction and Multituberculata. *Univ. Calif. Publ. Geol. Sci.* 48:1–105.
Clemens, W.A. Jr. 1986. On Triassic and Jurassic mammals. In: The beginning of the age of dinosaurs (Padian, K., ed.). Cambridge: Cambridge University Press, pp. 237–246.
Clemens, W.A. Jr., and Kielan-Jaworowska, Z. 1979. Multituberculata. In: Mesozoic mammals: The first two-thirds of mammalian history (Lillegraven, J.A., Kielan-Jaworowska, Z., and Clemens, W.A. Jr., eds.). Berkeley: University of California Press, pp. 99–149.
Cope, E.D. 1888. The Multituberculata monotremes. *Naturalist* 22:259.
Crompton, A.W., and Sun, A. 1985. Cranial structure and relationships of the Liassic mammal *Sinoconodon*. *Zool. J. Linn. Soc.* 85:99–119.
Fraser, N.C., Walkden, G.M., and Stewart, V. 1985. The first pre-Rhaetic therian mammals. *Nature* 314:161–163.
Gidley, J.W. 1909. Notes on the fossil mammalian genus *Ptilodus*, with descriptions of new species. *Proc. U.S. National Mus.* 36:611–626.
Gingerich, P.D. 1988. Cladistic futures: Review of: *The phylogeny and classification of the tetrapods, vol. 1, Amphibians, reptiles, birds;* vol. 2, Mammals (Benton, ed. M.J.,), Oxford: Oxford University Press, 1988. *Nature* 336:628.
Granger, W., and Simpson, G.G. 1929. A revision of the Tertiary Multituberculata. *Bull. Amer. Mus. Nat. Hist.* 56:601–676.
Griffiths, M. 1978. *The biology of the monotremes*. New York: Academic Press.
Hahn, G. 1969. Beiträge zur Fauna der Grube Guimarota Nr. 3, die Multituberculata. *Palaeontographica, Abt.* A133:1–100.
Hahn, G. 1978. Neue Unterkiefer von Multituberculaten aus dem Malm Portugals. *Geol. et Palaeontol.* 12:177–212.
Hahn, G. 1981. Zum Bau der Schädel-Basis bei den Paulchoffatiidae (Multituberculata; Ober-Jura). *Senckenberg. Lethaea* 61:227–245.
Hahn, G. 1985. Zum Bau des Infraorbital-Foramens bei den Paulchoffatiidae (Multituberculata, Ober-Jura). *Berliner Geowissenschaften Abhandlung* A60:5–27.
Hahn, G. 1987. Neue Beobachtungen zum Schädel-und Gebiss-Bau der Paulchoffattiidae (Multituberculata, Ober-Jura). *Palaeovert.* 17:155–196.
Hahn, G. 1988. Die Ohr-Region der Paulchoffatiidae (Multituberculata, Ober-Jura). *Palaeovert.* 18:155–185.
Hill, J.E. 1935. The cranial foramina in rodents. *J. Mammal.* 16:121–129.
Hopson, J.A. 1987. Synapsid phylogeny and the origin of mammalian endothermy (abstr.). *J. Vert. Paleontol.* 7:18A.
Hopson, J.A., and Crompton, A.W. 1969. Origin of mammals. *Evol. Biol.* 3:15–72.
Hopson, J.A., Kielan-Jaworowska, Z., and Allin, E.F. 1989. The cryptic jugal of multituberculates. *J. Vert. Paleontol.* 9:201–209.
Jenkins, F.A. Jr. 1990. Monotremes and the biology of Mesozoic mammals. *Netherlands J. Zool.* 40:5–31.
Kemp, T.S. 1983. The relationships of mammals. *Zool. J. Linn. Soc.* 77:353–384.
Kemp, T.S. 1988. A note on the Mesozoic mammals, and the origin of therians. In: *The phylogeny and classification of the tetrapods, vol. 2* (Benton, M.J., ed.). Oxford: Clarendon Press, pp. 23–29.
Kermack, D.M., and Kermack, K.A. 1984. *The evolution of mammalian characters*. Sydney, Australia: Croom Helm Ltd.
Kermack, K.A. 1967. The interrelations of early mammals. *J. Linn. Soc. (Zool.)* 47:244–249.
Kermack, K.A., and Kielan-Jaworowska, Z. 1971. Therian and non-therian mammals. In: Early mammals (Kermack, D.M., and Kermack, K.A., eds.). *Zool. J. Linn. Soc.* 50 (suppl. 1): 103–115.
Kielan-Jaworowska, Z. 1971. Skull structure and affinitites of the Multituberculata. In: Results of the Polish-Mongolian palaeontological expeditions, Part III (Kielan-Jaworowska, Z., ed.). *Palaeontol. Polonica* 25:5–41.
Kielan-Jaworowska, Z. 1974. Multituberculate succession in the late Cretaceous of the Gobi Desert (Mongolia). In: *Ibid.*, Part V. 30:23–44.
Kielan-Jaworowska, Z., Crompton, A.W., and Jenkins, F.A. Jr. 1987a. The origin of egg-laying mammals. *Nature* 326:871–873.

Kielan-Jaworowska, Z., Dashzeveg, D., and Trofimov, B.A. 1987b. Early Cretaceous multituberculates from Mongolia and a comparison with late Jurassic forms. *Acta Palaeontol. Polonica* 32:3–47.

Kielan-Jaworowska, Z., Presley, R., and Poplin, C. 1986. The cranial vascular system in taeniolabidoid multituberculate mammals. *Phil. Trans. Royal Soc. London.* B313:525–602.

Krause, D.W. 1982. Jaw movement, dental function, and diet in the Paleocene multituberculate Ptilodus. *Paleobiol.* 8:265–281.

Kuhn, H.-J., and Zeller, U. 1987. The cavum epitericum in monotremes and therian mammals. In: *Morphogenesis of the mammalian skull* (Kuhn, H.-J., and Zeller, U., eds.). Hamburg: Verlag Paul Parey, pp. 51–70.

Luo, Z. 1989. Observation of the acousto-vestibular structures of multituberculate mammals with computerized tomography (abstr.). *J. Vert. Paleontol.* 9:30A.

Maier, W. 1987. The ontogenetic development of the orbitotemporal region in the skull of Monodelphis domestica (Didelphidae, Marsupialia), and the problem of the mammalian alisphenoid. In: *Morphogenesis of the mammalian skull* (Kuhn, H.-J. and Zeller, U., eds.). Hamburg: Verlag Paul Parey, pp. 71–90.

McKenna, M.C. 1987. Molecular and morphological analysis of high-level mammalian interrelationships. In: *Molecules and morphology in evolution: Conflict or compromise?* (Patterson, C., ed.). Cambridge: Cambridge University Press, pp. 55–93.

Miao, D. 1988. Skull morphology of Lambdopsalis bulla *(Mammalia, Multituberculata) and its implications to mammalian evolution, contr. geol.* (Univ. of Wyoming), Special Paper 4.

Miao, D. 1991. On the origins of mammals. In: *Origins of major groups of tetrapods: Controversies and consensus* (Schultze, H.-P., and Trueb, L., eds.). Ithaca, NY: Cornell University Press, pp. 579–597.

Miao, D., and Lillegraven, J.A. 1986. Discovery of three ear ossicles in a multituberculate mammal. *National Geogr. Res.* 2:500–507.

Moore, W.J. 1981. *The mammalian skull.* Cambridge: Cambridge University Press.

Novacek, M.J. 1990. Morphology, paleontology, and the higher clades of mammals. In: *Current mammalogy, vol. 2.* (Genoways, H.H., ed.). New York: Plenum Publishing Corporation, pp. 507–543.

Novacek, M.J., and Wyss, A. 1986. Origin and transformation of the mammalian stapes. In: *Vertebrates, phylogeny, and philosophy* (Flanagan, K.M., and Lillegraven, J.A., eds.). *Contr. Geol.* (Univ. of Wyoming), Special Paper 3, pp. 35–53.

Parrington, F.R. 1971. On the Upper Triassic mammals. *Phil. Trans. Royal Soc. London* B261:231–272.

Patterson, B., and Olson, E.C. 1961. A triconodontid mammal from the Triassic of Yunnan. *Internat. Colloq. on the Evol. of Lower and Non-specialized Mammals, pt. I*, pp. 129–191.

Patterson, C. 1981. Methods of paleobiogeography. In: *Vicariance biogeography: A critique* (Nelson, G. and Rosen, D.E., eds.). New York: Colubmia University Press, pp. 446–489.

Patterson, C. 1982. Morphological characters and homology. In: *Problems of phylogenetic reconstruction* (Joysey, K.A., and Friday, A.E., eds.). London and New York: Academic Press, pp. 21–74.

Presley, R. 1981. Alisphenoid equivalents in placentals, marsupials, monotremes and fossils. *Nature* 294:668–670.

Presley, R., and Steel, F.L.D. 1978. The pterygoid and ectopterygoid in mammals. *Anat. and Embryol.* 154:95–110.

Pridmore, P.A. 1985. Terrestrial locomotion in monotremes (Mammalia: Monotremata). *J. Zool., London* A 205:53–73.

Rowe, T. 1986. *Osteological diagnosis of Mammalia, L. 1758, and its relationship to extinct Synapsida.* Ph.D. Dissertation, University of California, Berkeley.

Rowe, T. 1988. Definition, diagnosis, and origin of Mammalia. *J. Vert. Paleontol.* 8:241–264.

Rowe, T., and Greenwald, N.S. 1987. The phylogenetic position and origin of Multituberculata (abstr.). *J. Vert. Paleontol.* 7:24A.

Sigogneau-Russell, D. 1989. Haramiyidae (Mammalia, Allotheria) en provenance du Trias superieur de Lorraine (France). *Palaeontographica Abt.* A 206:137–198.

Simpson, G.G. 1928. *A catalogue of the Mesozoic Mammalia in the geological department of the British Museum.* London: Oxford University Press,

Simpson, G.G. 1937. Skull structure of the Multituberculata. *Bull. Amer. Mus. Nat. Hist.* 73:727–763.

Simpson, G.G. 1938. Osteography of the ear region in monotremes. *Amer. Mus. Novitates* 978:1–15.

Sloan, R.E. 1979. Multituberculata. In: *The encyclopedia of paleontology* (Fairbridge, R.W., and Jablonski, D., eds.). Stroudsburg: Dowden, Hutchinson & Ross, Inc., pp. 492–498.

Tyndale-Biscoe, H., and Renfree, M. 1987. *Reproductive physiology of marsupials.* Cambridge: Cambridge University Press.

Ulinski, P.S. 1986. Neurobiology of the therapsid-mammal transition. In: *The ecology and biology of mammal-like reptiles* (Hotton, N. III, MacLean, P.D., Roth, J.J., and Roth, E.C., eds.). Washington, D.C.: Smithsonian Institution Press, pp. 149–171.

Watson, J.M., and Graves, J.A.M. 1988. Monotreme cell cycles and the evolution of homothermy. *Aust. J. Zool.* 36:573–584.

Whiteside, D.I., and Marshall, J.E.A. 1985. Pre-Rhaetic therian mammals. *Nature* 318:81–82.

Wible, J.R. 1991. Origin of Mammalia: The craniodental evidence reexamined. *J. Vert. Paleontol.* 11:1–28.

Wible, J.R., Miao, D., and Hopson, J.A. 1990. The septomaxilla of fossil and recent synapsids and the problem of the septomaxilla of monotremes and armadillos. *Zool. J. Linn. Soc.* 98:203–228.

Winge, H. 1923. *Pattedyr-Slaegter, vol. 1.* Copenhagen: C.A. Reitzels Forlag.

CHAPTER 7

Reconsideration of Monotreme Relationships Based on the Skull and Dentition of the Miocene *Obdurodon dicksoni*

MICHAEL ARCHER, PETER MURRAY, SUZANNE HAND, AND HENK GODTHELP

Overview

In this chapter, we (1) describe and illustrate the anatomy of a complete skull, posterior half of a dentary, and cheektooth dentition of a Miocene ornithorhynchid, *Obdurodon dicksoni* from Riversleigh, Queensland, Australia; (2) use this new information to reevaluate understanding of the anatomy of the living *Ornithorhynchus anatinus*; (3) revise previous understanding about the monotreme dentition; and (4) reconsider monotreme relationships.

The skull differs from that of *Or. anatinus* in many features but demonstrates that by the Miocene much of the pattern evident in *Or. anatinus* had already been attained. However, in its more primitive state, the Miocene taxon reveals details of morphology unclear in the living taxon such as the undoubted presence of an enormous septomaxilla at the front of the skull. Contrary to recent suggestions, there is no evidence of postdentary bones in the mandible. On balance, *Obdurodon* could have been ancestral to *Ornithorhynchus*.

The dentition of *Ob. dicksoni* is more plesiomorphic than that of *Or. anatinus*, but the tooth formula agrees in general with the ontogenetic pattern established for the living Platypus. Molar morphology is highly autapomorphic and mimics, with half as many teeth, the serial pattem of four tribosphenid molars.

The evidence regarding monotreme relationships is reviewed. Although it is contradictory, on balance it does not support a close relationship between monotremes and any group of tribosphenid mammals. On the one hand, although the contention of Kielan-Jaworowska et al. (1987) that monotremes are descendants of pre-tribosphenid eupantotheres such as peramurids (or vincelestids as discussed here) is possible, there are as many similarities to molars of South American mesungulatid dryolestoids (Archer et al., 1979). Although similarities to these particular late Cretaceous dryolestoids are probably convergent (aspects of the mesungulatid pattern probably being secondarily monotreme-like), the dentition of the early Cretaceous monotreme *S. galmani* is at least superficially annectant between that of dryolestoids and Oligo-Miocene ornithorhynchids of the genus *Obdurodon*.

Contents

Introduction, 76
The Dentition of *Obdurodon dicksoni*, 76
   Source for information about the dentition, 77
   Crown and root morphology, 79
The Dental Formula of Ornithorhynchids, 79
Molar Structure of Ornithorhynchids, 80
   Previous suggestions, 80
   Dryolestoid similarities, 81
   Ontogenetic falsification?, 82
   Adaptive advantage in elaborating the dryolestoid molar pattern?, 82
Occlusion and Dental Function of Toothed Ornithorhynchids, 82
   Steropodon galmani, 82
   Obdurodon, 83
   Functional analogues to the therian molar row, 83
   Molar stability during occlusion, 83
   Enamel surface structures, 84
The Cranium and Dentary of *Obdurodon dicksoni*, 84
   Description, 84
   Discussion of cranial morphology, 86
   Conclusions based on cranial morphology, 87
   Dentary, 88
   Overview and limitations of cranial comparisons, 88
Phylogenetic Relationships of Monotremes, 89
   Plesiomorphic osseous cranial/dental features certainly or possibly unique within Mammalia, 89
   Mammalian synapomorphic osseous cranial/dental features, 89
   Autapomorphic osseous cranial/dental features, 89
   Synapomorphic cranial/dental features that appear to relate monotremes to specific subgroups of mammals, 89
   Position of monotremes with respect to mammals in general, 89
   Position of monotremes with respect to tribosphenid mammals, 90
   Position of monotremes with respect to non-tribosphenid therian mammals, 90
   Relationships within Monotremata, 90

Overview, *continued*

As suggested by Bonaparte (1990), monotremes could be a specialized part of an endemic southern eupantothere if not dryolestoid radiation.

The evidence, not wholly negative, thus tends to emphasize the sharp separation from all other mammals seen throughout the whole organization of *Ornithorhynchus*. For what it is worth, it tends to remove the group from any central position in therian phylogeny, to emphasize the rather widespread

Contents, *continued*

Biogeographic Considerations, 91
Conclusions, 91
Acknowledgments, 91
Emendation, 92
References, 92

opinion that *Ornithorhynchus* is not merely a more primitive therian, but something quite distinct (Simpson 1929, p.14).

Introduction

The relationships of monotremes have been vigorously debated ever since they were first known. Along the way, every conceivable sistergroup relationship has at one time or another been proposed from birds to marsupials and even turtles. Recent reviews or key examinations of monotreme relationships have most recently been given by Griffiths (1968, 1978); Woodburne and Tedford (1975); Clemens (1979); Kemp (1982, 1983); Murray (1984); Archer, Flannery, Ritchie, and Molnar (1985); Kielan-Jaworowska, Crompton, and Jenkins (1987); Pascual, Archer, Ortiz Jaureguizar, Prado, Godthelp, and Wand (1992a,b); and Archer, Hand, Jenkins, Murray, and Godthelp (1992). Despite a great deal of attention, there is little in the way of a consensus view.

Principal reasons for much of the uncertainty include the paucity of information about dental morphology of the group; dental ontogeny of this and most Mesozoic mammal groups; plesiomorphic cranial morphology of monotremes; skulls of potentially related Mesozoic mammal groups such as peramurids, dryolestoids, symmetrodonts, and docodonts; and the fossil record of monotremes.

Modern studies of dental ontogeny (Luckett and Zeller, 1989) and enamel ultrastructure (Lester and Archer, 1986; Lester and Boyde, 1986) are providing new, phylogenetically useful information. Recent discovery of skulls of vincelestid peramurids (Bonaparte and Rougier, 1987) have significantly increased understanding about the cranial morphology of this key group of Cretaceous mammals. Discovery of two isolated teeth of the Oligo-Miocene ornithorhynchid *Obdurodon insignis* (Woodburne and Tedford, 1975) provided the first look at non-vestigial monotreme teeth, and discovery of the Early Cretaceous monotreme *Steropodon galmani* (Archer et al., 1985; Kielan-Jaworowska et al., 1987) provided the first look at a presumably plesiomorphic lower molar row of a monotreme.

Discovery of a complete skull (Archer, Hand and Godthelp, 1986, 1991; Archer et al., 1992) and a nearly complete dentition of the Miocene *O. b. dicksoni* provided the first opportunity to examine non-deciduous upper and lower tooth rows as well as a relatively plesiomorphous monotreme cranium. A minimal description of the dentition of this, the only well-preserved Tertiary monotreme, was given by Archer et al. (1992). Here we review the evidence to provide an adequate basis for reconsidering the phylogenetic affinities of monotremes.

In this chapter, terminology of crown morphology is based on the structural/functional thegotic nomenclature devised by Every (1972, 1974). Cranial nomenclature used follows works cited and is illustrated in Figures 7.2–7.5. Use of the taxonomic term *tribosphenid* refers to Tribosphenida McKenna, 1975. Use is made here (following the rationale of Aplin and Archer, 1987) of the term *Eutheria* in its original unambiguous sense as proposed by Gill (1872 i.e., Marsupialia plus Placentalia), rather than as subsequently and inappropriately used by Huxley (1880) as a senior synonym of Placentalia (Owen, 1837).

The Dentition of *Obdurodon dicksoni*

Because structural homologies with tribosphenid molars are uncertain, crown morphology is described here primarily in terms of topography and apparent function. The dental nomenclatural system best suited for this purpose is Every's thegotic terminology (Every 1972, 1974). Thus 'crests' and 'lophs' of more conventional descriptions are functionally blades and are called *drepanons* (*drepanids* in lower teeth; sing. drepanon, -id). A 'cusp' is called an *akis* (akid; pl. akises, akids) and akises commonly occur at the ends of the blades they define. Akises that occur on the cingulum or basal portions of the crowns may serve either as puncturing structures or, following faceting, as accessory blades. A blade subtended by two akises is a diakidrepanon (-id; pl. diakidrepanons, -ids). Two blades sharing an akis is a *triakididrepanon* (-id), functional units of this kind being the dominant features of ornithorhynchid molars.

Because we doubt that the akises at the buccal ends

Chapter 7. Reconsideration of Monotreme Relationships

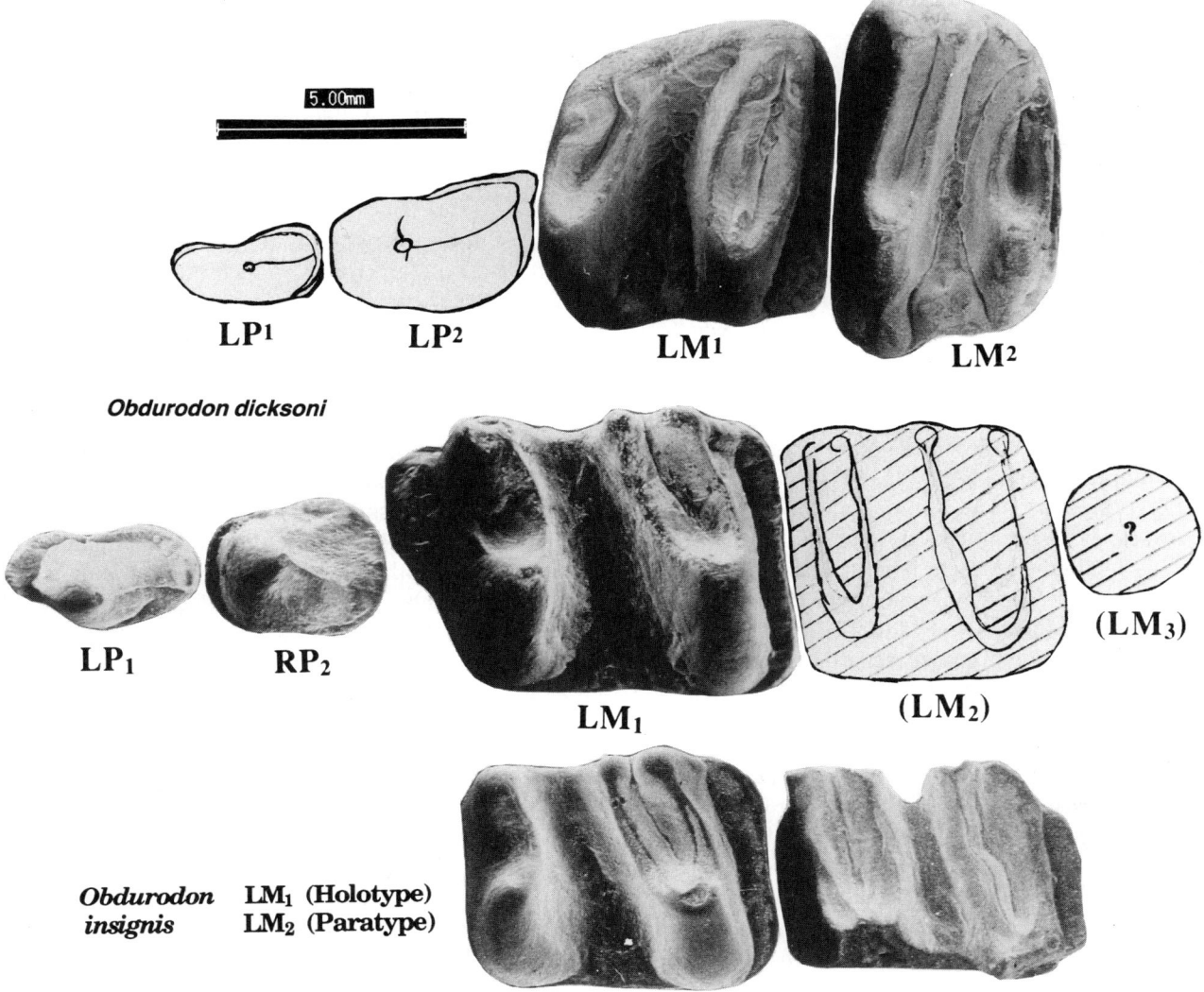

FIGURE 7.1. Composited SEM photographs of the upper left and lower left dentitions of *Obdurodon dicksoni* (upper two rows) and the holotype and paratype of *Ob. insignis* (bottom row) as reinterpreted here. Anterior is to the left. For upper teeth, buccal is toward the top. For lower teeth, buccal is toward the bottom. Because we do not have an LP_2, an RP_2 is shown in its place. The upper premolars are drawn in outline because they are bound into the skull. Although the upper molars were isolated when found, they are compatible in all regards and may represent the same individual.

of each primary blade on the upper molars are homologous with stylar cusps of therian molars, we have numbered them sequentially for each molar as buccal blade akises 1–4 (BBA1–4). Similarly, because of doubt about the homology of the lingual akises, we have referred to them as the anterolingual and posterolingual akises. Although we are reasonably confident that the anterior half of the lower molars is the homologue of the symmetrodont, dryolestoid, and tribosphenid trigonid with a protakid (=protoconid), metakid (=metaconid), and in some molars a parakid (=paraconid), the interordinal homologies (if any) of the lingual akids of the posterior triakididrepanid are unknown. They are therefore referred to as lingual blade akids 3–4 (LBA3–4), 1 being the parakid and 2 the metakid. Although it might be a hypakid (=hypoconid), we are not confident about the homology of the posterobuccal akid of the posterior triakididrepanid; hence it is referred to here as the primary posterobuccal akid.

SOURCE FOR INFORMATION ABOUT THE DENTITION. The premaxillae and maxillae of the skull of *Ob. dicksoni* reveal no alveoli for any teeth anterior to the two premolariform teeth, designated here P^1 and P^2. Behind these are alveoli for two molariform teeth, designated here M^1 and M^2. There is no indication of a third upper molar, although there appears to be one in the lower dentition.

Material contributing to the interpretation of the lower dentition includes the posterior portion of an edentulous dentary, two isolated premolars, and two molars.

Crown and Root Morphology. Figure 7.1 provides occlusal views of upper and lower teeth. Here we give a description that supplements these photographs. These teeth were all found as isolated specimens. Association of the two upper molars shown together is based on their similar small size (in contrast to a considerably larger M^1 and M^2 from the same deposit). The upper premolars are shown in outline because of the difficulty of obtaining SEM photographs of these teeth, which are fixed in the skull. The association of lower premolars shown is hypothetical and based on the presumption that, as in the upper premolar row, P_2 is larger than P_1. Teeth interpreted as lower premolars differ in morphology from upper premolars.

Upper premolars. The identity of the second upper cheektooth as a premolar, despite its premolariform crown, is in doubt because it is at least three-rooted. However, the abrupt change in crown form and shape between this tooth and that identified here as M^1 is so striking that we are urged to regard its "polyfangled" condition as autapomorphic. Luckett and Zeller (1989) have also raised doubts (see Simpson, 1929) about the homology of a tooth anlage in this position (see below).

The crowns of both P^1 and P^2 are basically simple with a primary akis (presumably the parakis). The P^2 exhibits an area of wear on the crown that starts at the tip and descends down the posterior flank onto the posterior shelf. The angle and position of this area corresponds with the diakidrepanid of the M_1 and provides some information about dental function (see below). On P^1 there is no conspicuous trailing blade. Rugose basal cingula are well-developed on both teeth except below the lingual base of the parakis.

Upper molars. Although the size range of each upper molar is considerable (see Fig. 7.1), the morphology in our albeit small sample is remarkably constant considering the variation well-known for molar rudiments found in *Or. anatinus* (e.g., Simpson, 1929).

In essence, each upper molar is dominated by two transverse triakididrepanons on either side of a deep central transverse basin. The buccal and lingual margins of the crowns are higher than the central basins between the triakididrepanons. All blades are concave occlusally and arcuate such that the two of each triakididrepanon bucally converge toward each other but do not share a buccal akis. There are two basal lingual akises between the two primary akises and another on the posterolingual margins of M^1 and anterolingual margin of M^2, thereby providing an interdental set analogous to that between the primary lingual akises of each individual molar. The anterior triakididrepanon of M^1 and posterior triakididrepanon of M^2 are asymmetric, with the anterior blade of the former and the posterior blade of the latter being shorter than their adjacent counterparts. A large posterior cingulum on M^1 and anterior cingulum on M^2 provide a large transverse interdental basin as large as that developed between the two triakididrepanons of M^1. There is a smaller anterior cingulum on M^1 and a much smaller posterior cingulum on M^2. The buccal and lingual flanks of all akises are inclined toward the center of the crowns, making the blade lengths considerably narrower than the basal width of the crowns. This may reflect the very short roots of all molar crowns, which could have provided only limited stability to offset normal pressures of transverse shear (see functional discussion below).

In M^1, the arcuate posterior blade sweeps posterobuccally out to the buccal margin of the crown and the posterobuccal corner of the crown. Halfway along its length it is punctuated by a buccal akis (BBA2) that in effect divides the blade into a transverse and longitudinal section. The basal midbuccal region of M^2, between BBA2 and 3, has a prominent akis rather than a longitudinal blade, although with faceting or wear, it could function as a blade. The four triakididrepanons define three deep embrasures, presumably for the corresponding structures of the lower molars (but see below under functional discussion).

Within the posterior triakididrepanon of M^1 there is a transverse ridge similar in appearance to the 'mesocrista' of mesungulatid and dryolestid dryolestoids, although there is no clear 'stylocone' unless it is homologous with BBA3. This is conceivable because the mesocrista actually rises to contact the apex of BBA3. To see this triakididepanon system as derived from the pattern evident in mesungulatids, the anterior blade's contact with BBA3 would have to be derivative, a posteriad inflexion of the buccal margin of the blade. None of the other triakididrepanons have any trace of a mesocrista.

Lower premolars. Two isolated teeth appear to represent an LP_1 and RP_2. Neither has the morphology of the upper premolars but both are basically ornithorhynchid-like in terms of the relationship of the primary cusp to the basal cingulum. The P_1 is relatively long and narrow with a complete basal cingulid and prominent but low anterior and posterior cingular akids. The P_2 is relatively short and wide with a complete buccal but incomplete lingual cingulid. Both are relatively unspecialized teeth with a protakid and slightly developed trailing blades. Basal cingula are reasonably well-developed.

Lower molars. There are two lower molars of *Ob. dicksoni* and both appear to represent LM_1. The most complete of the two (Fig. 7.1) is so identified for three reasons: (1) it is basically similar in distinctive structure to the M_1 of *Steropodon galmani*, whose entire molar row is in situ; (2) the anterior transverse blade is only a diakidrepanid with one lingual and one buccal akid, which suggests that this blade passed anterior to M^1 rather than between two triakididrepanons of the upper dentition; and (3) it is distinctly different in morphology from the paratype of Ob. insignis, which we are now confident is LM$_2$ (rather than a right upper molar as originally determined by Woodburne and Tedford, 1975; see below) and very similar to the holotype of Ob. insignis, which we are similarly confident is LM$_2$ (rather than a posterior right upper molar as originally interpreted).

Our only doubts concern the less complete of the two because, although both teeth are from the same locality (Ringtail Site), it reveals a somewhat simpler anterior half (see below). We conservatively presume that this is an indication of intraspecific variation not otherwise demonstrated by the sample of upper molars rather than an indication of a second species.

The anterior half of the relatively complete crown consists of an anteriorly concave diakidrepanid with a steeply inclined subsidiary drepanid extending anterolingually from the anterobuccal akid (probably a 'protakid'). The basin defined between these two blades is shallow. In this regard, the incomplete M$_1$ is simpler in lacking any trace of the subsidiary anterior drepanid. The posterior half of both first molars is a diakididrepanid with a very tall buccal akid whose homology is unclear. Similarly, the homology of the lingual akids is unclear. We suspect, for reasons noted below, that they have no homologues with the structures of therian molars. The basic structure of this posterior half is similar to but a mirror image of the triakididrepanons of the upper molars, between which it presumably interdigitated. The basin within the posterior triakididrepanid is well-defined and enclosed on all sides, as are the comparable basins of the upper molars. There are no mesocristae in any of the triakididrepanids. Between the anterior and posterior halves, there is a deep transverse midvalley with small akids near its buccal margin and poorly defined, steeply inclined longitudinal blades on its lingual margin. All primary buccal and lingual akids curve medially toward the center of the crown such that the transverse blades, as in the upper molars, are narrower than the basal crown width. The anterior cingulid is damaged except on its anterolingual end but was clearly well-developed. The posterior cingulid is also well-developed and presumably functioned with the anterior cingulid of M$_2$ to provide an occlusal counterpart for the posterior triakididrepanon of M^1. The posterior cingulid ascends the flank of the most posterior akid in the more complete tooth but fails to do so in the incomplete molar.

Although we lack an M$_2$ for Ob. dicksoni, the paratype of Ob. insignis (see Fig. 7.1 and the figures in Woodburne and Tedford, 1975) appears to be an incomplete LM$_2$. From what is preserved, it differs from M$_1$ primarily in the fact that its anterior half is a triakididrepanid as are the anterior halfs of the upper molars.

Roots. P^1 has two roots, the anterior of which appears to enlarge towards its distal end (which is hidden in the alveolus). P^2 has at least three roots. There are two primary roots: a large, wide posterior root that extends transversely below the entire width of the tooth's posterior shelf; and a large but smaller anterior root below the parakis. A small accessory root, sharing its base with the anterior root, is distinguished from it on the lingual side of the tooth, in a position effectively between the anterior and posterior root. If there are additional roots on the root platform (although we suspect there are not), they are not visible from the alveolar margins. Alveoli for the upper molars indicate that M^1 had six and the M^2 four roots, these being arranged more or less in transverse rows of three and two, respectively. The largest M^1 preserves some of its roots. These are basically of the same kind as those described for M$_1$ by Woodburne and Tedford (1975) for Ob. insignis. They are all short, stumpy, with tiny to possibly blocked root canals. In most cases the roots are less than a third the height of the crown. They do not narrow toward the distal ends, but their distal width is not sufficient to ensure that they remain in the alveoli after death as the skull clearly demonstrates. On the undersides of some teeth there is a calcified layer of rugose material that looks almost like cementum in the way that it appears to adhere to the base of the crown.

The lower premolars are represented by two isolated teeth (see above). Both P$_1$ and P$_2$ had two roots. None of the molars or dentary fragments give any information about root number for M$_1$ or M$_2$. However, M$_3$ almost certainly had only a single root positioned near the posterolingual margin of the alveolar row. The M$_1$ of Ob. insignis, described by Woodburne and Tedford (1975), rests on six roots. Their diminutive size and shape are basically similar to those below the M^1 of Ob. dicksoni.

The Dental Formula of Ornithorhynchids

There have been interminable debates (e.g., see Luckett and Zeller, 1989) about the number of molars and premolars in monotremes. Part of the reason has been Kühne's persistence (e.g., Kühne, 1973, 1987) that the cheektooth number and pattern of apparent diphyodonty of monotremes are basically the same as those of marsupials, one of the mainstays in his argument for the monophyly of Marsupionta. Kühne's contention about the dentition is based mainly on the work of Green (1937), which concluded that there were five distinct upper and lower postcanine cheekteeth: *v* or *dv* (our P1), *w* (our P2), *x* (our M1), *y* (our M2), and *z* (our M3). The first appeared to be an unreplaced deciduous premolar (*dv*), the replacement tooth (*v*) developing in at least some lower dentitions but subsequently resorbing. Kühne has interpreted this pattern to be the same as that of marsupials, in which the ephemeral fifth tooth from the rear of the molar row *appears* to be a deciduous predecessor to P3.

If v is a deciduous tooth for which a genuine successor starts to develop in *Ornithorhynchus anatinus*, as Wilson and Hill (1907) and Green (1937) suggest, the situation is *not* clearly like that of marsupials. Berkovitz (1967), Kirkpatrick (1969), and Archer (1974a, 1978) have given reasons for regarding the replaced tooth—at least in the marsupials they ontogenetically examined (didelphids, dasyurids, and macropodids)—as an ephemeral M1 evicted at the time P3 erupts. In any case, Luckett and Zeller (1989) dispute Kühne's claim that v is a deciduous tooth, having seen no replacement anlage develop in the ontogenetic series they studied. But Green also indicated that some individuals he examined failed to develop the replacement tooth, so its appearance could be variable.

Clearly there is unresolved conflict here. If Green and Wilson and Hill did see a successional tooth developing from the dental lamina lingual to the developing v, their interpretation of v as a deciduous tooth, and hence part of a premolar family, may be correct. Certainly in terms of crown morphology, the tooth we interpret in *Ob. dicksoni* to be P1 is distinctly premolariform. But it would not necessarily mean that every tooth posterior to this successional family was a molar as Kühne would argue, because tooth w could be a deduous tooth that does not develop a successional anlage in the same way that the free edge of the dental lamina lingual to tooth v in some individuals fails to develop a replacement tooth.

Jenkins and Schaff (1988) have described an early Cretaceous triconodont (*Gobiconodon ostromi*) that appears to demonstrate successional tooth families well within the molariform series, thereby raising doubts about the distinction between molars and premolars simply on the basis of patterns of tooth replacement. Owen's original criterion for distinguishing between premolars and molars was based on diphyodonty. Determination of a postcanine as a molar or premolar depended on whether it was in front of another that was replaced (and hence a premolar like the tooth behind it) or behind the last one that was replaced (and therefore a molar). Although this procedure appears simple in theory, it is often difficult to apply in practice. As Archer (1978) and Luckett and Zeller (1989) point out, even embryological investigations can fail to resolve the problem because an original condition of a successional relationship can be obfuscated by differential growth of the oral epithelium, dental lamina, and the tooth buds themselves. Interpretation of embryological relationships sometimes seems to depend more on serendipity and preconception than on 'simple' observation.

Our only doubts about the nature of the cheekteeth in *Ob. dicksoni* focus on the tooth we have interpreted as P². Its crown appears to be supported by three roots. Because the number of roots a mammalian tooth has appears in general to roughly correlate with the size and complexity of the crown, several roots beneath a simple single-cusped premolariform tooth might be an indication that the crown has undergone secondary simplification. Considering that Green, Wilson and Hill, and Luckett and Zeller found no evidence for a successional tooth to w, and Luckett and Zeller noted that developmentally it appeared more like a molar than a premolar, we suggest serious consideration of the possibility that it (and presumably its occlusal counterpart in the lower tooth row, our P_2) is in fact a molar. However, we have been persuaded by the abrupt change in crown morphology between P² and M¹, as well as the abrupt increase from three roots to six, to conclude that P² is more appropriately regarded as an autapomorphically multirooted posterior premolar.

In summary, we suggest that our paleontological data corroborate a subset of the ontogenetic data. The two anlagen suggested by Green (1937) to be presumptive premolars (v,w) are indeed represented in *Obdurodon dicksoni* by premolariform crowns, although the three-rooted condition of the upper posterior premolariform tooth does not contradict the possibility that this tooth is a secondarily simplified molar. Occlusal relationships would then suggest that the corresponding tooth in the lower dentition might also be a molar even though it has only two roots, as do P¹ and P_1. Paleontology provides evidence for two upper and three lower molariform teeth. There appears to be no homologue in the fossil dentition for the posterior upper anlage reported by Green (1937) and others.

Molar Structure of Ornithorhynchids

PREVIOUS SUGGESTIONS. The vestigial molar rudiments of *Ornithorhynchus anatinus* have been used in the past to interpret the relationships of monotremes (e.g., Simpson, 1929; Hobson and Crompton, 1969; Mills, 1971). Recent suggestions for relationships on the basis of the dentition of the living platypus have included marsupials and morganucodontids. Although the two teeth of *Obdurodon insignis* were the first well-formed monotreme molars found, Woodburne and Tedford had difficulty determining the orientation of these isolated teeth. Nevertheless, they noted similarities to peramelid marsupials, symmetrodonts, and morganucodontids. On the basis of ultrastructure of molar enamel of *Ob. dicksoni* (then known but unnamed), Lester and Archer (1986) concluded that monotremes represented a unique and possibly plesiomorphic group of mammals, albeit one somewhat similar among other groups examined to multituberculates. Archer et al. (1985) suggested that the lower molars of *Steropodon galmani* are tribosphenid-like and suggested that monotremes might be the plesiomorphic sistergroup of aegialodontids plus eutherians. Kielan-Jaworowska, Crompton, and Jenkins (1987) agreed on the basis of *S. galmani* that monotremes were therian but not tribosphenid and perhaps the plesiomorphic sistergroup of tribosphenid mammals. Van Valen (1988a, 1988b) suggested that monotremes may have given rise to the Early Cretaceous *Gondwanatherium patagonicum* (regarded by Bonaparte (1990) to be a paratherian). Archer, Hand and Godthelp (1989) suggested, mainly on the basis of the then undescribed dentition of *Ob. dicksoni*, that monotremes shared features with eupantotheres such as the Early Cretaceous vincelestid peramurids and Late Cretaceous Los Alamitos mesungulatid dryolestoids. Bonaparte (1990), on the basis of comparisons of the teeth of *S. galmani* with those of the Los Alamitos mammals, independently suggested the possibility that monotremes might be derivatives of Gondwanan dryolestoids. Archer et al. (1992) considered the viewpoint we develop here, that monotremes may have descended from eupantotheres that had a molar morphology similar to that of dryolestoids but a cheektooth formula similar to that of vincelestid peramurids.

EUPANTOTHERE (SPECIFICALLY DRYOLESTOID) SIMILARITIES. Considering the almost complete dentition of *Ob. dicksoni* described above, it is apparent that the cheekteeth share seemingly distinctive features with eupantotheres in general and, perhaps convergently, mesungulatid dryolestoids in particular. These latter include (1) mildly crenulated enamel with vertical fluting; (2) seeming dominance of apical blade abrasion without conspicuous thegotic facets otherwise characteristic of many dryolestoids (Krebs, 1971); (3) an abrupt change in morphology between premolariform and molariform teeth; (4) metakids with an anteriorly extended flange or swelling; (5) transverse mesocrista within buccal midvalley of triakididrepanon on upper molars (most dryolestoids and posterior halves of molars of *Ob. dicksoni*); (6) primary buccal akis of dryolestoids ('stylocone') may be homologous with BBA3 in *Ob. dicksoni*; (7) derived (for dryolestoids) shapes and relationships of basal lingual akises, primary axis, and basal cingula of mesungulatids (and Los Alamitos dryolestoids such as *Grobertherium novasi*) and posterior half of upper molars of *Ob. dicksoni* basically similar; (8) same basic relationships shared in comparable portions of lower molars; and (9) a reduced molar number (most dryolestoids have eight fide Krebs, 1971; mesungulatids have four to five fide Bonaparte, 1986b; ornithorhynchids have three). Further, monotremes and Cretaceous dryolestoids share loss of the coronoid and splenial from the lower jaw, these bones being present in at least some Jurassic dryolestoids (Krebs, 1971).

Allowing that neither the 'dimer theory' (fusion of primary tooth buds as an embryological mechanism for increasing the complexity of teeth in evolving lineages) nor an embryological doubling and subsequent fusion of the presumptive cells destined to produce individual teeth is a plausible way of deriving a 'double-mesungulatid' pattern in ornithorhynchids, we suggest three alternative testable explanations of how the ornithorhynchid pattern might have evolved from a dryolestoid with a mesungulatid-like molar pattern:

1. If the anterolingual akis of the upper molars of *Ob. dicksoni* is the parakis (=paracone), as most mammalogists familiar with therian mammals would initially presume, its associated blades and basal cingula could be the homologue of a mesungulatid-like upper molar. The posterior half of the *Ob. dicksoni* molar could then be neomorphic, an elaboration of the original posterior cingulum. The same could be true of the lower molars if the anterobuccal akid is the protakid (=protoconid).

2. If the posterolingual akis is the parakis, its triakididrepanon could be the homologue of a mesungulatid-like molar and the anterior portion of the tooth the elaboration of the original anterior cingulum. The same could be true for the lower if the posterobuccal akid is the protakid.

3. The hypothesis we cautiously suggest is that the mesungulatid-like portion of the upper molars of ornithorhynchids is the posterior triakididrepanon and its basal structures, while that of the lower molars may be the anterior triakididrepanid, the adjacent portions of the molars having developed as occlusal counterparts within what was formerly the interdental embrasure between the upper and lower crowns of the ancestor.

Our reasons for supporting the third hypothesis include the following. In common with the upper molars of mesungulatid-like dryolestoids, the posterior triakididrepanon retains a transverse mesocrista and posterior and anterior cingula with basal lingual akises. In common with lower mesungulatid molars, the lingual akid (presumed metakid) of the anterior diakidrepanid has a pronounced anterior process or crest. Further, on the presumption that the Early Cretaceous *Steropodon galmani* would demonstrate an annectant condition between *Ob. dicksoni* and mesungulatid-like dryolestoids, its posterior blade system is not as mesungulatid-like as its anterior triakididrepanid—a contrast particularly evident on M_2. The anterior blade of this posterior system is relatively poorly defined, inclines ventrolingually, and lacks a terminal lingual akid (the LBA3 of *Obdurodon* species). We therefore suggest that the pattern of the posterior half of the lower molars of ornithorhynchids may be convergent on that of dryolestoid trigonids.

Consequent on this hypothesis, we would posit a late Jurassic to Early Cretaceous ancestor of monotremes that may have been a dryolestoid (or similar eupantothere) with three molars having a mesungulatid-like crown pattern separated by interdental embrasures, perhaps similar to the dryolestoid *Leonardus cuspidatus* (Bonaparte, 1990). Then, as "simply" as the protakis/talonid system gradually developed in therian mammals from peramurids to tribosphenids, or as neomorphic pre-talonids developed in shuotheriid yinotheres (Chow and Rich, 1982), opposing anterior upper and posterior lower basal cingula may have elaborated a second set of interdigitating triakididrepanon/id structures. All stages of this elaboration could have developed with occlusal advantage by anterolingual development of the anterobuccal corner of the anterior cingulum of the upper molar simultaneously with posterobuccal elaboration of the posterolingual corner (adjacent to metakid) of the lower molar. The result would be a mirror-image "cloning" of the dryolestoid primary structures and a doubling of the number of transverse blades.

For these reasons, while the anterior halves of the lower molars of monotremes may be homologues of the therian trigonid, the posterior halves may be better regarded as "pseudotalonids". As Kielan-Jaworowska et al. (1987) pointed out, it is highly probable that at least the lingual half of the talonid of *Steropodon galmani* is convergent on that portion of the tribosphenid molar. Here we are going one step further in suggesting that the whole of the posterior half of the lower molars of monotremes may be convergent.

ONTOGENETIC FALSIFICATION? A challenge to our preferred hypothesis of character transformation comes from the fact that Green (1937) found that the principal anterior cusp of the upper and lower molars of *Or. anatinus* developed first. This is not surprising in terms of the lower molars, where we presume the analogue of the protakid is the principal anterobuccal akid. However, this is not the case on the upper molars. If we were to rationalize this incongruity, we would point out that on both the M1 and M2, the anterolingual akis is the largest cusp and, as others have demonstrated, relative size can influence the sequence of ontogenetic development and calcification. In *Antechinus flavipes*, for example, a dasyurid marsupial with a metacone larger than the paracone, the metacone appears and calcifies first (Archer, 1974b).

ADAPTIVE ADVANTAGE IN ELABORATING THE DRYOLESTOID MOLAR PATTERN? Allowing that the hypothesis developed above for how a mesungulatid-like dryolestoid pattern could have given rise to the ornithorhynchid pattern, why might it have happened? We suggest that it could have been an adaptive response to pressure to elaborate the number of shearing blades in a lineage that had already undergone, for whatever reason, a reduction in the number of molars from eight (normal dryolestoid number) to three. Just as tribosphenid mammals elaborated the number and width of their shearing blades on a reduced number of teeth, ancestral monotremes may have done the same.

The large size of the M3 in *Steropodon galmani* compared with the apparently single-rooted M_3 and missing M^3 in species of *Obdurodon* and virtually edentulous condition of *Ornithorhynchus anatinus* suggests that the tendency for molar reduction is entrenched in this lineage. And yet at least in so far as the species of *Obdurodon* had relatively wider and more elaborate transverse shearing blades, they appear to have compensated for reduction of their posterior molars. Why *Or. anatinus* took the path towards 'gummyhood', however, is unclear.

Occlusion and Dental Function of Toothed Ornithorhynchids

Interpretation of ornithorhynchid dental function based on living monotremes is difficult because they operate either with degenerate cheektooth rudiments (juvenile platypuses) or with no teeth at all. Adult Platypuses use horny pads near the rear of their mouths to masticate mainly small aquatic crustaceans and insects. Short-beaked Echidnas (*Tachyglossus aculeatus*) use elaborate structures on the posterior area of their tongues to crush insects against specialized areas of the palate (underlain by the "echidna pterygoid"). Long-beaked echidnas (*Zaglossus* spp.) use the same type of structures to consume (if not crush) worms.

STEROPODON GALMANI. The M_{1-3} of *S. galmani*, the only teeth known of this taxon, are opal casts. Consequently surface detail is obscure and possibly misleading. However, it is apparent that there is loss of dental material on the apices of all akids and most blades and seemingly from the flanks and occlusal basins of most blades. What looks like relatively precise thegotic faceting occurs on the buccal cingulum between the bases of the two buccal akids of M_1, presumably caused by tooth/tooth contact with the apex of the primary posterolingual akid of the M^1, in the occlusal basins of the cingula and mid-crowns and, less certainly, on the vertical flanks of the primary blades. The anterior blade of the pseudotalonid, which is topographically analogous to the cristid obliqua of tribosphenid mammals, forms the posterolingual rim in a blade system that includes a steep blade posterior to the metakid and the posterior blade of the trigonid. These, and the basal cingulum on the buccal margin of the midvalley, have been thegosed by the lingual half of the anterior triakididrepanon of the unknown upper molars. Facing blades and the basal buccal cingula between the lower molars have been thegosed by the lingual half of the posterior triakididrepanon of the upper molars.

Abrasion on most of the interblade areas suggests

that when the lower teeth were brought up toward the upper teeth in anticipation of mastication, the point of initial contact (with or without food between) may have been less exact than is normally the case in therians. Perhaps the positions of the blades were adjusted as the masticatory stroke proceeded following first contact. Such a pattern might result if, during mastication, hard-shelled foods (e.g., beetles or crustaceans) were first "punctured" by akis-akid incussion (producing the apical subhorizontal areas of wear seen on all akises and akids) much as stylar cusps of marsupials are sometimes used to "tenderize" insect cuticle prior to segmentation. Subsequent efforts to shear the tenderized exoskeleton may have dragged abrasive materials over the akises/akids and blades rounding their cutting edges and trailing surfaces.

OBDURODON. The Riversleigh sample of *Ob. dicksoni* teeth includes none that are more than very lightly worn. This should not, however, be taken as an intimation that they were deciduous because, curiously, most of the thousands of Riversleigh mammal specimens so far recovered exhibit little-worn teeth. Only the paratype of *Ob. insignis* from central Australia is heavily worn, but it is also, unfortunately, severely damaged. Under low-angle lighting, some of the *Ob. dicksoni* molars reveal signs of initial thegotic faceting. In particular, this is evident on the flank of the anterior blade of M^1 and the posterior flank of the anterior blade of M_1. Most teeth show small abrasion areas on the tips of the akises/akids and several show very slight abrasion on the apical margins of the blades. In their unworn state, some blades reveal overturned apices above gently concave trailing flanks, suggesting that they would facilitate thegosis as the means of restoring the sharpness of the blades.

The worn M_2 of *Ob. insignis* reveals extensive wear on the apices of the blades, presumably from initial mastication of hard materials prior to finer segmenting between opposing blades. The enamel has been breached on the buccal half of each blade such that there are now blade pairs. However, the secondary blades on the internal flanks of the triakididrepanids are not serviced by thegosis and so have structural sharpness only. All of the blades stand proud of the dentine lakes that they surround, indicating that mastication is removing dental material and enhancing the efficacy of the blades.

The orientation of the dentine lakes on the blades is significantly different for the four blades. On those of the trigonid, the abrasion planes are at right angles to the sides of the blades. On the anterior blade of the pseudotalonid, the abrasion surface is inclined at about 45 degrees to the horizontal. Although it is rampant speculation, it is possible that these lakes are being produced in two different ways. For those developing at right angles to the blade flanks, mastication prior to interdigitation of the blade systems may be the cause. In the case of the inclined dentine lake, attempts to match structures on the upper molars suggest that the lingual basal akis adjacent to the posterolingual base of the primary anterolingual akis of the M^1 actually furrows the dentine lake of M_2. Because the same relationship of abrasion surfaces occurs on *S. galmani*, we suggest that its unknown upper molars also had basal cingular cusps of the type that occur in species of *Obdurodon* (and some dryolestoids).

FUNCTIONAL ANALOGUES TO THE TRIBOSPHENID MOLAR ROW. On the lower molars, an analogue to the wide tribosphenid talonid basin is the basin between the trigonid and posterior triakididrepanid of the same tooth. A second, quite different analogue is the basin constructed by the facing posterior cingula of adjacent molars. The combined effect is the production of four alternating sets of shearing blades and trailing basins—i.e., a system not unlike that seen in many tribosphenids (e.g., marsupials) where four trigonids alternate with four talonid basins. In the upper molars, the corresponding pattern is the same with the four triakididrepanons of two molars functioning as analogues of four therian upper molars with the intratooth basins alternating as analogues of abutting basal cingula enclosing interdental embrasures. In short, two highly autapomorphic monotreme molars at least in part mime, albeit in a less complex manner, the vertical shearing and horizontal segmentative functions of four tribosphenid molars.

MOLAR STABILITY DURING OCCLUSION. The arcuate intra- and intermolar basins suggest that some degree of rotational interdigitation might occur with rotation of the lower molars against the fixed maxillary teeth. Murray (1981) reported that echidnas rotate their lower jaws as their primary method of mastication. One possible correlate of this in ornithorhynchids might be their very short upper and lower molar roots, which seem somewhat insubstantial (viz. the lack of molars in the skull of *Ob. dicksoni* while the premolars remain in situ). In more conventional therian mammals, the long roots of molars undoubtedly help to resist pressures induced by transverse shear. Perhaps if the blades and akids of the lower molars could be "rolled" against their occlusal counterparts in the upper molars, the risks of molar evulsion would be reduced. On the other hand, the only well-worn tooth we have seen, the paratype of *Ob. insignis*, seems to have reduced, via abrasion and thegosis, the curvature of its intra- and interdental basins, which suggests that the elevated edges of the concave basins are being lost at a faster rate than the central regions, a process that does not seem to support a suggestion of rotational occlusion.

ENAMEL SURFACE STRUCTURES. The unworn flanks of the molar blades of *Ob. dicksoni* have a slightly crenulated appearance. The orientation of the crenulations is not consistent, but there is a clear tendency for them to be at right angles to the long axes of the blades. As in docodontids, perhaps with thegotic honing of the flanks, these crenulations develop microscopic blade edges of their own and act a bit like a cheese grater during the process of communition of food. The basins have a mammilary rather than crenulated appearance but, as with the crenulations, thegosis may convert these into a floor of tiny accessory horizontal blades.

The Cranium and Dentary of *Obdurodon dicksoni*

The skull (Figs 7.2–7.4; Table 7.1) was illustrated but not described in Archer, Hand, and Godthelp (1986, 1989), and key features of the snout were noted by Archer et al. (1991).

DESCRIPTION. In addition to its possession of a permanent cheek dentition, the cranium of *Ob. dicksoni* is readily distinguished from *Or. anatinus* by its much larger size (by a factor of approximately 0.33 to 0.45 in most dimensions); its overall robustness, including a much greater thickness of the neurocranial laminae; the more rounded contours of the anterior rostral margin; the fully enclosed interpremaxillary fenestra, and the much wider fossae for the adductors of the dentary. Concomitant with its large adductor complex, developed in association with the presence of a functional permanent dentition, *Ob. dicksoni* has deeper, more complex glenoid fossae, more robust zygomatic arches, and more prominent temporalis crests located higher on the neurocranium. *Obdurodon dicksoni* differs from *Or. anatinus* in its shorter, narrower posterior palatine processes, resulting in the exposure of more of the presphenoid; its higher palatal arch, and in possessing large, deep, oval fossae on either side of the maxillary palate that commences a short distance anterior to the first premolar. The basicranium of *Or. anatinus* is more deflected than that of *Ob. dicksoni* and, although the splanchnocranial portion of the cranium is much larger than that of *Or. anatinus*, the braincase proportions of the two forms are very similar (Table 7.2).

The primary differences between the crania of *Ob. dicksoni* and *Or. anatinus* are in their respective sizes, allometric proportioning, and trophic specializations. The basic anatomical arrangement of the cranium of *Ob. dicksoni* is essentially the same as in *Or. anatinus*, although many finer anatomical details distinguish the two taxa. Many features relating to the masticatory complex are obviously degenerate in *Or. anatinus*.

The bill structure of *Ob. dicksoni* departs from that of *Or. anatinus* in several important respects. The most obvious feature is in the midline contact of the anterior processes of the premaxillae ventrally and the septomaxillae dorsally, which fully enclose the interpremaxillary fenestra in contrast to the condition in *Or. anatinus*, in which the anterior processes are separated by a wide gap. The contour of the anterior portion of the rostrum of *Ob. dicksoni* and the proportions of its enclosed fenestra also differ from the conditions in *Or. anatinus*. In *Ob. dicksoni*, the rostral termination is rounded as

TABLE 7.1. Abbreviations

ae	articular eminence
af	adductor fossa
alm	anterolateral maxillary process
aof	anteorbital fossa
apf	anterior palatal fenestra
apm	anteromedian maxillary process
aps	accessory process septomaxilla
cpt	pterygoid canal
crp	crista parotica
fgp	fossa for gripping pad
fmr	fenestra molar root
fn	facial nerve canal
fpo	foramen pseudovale
fpr	foramen pseudorotundum
fr	frontal
fv	vestibular foramen
fve	foramen vasculosum externa
gf	glenoid fossa
gin	groove for infraorbital nerve
gmc	groove for infraorbital nerve
gpf	greater palatine foramen
ica	internal carotid canal
ioc	infraorbital canal
iof	infraorbital foramen
jf	jugular foramen
ju	jugal
M1–3	molars 1–3
mx	maxilla
na	nasal
npf	nasopremaxillary fenestra
oc	occipital bone
occ	occipital condyle
P3	third premolar
P4	fourth premolar
pal	palatine bones
pgn	preglenoid notch
pic	pit for M. longus capitus and M. rectus capitus
plm	posterolateral maxillary process
pmx	premaxilla
poc	postorbital crest
pot	postorbital tuberosity
ppm	palatal process of maxilla
prh	processus hyoideus
psp	presphenoid
ptc	posterior temporal canal
smx	septomaxilla
sof	sphenorbital foramen
spf	sphenopalatine foramen
sq	squamosal
tc	temporalis crest
vo	vomer

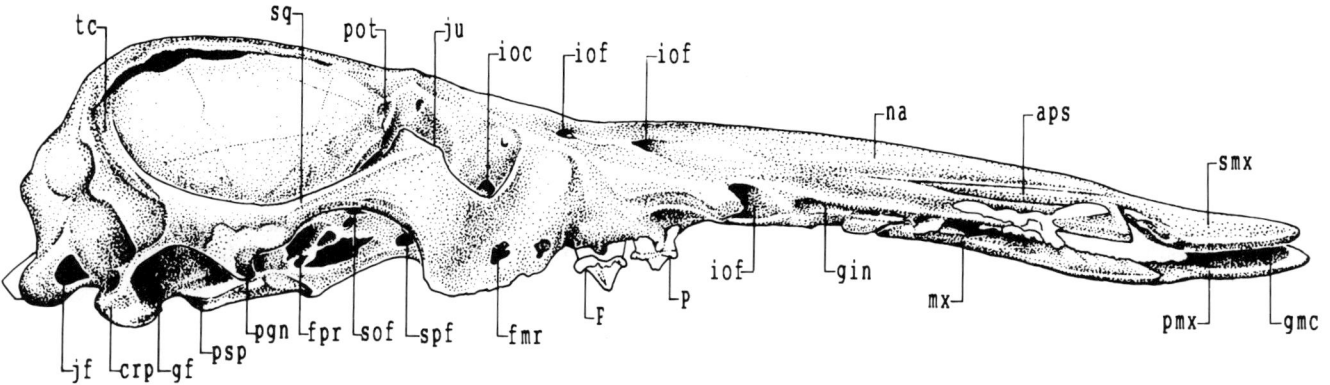

FIGURE 7.2. Skull of *Obdurodon dicksoni*, lateral aspect.

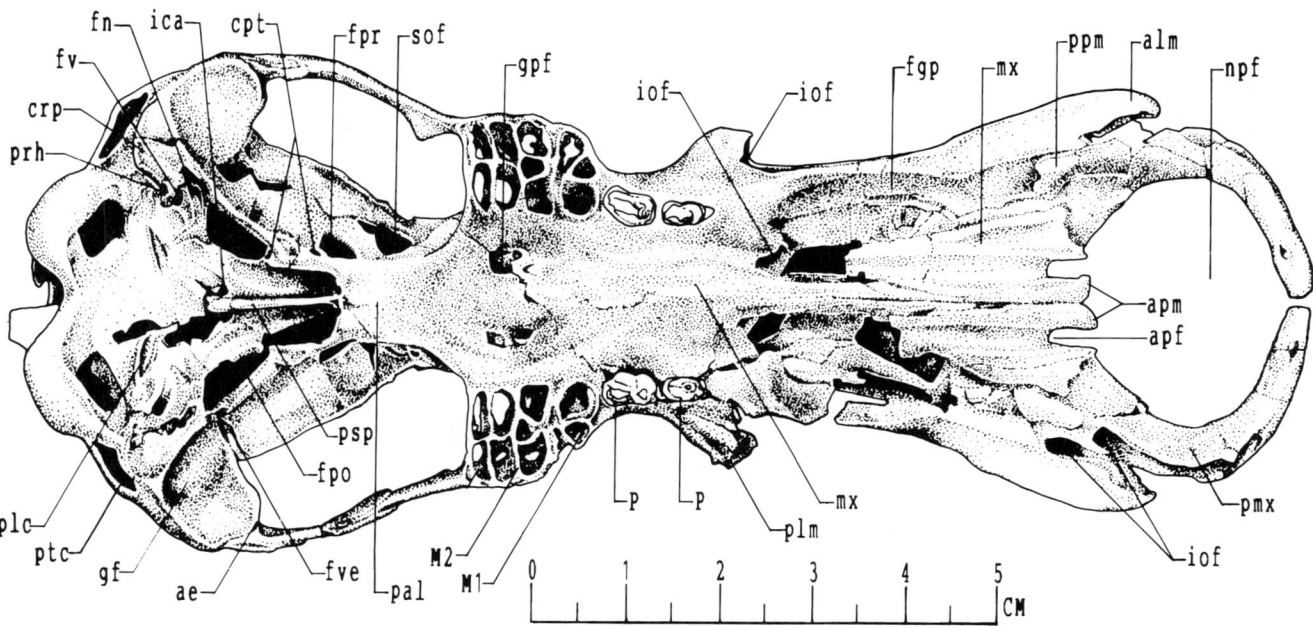

FIGURE 7.3. Skull of *Obdurodon dicksoni*, ventral aspect.

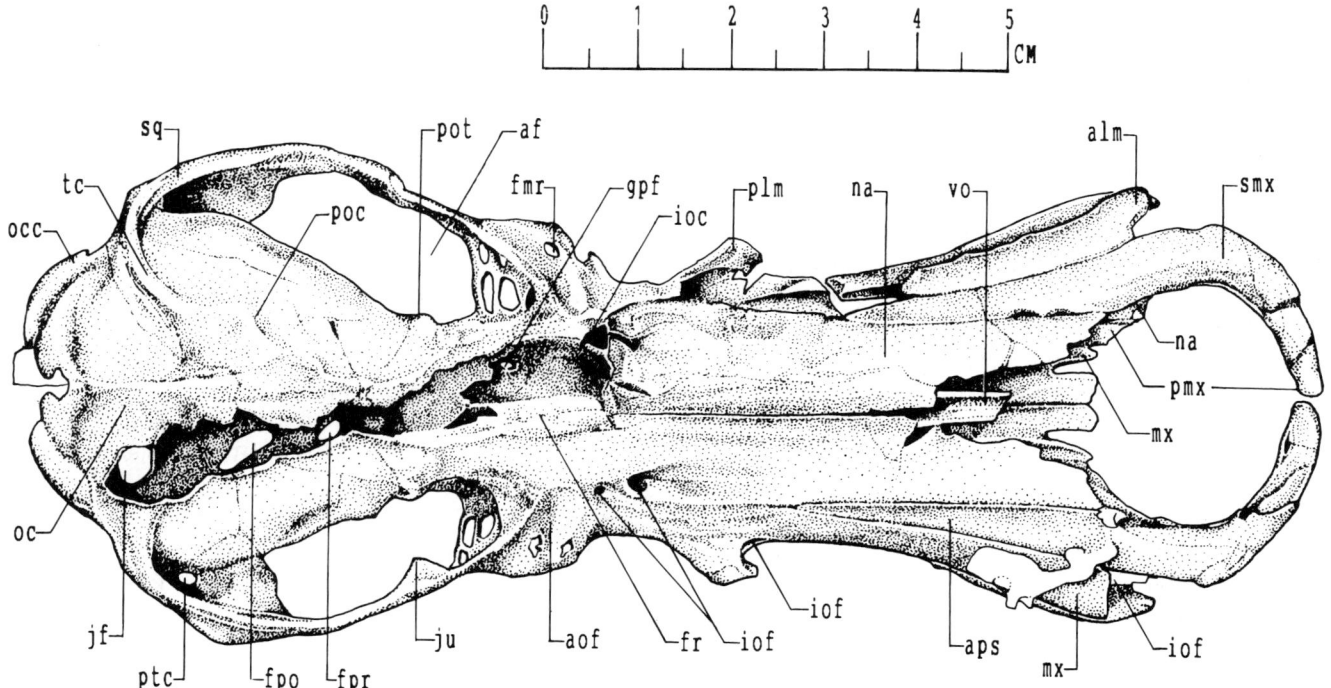

FIGURE 7.4. Skull of *Obdurodon dicksoni*, dorsal aspect.

TABLE 7.2. Comparison of cranial features of *Ob. dicksoni* with those in *Or. anatinus*

Feature	Ob. dicksoni	Or. anatinus
Cranium	large, robust	small, gracile
Rostrum	straight	deflected
Premaxilla/septomaxilla	separate	fused
Anterior processes	meet in midline	widely separated
Interseptomaxillary fenestra	enclosed, oval	open, V-shaped
Anteropremaxillary processes	prominent, beak-like	absent
Lateral margin palatine process	concave	convex
Posterior palatine process	short, narrow	long, wide
Palatal arch	deep	shallow
Posterior margin molar alveoli	straight	rounded
Postalveolar palatal notch	absent	present
Anterolateral infraorbital foramen	large	small
Palatal infraorbital foramina	small, slit-like	large
Anterolateral palatal fossae	deep, wide, oval	shallow, narrow
Articular facet ectopterygoid	conspicuous, triangular	weak
Zygomatic arches	bowed laterally	straight
Glenoid fossa	broad, deep, oblique	narrow, transverse
Occipital fenestrae	rhomboidal, oblique	oval, transverse
Hyoid pits	small	large
Posterior temporal canal	large	small
"Nuchal" crest	prominent	weak

opposed to being transversely straight. The interpremaxillary fenestra of *Ob. dicksoni* is oval and wider than it is deep, as opposed to being triangular and deeper lengthwise.

Compared to *Or. anatinus*, the sutures of the rostrum of *Ob. dicksoni* appear to show a lesser degree of fusion. In *Ob. dicksoni*, the patent bony contacts reveal the complex relationship of the nasals, septomaxillae, premaxillae, and vomer within the interpremaxillary fenestra and anterior palatal region, which are often obscure in mature *Or. anatinus*. The dorsal and ventral laminations of the anterior portion of the rostrum, usually designated premaxillae, are deeply divided by the wide groove for the marginal cartilage, so much so that only a thin lamination of bone connects the two plates mesially. Moreover, on the medial aspect of the process, a definite suture line entirely separates the two processes over much of their visible extent (Figs 7.4, 7.5).

DISCUSSION OF CRANIAL MORPHOLOGY. The morphology of the rostrum of *Ob. dicksoni* clearly expresses the retention of a large septomaxilla that constitutes most, if not all, of the dorsal surface of the anterior portion of the rostrum in adult monotremes. The premaxilla, with its posteriorly directed processus accessorius, is confined to the ventral surface of the rostrum. Embryological investigations of *Or. anatinus* and the echidna, *Tachyglossus aculeatus*, by Kuhn (1971) and Wible et al. (1990) fully support the retention of a large septomaxilla in monotremes. Although the presence of the septomaxilla in monotremes has been previously suggested, particularly in relation to the dorsal midline contact of the "premaxillae" in *T. aculeatus*, and sometimes with regard to the egg-tooth (Grasse, 1955), there is little direct evidence of its presence from the comparative anatomy of living monotremes.

The immediate functional significance of the septomaxilla-premaxilla lamination in ornithorhynchids is to retain the large cartilago marginalis, which is sandwiched between the two layers of bone. It is therefore of some importance that the marginal cartilage is also present in embryonic tachyglossids, eventually to be represented only by a faint marginal sulcus in adults. This may indicate that the beak of *T. aculeatus* (and other echidnas) was derived from an ancestor with structural relations more like those of ornithorhynchids.

The structure of the marginal cartilages of the bill of *Ob. dicksoni* was similar to that of *Or. anatinus*, differing primarily in proportions, which if anything indicated that they were greatly expanded. This is suggested by the exceptionally wide, deep marginal sulcus formed between the oral processes of the septomaxillae and premaxillae. In a formalin-preserved *Or. anatinus*, the marginal cartilage is about 6.0 mm thick within the peripheral groove. The groove in a macerated *Or. anatinus* specimen is between 1.0 mm and 0.5 mm wide. The marginal sulcus in *Ob. dicksoni* is about twice as wide and deep as the maximum dimensions of *Or. anatinus*, in which the marginal cartilage tapers out to a maximum width of about 8.0 mm. By analogy, the marginal cartilage of *Ob. dicksoni* would have been at least twice as wide (i.e., about 16.0 mm) as that of *Or. anatinus*. This linear expansion of the cartilages around *Ob. dicksoni*'s comparatively large bill would result in an even greater apparent disproportion between the neurocranium and splanchnocranium of *Ob. dicksoni* than is apparent in *Or. anatinus*.

The anterolateral infraorbital foramina (IOF-3) in *Ob. dicksoni* are proportionately much larger than in *Or. anatinus*. In contrast, the palatal infraorbital foramina (IOF-4) of *Ob. dicksoni* are represented by small, inconspicuous slits. The fibers of V^2 that emanate from the anterolateral foramina predominantly course ventrad. It appears that the palatal branches in *Ob. dicksoni* were reduced, possibly in relation to the large anterolateral palatal fossae, and the anterolateral fibers provided most of the sensory fibers to the palatal region otherwise supplied by the large palatal branch in *Or. anatinus*.

Obdurodon dicksoni's conspicuous anterolateral palatal fossae, which partially correspond to the position of the secateuring ridges of *Or. anatinus* (Burrell, 1927), may have accommodated larger, more complex cornified crushing pads for gripping its prey. In *Or. anatinus*, the palatal infraorbital nerve lies in a deep furrow, which also partially corresponds to the palatal fossae of *Ob. dicksoni*. However, because the palatal infraorbital foramina are small, slit-like openings in *Ob. dicksoni*, it is unlikely that they were specifically associated with the nerve.

In *Or. anatinus*, a shallow, trough-like fossa in the maxillary root of the zygomatic arch represents the connective tissue support and origin of the cheek pouch retractor muscles, the inserting upper and middle portion of the platysma, and the m. sphincter bursae buccales (Huber, 1931). A much larger and more clearly defined fossa is present in *Ob. dicksoni*, and it is reasonable to assume that this extinct taxon had well-developed cheek pouches.

The proportionally larger bill of *Ob. dicksoni*, relatively both wider and longer than that of *Or. anatinus*, and its more robust construction, entailing a marked thickening of the frontal bones to assist bending stresses at the base of the rostrum (Figs 7.2, 7.4), combined with its less decurved cranial axis, imply a difference in feeding habits between *Ob. dicksoni* and *Or. anatinus*. The straighter basicranium of *Ob. dicksoni* may indicate that it foraged in a more horizontal attitude than *Or. anatinus*. The stout construction of the beak of *Ob. dicksoni* would have allowed it to lift substantial cobbles and substrate debris in seeking out larger, solitary prey items.

CONCLUSIONS BASED ON CRANIAL MORPHOLOGY. The primary observation that can be extracted from the cranium of *Ob. dicksoni* is that the ornithorhynchid monotremes had fully attained their highly derived status by the early to middle Miocene and, by inference, well before this time. If *Or. anatinus* is a descendant of the *Obdurodon* lineage, its principal morphological trend has been one of structural reduction since the mid-Tertiary through the loss of its permanent denition, associated reduction of its jaw adductors, and a marked decrease in overall body size. There are no features of *Ob. dicksoni* that appear to be more derived than the corresponding character states in *Or. anatinus*. The main difference between the two is the seemingly degenerate nature of the latter, and some of the specializations of the modern animal (e.g., its down-turned snout) may have been autapomorphically derived in response to its preference for small prey.

However, the larger size of *Ob. dicksoni*'s cranium and the comparatively unfused state of the sutures between its rostral elements offer some clear advantages in the interpretation of its anatomy. We have therefore been able to verify the presence and precisely delineate the extent of the septomaxilla in an adult monotreme. The septomaxilla was identified as a large anterolateral rostral element in the echidna embryo by Gaupp (1908), subsequently by Griffiths (1968) and Kuhn (1971), and most recently by Wible et. al. (1990), but

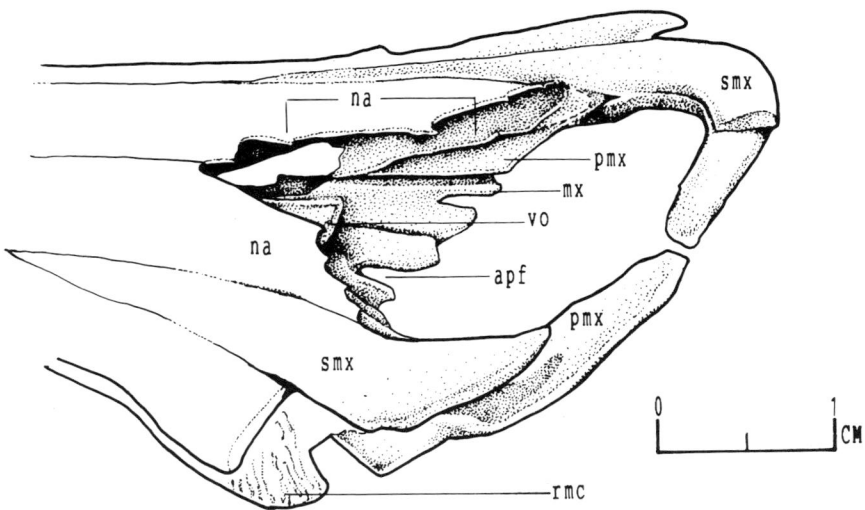

FIGURE 7.5. Morphological details of the cranium of *Obdurodon dicksoni*, anterior aspect of rostrum.

its early fusion to the premaxilla apparently obfuscated the extent of its contribution to the anterodorsal portion of the rostrum in adults. Consequently, most anatomists have long considered the entire bone to be the premaxilla, vaguely relegating the septomaxillary portion to the fused dorsal midline element in echidnas and to the bony structure underlying the egg-tooth, or as Goodrich (1958, p. 317) commented: "In monotremes it [the septomaxilla] fuses early with the premaxillary forming an important part of its outer surface." A bone commonly interpreted to be the septomaxilla in the armadillo (Broom, 1897) has been determined by Wible et al. (1990) to be a neomorphic structure, making monotremes the only living mammals to retain a true septomaxilla.

The hypertrophy of the septomaxilla in ornithorhynchids appears to be related to supporting the cartilago marginalis, which becomes deeply lodged between it and the similarly constructed premaxillary processes below. The enormous extent of the septomaxilla in monotremes implies that the dorsal surface of the premaxilla either has been entirely lost or is represented by another element that, by virtue of its intimate relationship and position relative to the two bony laminae, is probably the marginal cartilage itself.

The septomaxilla is externally exposed as a separate element in morganucodontids (Kermack et al., 1981). The bone is also present in *Sinconodon* (Patterson and Olsen, 1961) and *Oligokyphus* (Kühne, 1956). The recently described peramurid *Vincelestes neuquenianus* from the early Cretaceous of South America also retains the facial process of a septomaxilla (Bonaparte and Rougier, 1987). In mammal-like reptiles (e.g., *Procyanosuchus* and *Thrinaxodon*), the septomaxilla retains the intimate relationship with the vomeronasal organ (Duvall, 1986) as it does in modern lepidosaurs, or it was reduced to a small splint or not exposed externally (e.g., *Trilobodon*). In contrast, in monotremes the large septomaxillae are no longer anywhere near the vomeronasal organ, which implies that their ancestors may have been derived from synapsids in which the relationship between this bone and organ had ended. The profile shape of the septomaxilla in *Ob. dicksoni* is strikingly similar to that of certain cynodonts and it is tempting to speculate that the marginal notch of the ornithorhynchid bill is a homologue of the vomeronasal notch in mammal-like reptiles. However, the vomeronasal organ is associated with specialized olfactory functions and is usually lost or reduced in secondarily aquatic tetrapods. Because the sensory modalities relating to the vomeronasal organ project to the accessory olfactory bulb, relationship of this organ to the electroreceptive-tactile sensorium of the bill, apparently supplied wholly by the trigeminal nerve, seems very remote. Although functionally autapomorphic, the high definition and large size of the septomaxilla in monotremes might relegate this character to the primitive end of the mammalian morphocline, more so perhaps than in morganucodontids, in which the bone was already much reduced.

DENTARY. A left dentary fragment of *Ob. insignis* was described by Archer et al. (1978). A slightly more complete portion of a dentary of *Ob. dicksoni* was collected from Quentin's Quarry Site at Riversleigh. In the dentary of *Ob. insignis*, the region of the ascending process was broken. In *Ob. dicksoni*, this is preserved and reveals a distinctive ascending process, much better developed (about 5 mm high) than in any specimen of *Or. anatinus*. Its large size might be expected in response to the better-developed teeth and presumably greater occlusal pressures required if relatively large foods were eaten. The posterior margin ascends vertically, while the anterior margin inclines posteriad at an angle of about 70 degrees to the horizontal. The apex is slightly damaged but appears to have been subrounded. The lingual flank of this process clearly did not carry a coronoid bone. Posterior to the ascending process, the dorsal edge of the ramus anterior to what would have been the articular condyle is nearly horizontal and about 2 mm above the level of the alveolar margin. There is a distinct, elongate posteromedial angular process, clearly better developed than the rudiment of that structure in *Ob. insignis* and much better developed than any in *Or. anatinus*. The internal coronoid process behind the alveolar shelf has a large down-turned eminence that, although broken at its distal end, clearly served as a significant muscle attachment. The mandibular canal was larger than that structure in *Or. anatinus* and much larger than in *Ob. insignis*. Because it was earlier concluded that the size of this canal probably reflected the size of the bill and sensory structures at the front of the head, we would have been inclined to suggest what we already know, that the bill of *Ob. dicksoni* was larger than that of the living species. Within the vestibule of this canal is a mandibular fenestra that connects to the masseteric fossa on the lateral side.

The alveolar pattern and morphology are basically as previously described for *Ob. insignis* (Archer et al., 1978). The allocation of alveoli to individual lower molars is difficult and ultimately will require recovery of lower molars that retain roots. Tentatively, we would allocate the last alveolus to a small single-rooted M_3 and possibly the next five (three lingual and two buccal) to M_2, although this seems to leave very little room, in terms of upper molar row length, for a five- or six-rooted M_1 of the kind that characterizes the holotype of *Ob. insignis*.

OVERVIEW AND LIMITATIONS OF CRANIAL COMPARISONS. The cranium of *Obdurodon dicksoni* offers some new insights into the relationships of the Monotremata with other mammalian groups in that its more definitive structural relationships have served to clarify important aspects of monotreme anatomy. The problem with the monotreme cranium is its lack of undoubted synapomorphic features with any other mammalian group below the level of subclass for which crania are known. Certainly it reveals numerous symplesiomorphic features shared with morganucodontids, *Sinconodon*, and vincelestids.

Perhaps we seek an appropriate plesiomorphic sister-group for monotremes somewhere between the structural grades of mesungulatid dryolestoids and vincelestid eupantotheres. Although the latter are known from skulls, these have yet to be adequately described.

Phylogenetic Relationships of Monotremes

Two aspects of the phylogenetic relationships of monotremes concern us here: the interordinal position of Monotremata within (or without?) Mammalia; and the intraordinal relationships of monotremes to each other.

To use information provided by the skull and dentition of *Obdurodon dicksoni* to clarify the first, we need to review current understanding about the distribution of key cranial and dental features of monotremes. These may be placed in one of the four categories noted below. Comparable categorization of postcranial anatomy (e.g., Griffiths, 1968, 1978; Kemp, 1982, 1983; Jenkins, 1969, 1973; Jenkins and McClearn, 1984; Murray, 1984) provides a basically similar distribution pattern of significant features but is made more difficult by the relative lack of information about the postcranial anatomy of most Mesozoic mammal groups (see Szalay, this volume, chapter 9).

PLESIOMORPHIC OSSEOUS CRANIAL/DENTAL FEATURES CERTAINLY OR POSSIBLY UNIQUE WITHIN MAMMALIA. Monotremes appear to be unique among at least living mammals in retaining the following bony or pre-bony cranial features: pila antotica and osseus derivatives (although the condition in nonliving mammals is unclear); sclerotic cartilages (although they do not develop bone); curved but uncoiled cochlea; horizontal posture of ectotympanic (but posture may be derived).

MAMMALIAN SYNAPOMORPHIC OSSEOUS CRANIAL/DENTAL FEATURES. Because the definition of mammals is in a constant state of flux, features regarded as definitive are similarly inconstant (e.g., Kemp, 1982; Rowe, 1988; Wible et al., 1990). The following features present in monotremes are among those that have been cited as synapomorphies of Mammalia: absence of median shelf of septomaxilla (Wible, 1989); absence of prefrontal and postorbital (also in tritylodontids and tritheledontids); lamina cribosa (Zeller, 1988); three auditory ossicles; stapedial artery and lateral head vein run through middle ear (Wible, 1989); dentary-squamosal jaw joint (also in chiniquodontids and tritheledontids); articular condyle on the dentary; mammalian molar pattern (derived from primary three cusps—viz the trigonids of the lower molars); triangular masticatory stroke (Crompton and Jenkins, 1973); prismatic enamel (Lester and Archer, 1986) (also in tritheledontids); diphyodont(?) dentition (only rudiments left of second-generation teeth).

AUTAPOMORPHIC OSSEOUS CRANIAL/DENTAL FEATURES. Although of no value per se in determining the interordinal relationships of monotremes, their autapomorphic conditions are a reservoir of potential synapomorphies when skulls of other Mesozoic mammals are finally found. Cranial and dental autapomorphies include (there are many): loss of lachrymal and frontal bones; bony bill; hypertrophied septomaxilla; loss of facial process of premaxilla as such (may be marginal cartilage); reduction of alisphenoid contribution to sidewall of braincase (fide Hopson, Bonaparte, and Rougier, 1989, compared with the contribution in peramurids—i.e., *Vincelestes*); hypertrophy of lamina obturans in sidewall of braincase; loss of pila metopica and osseous derivatives; fused parietals (such as "parietale mediale" fide Wible et al., 1990); loss of interparietal (ibid.); hypertrophied dryolestoid? molar pattern; increase in number but reduction in size of molar roots; many other dental and cranial features (noted in this chapter).

SYNAPOMORPHIC CRANIAL/DENTAL FEATURES THAT APPEAR TO RELATE MONOTREMES TO SPECIFIC SUBGROUPS OF MAMMALS. Here again, the lack of cranial information about symmetrodonts, mesungulatids, and dryolestoids makes the search for synapomorphies shared by monotremes and other mammal groups, particularly eupantotheres, a very frustrating task. However, among those so far previously discussed or proposed here are: loss of coronoid and splenial (shared with derived eupantotheres, multituberculates, and therians; contra Greenwald's personal communication in Wible et al., 1990); apparent trigonids on lower molars (with symmetrodonts, eupantotheres, and tribosphenids); mesocrista on upper molars (with dryolestoids); basal cingular akises/akids adjacent to main akises/akids (with mesungulatid dryolestoids); reduction of cheektooth formula to P1–2 and M1–3 (with many "post-dryolestoid" groups such as vincelestid peramurids, but reduction commonly convergent); apparent loss of cheektooth diphyodonty (shared with marsupials) (fide, e.g., Berkovitz, 1967; Archer, 1978; but see Luckett, this volume).

POSITION OF MONOTREMES WITH RESPECT TO MAMMALS IN GENERAL. It is possible that monotremes are descendants of an advanced cynodont that independently acquired "mammalian" features such as a dentary-squamosal jaw articulation and three auditory ossicles. However, we are not persuaded that this hypothesis is the most parsimonious. The bony floor for the facial and trigeminal ganglia (Wible & Hobson, this volume), prootic canal (ibid.), bulbous promontorium (ibid.), grove for the perilymphatic duct (ibid.), enclosure of mandibular nerve by the anterior lamina of the petrosal (ibid.), dryolestoid-like features of the molars (above) and reduced dentition (above) are all features of groups generally regarded to be within Mammalia (but see Rowe (1988) for a different point of view).

POSITION OF MONOTREMES WITH RESPECT TO TRIBOSPHENID MAMMALS. Our data supports the contention of Kielan-Jaworowska et al. (1987) that monotremes represent a pre-tribosphenid but therian group; that is, they lack a protocone and a 'true' talonid but have a well-developed trigonid. The only structure in the dentition that appears to be shared with tribosphenid mammals is the trigonid but this is a symplesiomorphy within tribosphenids being found in symmetrodonts, eupantotheres and yinotheres. The structures most similar to talonids in monotremes are almost certainly neomorphic 'pseudotalonids.'

However, although not homologous to tribosphenid molars, a partial analogue to the tribosphenid talonid basin can be found in the basin between the trigonid and posterior triakididrepanid of the same tooth. A second set of occlusal basins is provided by the occlusal surfaces of the facing posterior and anterior cingula of adjacent molars. The net effect is that two monotreme lower molars mime four therian molars. In the upper molars, the corresponding pattern is the same with the four triakididrepanons of two molars functioning as analogues of four therian upper molars with the intratooth basins alternating as analogues of abutting basal cingula enclosing interdental embrasures. In short, two highly autapomorphic monotreme molars mimic four tribosphenid molars.

It is interesting to consider that at least four different groups of mammals have experimented with the production of therian-like molars: docodonts, shuotheriids, tribosphenids and now, it would appear, monotremes.

POSITION OF MONOTREMES WITH RESPECT TO NON-TRIBOSPHENID THERIAN MAMMALS. In terms of molar structure, the non-tribosphenid therian mammal group that monotremes most resemble are eupantotheres and, in particular, dryolestoids. Unfortunately, most eupantothere groups are not yet known from skulls, thereby restricting comparisons mainly to the dentary and the dentition. The dentaries of dryolestoids show no striking similarities to those of monotremes other than the loss (by the Cretaceous) of all postdentary bones from the mandible. However, loss of these bones in monotremes and dryolestoids could as easily be the result of convergence, considering how often this seems to have happened independently in various groups of Mesozoic mammals.

Although Archer et al. (1989) noted the dryolestoid affinities of monotremes, they also noted similarities to the skull of the Cretaceous peramurid *Vincelestes*. The two most striking similarities were the presence of a septomaxilla and a similarly reduced cheektooth number. However, molar morphology is quite different, the vincelestid pattern being more like that of *Peramus*, which is commonly regarded as the sistergroup of tribosphenid mammals (aegialodontids plus eutherians). Hopson et al. (1989) further noted that the sidewall of the braincase of *Vincelestes* is not like that of monotremes, although this in itself would not exclude the vincelestid pattern from being ancestral to that of monotremes.

Wible et al. (1990) suggested that monotremes are the sistergroup of multituberculates plus vincelestids-marsupials-placentals on the basis that monotremes lack four synapomorphies of the middle ear, including a cochlea that spirals at least one full turn, periotic enclosure of the perilymphatic duct, a post-promontorial tympanic sinus, and enlarged alisphenoid. They suggested that although multituberculates and the other members of their sistergroup lack septomaxillae, multituberculates independently lost this bone. Further, the very large facial wing of the premaxillae that extends to the nasals in all of these groups except monotremes (which have no facial wing as such) may also be the result of convergent gain in multituberculates.

We support Wible et al.'s (1990) hypothesis, but would add that where monotremes fit into the tree, so too probably fit at least dryolestoid eupantotheres. How these relate to the remaining eupantotheres is less clear.

RELATIONSHIPS WITHIN MONOTREMATA. Monophyly of the Monotremata has been frequently defended (e.g., Murray, 1984). Apparent synapomorphies include the bill, the spur on the inside of the hind foot, the detrahens muscle as the jaw adductor, and electroreceptors on the rhinarial surface. Unfortunately, the fossil materials of species of *Steropodon* and *Obdurodon* have not revealed anything of phylogenetic value about the relationship between tachyglossids and ornithorhynchids. Further, despite considerable searching, no pre-late Miocene or any toothed tachyglossids are known.

Considering the functional significance of the highly distinctive septomaxilla-premaxilla lamination in ornithorhynchids, which appears to relate to retention and support of the large cartilago marginalis sandwiched between their rostral processes, it is interesting to note that the marginal cartilage is also present in embryonic tachyglossids, although it eventuates only as a faint marginal sulcus in adults. This would suggest that the common ancestor of both echidnas and platypuses, whenever it occurred, had a bill or at least a marginal cartilage, which would then be another synapomorphy uniting Monotremata.

Considering patristics, no dental or mandibular feature of the early Cretaceous *Steropodon galmani* known to us would prohibit it having given rise to the species of *Obdurodon*. Similarly, there are no features in the skull of *Ob. dicksoni* that would prohibit it from being an ancestor of *Ornithorhynchus anatinus*. The primary differences between the species of *Obdurodon* and *Ornithorhynchus* include features that either are more degenerate in the living taxon (e.g., loss of teeth, reduced size and reduction and fusion of medial premaxillary and septomaxillary processes) or reflect a different mode of feeding. In *Ob. dicksoni*, the snout is more horizontal, the teeth better formed, the masseteric and temporalis muscles better developed, and the pterygoid fossae and ascending ramus of the dentary larger—features that combine to suggest this species foraged for larger prey on the surface of its underwater substrate. In contrast, the living species, with its down-turned snout, ephemeral degenerate teeth, and reduced strength of its masticatory apparatus, feeds on small invertebrates recovered by furrowing through rather than gathering on the surface of the substrate of its feeding environment.

Biogeographic Considerations

Our conclusions lead us to the following interpretation of paleobiogeographic events. Until the late Cretaceous, Gondwana was occupied by a wide variety of dryolestoids and other pre-tribosphenid mammals, as the diverse Late Cretaceous Los Alamitos fauna demonstrates (Bonaparte, 1990). At least two of these pre-tribosphenid groups, dryolestoids and peramurids, had populations in the presumptive area of South America as well as possibly Antarctica and Australia.

By at least 110 MYBP, the age of *Steropodon galmani*, some pre-tribosphenid group had produced monotremes in at least the Australian portion of Gondwana. It is worth considering that at this time Lightning Ridge (the type locality for *S. galmani*) was within the Antarctic Circle, at which time and place mean annual temperatures varied between −5 and +8°C (Rich and Rich, 1989). Such low temperatures may help to explain the high metabolic scope of modern monotremes (e.g., Augee, 1978) and perhaps the relatively large size of *S. galmani* (within the range of the smaller dinosaurs and comparable to the largest stagodontid marsupials) (Archer et al., 1985) compared with contemporary mammals on other continents.

By 45 MYBP its descendants, with microbiotheriids (or microbiotheriid derivatives), paleochiropterygoid-like bats, and other groups such as madtsoiid snakes and meiolaniid turtles, were within the Australian land mass when it finally broke away from Antarctica. Although our current investigations of the 54 million years old Tingamarra Local Fauna of southeastern Queensland are at an early stage, we have already recovered representatives of a number of Gondwanan mammal groups (Godthelp et al., 1992) but, so far, no representatives of the Monotremata. But clearly monotremes continued to diversify in Australia producing at least two families by the Miocene: ornithorhynchids and tachyglossids. In apparent contrast to tachyglossids, ornithorhynchids appear to have undergone a severe decline since the middle Miocene and may now be teetering on the brink of extinction. The reasons for this decline are undoubtedly complex but certainly included the late Miocene loss of the inland and northern rain forests and associated waterways where *Ob. dicksoni* lived (Archer et al., 1989). A further factor may well have been the arrival in Australia, by at least the Pleistocene, of the aquatic hydromyine murids. It is probably significant that the similar-sized, predatory water rat (*Hydromys chrysogaster*) occurs in the same waterways as the modern Platypus throughout its range and beyond.

Conclusions

We began this examination noting Simpson's (1929) view that whatever kind of mammal the Platypus was, it was unlike any other. Following the discovery of the Early Cretaceous *Steropodon galmani* and the skull and dentition of the early to middle Miocene *Obdurodon dicksoni*, we are urged to conclude that monotremes lie somewhere within the paraphyletic cluster of pre-tribosphenid mammals known as eupantotheres and *perhaps* closer to dryolestoids than to any other group normally included here (e.g., Kraus, 1979).

The dentition indicates that monotremes are pre-tribosphenid as Kielan-Jaworowska et al. (1987) concluded. There are resemblances between Cretaceous/Tertiary monotremes and South American peramurids and mesungulatid dryolestoids as noted by Archer et al. (1992) and Bonaparte (1990). However, although distinctive resemblances between the molars of ornithorhynchids and mesungulatids urge us to consider monophyly, because the latter are late Cretaceous in age they are at least as likely to be convergent on ornithorhynchids as they are to be dentally conservative descendants of unknown Jurassic to early Cretaceous dryolestoids that could have given rise to monotremes. Nevertheless, mesungulatid molar structure provides an interesting eupantotherian starting point from which the monotreme molar pattern could potentially have been derived.

Nothing about the molar morphology of the relatively plesiomorphic ornithorhynchid *Ob. dicksoni* gives support to the concept of Marsupionta (Kühne, 1973, 1987) or, for that matter, to any other currently competing hypothesis (see also Szalay, this volume, chapter 9).

The skull provides incontrovertible corroboration of the view (most recently proposed by Wible et al., 1990) that monotremes are the only living mammals with a septomaxilla. This feature is a mammalian symplesiomorphy, although it is hypertrophied in ornithorhynchids apparently as an adaptation to supporting the elaborate bill and sensory organs characteristic of this group. Overall the cranium of the Oligo-Miocene *Ob. dicksoni* demonstrates the extent to which the living platypus has "degenerated", although it is clear that the "modern" ground plan for platypuses was established at least as early as the late Oligocene. Nothing about the structure of *S. galmani* argues against the possibility that it could be ancestral to the species of *Obdurodon*, and nothing about *Ob. dicksoni* suggests it could not be ancestral to the species of *Ornithorhynchus*.

We have been unable to clarify the relationships of ornithorhynchids to tachyglossids other than to note that the autapomorphic conditions of the latter are better seen as derivatives of the former than vice versa. The basic ornithorhynchid dental pattern seems to have been established since at least the Early Cretaceous, well prior to the oldest known (late Miocene) tachyglossids and molecular clock dates for the last common ancestor of tachyglossids and ornithorhynchids.

The relatively rapid "degeneration" of ornithorhynchids since the Miocene should be cause for serious concern about the future of this lineage, which is, whatever it is, most certainly something quite distinct.

EMENDATION. In Archer et al. (1992), the Holotype of *Obdurodon dicksoni* is incorrectly cited as QM F20564 on p. 16 although it is correctly cited elsewhere in that paper. The correct number is QM F20568 (formerly AR17170). Other paratypes are: from Ringtail Site, Gag Plateau, QM F18986 (= AR5383); QM F18988 (= AR5429); QM F18987 (= AR5544); QM F18985 (= AR8608); QM F18983 (= AR9406); QM F18982 (= AR9705); QM F18980 (= AR13015); QM F18979 (= AR13625); QM F18976 (= AR16376); QM F18973 (= AR16380); and QM F16888; from Quentin's Quarry, Gag Plateau, QM F18981 (= AR13401); QM F18975 (= AR16479); from Bob's Boulders Site, Gag Plateau, QMF QM F18984 (= AR6717); from Neville's Garden, D-Site Plateau, QM F19877-78 (= AR17007-08), QM F18974 (= AR17479). Some of the AR and QM F prefixes were inadvertently confounded in the original description (p. 16).

ACKNOWLEDGMENTS. We wish to acknowledge the vital financial support the Riversleigh Project has had from the Australian Research Grant Scheme (Grant PG A3 851506P); the National Estate Grants Scheme (Queensland); the Department of Arts, Sport, the Environment and Territories; University of N.S.W.; IBM Australia Pty Ltd; ICI Australia Pty Ltd; the Australian Geographic Society; Wang Australia Pty Ltd; the Queensland Museum; the Australian Museum; Mount Isa Mines Pty Ltd; and Surrey Beatty & Sons Pty Limited. Critical logistical support in the field and laboratory has been received from the Riversleigh Society, the Friends of Riversleigh, the Royal Australian Air Force, the Australian Defence Force, the Queensland National Parks and Wildlife Service, the Riversleigh Consortium (Riversleigh being a privately owned station), the Mount Isa Shire, the Northwest Queensland Tourism and Development Board, the Gulf Local Development Association, PROBE, and many volunteer workers and colleagues. Peter Murray's involvement was made possible by the Museums and Art Galleries of the Northern Territory. The SEM photographs were produced on equipment made available by the University of N.S.W. Electronmicroscopy Unit. Light photographs were taken by Ross Arnett of the Photographic Unit of the University of N.S.W. Drawings of skulls were by Peter Murray. We gratefully acknowledge the support made available to allow Archer to present the paper at the meeting in the American Museum. This support also enabled Archer to benefit from discussions with other participants, including A.W. Crompton, J.R. Wible, J.A. Hopson, U. Zeller, F. Szalay, R. Presley, M.C. McKenna, F. Jenkins, Z. Kielan-Jaworoska, W.P. Luckett, R.H. Tedford, D.R. Prothero, W. Maier, R.L. Cifelli, and T. Rowe.

CORRESPONDENCE ADDRESS. Michael Archer, School of Biological Science, University of New South Wales, P.O. Box 1, Kensington NSW, Australia 2033.

References

Aplin, K., and Archer, M. 1987. Recent advances in marsupial systematics with a new syncretic classification. In: *Possums and opossums: Studies in evolution.* Sydney: Surrey Beatty and Sons, pp. 15-72.

Archer, M. 1974a. The development of cheek-teeth in *Antechinus flavipes* (Marsupialia, Dasyuridae). *J. Proc. Roy. Soc. West. Aust.* 57: 54-63.

Archer, M. 1974b. The development of premolar and molar crowns of *Antechinus flavipes* (Marsupialia, Dasyuridae) and the significance of cusp ontogeny in mammalian teeth. *J. Proc. Roy. Soc. West. Aust.* 57:118-125.

Archer, M. 1978. The nature of the molar-premolar boundary in marsupials and a reinterpretation of the homology of marsupial cheekteeth. *Mem. Qd Mus.* 18:157-164.

Archer, M., Flannery. T.F., Ritchie, A., and Molnar, R.E. 1985. First Mesozoic mammal from Australia—An early Cretaceous monotreme. *Nature* 318: 363-366.

Archer, M., Godthelp, H., Hand, S.J., and Megirian, D. 1989. Fossil mammals of Riversleigh, northwestern Queensland: preliminary overview of biostratigraphy, correlation and environmental change. *Aust. Zool.* 25:29–65.

Archer, M., Hand, S.J., and Godthelp, H. 1986. *Uncovering Australia's Dreamtime.* Syndey: Surrey Beauty and Sons.

Archer, M., Hand, S.J., and Godthelp, H. 1989. Dentition of the Oligocene/Miocene ornithorhynchid genus *Obdurodon* and the phylogenetic relationships of monotremes. Conference and Australasian Vertebrate Evolution, Palaeontology and Systematics, Sydney, March 1989 (Abstracts): 1.

Archer, M., Hand, S.J., and Godthelp, H. 1991. *Riversleigh.* Sydney: Reed Books.

Archer, M., Jenkins, F.A. Jr., Hand, S.J., Murray, P., and Godthelp, H. 1992. Description of the skull and non-vestigial dentition of a Miocene platypus (*Obdurodon dicksoni* n.sp.) from Riversleigh, Australia and the problem of monotreme origins. In: *Platypus and echidnas* (Augee, M., ed.). Sydney: Royal Zoological Society of New South Wales, pp. 15–27.

Archer, M., Plane, M.D., and Pledge, N.S. 1978. Additional evidence for interpreting the Miocene *Obdurodon insignis* Woodburne and Tedford, 1975, to be a fossil platypus (Ornithorhynchidae: Monotremata) and a reconsideration of the status of *Ornithorhynchus agilis* De Vis, 1895. *Aust. Zool.* 20: 9–27.

Augee, M.L. 1978. Monotremes and the evolution of homeothermy. *Aust. Zool.* 20:111–119.

Berkovitz, B.K.B. 1967. The dentition of a 25-day pouch-young specimen of *Didelphis virginiana* (Didelphidae: Marsupialia). *Arch. Oral Biol.* 12:1211–1212.

Bonaparte, J.F. 1986a. A new and unusual late Cretaceous mammal from Patagonia. *J. Vert. Paleo.* 6: 264–270.

Bonaparte, J.F. 1986b. Sobre *Mesungulatum houssayi* y nuevos mamíferos cretácios de Patagonia, Argentina. *Actas IV Congreso Argentino de Paleontología y Bioestratigrafía (Mendoza)* 2:48–61.

Bonaparte, J.F. 1986c. History of the terrestrial Cretaceous vertebrates of Gondwana. *Acta Congreso Argentino de Paleontología y Bioestratigrafía (Mendoza)* 2: 63–95.

Bonaparte, J.F. 1990. New late Cretaceous mammals from the Los Alamitos Formation, northern Patagonia. *Nat. Geog. Res.* 6:63–93.

Bonaparte, J.F., and Rougier, G. 1987. Mamiferos del Cretacico inferior de Patagonia. IV *Congreso Latinoamericano de Paleontología, Bolivia* 1: 343–359.

Bonaparte, J.F., and Soria, M.F. 1985. Nota sobre el primer mamífero del Cretácico Argentino, Campaniano-Maastrichtiano (Condylarthra). *Ameghiniana* 2–4:177–183.

Broom, R. 1897. Nasal-floor bone in the hairy armadillo. *J. Anat. Physiol.* 31.

Burrell, H. 1927. *The platypus.* Adelaide: Rigby Ltd.

Chow, M., Rich, T.H. 1982. *Shuotherium dongi*, n. gen. and sp., a therian with pseudo-tribosphenic molars from the Jurassic of Sichuan, China. *Aust. Mammal.* 5: 127–142.

Clemens, W.A. 1979. Notes on the Monotremata. In: *Mesozoic mammals: The first two-thirds of mammalian history* (Lillegraven, J.A., Kielan-Jaworowska, Z., Clemens, W.A., eds.). Berkeley: University of California Press, pp. 309–311.

Duvall, D. 1986. A new question of pheromones: Aspects of possible chemical signaling and reception in mammal-like reptiles. In: *The ecology and biology of mammal-like reptiles* (Hotten, N. III, MacLean, P., Roth, J., Roth, E., eds.). Washington, D.C.: Smithsonian Institution Press, pp. 219–238.

Every, R.G. 1972. *A new terminology for mammalian teeth.* Christchurch: Pegasus Press for the Centre for the Study of Conflict.

Every, R.G. 1974. Thegosis in prosimians. In: *Prosimian biology* (R. D. Martin, G.A. Doyle, and A.C. Walker, eds.). London: Duckworth, pp. 579–619.

Fitsinger, L.J. 1826. *Neue Classification der Reptilien.* Wien: J.G. Heubner Verlag.

Gaupp, E. 1908. Zur entwickelungsgeschichte und vergleichenden Morphologie des Schadels von *Echidna aculeata* var. *typica*. *Denkschr. Med. Naturwiss. Gesellsch. Jena* 6:539–788.

Gill, T. 1872. Arrangement of the families of mammals with analytical tables. *Smithsonian Misc. Coll.* 11:1–98.

Godthelp, H., Archer, M., Cifelli, R., Hand, S.J., and Gilkeson, C. 1992. Earliest known Australian Tertiary mammal fauna. *Nature* 356:514–16.

Goodrich, E. 1958. *Studies on the structure and development of vertebrates.* New York: Dover.

Grassé, P. 1955. Order des Monotremes. In: *Traite de Zoologie T. XVII* (Grassé, P., ed.). Paris: Librares de L'Academie de Medcine 120, Blvd St.-Germaine, pp. 46–91.

Green, H.L.H.H. 1937. The development and morphology of the teeth of *Ornithorhynchus*. *Phil. Trans. Roy. Soc. London*, B228:367–420.

Gregory, J.E., Iggo, A., McIntyre, A.K., and Proske, U. 1987. Electroreceptors in the platypus. *Nature* 326:386–387.

Griffiths, M. 1968. *Echidnas.* Oxford: Pergamon Press.

Griffiths, M. 1978. *The biology of the monotremes.* New York: Academic Press.

Griffiths, M. 1988. The platypus. *Scientific American* 256:84–91.

Hopson, J.A., Bonaparte, J.F., and Rougier, G.W. 1989. Braincase structure of a non-tribosphenic therian mammal from the early Cretaceous of Argentina. *J. Vert. Paleo. Abstracts* 9: 25A.

Hobson, J.A., and Crompton, A.W. 1969. Origin of mammals. *Evol. Biol.* 3:15–72.

Huber, E. 1931. Studies on the organization of monotremes contrasted with the marsupials and placentals. *Morph. Jb.* 1931:66–67.

Huxley, T.H. 1880. On the application of the laws of evolution to the arrangement of the Vertebrata and more particularly of the Mammalia. *Proc. Zool. Soc. London* 1880:649–662.

Jenkins, F.A. Jr. 1969. The evolution and development of the dens of the mammalian axis. *Anat. Rec.* 164:173–184.

Jenkins, F.A. Jr. 1970. Limb movements in a monotreme (*Tachyglossus aculeatus*): A cineradiographic analysis. *Science* 168:1473–1475.

Jenkins, F.A. Jr. 1973. The functional anatomy and evolution of the mammalian humero-ulnar articulation. *Am. J. Anat.* 137:281–297.

Jenkins, F.A. Jr. 1984. A survey of mammalian origins. In:

Mammals (Broadhead, T.W., ed.). *University of Tennessee Dept Geol. Sci. Studies in Geology* 8:32–47.

Jenkins, F.A. Jr., and McClearn, D. 1984. Mechanisms of hind foot reversal in climbing mammals. *J. Morph.* 182:197–219.

Jenkins, F.A. Jr., and Schaff, C.R. 1988. The early Cretaceous mammal *Gobiconodon* (Mammalia, Triconodonta) from the Cloverly Formation in Montana. *J. Vert. Paleo.* 8:1–24.

Kemp, T.S. 1982. *Mammal-like reptiles and the origin of mammals.* London: Academic Press.

Kemp, T.S. 1983. The relationships of mammals. *Zool. J. Linn. Soc.* 77:353–384.

Kermack, K., Musset, F., and Rigney, S. 1981. The skull of *Morganucodon. Zool. J. Linn. Soc. London* 71:1–158.

Kielan-Jaworowska, Z., Crompton, A.W., and Jenkins, F.A. Jr. 1987. The origin of egg-laying mammals. *Nature* 326:871–873.

Kirkpatrick, T.H. 1969. *The dentition of the marsupial family Macropodidae with particular reference to tooth development in the grey kangaroo* Macropus giganteus *Shaw*. Ph. D. Thesis, University of Queensland, Brisbane.

Krebs, B. 1971. Evolution of the mandible and lower dentition in dryolestids (Pantotheria, Mammalia). In: *Early mammals* (Kermack, D.M., and Kermack, K.A., eds.). London: Academic Press, pp. 91–102.

Kuhn, H.-J. 1971. Die entwicklung und morphologie des Schadels von *Tachyglossus aculeatus. Abhand. Senchenberg. Naturforsch. Gesellsch.* 528:1–244.

Kuhn, H.-J., and Zeller, U. 1987. The cavum epiptericum in monotremes and therian mammals. In: Morphogenesis of the mammalian skull Kuhn, H.J., and Zeller, U., eds. Hamburg and Berlin: Verlag Paul Parey, pp. 51–70.

Kühne, W.G. 1956. *The Liassic therapsid* Oligokyphus. London: British Museum (Natural History).

Kühne, W.G. 1973. The systematic position of monotremes reconsidered (Mammalia). *Z. Morph. Tiere* 75: 59–64.

Kühne, W.G. 1987. Marsupionta contra Theria. In: *Deutsche Ges. Säugetierkunde, 61. Hauptversammlg., Kurzfassungen der Vorträge u. Posterdemonstr* Schmidt, C., ed. Hamburg: Paul Parey, pp. 26–27.

Lester, K.S., and Archer, M. 1986. A description of the molar enamel of a middle Miocene monotreme (*Oburodon*, Ornithorhynchidae). *Anat. Embryol.* 174:145–151.

Lester, K.S., and Boyde, A. 1986. Scanning microscopy of platypus teeth. *Anat. Embryol.* 174:15–26.

Lewis, O.J. 1963. The monotreme cruro-pedal flexor musculature. *J. Anat., Lond.* 97:55–63.

Luckett, W.P., and Zeller, U. 1989. Developmental evidence for dental homologies in the monotreme *Ornithorhynchus* and its systematic implications. *Z. Säugetierkunde* 54: 193–204.

Mills, J.R.E. 1971. The dentition of *Morganucodon*. In: *Early mammals* (Kermack, D.M., and Kermack, K.A., eds.). London: Academic Press, pp. 29–63.

Murray, P.F. 1981. A unique jaw mechanism in the Echidna, *Tachyglossus aculeatus* (Monotremata). *Aust. J. Zool.* 29:1–5.

Murray, P.F. 1984. Furry egg-layers: The monotreme radiation. In: *Vertebrate zoogeography and evolution in Australasia* (Archer, M., and Clayton, G., eds.). Perth: Hesperian Press, pp. 571–583.

Novacek, M.J., and Wyss, A.R. 1986. Higher-level relationships of the Recent eutherian orders: Morphological evidence. *Cladistics* 2:257–287.

Pascual, R., Archer, M., Ortiz Jaureguizar, E., Prado, J.L., Godthelp, H., and Hand, S.J. 1992. The first non-Australian monotreme: an early Paleocene South American platypus (Monotremata, Ornithorhynchidae). In: *Platypus and echidnas* (Augee, M., ed.). Syndey: Royal Zoological Society of New South Wales, pp. 1–14.

Patterson, B., and Olsen, E. 1961. A triconodontid mammal from the Triassic of Yunnan. In: *International colloquium on the evolution of lower and non-specialized mammals.* Brussels: Koninlijke Vlaamse Academie voor Wetenschappen, Letteren en Schone Kunsten van Belgie, pp. 129–191.

Presley, R. 1981. Alisphenoid equivalents in placentals, marsupials, monotremes, and fossils. *Nature* 294:668–670.

Rich, T.H., and Rich, P.V. 1989. Polar dinosaurs and biotas of the early Cretaceous of southeastern Australia. *Nat. Geog. Res.* 5:15–53.

Rowe, T. 1988. Definition, diagnosis, and origin of Mammalia. *J. Vert. Paleont.* 8:241–264.

Van Bemmelen, J.F. 1901. Der Schädelbau der Monotremen. *Semon's Zoologische Forschungsreisen in Australien, Denkschriften der medicinisch-naturwissenschaftlichen Gesellschaft zu Jena* 6:729–798.

Van Valen, L.M. 1988a. Faunas of a southern world. *Nature* 333: 113.

Van Valen, L.M. 1988b. Paleocene dinosaurs or Cretaceous ungulates in South America. *Evol. Monogr.* 10:1–79.

Wible, J.R. 1989. Differences in the petrosal bone among the major groups of Mesozoic and Recent mammals. *J. Vert. Paleo. Abstracts* 9:44A.

Wible, J.R., Miao, D., and Hopson, J.A. 1990. The septomaxilla of fossil and recent synapsids and the problem of the septomaxilla of monotremes and armadillos. *Zool. J. Linn. Soc.* 98:203–228.

Wilson, J.T., and Hill, J.P. 1907. Observations on tooth-development in *Ornithorhynchus. Quart. J. Microscop. Sci.* 51:137–165.

Woodburne, M.O., and Tedford, R.H. 1975. The first Tertiary monotreme from Australia. *Amer. Mus. Nat. Hist. Novitates* 2588:1–11.

Zeller, U. 1988. The lamina cribosa of *Ornithorhynchus* (Monotremata, Mammalia). *Anat. Embryol.* 178:513–519.

CHAPTER 8

Ontogenetic Evidence for Cranial Homologies in Monotremes and Therians, with Special Reference to *Ornithorhynchus*

ULRICH ZELLER

Overview

The ontogeny of the tympanic and otic regions of the head of *Ornithorhynchus anatinus* was studied in serial sections of fifteen stages and compared with *Tachyglossus* (Monotremata) and therian mammals. Recent monotremes retain a number of synapsid plesiomorphies in the middle ear. These include the absence of a caput mallei, the synostosis between praearticulare and tympanicum, the columelliform stapes, the presence of a secondary tympanic membrane at the lateral aperture of the recessus scalae tympani, and the absence of a bulla tympanica. In addition, *Ornithorhynchus* retains the plesiomorphic synapsid condition in that the foramen perilymphaticum and recessus scalae tympani are close to the cranial base; a processus recessus and an aquaeductus cochleae are lacking. Besides these plesiomorphies, *Ornithorhynchus* shares with *Tachyglossus* the following derived characters: The incus lies medial to the malleus, the tympanicum is nearly horizontal, the stapes is not penetrated by the a. stapedia, the ductus cochlearis is elongated rostromedially into the cranial base and, in contrast to the Theria, does not form a cochlea. Among mammals, these characters can be regarded as synapomorphic for the Monotremata. Monotremes resemble therians in having three middle ear ossicles, which, in the adults, together with praearticulare and tympanicum, are separate from the jaw apparatus. In both groups the praearticulare forms the anterior process (folii) of the malleus, and the tympanicum is the frame for the tympanic membrane. Therefore, a high degree of separation of the angulare, articulare, and praearticulare from the lower jaw between the praearticulare and dentale, significant in the transmission of airborne sound, is likely to have been present in the common ancestor of all Recent mammals. Other shared derived characters (synapomorphies) of monotremes and therians are the squamoso-dentary jaw joint and the m. tensor tympani. Recent Mammalia, therefore, form a monophyletic group. The morphogenesis of the head does not provide evidence for close phylogenetic relationships among the Monotremata, Triconodonta, Multituberculata, or Pantotheria. The available evidence suggests that the dichotomy of the phyletic lines leading to monotremes and therians occurred well before the origin of Eupantotheria.

Contents

Introduction, 96
Middle Ear, 96
Foramen Perilymphaticum and Recessus Scalae Tympani, 100
Conclusions, 102
Acknowledgments, 105
References, 106

Introduction

The early phase of mammalian phylogeny is characterized by a number of profound constructional changes that resulted in a new grade or plateau of organization (Starck, 1978a, 1978b). Of particular importance are the increase of relative brain size and the concomitant enlargement of the cranial cavity (Kuhn and Zeller, 1987). In addition, profound changes in the construction of the jaw apparatus and the oral cavity took place; these changes were functionally related to the development of milk glands, ingestion of milk by the newborn, chewing, and homoiothermy. Examples are the development of heterodonty, the secondary palate, muscular cheeks, and the reorganization of the jaws, leading to the separation of ear ossicles (Gaupp, 1913; Kuhn, 1971; Zeller, 1989a; Maier, 1990). In addition, specializations of the sense organs, including an enlargement of the cochlear duct and the remodeling of the otic capsule of the skull, are diagnostic characters of mammals.

The question of whether these constructional changes were already achieved in the common ancestor of all Recent mammals or took place in parallel, and independently, in several phyletic lines is still unsettled and remains a matter of controversy. In spite of the great variety of mammalian Mesozoic fossil remains already described, the common ancestor of Recent Mammalia is not documented in the fossil record. However, it can be reconstructed partly by comparing the morphogenesis of the living groups of mammals.

For this reconstruction the comparative anatomy of the skull is especially rewarding because the richness of its features provides detailed information about the biology of the organism. However, the osseous skull of the adult mammal does not display all characters on which comparative analyses have to be based. Only by studying ontogeny—the process of cranial morphogenesis, including the relationships among the skeletal, neural, sensory, muscular, and other features of the head—can homologies be assessed and an understanding of cranial morphology of the Recent Mammalia be achieved (Starck, 1967; Kuhn, 1987; Maier, 1987; Zeller, 1987). These ontogenetic data help to interpret cranial structures in fossil mammals.

In the interpretation of the cranial structures of Triconodonta and Multituberculata, the monotremes, especially *Ornithorhynchus*, are used primarily for comparisons (e.g., Clemens and Kielan-Jaworowska, 1979; Crompton and Sun, 1985). Until recently, due to lack of material, the morphogenesis of the skull of *Ornithorhynchus* was not well known. Only from *Tachyglossus* had the relevant stages been studied in detail on the basis of histological serial sections by Gaupp (1908) and later by Kuhn (1971).

I have studied skull development of *Ornithorhynchus* in fifteen serially sectioned heads of embryos, nest young, one subadult, and one adult (Zeller, 1989a). The following analysis of the tympanic and otic regions of the head is mainly drawn from this study. The comparative analysis follows Hennig's (1950) approach.

Middle Ear

In the reptilian ancestors of mammals, the quadrate and articular were separated from the jaw apparatus and employed in the transmission of airborne sound (Fig. 8.1). The separation of the articular from the lower jaw took place between the angular and prearticular (both dermal bones) on the one side, and the dentary on the other side. The original jaw bones prearticular and angular are maintained in mammals. During phylogeny, the prearticular (=goniale) fuses with the articular, forming the anterior process (processus folianus) of the adult malleus. The angular (=tympanic) is used as the

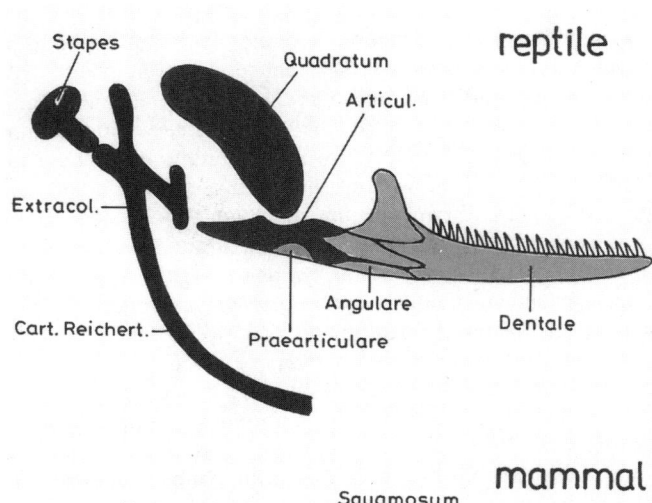

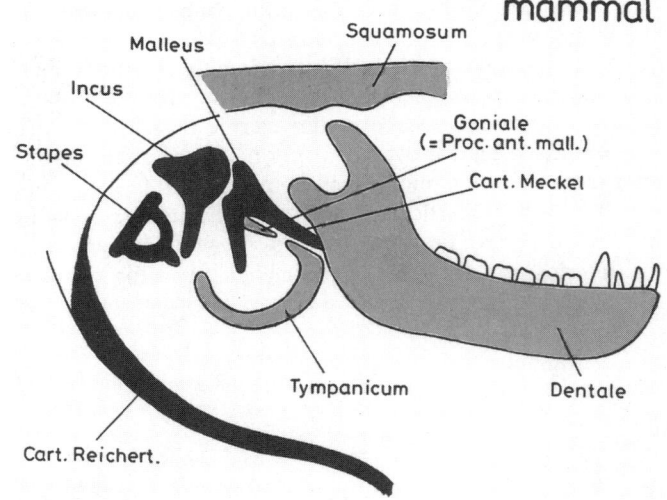

FIGURE 8.1. Schematic representation of the lower jaw and jaw articulations in a reptile (above) and mammal (below) to show the major features of remodeling of this region that occurred at the reptilian-mammalian boundary. Modified from Gaupp (1913).

frame for the tympanic membrane. The mammalian secondary jaw joint lies rostral to the reptilian primary jaw joint between the dentary and the squamosal. This, in essence, is the content of the Reichert-Gaupp theory (Reichert, 1837; Gaupp, 1913).

The reorganization of the jaw apparatus in therapsids is well documented in the fossil record (Kermack et al., 1973; Allin, 1975, 1986; Crompton and Jenkins, 1979; Crompton and Hylander, 1986; Moore, 1981). Nevertheless, the question of whether the separation of the postdentary accessory jaw bones (articular, prearticular, angular) from the lower jaw and their use in the transmission of airborne sound was already achieved in the common ancestor of monotremes and therians, or whether this took place in parallel and independently in the two phyletic lines, is not fully answered. Fleischer (1973, 1978), for example, suggested that the ear ossicles of monotremes and therians evolved independently in two phyletic lines, whereas Kemp (1983) regarded the sound-conducting system of monotremes and therians as synapomorphic. In order to assess the homologies and the character-state polarities of the mammalian sound-conducting system, therefore, the morphogenesis of the tympanic region of *Ornithorhynchus* was compared to that of *Tachyglossus* and the Theria.

Figure 8.2 shows the reconstruction of the otic region of the skull of a nest young platypus of 180 mm length viewed ventrally. At this developmental stage, most parts of Meckel's cartilage are still present; its rostral half lies in a medial groove of the dentary. As in other mammals, the caudal end of Meckel's cartilage forms the cartilaginous anlage of the malleus; it is already partially ossified, as is the articular bone. This ossification is in contact with the prearticular from the beginning, as in all other mammals. Laterally it is synchondrotically connected to the crista parotica. The cartilaginous manubrium of the malleus is well developed. The tympanic and prearticular are already developed in the 180 mm stage as dermal bones of Meckel's cartilage. Both develop in close contact to Meckel's cartilage and do not contact the dentary at any point. Compared to other mammals, the prearticular is relatively large. The tympanic, however, is small and horseshoe shaped. It lies in an almost horizontal position.

The proximal part of Reichert's cartilage of *Ornithorhynchus* forms the floor of the caudal part of the tympanic cavity. Medial to the tympanic cavity, it is synchondrotically connected to the ventral surface of the cochlear capsule. Also, other elements of the chondrocranium (e.g., incus-crista parotica; palatine cartilages) tend to fuse in nest young *Ornithorhynchus*. In addition to the extraordinary thickness of the cartilage, the fusion of neighboring elements of the chondrocranium increases its mechanical stability in both monotremes. This can be regarded as a specific adaptation to the mode of ontogeny in egg-laying mammals, because

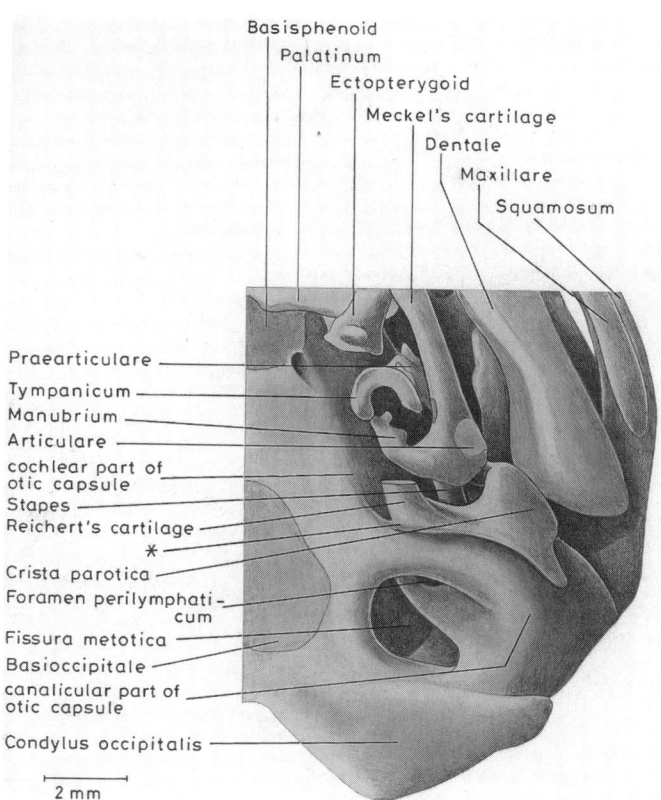

FIGURE 8.2. *Ornithorhynchus anatinus*, nest young of 180 mm dorsal contour length (DCL) (J.P. Hill collection MO 39, Hubrecht Laboratory, Utrecht, The Netherlands). Model of the otic and occipital regions of the skull in ventral view. The otic capsule is still entirely cartilaginous. The asterisk indicates the position where Reichert's cartilage is fused to the cochlear capsule. Distally to that fusion, Reichert's cartilage is removed.

nest or pouch young monotremes develop without protection of fetal membranes and, therefore, need a mechanically stable chondrocranium. The fusion of Reichert's cartilage with the cochlear capsule in monotremes does not provide any evidence that the tympanic processes of the petrosal or other constituents of the bulla tympanica of therians "may be possible to relate . . . , topologically, to the rather simple structure found in monotremes" (Presley, 1984, p. 192). Rather, the bulla is a neomorph of the Theria, correlated with the increasing inclination of the tympanic and the expansion of the tympanic cavity (van der Klaauw, 1931; Zeller, 1986). Several morphologically nonuniform skeletal elements contribute to the formation of the bulla (e.g., entotympanics, alisphenoid, basisphenoid, tympanic, periotic) (van Kampen, 1905), the ontogeny and topographical relations of which vary considerably among different taxa. Evidence for the homology of these skeletal elements with Reichert's cartilage of monotremes is lacking (Zeller, 1987, 1989a). Aside from Reichert's cartilage, a skeletal floor of the tympanic cavity and a bulla are lacking in monotremes.

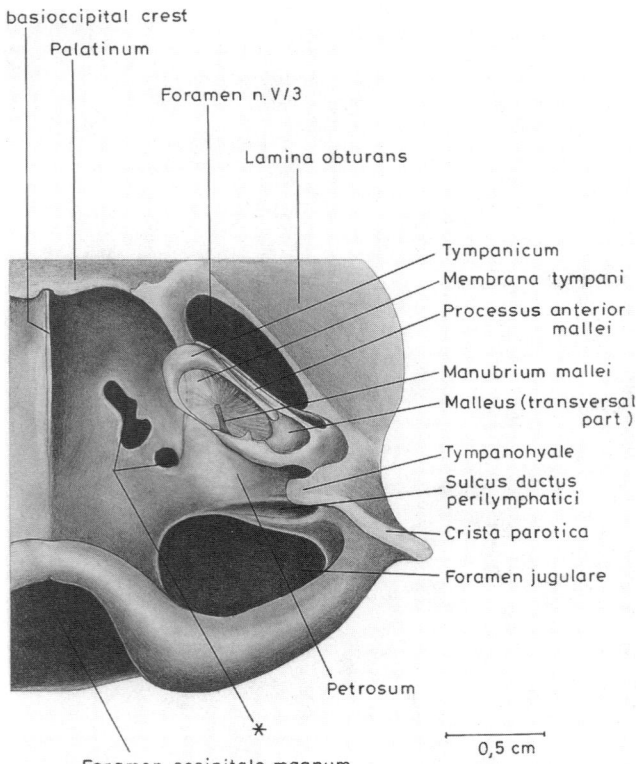

FIGURE 8.3. *Ornithorhynchus anatinus*, adult. Otic and occipital regions of the skull with ear ossicles and tympanicum in situ. The malleus is affixed to the crista parotica by dense connective tissue. The tympanicum is firmly connected to the anterior process of the malleus formed by the praearticulare. Tympanicum and tympanic membrane are in an almost horizontal position. The groove housing the perilymphatic duct is clearly recognizable; an aquaeductus cochleae is lacking. The asterisk indicates two positions where the basisphenoid is resorbed. The ectopterygoid is removed.

In the adult *Ornithorhynchus*, the position of the ear ossicles and the tympanic has not significantly changed compared to the 180 mm stage (Fig. 8.3). Meckel's cartilage, however, is ossified as part of the dentary or, in its caudal segment, entirely resorbed. The prearticular has fused with the articular and, as in all other mammals, forms the anterior process of the malleus. Laterally, the malleus and incus are closely connected to the crista parotica by dense connective tissue. The malleus is a frame-like construction, since its lateral and caudal parts are thicker than its medial part, which is only a thin osseous plate (Figs. 8.3, 8.4). It consists mainly of the phylogenetically ancient transverse part and of the manubrium. A caput mallei is lacking, just as in *Tachyglossus* (Kuhn, 1971; Starck, 1978a). As Meckel's cartilage is entirely resorbed, the anterior process is formed solely by the prearticular. It is synostotically connected to the tympanic and, as in most other synapsids, pierced by a foramen through which the chorda tympani pass (Fig. 8.4). Many Theria also possess a large transverse part of the malleus with a long anterior process (folii) firmly connected to the tympanic ("ancestral type") (Fleischer, 1978). Reduction of the transverse part and of the anterior process of the malleus in addition to the formation of a caput mallei are correlated to each other and occur in therians with a mobile malleus-incus complex (primates, some rodents, proboscideans, and tylopods) (Fleischer, 1973, 1978; Starck, 1979).

The incus is only a flattened platelet, medial to the transverse part of the malleus. A short process is barely recognizable. Figure 8.5 illustrates the unique position of the incus of monotremes in comparison with a therian (*Tupaia*). In both monotremes the incus is flat and lies medial to the malleus. This position of the incus has not been found in any other mammal. In Theria the incus lies caudal to the malleus. In both monotremes the incus is connected to the malleus by dense connective tissue and by cartilage. A quadrato-articular synovial joint is lacking in monotremes, as in many therians (Werner, 1960). Generally, malleus and incus form a functional unit in all mammals (Fleischer, 1978).

The stapes of the platypus is connected to the incus by a synchondrosis. In *Ornithorhynchus*, as in *Tachyglossus*, it is columelliform and, in contrast to most other synapsids, is not penetrated by the stapedial artery, contrary to earlier reports by Goodrich (1915). This is obviously a derived character.

The tympanicum of *Ornithorhynchus* supports the eardrum, is sickle shaped, and is firmly connected to the anterior process of the malleus. The latter character can be regarded as a synapsid plesiomorphy because, in therapsids, angular and prearticular are also tightly joined together. The shape of the tympanic of *Ornithorhynchus* resembles that of the angular of Mesozoic synapsids. In *Ornithorhynchus* it is only loosely connected by connective tissue to other surrounding elements of the skull, as is observed among marsupials and insectivores (Fleischer, 1978). This can be regarded as the primitive mammalian condition. The tympanic bone and the tympanic membrane are only slightly inclined; the angle between these and the horizontal measures only 25 to 30 degrees. This, contrary to earlier interpretations, is not a primitive mammalian character, because the angular of therapsids and that of early mammals, such as the triconodont *Morganucodon*, are in an almost vertical position (Kermack et al., 1973). The low inclination of the tympanic of monotremes was achieved secondarily by the enlargement of the cranial cavity, which is very wide in this region. The tympanicum, malleus, and incus were transferred phylogenetically into an almost horizontal position by the lateral and ventral displacement of the sidewall of the braincase and of the otic capsule. This is also true for the ancestors of many groups of living therians, and, among

FIGURE 8.4. *Ornithorhynchus anatinus*, isolated ear ossicles and tympanicum. **a** and **b**: nest young of 180 mm DCL; **c** and **d**: adult. **a** and **c**: ventral view; **b** and **d**: dorsal view. In the nest young the caudal end of Meckel's cartilage is partially ossified as the articulare, which is in contact with the praearticulare from the beginning. The manubrium, incus, and stapes are still cartilaginous. The synchondrosis between the incus and the crista parotica was cut. The anterior part of Meckel's cartilage is removed. In the adult, Meckel's cartilage is entirely resorbed. The anterior process of the malleus is solely formed by the praearticulare. The asterisks in **b** and **d** indicate a foramen in the praearticulare through which the chorda tympani pass.

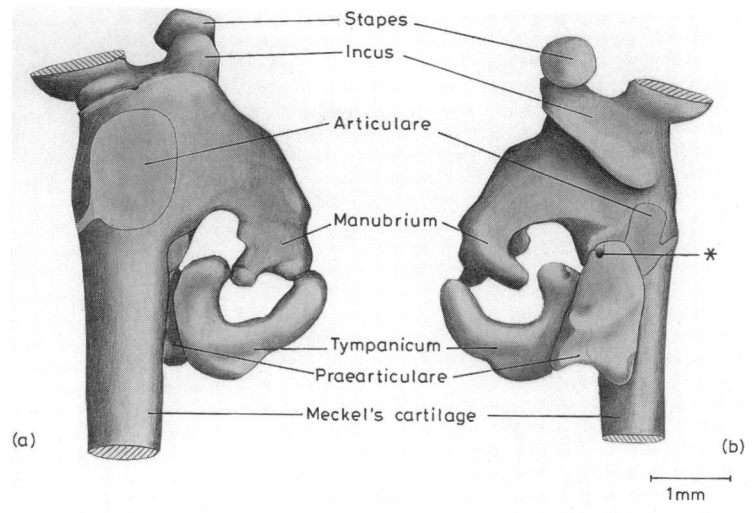

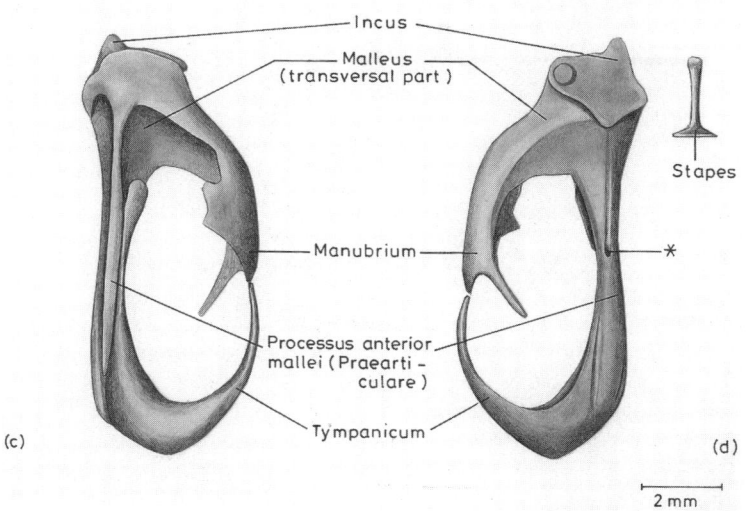

the Recent forms, in Didelphidae and Insectivora the tympanicum is still found in an almost horizontal position (Starck, 1967). However, this position of the tympanic must be regarded as independently derived in monotremes and therians, as the increase of relative brain size and the concomitant enlargement of the cranial cavity occurred independently in the two phyletic lines leading to Recent mammals (Kuhn and Zeller, 1987). Derived from the horizontal position ("tertiary"), the tympanic was inclined by the enlargement of the cochlear capsule and the expansion of the tympanic cavity during the phylogeny of most therians (Kuhn, 1971).

In the 180 mm nestling platypus, the tympanic cavity is pneumatized only to a small extent and is still partially filled with tympanic mucoid tissue. Later the tympanic cavity expands at the expense of this tissue. Corresponding to the lack of caput mallei, an epitympanic recess of the tympanic cavity is lacking in both monotremes. In *Ornithorhynchus*, the proximal part of the tympanic cavity (recessus tympanicus pharyngis) (Eschweiler, 1899a) communicates with the pharynx through a large opening. An auditory tube is lacking. The recessus tympanicus pharyngis is separated from the "atticus tympanicus" (Eschweiler, 1899a), which contains most parts of the ear ossicles, by a narrow constriction. As *Tachyglossus* and Theria possess an auditory tube supported by a small cartilage (Eschweiler, 1899a,b), its lack in *Ornithorhynchus* is regarded here as secondary and derived.

The m. tensor tympani is well developed in *Ornithorhynchus* (Zeller, 1989a, figs. 46–48) as in *Tachyglossus* (Eschweiler, 1899a; Schulmann, 1906; Kuhn, 1971). The mammalian tensor tympani muscle is innervated by the trigeminal nerve and is a derivative of the m. adductor posterior of other tetrapods. It developed in correlation with the transformation of the primary (reptilian) jaw joint into part of the sound-conducting system and the formation of the secondary (squamoso-dentary) jaw articulation (Barghusen, 1986). The m. stapedius (facial nerve), present in most therians, is lacking in both monotremes.

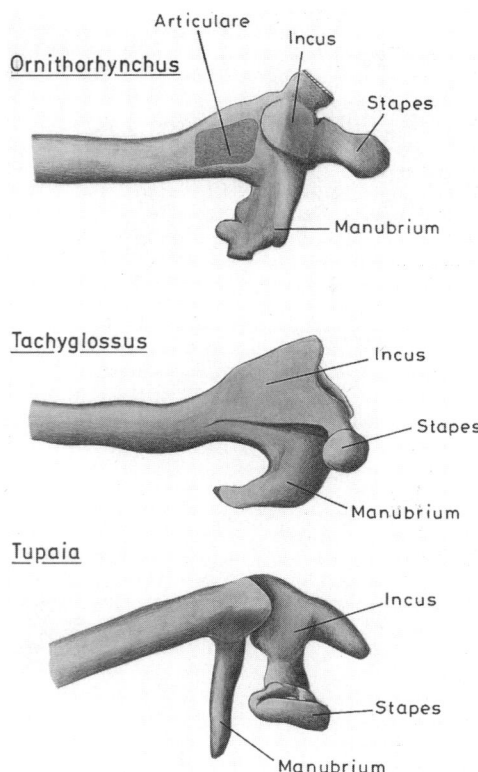

FIGURE 8.5. Caudal end of Meckel's cartilage together with the cartilaginous incus and stapes of the monotremes *Ornithorhynchus* (180 mm DCL) and *Tachyglossus* (53 mm crown-rump length) (Kuhn, 1971) and of the therian mammal *Tupaia* (34-day-old embryo) (Zeller, 1987). In both monotremes the incus is flat and lies medial to the malleus. In therians, however, the incus lies caudal to the malleus. The condition in monotremes is derived (see text). Not to scale.

Foramen Perilymphaticum and Recessus Scalae Tympani

Compared to their synapsid ancestors, the ductus cochlearis is lengthened to a considerable extent in mammals: in therians, in a spiral course forming a cochlea; in monotremes, in a straight course, extending rostromedially into the area of the cranial base. The lengthening of the cochlear duct has largely affected the configuration of the otic capsule and its relationships to the cranial base, including the position of the perilymphatic foramen through which the perilymphatic duct passes.

The perilymphatic space between the membranous labyrinth and the otic capsule of the skull is connected to the cranial cavity in all tetrapods. This connection is regarded as being necessary to release the pressure effected by the movements of the stapes within the fenestra ovalis (Wever, 1978). In all amniotes, the perilymphatic duct leaves the otic capsule through the perilymphatic foramen and enters a space, triangular in cross section, situated between the perilymphatic foramen and the cranial base (de Burlet, 1934; de Beer, 1937; Zeller, 1985a, 1985b), the recessus scalae tympani (Fig. 8.10). Medially this space faces the cranial cavity. In many groups of extant reptiles, at the lateral aperture of the recessus scalae tympani, the perilymphatic duct contacts the tympanic cavity with only the secondary tympanic membrane lying between them.

Among Theria, this contact is present at the fenestra rotunda, which is closed by the secondary tympanic membrane. In addition to the perilymphatic duct, this membrane functions to release the pressure originating in the perilymphatic space by the movements of the stapes (Arnold, 1974).

Monotremes were regarded by Presley (1980) as being anomalous among Recent mammals in that they supposedly lacked a contact between the perilymphatic space and the tympanic cavity at the secondary tympanic membrane. This would mean that they must possess only the perilymphatic duct for pressure release, resembling some reptiles (e.g., Amphisbaenia, *Aniella*). In *Ornithorhynchus* (Zeller, 1989a) the cochlear part of

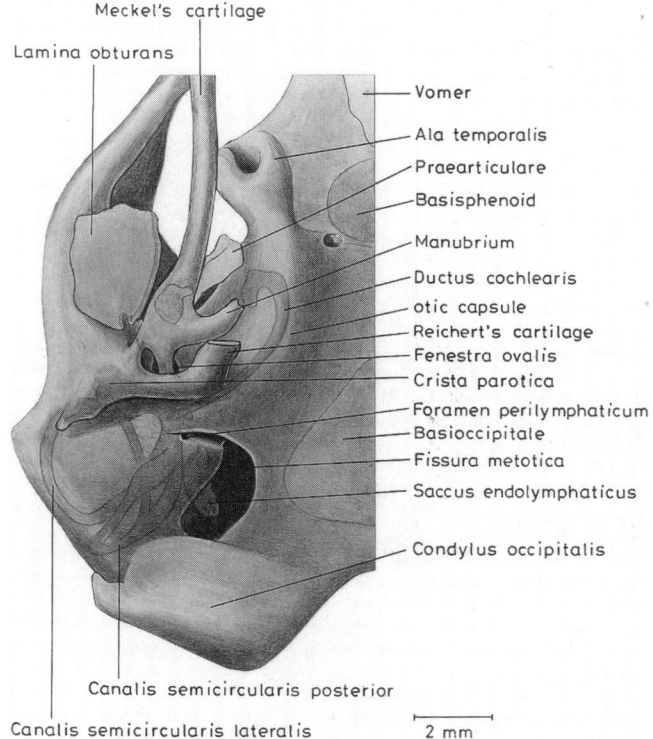

FIGURE 8.6. *Ornithorhynchus anatinus*, nest young of 180 mm DCL (MO 39). Model of occipital, otic, and parts of the orbitotemporal regions of the skull viewed ventrally (compare with Fig. 8.2). The cartilage of the otic capsule is shown as being transparent to demonstrate the shape of the membranous labyrinth. The perilymphatic foramen lies close to the cranial cavity at the anterior end of the fissura metotica. An aquaeductus cochleae is lacking. The distal segment of Reichert's cartilage is removed.

the otic capsule is elongated and flat and the perilymphatic foramen is situated at the anterior end of the fissura metotica (=foramen jugulare of the osteocranium) (Figs. 8.2, 8.6). This position of the perilymphatic foramen and the shape of the cochlear part of the otic capsule, including its relationship to the cranial base, correspond to the shape of the cochlear duct (Fig. 8.6). The ductus cochlearis of *Ornithorhynchus* is elongated compared to that of most reptiles, but, in contrast to therians, it is relatively short and uncoiled (Denker, 1901). The monotreme cochlear duct is only slightly curved; its rostral end, containing the papilla lagenae, is sickle shaped. Compared to that of other mammals, the perilymphatic foramen of *Ornithorhynchus* is situated far medially, resembling the perilymphatic foramen of many reptiles. In the 180 mm stage of *Ornithorhynchus*, the perilymphatic duct leaves the cartilaginous otic capsule through the perilymphatic foramen and enters the recessus scalae tympani, being triangular in cross section. From there, the perilymphatic duct runs medially in a shallow groove to enter the cranial cavity, where it communicates with the leptomeningeal space. Exactly at the area of this communication, the glossopharyngeal nerve leaves the cranial cavity. Ventrally it is not supported by skeletal tissue; therefore, a canal enclosed in cartilage or bone that would have housed the perilymphatic duct, i.e., a cochlear aqueduct, is lacking. Laterally the perilymphatic duct contacts tympanic mucoid tissue (Zeller, 1989a,b, fig. 51). At this stage the tympanic cavity is pneumatized only to a small extent. It does not reach its definitive extension until late in ontogeny. In the subadult *Ornithorhynchus*, the tympanic cavity still does not reach the recessus scalae tympani. The section of an adult *Ornithorhynchus* (Fig. 8.7) shows that, late in ontogeny, the tympanic cavity extends to the recessus scalae tympani. The lateral aperture of the recessus scalae tympani is closed by the secondary tympanic membrane (Zeller, 1991). The same has been observed in *Tachyglossus* (Eschweiler, 1899b; Kuhn, 1971). From this it is concluded that in all Recent mammals the perilymphatic space contacts the tympanic cavity at the secondary tympanic membrane, contrary to earlier reports. Moreover, the secondary tympanic membrane of all amniotes can be regarded as homologous.

The secondary tympanic membrane of *Ornithorhynchus* does not insert at a complete bony rim, because skeletal tissue ventral to the perilymphatic duct is lacking (Fig. 8.7). Therefore, a fenestra rotunda completely surrounded by bone is lacking. In addition, an aquaeductus cochleae, surrounding the perilymphatic duct, does not exist even in the adult *Ornithorhynchus*.

In therians, the cochlear part of the otic capsule is more prominent than in monotremes and extends far ventrally. It contains the cochlear duct, which is coiled to form a cochlea. The enlargement of the otic capsule also affects the position of the perilymphatic foramen, which in therians is situated more laterally in relation to the skull base. The perilymphatic duct is enclosed in a canal surrounded by skeletal tissue, the aquaeductus cochleae. The floor of the aquaeductus cochleae is formed by a lateral extension of the skull base, the processus recessus (Fig. 8.8). The secondary tympanic membrane inserts around a bony rim, the fenestra rotunda. In most therians the secondary tympanic membrane forms the medial boundary of a recess of the tympanic cavity, the fossula fenestrae rotundae (Frick, 1952; Zeller, 1985a, 1985b).

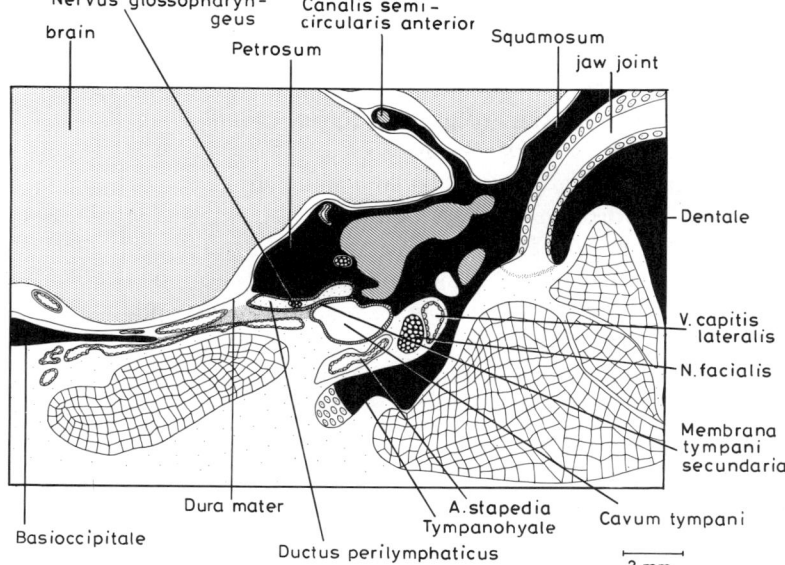

FIGURE 8.7. *Ornithorhynchus anatinus*, adult. Section through the head at the recessus scalae tympani. The caudal extension of the tympanic cavity contacts the perilymphatic duct at the secondary tympanic membrane. Skeletal tissue ventral to the perilymphatic duct is lacking, hence, an aquaeductus cochleae is absent. The glossopharyngeal nerve leaves the cranial cavity by passing through the perilymphatic duct.

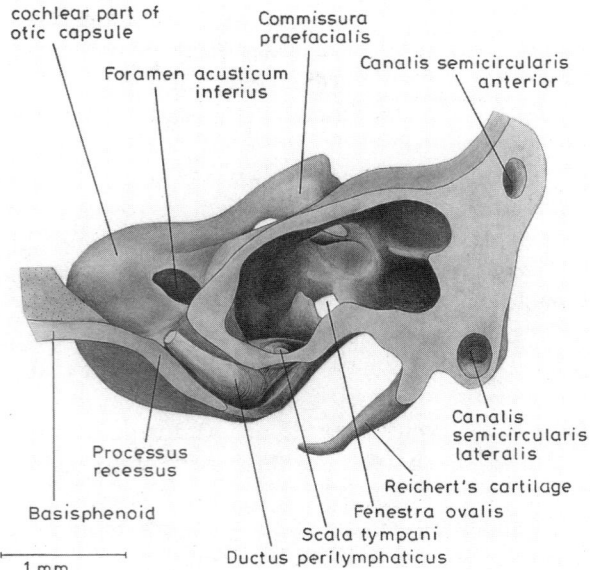

FIGURE 8.8. *Tupaia belangeri*, thirty-fourth day of ontogeny. Anterior part of cartilaginous otic capsule, scala tympani, and ductus perilymphaticus in posterior view. The cochlear capsule is greatly expanded and extends far dorsally and ventrally relative to the cranial base. The perilymphatic duct leaves the otic capsule by passing through the foramen perilymphaticum, enters the recessus scalae tympani, and from there turns medially to enter the cranial cavity. Ventrally the perilymphatic duct is supported by the processus recessus, a part of the cranial base that was shifted laterally through the expansion of the cochlear capsule. By the formation of the processus recessus the perilymphatic duct is enclosed into an aquaeductus cochleae. The anlage of the secondary tympanic membrane (hatched) develops at the lateral aperture of the recessus scalae tympani. From Zeller (1985a).

Conclusions

Compared to other synapsids, monotremes have retained a number of synapsid plesiomorphies in their tympanic region. These include (1) the absence of a caput mallei; (2) the synostosis between praearticulare and tympanicum, the columelliform stapes; and (3) the absence of a bulla tympanica. In addition, *Ornithorhynchus* shares with *Tachyglossus* the following derived characters: (1) the incus is flat and lies medial to the malleus; (2) the tympanicum is almost horizontal; (3) the stapes is not penetrated by the stapedial artery; and (4) the ductus cochlearis is elongated but does not form a cochlea. These characters can be regarded as synapomorphic for the Monotremata, that is, as having been apomorphies in their common ancestor. Among mammals, they are autapomorphic for the Monotremata.

Monotremes resemble therians in the following derived characters. Compared to the ancestral synapsid condition, the prearticular, articular, and quadrate are reduced in size and, together with the angular, are separated from the dentary. The prearticular forms the anterior process of the malleus. The tympanicum supports the tympanic membrane. The m. tensor tympani is present. The squamoso-dentary articulation is the functional jaw joint (Table 8.1).

These mammalian synapomorphies show that a high degree of separation of the postdentary accessory bones from the jaw as ear ossicles and tympanic must have already occurred in their common ancestor (Fig. 8.9). However, the close relationship of tympanic and prearticular to the dentary in the early ontogeny of some Theria (e.g., *Monodelphis*, *Didelphis*, *Halmaturus*, *Orycteropus*) (Gaupp, 1913; Toeplitz, 1920; Maier, 1987) indicate that the malleus of the common ancestor of Recent mammals was not entirely free of the lower jaw but, in the adults, was still connected to it by a per-

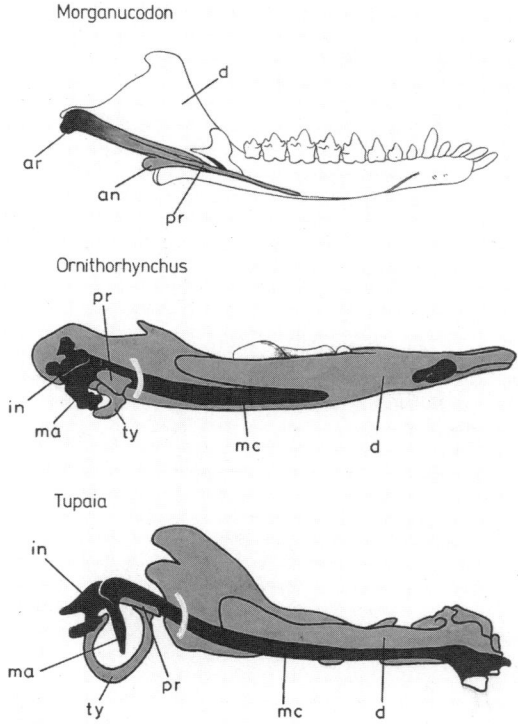

FIGURE 8.9. Comparison of left lower jaws of *Morganucodon* (adult; from Kermack et al., 1973), *Ornithorhynchus* (180 mm DCL), and *Tupaia* (thirty-fourth day of ontogeny). In *Morganucodon* the postdentary accessory jaw bones were still connected to the lower jaw, the slender praearticulare extends far rostrally, and, possibly, an ossified Meckel's cartilage persisted in the adult. In all extant mammals, articulare, praearticulare, and angulare (=tympanicum) are separated from the lower jaw between the praearticulare and dentale (white crescent); the praearticulare is fused to the articulare and forms the anterior process of the malleus. The tympanicum of all living mammals supports the tympanic membrane. The condition of these can be derived from a state represented by *Morganucodon*. Not to scale. For abbreviations, see Table 8.2.

Chapter 8. Ontogenetic Evidence for Cranial Homologies

TABLE 8.1. Character analysis of cranial features in monotremes and therians

Monotremata .... plesio- and ——— apomorphic features in comparison with the supposed ancestral condition of Mammalia	Supposed ancestral condition for Mammalia .... plesio- and ——— apomorphic features in comparison with ancestral Synapsida	Theria .... plesio- and ——— apomorphic features in comparison with the supposed ancestral condition of Mammalia
◀··	Foramen perilymphaticum and recessus scalae tympani close to cranial base ——————▶	displaced laterally by expansion of cochlear capsule
(*Tachyglossus* autapomorphic)		
◀··	Processus recessus and aquaeductus cochleae absent ————————————▶	present
◀··	Ductus perilymphaticus relatively short ————————▶	elongated
◀··	perilymphatic space contacts tympanic cavity at the secondary tympanic membrane ·················▶	
elongated in straight course, cochlea absent ◀———	Ductus cochlearis short ————————▶	elongated in spiral course to form a cochlea
◀··	Caput mallei absent — — — — — — — — — — — —▶	present
◀··	Praearticulare firmly connected to tympanicum — — — — — —▶	separated from tympanicum
◀··	Stapes columelliform ————▶	with wide spatium intercrurale
◀··	Bulla tympanica absent — — — — — — — — — — — —▶	present
incus medial to malleus ◀———	Incus located caudal to malleus ·······················▶	
horizontal ◀———	Tympanicum in upright position ————————▶	horizontal
not penetrated ◀———	Stapes penetrated by a. stapedia ·······················▶	
◀··	Articulare, praearticulare and angulare separated from the lower jaw between praearticulare and dentale ·······················▶	
◀··	Praearticulare fused to articulare forming anterior process of malleus ·······················▶	
◀··	Tympanicum supporting tympanic membrane ·······················▶	
◀··	squamoso-dentary articulation is functional jaw joint ·······················▶	
◀··	M. tensor tympani present ·······················▶	

sisting and possibly ossified Meckel's cartilage, as was the case in *Morganucodon* (Kermack et al., 1973) and Pantotheria (e.g., Upper Jurassic Dryolestidae) (Krebs, 1971, 1988). Also in nest young platypuses which actively open their mouths to ingest milk, and in the adults of some Theria (e.g., *Micropotamogale*) (Kuhn, 1987), the malleus is connected to the dentary by a persisting Meckel's cartilage. Functionally, because of the elasticity of the cartilage, this is of little importance, and it is reasonable to assume that in the adults of the common ancestor of Recent mammals the angular, prearticular, and articular were used in the transmission of airborne sound, although they were not yet completely separated from the lower jaw. In addition, the

TABLE 8.2. Abbreviations

an	angulare
ar	articulare
cb	cranial base
cc	cranial cavity
d	dentale
in	incus
ma	manubrium
mc	Meckel's cartilage
N.IX	n. glossopharyngeus
oc	otic capsule
pc	processus recessus
pr	praearticulare
prm	promontorium
st	stapes
ty	tympanic

shape and relative position of the tympanic of monotremes and some Theria closely resemble the angular of *Cynognathus* and *Morganucodon* (Kermack et al., 1973, 1981; Kermack and Mussett, 1983). Its lower part can be traced back to the reflected lamina of the angular of many therapsids (Allin, 1975). This resemblance in size and position suggests a similarity in function and further corroborates the theory proposed by Allin (1975) that the angular of advanced therapsids, early mammals, and, most probably, the common ancestor of Recent mammals already supported a layer of dense connective tissue situated at the floor of a laterally open depression of the surface of the head (Allin, 1986) and transmitted airborne sound to the articular and from there to the quadrate and the adjacent stapes (Allin, 1975; Kermack et al., 1981). The tympanic membrane of Recent mammals can be derived from this layer of connective tissue. The quadrate of many cynodonts already contacted a process of the stapes (Crompton, 1972; Allin, 1975, 1986; Chatterjee, 1982), therefore, it is likely that it was part of the sound-conducting system in the common ancestor of Recent mammals, as in *Morganucodon* (Kermack et al., 1981). The quadrate of therapsids and of *Morganucodon* lies dorsal and caudal to the articular. In therians, the incus has retained this position relative to the articular. The incus of monotremes was independently transformed into a small plate medial to the articular by the lateral displacement of the cranial sidewall and otic capsule due to the considerable widening of the cranial cavity in the otic region (Zeller, 1989a). This position of the incus of monotremes must be regarded as autapomorphic.

In Multituberculata (e.g., *Lambdopsalis*) (Miao and Lillegraven, 1986) malleus, incus, and tympanicum (Allin, 1986) closely resemble those of many Theria. This, together with the evidence from the morphogenesis of monotremes and therians, suggests that mammals can be regarded as a monophyletic group of synapsids with a squamoso-dentary jaw articulation and a specialization of the original jaw bones angular, prearticular, articular, and quadrate for sound conduction.

In Theria, the pressure originating through the minute vibrations of the stapes at the fenestra ovalis is suggested as being released at the fenestra rotunda. Monotremes were in the past (Presley, 1980) regarded as being unique among mammals in lacking such a membrane and, as in some reptiles, using only the perilymphatic duct for pressure release. However, the ontogenetic evidence of this study clearly indicates that in *Ornithorhynchus*, late in ontogeny, the tympanic cavity expands to the recessus scalae tympani and, in the adult, contacts the perilymphatic duct at the membrana tympani secundaria in the plane of the lateral aperture of the recessus scalae tympani, just as in *Tachyglossus* (Kuhn, 1971). The secondary tympanic membrane of all Recent mammals can be regarded as homologous.

In *Ornithorhynchus*, as in most reptiles, the perilymphatic foramen and the recessus scalae tympani are situated near the skull base and close to the cranial cavity. The perilymphatic duct is comparatively short. An aquaeductus cochleae and a fenestra rotunda are lacking. The comparison with reptiles and fossil synapsids indicates that the medial position of the recessus scalae tympani of *Ornithorhynchus* is a primitive synapsid character (Fig. 8.10).

In therians the perilymphatic foramen and the recessus scalae tympani are displaced laterally through the formation of a spiral cochlea and the concomitant enlargement of the otic capsule; the perilymphatic duct is enclosed in an aquaeductus cochleae (Zeller, 1985a, 1985b).

The recessus scalae tympani of the monotreme *Tachyglossus* is derived within the Monotremata in that the "aquaeductus cochleaeal," in addition to the perilymphatic duct, contains the glossopharyngeal nerve (Kuhn, 1971). This indicates that the "aquaeductus cochleae" of *Tachyglossus* is not homologous, but evolved independently of that of therians.

The common ancestor of all Recent mammals must have had a comparatively small cochlear capsule of the skull containing an uncoiled ductus cochlearis. The elongated and slightly curved cochlear duct of monotremes and the expanded cochlea of therians can be derived only from such an ancestral condition. In particular, the monotreme cochlear duct cannot be derived from the cochlea of therians. In the common ancestor of Recent mammals, the position of the perilymphatic foramen and of the recessus scalae tympani must have been similar to that of therapsids and early mammals in which the cochlear capsule was comparatively small and contained an uncoiled cochlear duct. Such a condition is found in cynodonts (Allin, 1986), for instance in *Diademodon* (Watson, 1916) and in *Morganucodon* (Kermack et al., 1981). The cochlear cavity in the latter was straight and even shorter than in any extant mammal

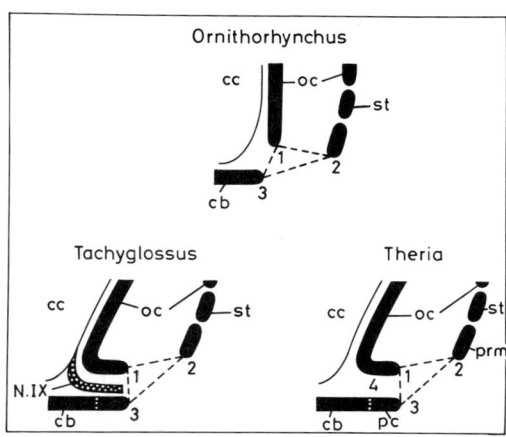

FIGURE 8.10. Schematic cross sections through the right otic region of *Ornithorhynchus*, *Tachyglossus*, and Theria (compare with Fig. 8.8). Midline of skull is to the left. 1–2 = foramen perilymphaticum; 2–3 = lateral aperture of the recessus scalae tympani; 1–2–3 = recessus scalae tympani; 4 = aquaeductus cochleae. *Ornithorhynchus* resembles living reptiles (e.g., Lacertilia) in that the recessus scalae tympani lies close to the cranial base; the aquaeductus cochleae is absent. In Theria the recessus scalae tympani was displaced laterally by the expansion of the cochlear capsule. The floor of the aquaeductus cochleae is formed by the processus recessus, a part of the cranial base that was shifted laterally together with the recessus scalae tympani. The "aquaeductus cochleae" of *Tachyglossus* is uniquely derived in that it contains, in addition to the perilymphatic duct, the glossopharyngeal nerve (Kuhn, 1971). The conditions of *Tachyglossus* and of Theria can be derived, independently, from a state represented by *Ornithorhynchus* (see text). For abbreviations, see Table 8.2.

(Graybeal et al., 1989). In both of these synapsids the perilymphatic foramen was close to the cranial base and a cochlear aqueduct was lacking, as is the case in Recent Lacertidae. Among living mammals *Ornithorhynchus*, which has an uncoiled cochlear duct shorter than that of *Tachyglossus* (Denker, 1901), retains the medial position of the recessus scalae tympani as a primitive character. The respective autapomorphic conditions of *Tachyglossus* and of Theria can be derived from a structural stage represented by *Ornithorhynchus* (Fig. 8.10).

In all living mammals, the perilymphatic space contacts the tympanic cavity at the secondary tympanic membrane. The available evidence suggests that this character was already present in the last common ancestor of all living Mammalia.

The evidence obtained from the morphogenesis of the tympanic and otic regions of *Ornithorhynchus* and the comparison with other living and fossil synapsids is relevant for the assessment of the phylogenetic relationships of the Monotremata. Monotremes and therians have a number of well-corroborated synapomorphies (Table 8.1) (Zeller, 1989a) and, therefore, are a monophyletic group. In addition, monotremes and therians display their own mosaic of synapsid plesiomorphic and group-specific apomorphic characters (Zeller, 1989a). Considering the profound differences in the organization of the brain, the jaw muscles, and the lateral wall of the braincase (Kuhn and Zeller, 1987; Zeller, 1989a, 1989b), the respective autapomorphies of monotremes and therians (Table 8.1) indicate an early separation of the two phyletic lines leading to Recent monotremes and therians from their common ancestor (see also Miao, this volume; Archer, this volume; Szalay, this volume, chapter 9).

The comparative analysis of the morphogenesis of the skull indicates that the common ancestor of monotremes and therians possessed a short ductus cochlearis and a small cochlear capsule of the skull, a foramen perilymphaticum close to the cranial base, and no aquaeductus cochleae around the perilymphatic duct. The early Cretaceous Eupantotheria (e.g., *Vincelestes*) (Bonaparte and Rougier, 1987) already had derived mammalian characters shared with Theria. These are, for example, a spiral cochlea, an aquaeductus cochleae, and a fenestra rotunda (Wible, 1991; Wible and Hopson, this volume). It is unlikely, therefore, that the common ancestor of monotremes and therians was closely related to these eupantothere mammals. Such a relationship, however, was suggested by Kemp (1983), Archer et al. (1985), and Kielan-Jaworowska et al. (1987). I believe that the available evidence drawn from the morphogenesis of the head (see also Szalay, this volume, chapter 9, on the tarsus) suggests that the dichotomy of the phyletic lines leading to monotremes and therians occurred well before the origin of Eupantotheria or Zatheria. In addition, the comparative analysis of the braincase (Kuhn and Zeller, 1987; Zeller, 1989a, 1989b) does not provide evidence for a close phylogenetic relationship between monotremes and Triconodonta or Multituberculata.

New discoveries of fossil mammals from the Mesozoic and their interpretation in the light of the anatomy of the living forms will further elucidate the phylogenetic relationships of the monotremes.

ACKNOWLEDGMENTS. The author thankfully acknowledges the availability of material from the Embryological Collection of the Hubrecht Laboratory, Netherlands Institute for Developmental Biology, Utrecht, The Netherlands, viz. the specimen *Ornithorhynchus anatinus*, J.P. Hill Collection MO 39 (180 mm DCL). I would like to thank Dr. Guy G. Musser (American Museum of Natural History, New York), Dr. R.W. Thorington, Jr. (National Museum of Natural History, Washington, D.C.), and Dr. A. Lechleuthner (Köln) for the loan of specimens of *Ornithorhynchus*. Ms. Jutta Zeller prepared the serial sections and the photographs, Mr. Egbert von Bischoffshausen prepared the drawings, and Ms. Cyrilla Maelicke corrected the English. The help of all is gratefully acknowledged.

CORRESPONDENCE ADDRESS. Ulrich Zeller, Institute of Anatomy, University of Göttingen, Kreuzbergring 36, 3400 Göttingen, Germany.

References

Allin, E.F. 1975. Evolution of the mammalian middle ear. *J. Morph.* 147:403–438.

Allin, E.F. 1986. The auditory apparatus of advanced mammal-like reptiles and early mammals. In: *The ecology and biology of mammal-like reptiles* (Hotton, N. III, Maclean, P.D., Roth, J.J., Roth, E.C., eds.). Washington, D.C., and London: Smithsonian Institution Press, pp. 283–294.

Archer, M., Flannery, T.F., Ritchie, A., Molnar, R.E. 1985. First Mesozoic mammal from Australia—An early Cretaceous monotreme. *Nature* 318:363–366.

Arnold, W. 1974. Zur Frage der Produktion und Resorption der Perilymphe (Lymphabfluβ des Innenohres). *Z. Laryng. Rhinol.* 53:774–790.

Barghusen, H.R. 1986. On the evolutionary origin of the therian tensor veli palatini and tensor tympani muscles. In: *The ecology and biology of mammal-like reptiles* (Hotton, N. III, Maclean, P.D., Roth, J.J., Roth, E.C., eds.). Washington, D.C., and London: Smithsonian Institution Press, pp. 253–262.

Bonaparte, J.F., Rougier, G. 1987. Mamiferos del Cretacio inferior de Patagonia, *IV Congreso Latinoamericano de Paleontologia, Bolivia* I:343–359.

Chatterjee, S. 1982. A new cynodont reptile from the Triassic of India. *J. Paleontol.* 56:203–214.

Clemens, W.A., Kielan-Jaworowska, Z. 1979. Multituberculata. In: *Mesozoic mammals* (Lillegraven, J.A., Kielan-Jaworowska, Z., Clemens, W.A., eds.). Berkeley: University of California Press, pp. 99–149.

Crompton, A.W. 1972. The evolution of the jaw articulation of cynodonts. In: *Studies in vertebrate evolution* (Joysey, K.A., Kemp, T.S., eds.). Edinburgh: Oliver & Boyd, pp. 231–251.

Crompton, A.W., Hylander, W.L. 1986. Changes in mandibular function following the acquisition of a dentary-squamosal jaw articulation. In: *The ecology and biology of mammal-like reptiles* (Hotton, N. III, Maclean, P.D., Roth, J.J., Roth, E.C., eds.). Washington, D.C., and London: Smithsonian Institution Press, pp. 263–282.

Crompton, A.W., Jenkins, F.A. 1979. Origin of mammals. In: *Mesozoic mammals* (Lillegraven, J.A., Kielan-Jaworowska, Z., Clemens, W.A., eds.). Berkeley: University of California Press, pp. 59–73.

Crompton, A.W., Sun, A.-L. 1985. Cranial structure and relationships of the Liassic mammal *Sinoconodon*. *Zool. J. Linn. Soc. London* 85:99–119.

de Beer, G.R. 1937. *The development of the vertebrate skull.* Oxford: Clarendon Press.

de Burlet, H.M. 1934. Vergleichende Anatomie des stato-akustischen Organs. In: *Handbuch der vergleichenden Anatomie der Wirbeltiere* (Bolk, L., Göppert, E., Kallius, E., Lubosch, W., eds.). Berlin, Wien: Urban & Schwarzenberg, pp. 1293–1432.

Denker, A. 1901. Zur Anatomie des Gehörorgans der Monotremata. *Semon Zool. Forschungsreisen in Australien* 3(1):635–662.

Eschweiler, R. 1899a. Zur vergleichenden Anatomie der Muskeln und der Topographie des Mittelohres verschiedener Säugethiere. *Arch. mikr. Anat. Entwickl.-Gesch.* 53:558–622.

Eschweiler, R. 1899b. Die Fenestra cochleae bei *Echidna hystrix*. *Anat. Anz.* 16:584–590.

Fleischer, G. 1973. Studien am Skelett des Gehörorgans der Säugetiere, einschließlich des Menschen. *Säugetierk. Mitt.* 21:131–239.

Fleischer, G. 1978. Evolutionary principles of the mammalian middle ear. *Adv. Anat. Embr. Cell. Biol.* 55:1–70.

Frick, H. 1952. Über die Aufteilung des Foramen perilymphaticum in der Ontogenese der Säuger. *Z. Anat. Entwickl.-Gesch.* 116:239–279.

Frick, H., Starck, D. 1963. Vom Reptil-zum Säugerschädel. *Z. Säugetierk.* 28:321–341.

Gaupp, E. 1908. Zur Entwickelungsgeschichte und vergleichenden Morphologie des Schädels von *Echidna aculeata* var. typica. *Semon Zool. Forschungsreisen in Australien* 3(2):539–788.

Gaupp, E. 1913. Die Reichertsche Theorie, (Hammer-, Amboss- und Kieferfrage). *Arch. Anat. Entwickl.-Gesch.*, Suppl. 1912:1–416.

Goodrich, E.S. 1915. The Chorda tympani and middle ear in reptiles, birds, and mammals. *Quart. J. micr. Sci.* 61:137–160.

Graybeal, A., Rosowski, J.J., Ketten, D.R., Crompton, A.W. 1989. Inner-ear structure in *Morganucodon*, an early Jurassic mammal. *Zool. J. Linn. Soc.* 96:107–117.

Hennig, W. 1950. *Grundzüge einer Theorie der phylogenetischen Systematik.* Berlin: Deutscher Zentralverlag.

Kemp, T.S. 1983. The relationships of mammals. *Zool. J. Linn. Soc. London* 77:353–384.

Kermack, K.A., Mussett, F. 1983. The ear in mammal-like reptiles and early mammals. *Acta Palaeont. Pol.* 28:147–158.

Kermack, K.A., Mussett, F., Rigney, H.W. 1973. The lower jaw of *Morganucodon*. *Zool. J. Linn. Soc. London* 53:87–175.

Kermack, K.A., Mussett, F., Rigney, H.W. 1981. The skull of *Morganucodon*. *Zool. J. Linn. Soc. London* 71:1–158.

Kielan-Jaworowska, Z., Crompton, A.W., Jenkins, F.A. 1987. The origin of egg-laying mammals. *Nature* 326:871–873.

Krebs, B. 1971. Evolution of the mandible and lower dentition in dryolestids. In: *Early mammals* (Kermack, D.M., Kermack, K.A., eds.). *Zool. J. Linn. Soc. London* 50, Suppl. 1:89–102. New York, London: Academic Press.

Krebs, B. 1988. Mesozoische Säugetiere—Ergebnisse von Ausgrabungen in Portugal. *Sber. Ges. Naturf. Freunde Berlin* (N.F.) 28:95–107.

Kuhn, H.-J. 1971. Die Entwicklung und Morphologie des Schädels von Tachyglossus aculeatus. *Abh. senckenb. naturforsch. Ges.* 528:1–192.

Kuhn, H.-J. 1987. Introduction. In: *Morphogenesis of the*

mammalian skull (Kuhn, H.-J., Zeller, U., eds.). *Mammalia depicta* 13:9–15. Hamburg, Berlin: Paul Parey.

Kuhn, H.-J., Zeller, U. 1987. The cavum epiptericum in monotremes and therian mammals. In: *Morphogenesis of the mammalian skull* (Kuhn, H.-J., Zeller, U., eds.). *Mammalia depicta* 13:51–70. Hamburg, Berlin: Paul Parey.

Maier, W. 1987. Der Processus angularis bei *Monodelphis domestica* (Didelphidae; Marsupialia) und seine Beziehungen zum Mittelohr: Eine ontogenetische und evolutionsmorphologische Untersuchung. *Morph. Jb.* 133:123–161.

Maier, W. 1990. Phylogeny and ontogeny of mammalian middle ear structures. *Proc. Third Intern. Congr. Vertebrate Morphology, Antwerp., Netherlands J. Zoology* 40:55–74.

Miao, D., Lillegraven, A. 1986. Discovery of three ear ossicles in a multituberculate mammal. *National Geographic Research* 2:500–507.

Moore, W.J. 1981. *The mammalian skull.* Cambridge, New York, Sidney: Cambridge University Press.

Presley, R. 1980. The braincase in recent and mesozoic therapsids. *Mem. Soc. Géol. Fr. N. S.* 139:159–162.

Presley, R. 1984. The tympanic cavity of Mesozoic mammals. In: *Third symposium on Mesozoic terrestrial ecosystems Tübingen 1984* (Reif, W.-E., Westphal, F., eds.). Tübingen: Attempto Verlag, pp. 187–192.

Reichert, C. 1837. Über die Visceralbogen der Wirbeltiere im Allgemeinen und deren Metamorphosen bei den Vögeln und Säugetieren. *Müllers Arch. Anat., Physiol., wiss. Med.* 1837:120–222.

Schulmann, H. 1906. Vergleichende Untersuchungen über die Trigeminus-Musculatur der Monotremen, sowie die dabei in Betracht kommenden Nerven und Knochen. *Semon Zool. Forschungsreisen in Australien* 3(2):297–400.

Simpson, G.G. 1945. The principles of classification and a classification of mammals. *Bull. Am. Mus. Nat. Hist.* 85:1–350.

Starck, D. 1967. Le crâne des Mammiféres. In: *Traité de Zoologie* (Grassé, P. P., ed.). Paris: Masson, Cie, pp. 405–549.

Starck, D. 1978a. Das evolutive Plateau Säugetier. *Sonderbd. naturwiss. Ver. Hamburg* 3:7–33. Hamburg, Berlin: Paul Parey.

Starck, D. 1978b. *Vergleichende Anatomie der Wirbeltiere auf evolutionsbiologischer Grundlage*, vol. 1. Berlin, Heidelberg, New York: Springer.

Starck, D. 1979. *Vergleichende Anatomie der Wirbeltiere auf evolutionsbiologischer Grundlage*, vol. 2. Berlin, Heidelberg, New York: Springer.

Toeplitz, Ch. 1920. Bau und Entwicklung des Knorpelschädels von *Didelphis marsupialis*. *Zoologica, Stuttg.* 27:1–84.

van Kampen, P.N. 1905. Die Tympanalgegend des Säugetierschädels. *Morph. Jb.* 34:321–722.

van der Klaauw, C.J. 1931. On the auditory bulla in some fossil mammals, with a general introduction to this region of the skull. *Bull. Am. Mus. Nat. Hist.* 62:1–340.

Watson, D.M.S. 1916. The monotreme skull: A contribution to mammalian morphogenesis. *Phil. Trans. roy. Soc. London* B207:311–374.

Werner, C.F. 1960. *Das Gehörorgan der Wirbeltiere und des Menschen.* Leipzig: Thieme.

Wever, E.G. 1978. *The reptile ear.* Princeton: Princeton University Press.

Wible, J. 1991. Origin of Mammalia: The craniodental evidence reexamined. *J. Vert. Paleont.* 11:1–28.

Zeller, U. 1985a. The morphogenesis of the fenestra rotunda in mammals. In: *Functional morphology in vertebrates* (Duncker, H.-R., Fleischer, G., eds.). *Fortschritte der Zoologie* 30:153–157. Stuttgart, New York: G. Fischer.

Zeller, U. 1985b. Die Ontogenese und Morphologie der Fenestra rotunda und des Aquaeductus cochleae von *Tupaia* und anderen Säugern. *Morph. Jb.* 131:179–204.

Zeller, U. 1986. Ontogeny and cranial morphology of the tympanic region of the Tupaiidae, with special reference to *Ptilocercus*. *Folia Primatol.* 47:61–80.

Zeller, U. 1987. Morphogenesis of the mammalian skull with special reference to *Tupaia*. In: *Morphogenesis of the mammalian skull* (Kuhn, H.-J., Zeller, U., eds.). *Mammalia depicta* 13:17–50. Hamburg, Berlin: Paul Parey.

Zeller, U. 1989a. Die Entwicklung und Morphologie des Schädels von *Ornithorhynchus anatinus* (Mammalia: Prototheria: Monotremata). *Abh. senckenberg. naturforsch. Ges.* 545:1–188.

Zeller, U. 1989b. The braincase of *Ornithorhynchus*. In: *Trends in vertebrate morphology* (Splechtna, H., Hilgers, H., eds.). *Fortschritte der Zoologie* 35:386–391. Stuttgart, New York: G. Fischer.

Zeller, U. 1991. Foramen perilymphaticum und Recessus scalae tympani von *Ornithorhynchus anatinus* (Monotremata) und anderen Säugern. *Verh. Anat. Ges.* 84:441–443.

CHAPTER 9

Pedal Evolution of Mammals in the Mesozoic: Tests for Taxic Relationships

FREDERICK S. SZALAY

Overview

Similar characters are often the result of different causes or similar causes in different lineages. The danger of incorrectly considering them synapomorphies is minimized when transformations are considered in the complex context of development, function, and ecological morphology. Thus the analysis of homologies (which are either plesiomorphous, apomorphous, transformational, synapomorphous, or parallel similarities) requires causal research encompassing all of evolutionary morphology, and not only distribution analysis. The apomorphous constraints in the common ancestry, which are the bases of grouping organisms, can be deciphered even when synapomorphies cannot be clearly recognized. Ontogenetic, functional, and ecological factors, i.e., numerous specific evolutionary causes, and some relevant processes affect the *individual organism* and are therefore responsible for the patterns of phena. The selection and analysis of characters (and therefore the cause and process-based interpretations of distributional data) and hypotheses of evolution, ranging from those of characters to those of species, should be tested against all relevant observations within the noted context. This is a methodological tenet that falls out of evolutionary theory, not from a taxonomic one. *Null-group comparisons* suggest a number of probable apomorphies, transformation sequences, synapomorphies, and parallel acquisitions.

The joints of the crus and tarsus of mammals offer features with critical and functionally well understood characters that form the bases of complex, and therefore highly weightable, taxonomic properties. The cruropedal traits of monotremes, selected cynodonts, tritylodontids, morganucodontids, multituberculates, and therians were analyzed. Aspects of the homologies of the monotreme foot have been misinterpreted both functionally and phylogenetically. Monotremes, while highly modified in their tarsus, retain some clearly synapsid characters (also retained in morganucodontids). The sharing of probably noncynodont, apomorphous similarities of tritylodontids and morganucodontids supports tritylodontid and mammalian sistergroup affinities. Although multituberculates and ancestral therians also share some derived mammalian calcaneal traits, the details of some of the pedal joints do not support a uniquely shared common ancestry between therians (Tribotheria, Metatheria, and Eutheria) and multis. Lack of adequate tarsal data of docodonts and numerous "therians" (and "pantotheres") make the suggestion of any synapomorphy (as opposed to parallelism) common to the Allotheria and Theria as yet impossible to test. These exclusive similarities must go back at least 155 MYBP to an ancestry structurally not significantly removed from the known morganucodontids. Some of the primitive pre-mammalian traits present in stem mammals were lost and others transformed into either synapomorphous or parallel (yet also divergent) similarities seen in multituberculates and therians. Based on pedal and other postcranial evidence, however, the multituberculates and therians appear more recently related to one another than either group is to monotremes. This does not mean that multis and therians can be considered to be close relatives with any degree of confidence.

Contents

Introduction, 109
Some Methodological Issues Related to Character
    Analysis, 109
Establishing the Evidence: Homologies, 111
    Early cynodonts, 113
    Tritylodontidae, 113
    Morganucodontidae, 114
    Cloverly triconodont, 115
    Monotremata, 115
    Multituberculata, 121
    Theria, 122
Character Transformations, 122
Pedal Evolution and Mesozoic Mammal Phylogeny, 125
    Pedal evolution, 125
    Taxon phylogeny, 126
Acknowledgments, 126
References, 127

Introduction

Pedal evidence for understanding early mammalian evolutionary history has not been rigorously utilized in the past. Peripheral exceptions have been Schaeffer's (1941) monograph on tarsal evolution of lower vertebretes, touching on the Mammalia, and Lewis's (1963, 1964, 1983, 1989, etc.) numerous studies in which he has placed particular emphasis on muscle homologies, subsequently proposing structural bone morphology hypotheses. Yet whenever new Mesozoic and Tertiary tarsal evidence has become known, excellent descriptions and functional accounts have usually been provided (e.g., Jenkins, 1970a, 1970b, 1971; Krause and Jenkins, 1983). Several outstanding functional analyses (such as those of Jenkins and McClearn, 1984; Jenkins and Krause, 1983) probed specific issues of foot mechanics and successfully related these to articlar joint morphology and function on the one hand, and biological roles on the other, in Mesozoic mammals or their common ancestries. I have (Szalay, 1984) attempted an interim understanding of aspects of arboreal adaptations in mammalian feet within a phylogenetic context, and proposed stages of evolutionary change that characterized some crural and tarsal aspects of mammalian groups.

In this chapter I compare in some detail the known or available evidence from some cynodont reptiles and Mesozoic and modern mammals and choose specific characters as they bear on the taxonomic properties of the pedal complex. Following this I test a number of proposed phylogenetic hypothese dealing with the early branching of mammals against the taxonomic properties that are proposed. All of these tests are ultimately made against observations relating to all aspects of this character complex of living and fossil mammals.

Taxonomic usage must be clearly understood in any paper that deals with phylogenetic history. Here I restrict the name Theria to the concept of Parker and Haswell (1897) to include Metatheria, Eutheria, and the much less well understood Tribotheria (see footnote of Gregory, 1910, p. 230, concerning usage and priority over Eutheria Gill, 1872). The tribotherians, such as the pappotheriids, were the dentally known tribosphenic mammals whose reproductive attributes may have been viviparous, marsupial-like, rather than oviparous, monotreme-like. Inclusion of monotremes (and a number of Mesozoic groups) in the concept of Theria makes that group a virtual equivalent of the Mammalia, as matters stand currently (as also recognized by Rowe, 1988, who used a similarly restricted concept of the Theria as employed here). The Mammaliamorpha Rowe, 1988, although deserving of serious consideration, is a synonym of the concept of Mammalia long in usage among students working on primary evidence. In the concept of Mammalia used here, I therefore include the last common ancestor of Morganucodontidae and other undoubted sistergroups or direct descendants with attributes that are derivable from an *Eozostrodon*-like skeletal system. Monotreme and therian holophyly, with the exclusion of any other Mesozoic group, is certainly unsupported not only by the pedal evidence probed here, but also by several other well-studied homologies. The Multituberculata and Monotremata are unambiguously understood concepts and, similarly, the Tritylodontidae and Morganucodontidae have as yet no confusing connotations (see also Crompton and Luo, this volume).

Some Methodological Issues Related to Character Analysis

The methodology employed in constructing the formal structure of this chapter is a phylogenetic (evolutionary) one, distinct from a cladistic approach. I define the latter as one in which (a) all uniquely shared similarities are recognized as synapomorphies, (b) such designated synapomorphies are obtained from the dictates of a taxic scheme defined by an outgroup, and (c) character polarities are weighted according to an *a posteriori* perspective provided by the hypothesized cladistic relationships, all based on assumptions of parsimony. Works with large and causally unanalyzed character matrices that use consistency, which are parsimony based and algorithm driven, to support a hypothesis are also usually fully cladistic in their approach. My attempt to understand the evolutionary transformation of the pedal complex is decoupled from any taxon phylogeny-based outgroup comparison. The goals (see also Bock, 1981; Szalay and Bock, 1991; Szalay, 1993) are to (1) understand the functional-adaptive aspects of character complexes, (2) choose characters, (3) fit this understanding of characters into **taxonomic properties** (such as homologies, transformation series, apomorphies, and synapomorphies; see Fig. 9.1), and (4) construct a phylogenetic hypothesis of the organisms tested against the taxonomic properties. The evaluation of the transformation, and degree of relatedness, of the character complex examined is a character analysis within the conceptual confines described by Bock (1981) as **functional-adaptive analysis (F-A A)**.

An integral and fundamental part of phylogenetic analysis is a well-established *modus operandi*, which has grown out of comparative anatomy and paleontology and is defined formally by Szalay and Bock (1991) as **null-group comparison (NGC)**. The etymology of the term derives from the notion that no specific evolutionary relationship is assumed for the *taxa* whose character complexes are being compared. A **null group** consists of organisms that share a particular, most similar, homologous feature with that of the group under consideration. A null group is chosen, as a result of careful comparisons, for the purposes of understanding character transformation of the morphocline independent of taxon phylogeny. In spite of the character comparison, it does not assume the phylogenetic relationships between the *taxa* compared, as the cladistic outgroup method does. Nor does it automatically assume (derived) similarities to be synapomorphies without the appropriate **F-A A**. The null group, given the facts of mosaic evolution and an imperfect record of the lineages of organisms, can (and probably should) be expected to be different for different features of the groups studied.

It is from the results of such undertakings that taxonomic properties of varying probabilities are established. Homology is always a more confidently established hypothesis than the plesiomorphy and apomorphy of traits. A preexisting state obviously constrains those derived from it, and therefore apomorphies cannot be established without any confidence in the primitive conditions. This importance of establishing what is antecedent (primitive) to the apomorphic trait of the last common ancestor of a group is often overlooked by students accustomed to cladistic outgroup procedures. The reason is that such an approach is usually based on *a posteriori* weighting. The fossil record, however, is particularly invaluable in supplying the broad outlines of character transformations. Cartmill's (1981) views on parsimony analysis, and his reaffirmation of the long-held common-sense view and sound methodology concerning paleontological data, are particularly relevant here. A more detailed discussion on the nature of connections between evolutionary theory and taxonomic practice can be found in Szalay and Bock (1991) and Szalay (1993).

In order to illustrate the theoretical nature of character evolution, Figure 9.1 depicts a simple model of transformations and splittings that involve the character states of a single homologous trait (of hypothetical populations of organisms, and not taxa). It is from such characters, shared among many groups, that one attempts to derive those taxonomic properties against which the particular concepts of *taxa* and *evolutionary lineages* are tested (see Bock, 1981; Szalay and Bock, 1991). A brief examination of this figure shows that the identification of such taxonomic properties as apomorphies, transformation series, or synapomorphies must be sorted out from character reversals, convergences, or parallelisms (="apomorphic tendencies," "inside parallelism," "underlying synapomorphy," and "nonuniversal derived character state") before character polarities can be understood. This can be accomplished only with methods that derive from the underlying theory that accounts for the diverse evolutionary causes in light of what most probably happened in the evolution of the character states. Cladogram-based analysis in cladistics conflates (and ignores) this critical phyletic aspect of phylogenetics.

Sober (1989) in his recent book on parsimony as it relates to phylogenetic analysis all but admits that everything in phylogenetics depends on the understanding of character evolution and the causality and processes that affect them. In spite of this brief vision, he nevertheless goes on to concentrate on "logical strength," which he admits has nothing to do with plausability or evidential support. It is based on the roots of such analysis, so well unearthed and explicated (and quasi-approved, based on formal logic) by Sober, that many taxonomists (and some philosophers) dismiss confidence in research on character function, adaptation, and transformation, as if these had little to do with the understanding of phylogeny. It is disappointing to see Sober write that, "The less we need to know about evolutionary processes to make an inference about pattern, the more confidence we can have in our conclusions" (1989, p. 11). What produces character patterns, we may ask, if not the numerous specific evolutionary causes, based on the general processes of heredity, the chemistry of biological substances, development, tissue biology, mechanical constraints, or the diverse aspects of ecological morphology that shape character evolution? When can the taxonomist say that all of these operate equally in all of the characters and therefore they are ignorable? Attempts to understand features in the biological and paleontological context and the specifics of the causes that produced them give the only chance to decipher the actual historical meaning behind the data base of taxonomists, the characters themselves.

As may be obvious to many from the foregoing, the evaluation of the evidence, or rather the recovery or observation of the features from the specimens, is critically dependent on the perspective, methodology, and a host of assumptions about biology, mechanics, and evolutionary causality and the specific processes (Szalay, 1981). It is not only which characters are useful or "valid" and which are not, but also the taxonomist's ability to link features, and thus facilitate the testing of transformation hypotheses, that determines the success of understanding character evolution and the phylogenetic endeavors based on character analysis in taxonomy. Very often, students who claim that their phyletic and taxonomic efforts are independent from any functional-adaptive foundations in fact make extensive comparative use of characters and transformational perceptions that are the results of exhaustive non-taxic and character-oriented evolutionary research of others. These "process-free" taxonomists in such instances soundly (but inadvertently?) base their comparisons on significant character complexes.

Phylogenetic analysis is critically dependent on understanding the constraints that channel change. Unfortunately, most current understanding of these constraints is restricted to a semantic and theoretical realm. Research methodologies with testable ontology through

Chapter 9. Pedal Evolution of Mesozoic Mammals

FIGURE 9.1. Model of character transformation, derived from evolutionary theory, of *one homologous trait*, as opposed to models based either on taxonomic theory, or on an "evolutionary" theory based on taxonomic theory. Darwinian evolutionary theory recognizes both phyletic (patristic, anagenetic) and cladistic (branching, splitting) relationships which manifest themselves in temporal and phenetic distances of attributes. On the figure taxonomic properties are shown which can be derived from a model that accounts for both the phyletic and the cladistic components of evolutionary change: **a** is the homology manifested in a variety of states; **ai-aiv, a1-a4, ai-a1, ai-ax** are transformation series; **ai** is the diagnostic apomorphy of the ancestor of group delineated in the figure, and it also appears as a reversal; **ap** is the condition primitive *to* the ancestry of diagnosed group; **aii** and **aiii** are both synapomorphies and parallelisms in different lineages of the figure; **a2** is a parallel acquisition (sometimes designated as a "synapomorphy"; see text).

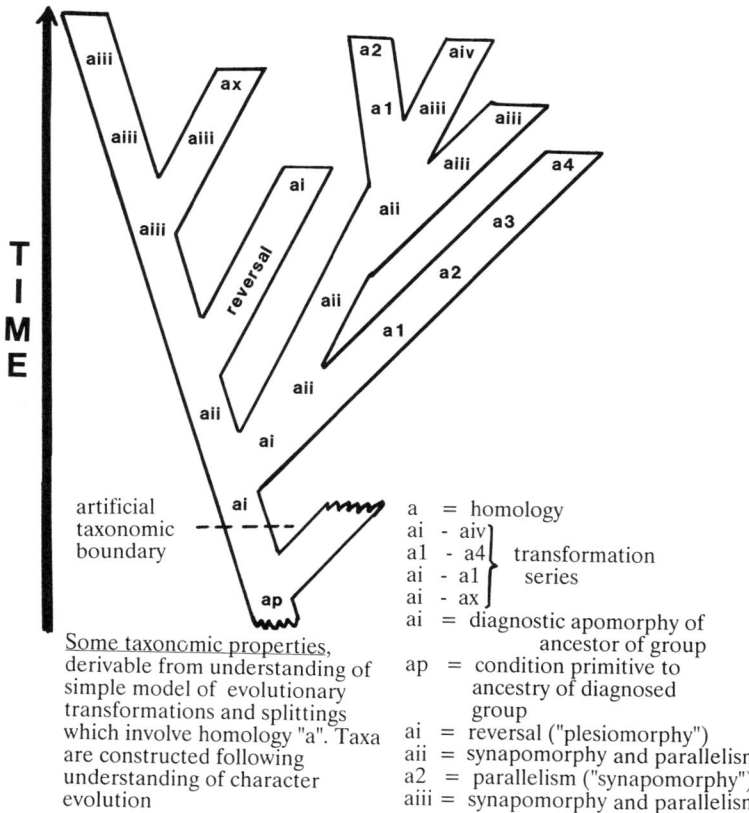

available epistemological means are sorely needed (Szalay and Bock, 1991). Meanwhile, a perspective that holds that aspects of a given area of form-function and behavior which direct a particular change (whether such transformations are perceived as either epiphenomenal or adaptational) is critical in the decision-making process of character analysis. Recognition of group-specific phylogenetic factors is the object of such research, these being the precious, sometimes incidental, results of sundry biological and evolutionary studies. Such focused studies, or a fallout of others, can be part of the causal explanatory framework of phylogenetics by establishing confidence in taxonomic hypotheses of all groups, ranging from homologous characters to taxic hypotheses. Such research is widely based in all areas of evolutionary biology, and the relevance of the latter to phylogenetics is supported by the objective evidence resulting from biologically broad-based studies, contrary to Cracraft's (1981) assertions. Taxonomy that attempts to sever its theoretical and empirical ties from these areas of biology returns to a form of scholasticism and the prevalent "rigor" of passimony-based algorithm-driven hypotheses. "Judgments" via such (but not all) algorithms are based on biologically unacceptable assumptions. Taxonomy so obtained cannot causally account for its basic currency—the characters. The value of characters, like that of cladograms, without an understanding of the evolutionary causality, can be highly questionable.

Establishing the Evidence: Homologies

Much of the material presented here has been characterized in some conventional anatomical-functional details. My aim is not to provide complete descriptions, but rather to focus on those aspects of known and new information that yield characters of reliable complexity (see Table 9.1 for definitions of the acronym abbreviations used in the text and figures). Analyses of a complex of characters often lead to robust taxonomic properties. Functionally and adaptively poorly understood characters often result in poor taxonomic properties because their evolutionary connections with putative homologues are suspect.

Before I discuss the specific results of my study, Lewis's (1983) critique of relevant previous work must be briefly addressed. According to Lewis, "The consensus of paleontological opinion would have us believe that cynodonts were already well on the way to achieving the essential characterisitcs of mammalian foot structure. . . . This view is, in fact, based on only the most general level of osteological analysis. . . . The morphological background needed for a more subtle and detailed analysis has simply not been adequately explored" (1983, pp. 28–29). He then proceeded, based on a few critical assumptions, to reinterpret Jenkins's (1971) comprehensive description of the Manda cynodont and the accompanying analysis of mammalian pedal evolution. In my view, Lewis has misjudged

TABLE 9.1. Abbreviations for morphology

	Bones, Joints, and Joint Facets
ACJ	astragalocuboid joint
ACu	astragalocuboid
AFi	astragalofibular
AN	astragalonavicular
ANJ	astragalonavicular joint
As	astragalus
ATa	astragalotibiale
ATi	astragalotibial
ATia	anterior astragalotibial
ATid	distal astragalotibial
ATil	lateral astragalotibial
ATim	medial astragalotibial
ATip	posterior astragalotibial
Ca	calcaneus
CaA	calcaneoastragalar
CaAd	distal calcaneoastragalar
CaCu	calcaneocuboid
CaCua	auxiliary (australidelphian) calcaneocuboid
CaCud	distal calcaneocuboid
CaCul	lateral calcaneocuboid
CaCum	medial calcaneocuboid
CaCup	proximal calcaneocuboid
CaFi	calcaneofibular
CaMt5l	calcaneal-**Mt5** ligament
CCJ	calcaneocuboid joint
CLAJP	continuous lower ankle joint pattern
CNJ	calcaneonavicular joint
Cu	cuboid
Ec	ectocuneiform
EMt1	entocuneiform-first metatarsal
EMt1J	entocuneiform-**Mt1** joint
EMt1l	lateral **EMt1**
EMt1m	medial **EMt1**
En	entocuneiform
Fe	femur
FFJ	femorofibular joint
Fi	fibula
LAJ	lower ankle joint
Mc	mesocuneiform
Mt	metatarsal
Na	navicular
NaCu	naviculocuboid
NCJ	naviculocuboid joint
Oc	os calcis
Ph	prehallux
SLAJP	separate lower ankle joint pattern
Su	sustentacular
Ta	tibiale
TFJ	tibiofibular joint, distal
Ti	tibia
TiFi	distal tibiofibular
TMTJ	tarsometatarsal joint
TTJ	transverse tarsal joint
UAJ	upper ankle joint

Topographical bony details, ligaments and tendons, and anatomical directions

ac	astragalar canal
adt	astragalar distal tuber
ampt	astragalar medial plantar tuberosity
at	anterior plantar tubercle
cflf	calcaneofibular ligament facet
cump	cuboidal medial process
gtpl	groove for tendon of peroneus longus

TABLE 9.1. *Continued*

lu	lunula
mc	meniscus
pfp	parafibular process
pp	peroneal process of calcaneus
ppl	process for peroneus longus on **Mt1**
ps	posterior shelf of distal tibia
sa	sulcus astragali
sc	sulcus calcanei
tc	tuber of calcaneus

Abbreviations used both in the text and on the figures for the cruropedal evidence. Specific joints are abbreviated by the combination of the first letters in capitals of the names of those units that contribute to the joint, and the letter *J* for joint. Abbreviations entirely in lower case designate landmarks of specific bones, anatomical directions, or muscles. Abbreviations are listed under two separate headings in order to facilitate retrieval of information; characters abbreviated in this table are specific homology designations rather than only topographical descriptive terms.

Jenkins's descriptive, functional, analytical, and evolutionary account. In fact, Lewis's assumption of the path of the tendon of the flexor fibularis based on his osteological misidentification of the homologies of the tuber and the peroneal process (discussed below) in monotremes (specifically echidnas) has even prompted him to postulate that crushing is the cause of the tuber orientation in the Manda specimen. He has also failed to note critical aspects of the **UAJ** in that specimen. Lewis's (1983) views are important to note here because his is the most recent opinion by an anatomist of extant forms on a topic related to my study.

The initial recognition of characters (and the construction of taxonomic properties) of significance ideally should depend on relatively phylogenetic assumption-free form-functional accounts. Yet the methodology and unexplicated taxic assumptions involved in Lewis's account overlooked joint kinematics and the functionally mediated probabilities of carpal and tarsal transformations. He simply chose echidna foot morphology as the "primitive stage of evolution" for understanding living mammals (". . . the foot of echidna is most informative, for although it possesses obviously aberrant specializations, these are *less extreme* than those of the platypus" [emphasis added] (1983, p. 23)). No explanation was provided as to what features in the monotremes promted his choice and why he made this choice. This is particularly curious because Gregory (1951, p. 365), in spite of his mistaken views concerning the "palimpsest theory" (Gregory, 1947) of monotreme origins, has given, for its time, a remarkably well argued functional-adaptive account as to why echidnas are derivatives of platypus-like animals. Echidnas, nevertheless, were Lewis's cladistic outgroup, although this was not formally stated (". . . the morphological features of monotreme can be readily conceived as preadaptive to the typical therian condition" (1983, p. 30)). Added to this was Lewis's conviction that an assumed taxic hierarchy of nature can supply character transformation series (" . . . modern monotremes, marsupials and placentals should present, in many anatomical features, an ascending scale of specialization" (p. 23)). Virtually all of Lewis's (1983) views of transformation are based on his assumtions, which I believe ultimately rest on some (but not all; see Renfree, this volume, chapter 2) corroborated aspects of another system, namely the protomarsupial reproductive pattern being derived from a monotreme-like ancestry. This unawareness of mosaic evolution, the total available fossil record, the nature of evolutionary causality, and the subsequent derivation of character transformation reconstruction from such a perspective is quite different from the null-group comparisons and functional-adaptive analyses practiced by others in comparative biology for a long time.

How powerful these beliefs were is illustrated by Lewis's view of the similarity of the echidna's calcaneal tuber and the general shape of the astragalus with those of marsupials: "The resemblance can be appreciated most easily by comparison with the marsupial generalized condition, where the heel is incompletely bent backwards, thus providing a transitional link to the characteristic eutherian pattern. . . . [and] . . . The marsupial talus which lacks a really clear-cut head and neck, also represents an ideal transitional form between the hemispherical echidna bone and the typical eutherian condition." (1983, p. 26). In fact, Lewis's approach is that of "scenario reconstruction," outlined by Eldredge and Cracraft (1980), who view it as a mere exegesis of cladistic "pattern."

EARLY CYNODONTS. Jenkins (1971) has given a monographic review of the known cynodont pedal remains, and in particular he has described and functionally analyzed an African middle Triassic right foot (BMNH TR.8), the Manda cynodont. Lewis (1983), who restudied this specimen along with monotremes, has critically commented on paleontological biases in assessing bone form (as noted above) and, according to him, the subsequent inherent misperceptions toward bony features. Yet he failed to better Jenkins's analytical account in any significant way. In fact, in addition to a methodologically unacceptable procedure for the evaluation of transformation (see above), Lewis has made several fundamental mistakes regarding monotreme foot anatomy and homologies, as I detail below.

I have restudied the Manda specimen, reconstructed the articular connections of the foot, and made comparisons with the other material reported here. This pes (Figs. 9.2, 9.3, and 9.11) shows a particularly interesting mosaic of traits. The tricontact **UAJ** (Szalay, 1984) or the phalangeal formula of 2,3,3,3,3 cannot be said to be stem mammalian apomorphies as they are present in at least this cynodont. Similarly, the well-defined calcaneal tuber, the articular contact between the calcaneus and astragalus (**CaA** and **Su** articulations), between the calcaneus and cuboid (**CaCu** articulation), and the well-defined peroneal process are obviously well-established before the last common ancestry of the Mammalia. The largely distal rather than distomedial orientation of the calcaneal **CaCu** facet is unusual in the context of these comparisons. The **LAJ** articulations of the calcaneus and astragalus show that the movements of the calcaneus involved a slightly helical rotation of the calcaneus not only on the **CaA** facets but primarily on the convex proximomedial portion of the calcaneal **Su** facet.

The astragalus unequivocally shows the clear (ancient synapsid) separation between the **ATi** and **AFi** facets at the area distal to the astragalar canal. Furthermore, I cannot differentiate between the medial and lateral tibial (**ATim** and **ATil**) facets, which are relatively well delineated in *Oligokyphus* and mammals. There is the slightest hint of a neck-like narrowing of the distal end, but the morphology of the **UAJ** facets does not suggest significant load and movement differences from the tibia and fibula; the articular surfaces are nearly uniformly convex. Yet it is the astragalus and its narrow contact with the navicular that is one of the most significant aspects of the foot. Articulation with the navicular is perceptibly sellar. The "nail-bearing" (exceptionally widened) unguals and the greatly widened distal ends of the metatarsals suggest powerful digging.

TRITYLODONTIDAE. *Oligokyphus* from the Liassic of England has been studied by Kühne (1956), who has described tibiae, fibulae, calcanea, and astragali in considerable detail and made comparisons with some cynodonts, multituberculates, and didelphids. He particularly noted what he considered the didelphid-like morphology of the astragalus.

There are significant differences in the astragalus and calcaneus (Fig. 9.4) from those of the Manda cynodont. These consist of the more highly differentiated body and head sections, which are reminiscent of mammals, and a clearly demarcated **ATim** facet of the astragalus. The large **ATim** facet suggests far more extensive movement in the **UAJ** than could be deduced from the early therapsids. The astragalar **CaA** articulation is unmistakably sellar, suggesting that considerable mobility could be achieved in the **LAJ** by the calcaneus, and therefore the whole foot distal to it. The astragalar **AN** articulation is slightly concave (sellar ?), and the separation of the **UAJ** articulations with the tibia and fibula persists. Although there is no clear demarcation of the lateral border of the astragalar **ATil** facet, I believe that the groove between the **AFi** facet and the more medial ti-

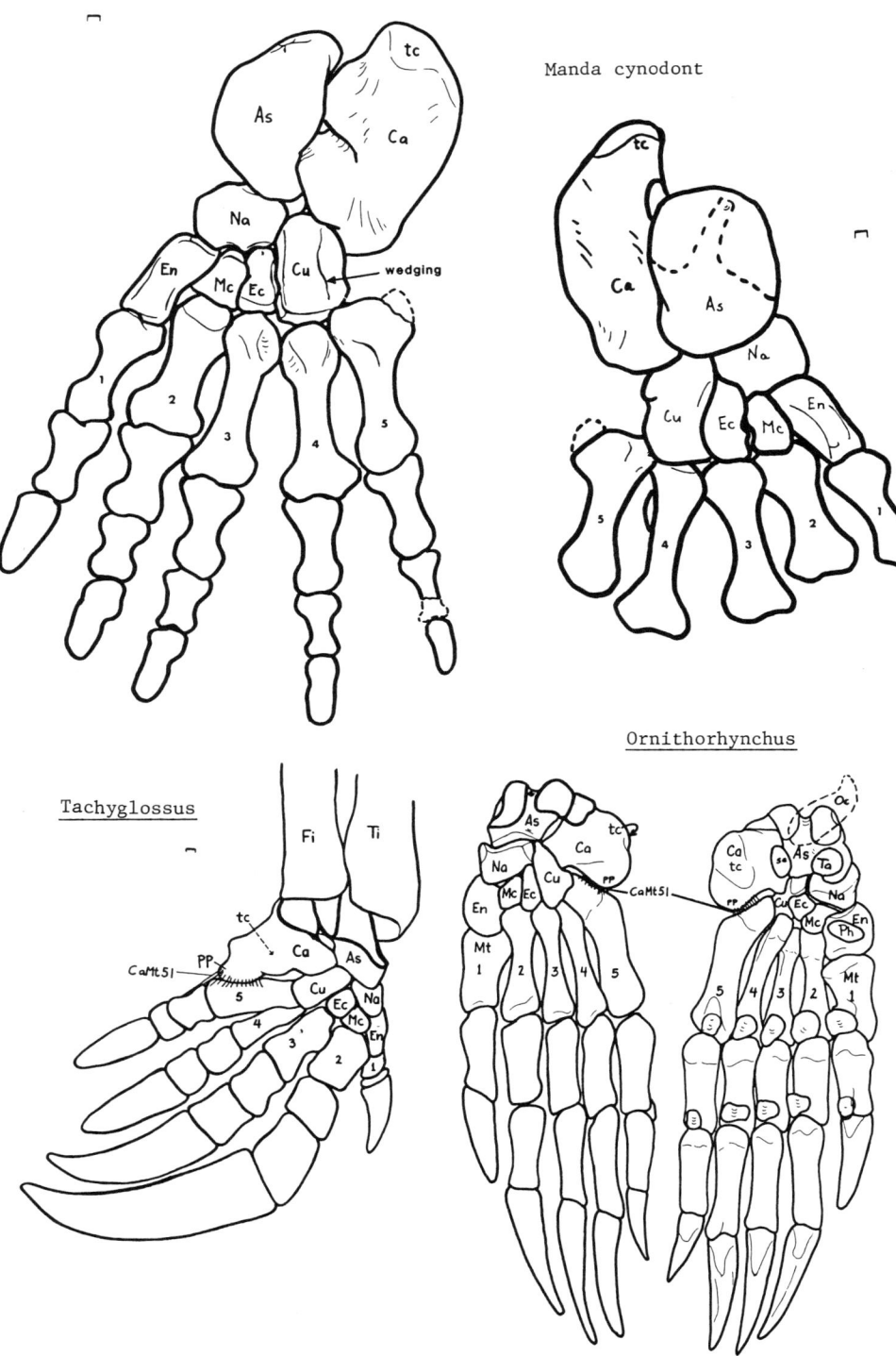

FIGURE 9.2. Above: Dorsal (right; phalanges not shown) and plantar (left) view of the right foot of the African middle Triassic Manda cynodont (BMNH Tr 8). Below left: Right foot and distal crus of *Tachyglossus* in dorsal view. Below right: Left foot of *Ornithorhynchus* in dorsal (left) and plantar (right) view; dashed outline is that of the os calcis. For abbreviations see Table 9.1. Scales represent 1 mm.

bial articulation is so similar to *Ornithorhynchus* that it was almost certainly nonarticular. The **CaCu** facet is largely medially oriented, and, as above, the calcaneal **Su** facet is convex with only its distal portion concave. The arc of **UAJ** motion, as in the Tanzanian cynodont, is remarkably small, suggesting **LAJ** adjustments.

MORGANUCODONTIDAE. Jenkins and Parrington (1976), in their descriptions of the postcrania of late Triassic mammals, gave an account of the proximal tarsals (*Eozostrodon*) and of a nearly complete left pes (South African *Megazostrodon*), the latter with damaged calcaneus and astragalus articulated in a block. They described the astragalus as having a distinct head and the navicular with a concave facet to receive the head. The collections from the Welsh Pont Alun quarry contain two astragalar morphs (Fig. 9.4). My account and emphasis differ from the original in some significant details.

The similarities to *Oligokyphus* of both the calcaneus

Chapter 9. Pedal Evolution of Mesozoic Mammals

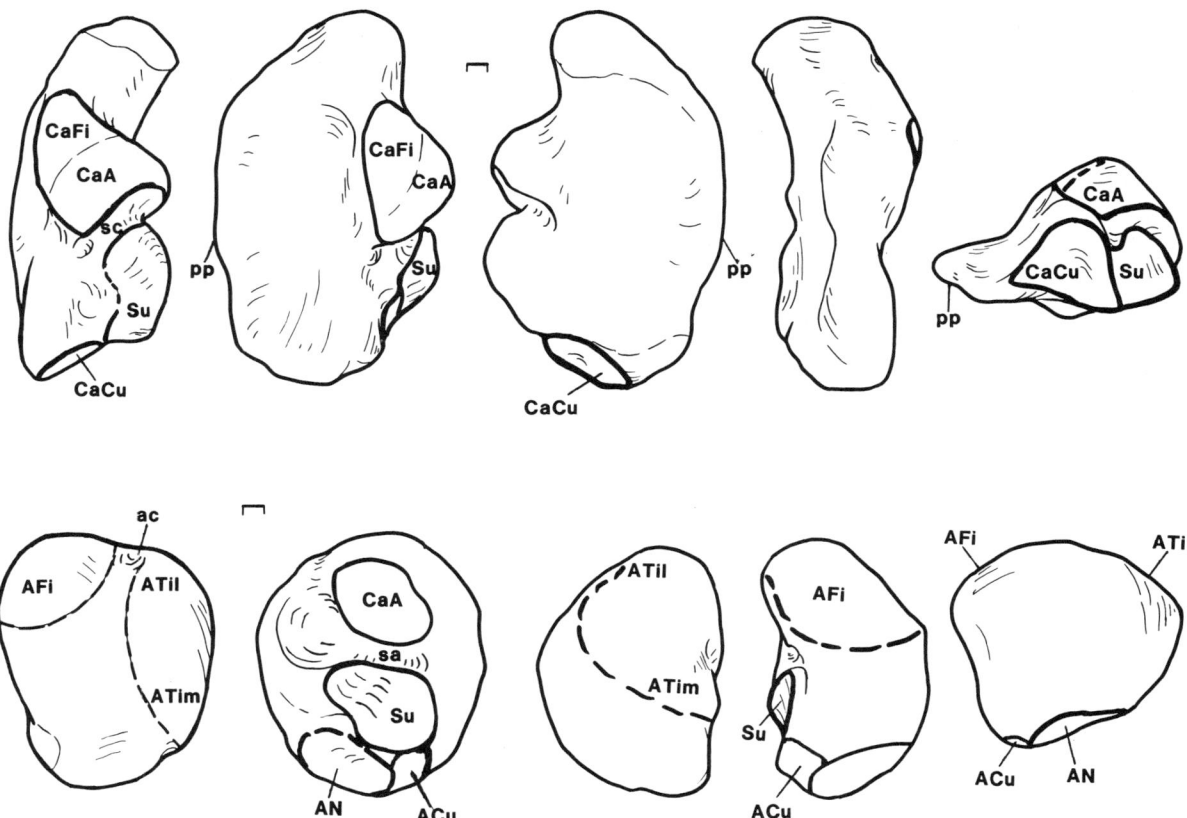

FIGURE 9.3. Manda cynodont, BMNH Tr 8, middle Triassic. Right calcaneus (above) and astragalus (below). From left to right for the calcaneus: surface with facets of the **LAJ** and **UAJ**, dorsal, plantar, lateral, and distal views. From left to right for the astragalus: surface with facets of **UAJ**, surface with facets of **LAJ**, medial, lateral, and distal views. For abbreviations see Table 9.1. Scales represent 1 mm.

and the two types of astragali are very great indeed (Fig. 9.4). The convex portion of the calcaneal **Su** facet persists to be the more extensive part of the sustentaculum compared to the distal concave portion. The **CaA** articulation is clearly sellar. On the larger astragalus the astragalar canal is more dorsal than in *Oligokyphus*, and there is contact with the cuboid, a condition that may or may not have been present in the tritylodontid. A neck and a head are well differentiated with a sellar (or at least concave) facet for navicular contact. As in the previously discussed groups, sustentacular support (i.e., a flange-like projection of the calcaneus under the astragalus) is minimal, probably less than that seen in the Manda cynodont.

CLOVERLY TRICONODONT. Jenkins and Schaff (1988) have described a nearly complete skeleton of an early Cretaceous triconodont, *Gobiconodon*. Of the tarsus, only a slightly broken left calcaneus and a bone, identified by the describers as an extratarsal spur, are known. The degree of astragalar superposition inferred is not more than in morganucodontids. The broken **CaFi** facet is extremely similar to that seen in tritylodontids and morganucodontids—a curved continuation of the **CaA** facet onto the base of the tuber. The tuber, albeit broken, was well developed and even more so than in the Triassic stem mammals. The peroneal process is somewhat reduced (i.e., derived) compared to the latter, retracted from the proximity of the **CCJ**, and what may have been the groove for the tendon of the peroneus longus is well accentuated on the lateral side of the bone. This condition is similar, although probably achieved independently, to that seen in the probably aquatic stagodontid marsupials of the Cretaceous (Szalay, 1993). This retraction of the peroneal process is common in most mammalian lineages.

MONOTREMATA. Since their discovery, these treasured living relics have been intensively studied from all aspects of their anatomy. Yet there is still new information available even from the descriptive osteology of the extremities, and this information has an altered significance in light of both the paleontological record of mammals and the new understanding derivable from ecomorphology. The most recent account is that of Lewis (1983), using monotreme morphology as a stage (set up on different criteria than the foot; in essence a cladistic outgroup of the Theria). This stage of evolution, according to him, must have preceded therian attributes in the foot. This theoretical perspective has led directly to a misidentification of some characters and resulted in an unsupported view of the evolution of

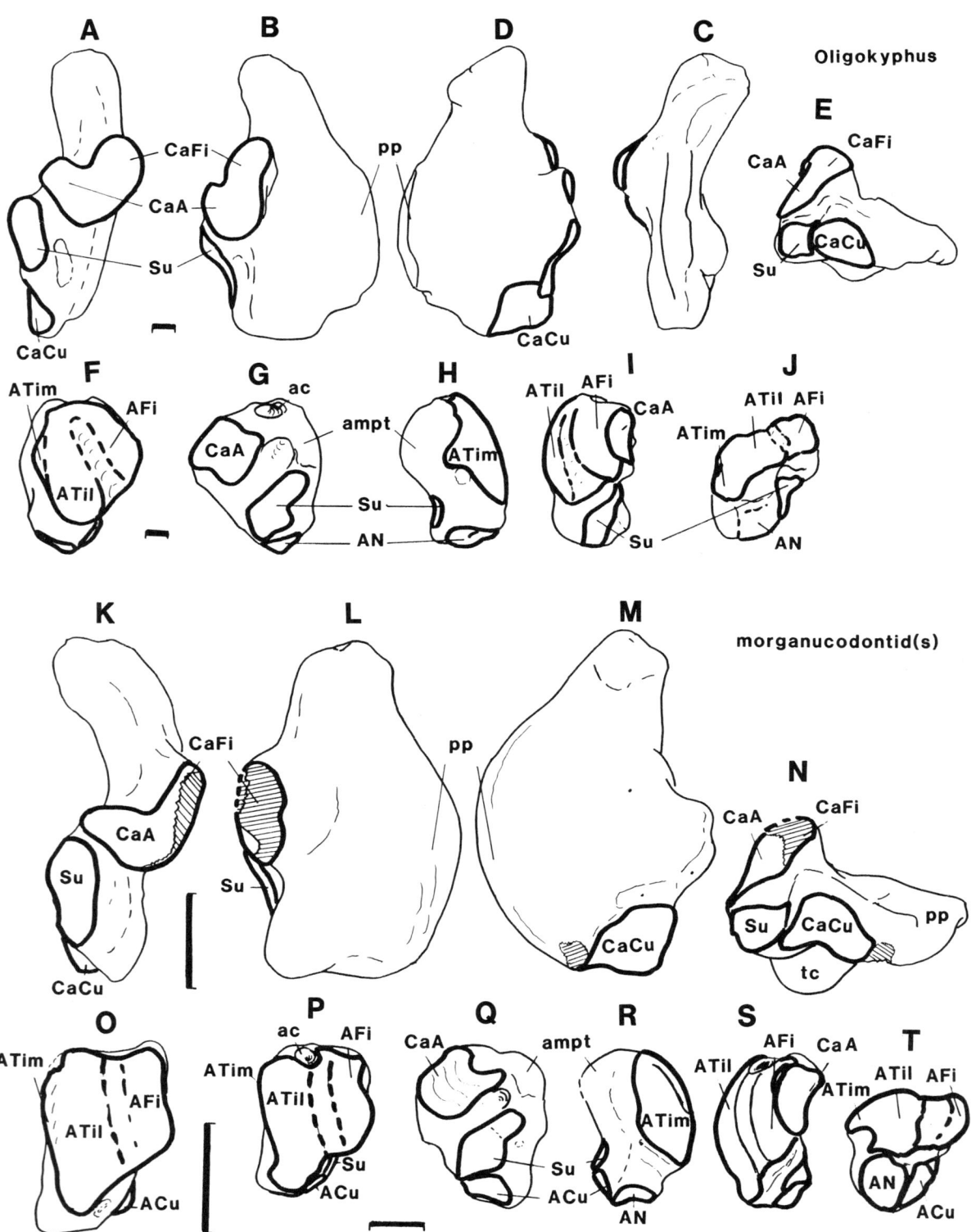

FIGURE 9.4. Comparison of the left calcanea and astragali of *Oligokyphus* and morganucodontid(s). *Oligokyphus* is based on BMNH R7503, 7504, 7500, and 7505, early Jurassic, England. From left to right for the calcaneus: surface with facets of the **LAJ** and **UAJ** (A), dorsal, (B), plantar (C), lateral (D), and distal (E) views. From left to right for the astragalus: surface with facets of **UAJ** (F), surface with facets of **LAJ** (G), medial (H), lateral (I), and distal (J) views.

Morganucodontid(s) based on Pont Alun quarry specimens (University Museum of Zoology, Cambridge), Wales, Triassic. From left to right for the calcaneus: surface with facets of the **LAJ** and **UAJ** (K), dorsal (L), plantar (M), and distal (N) views. From left to right for the astragali: surface with facets of **UAJ** (O and T), surface with facets of **LAJ** (P), medial (Q), lateral (R), and distal (S) views. For abbreviations see Table 9.1. Scales represent 1 mm.

FIGURE 9.5. *Tachyglossus aculeatus*, personal collection. A: Left femur and crus in slightly posterior and distal view. B: Left astragalus and calcaneus in distal view to show path of gastrocnemius tendon (stippled) and its insertion on tuber of calcaneus. C and D: Distal end of left crus in distal (C) an anterior (D) views, respectively. E: Left calcaneus and astragalus in dorsal (flexad) view to show path of tendon of gastrocnemius (stippled). For abbreviations see Table 9.1. Scale represents 1 mm.

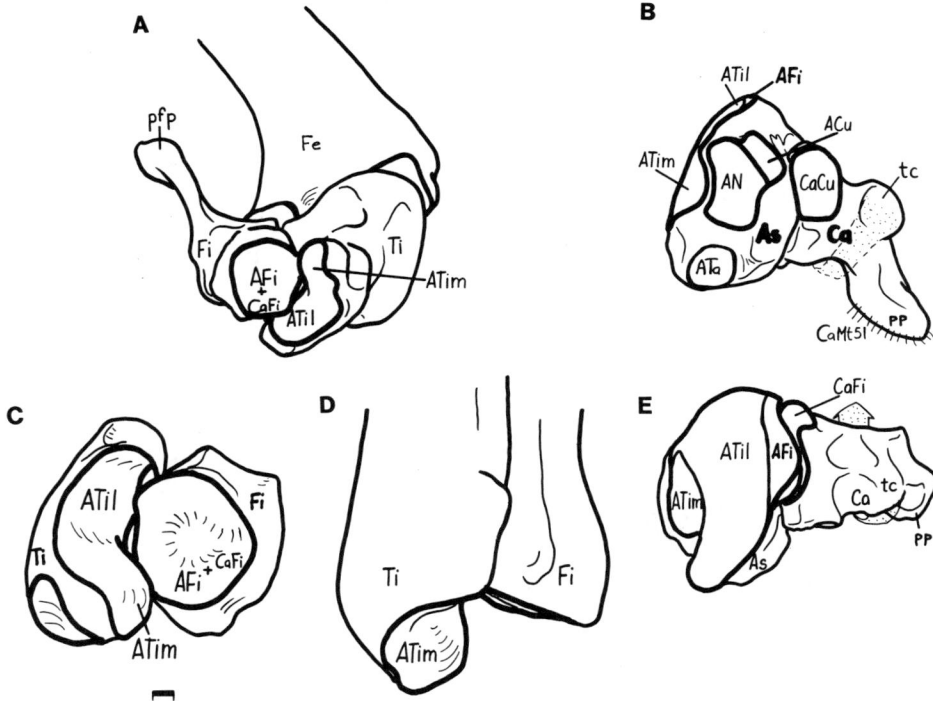

the pedal complex. There are some glaring differences between the foot structures of *Ornithorhynchus* and echidnas (see Figs. 9.2, 9.5–9.7, 9.10, 9.11). Nevertheless, there are uniquely shared attributes in this region of the skeleton that unequivocally support their holophyletic ties. The differences, as well as their unique similarities, become significant, however, when functional and ecological contexts are used in the analysis of their characters.

The account of some of the features must necessarily begin with the reasons for their correct identifications as homologues, as emphasized above. In living monotremes the calcaneal process previously identified by Lewis (1983) as the tuber is strongly bound by ligaments (**CaMt5l**) to a laterally and proximally attenuated process of **Mt5** (Figs. 9.2 and 9.5). In echidnas this connection extends to the proximal phalanx of the fifth ray as well. This area of the calcaneus in both groups of

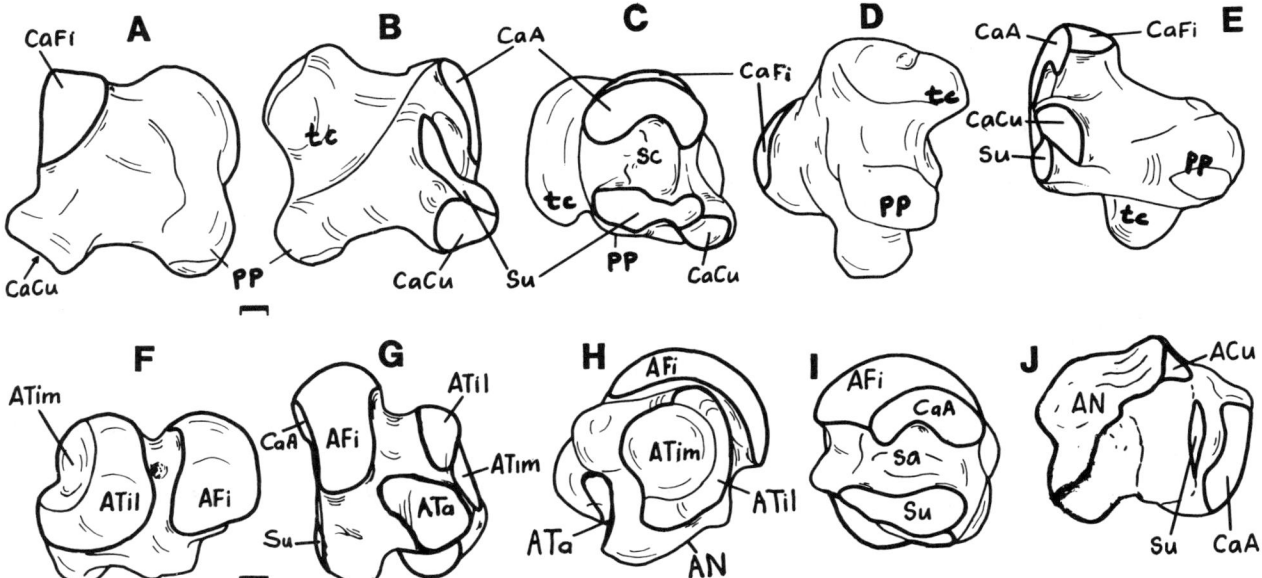

FIGURE 9.6. *Ornithorhynchus anatinus*, personal collection. Left calcaneus (above) and astragalus (below). From left to right: dorsal, plantar, medial, lateral, and distal views. For abbreviations see Table 9.1. Scales represent 1 mm.

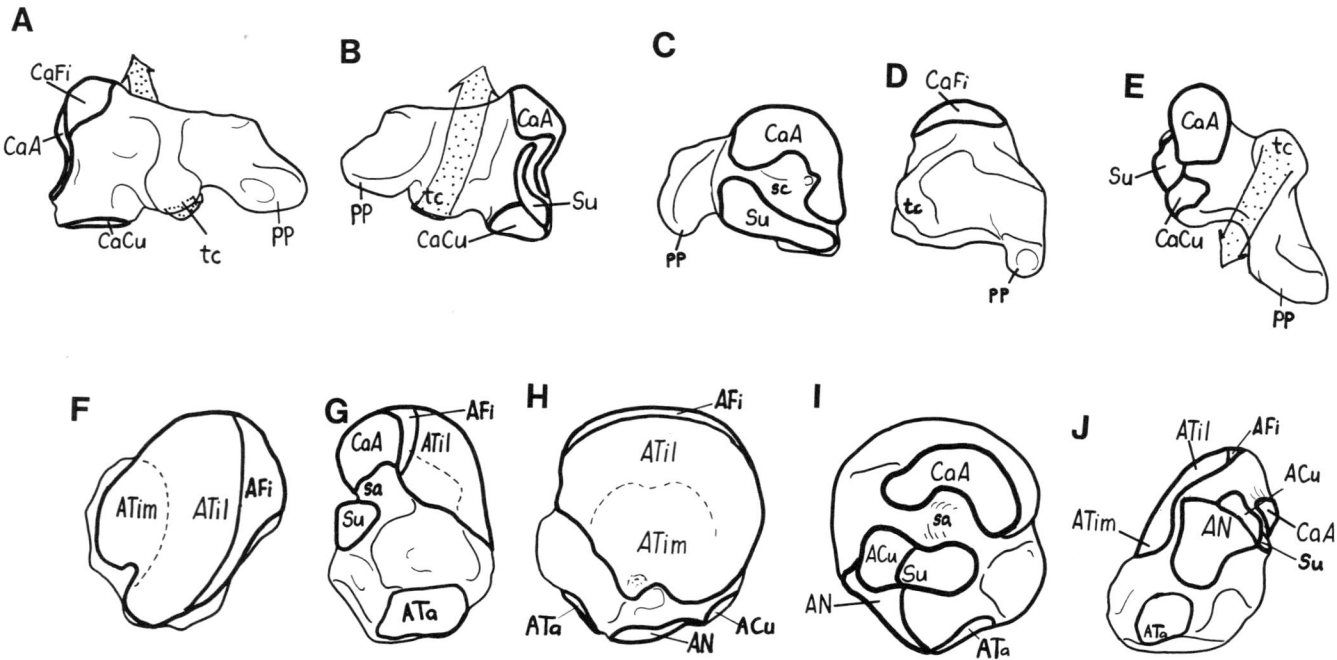

FIGURE 9.7. *Tachyglossus aculeatus*, personal collection. Left calcaneus (above) and astragalus (below). From left to right: dorsal, plantar, medial, lateral, and distal views. For abbreviations see Table 9.1.

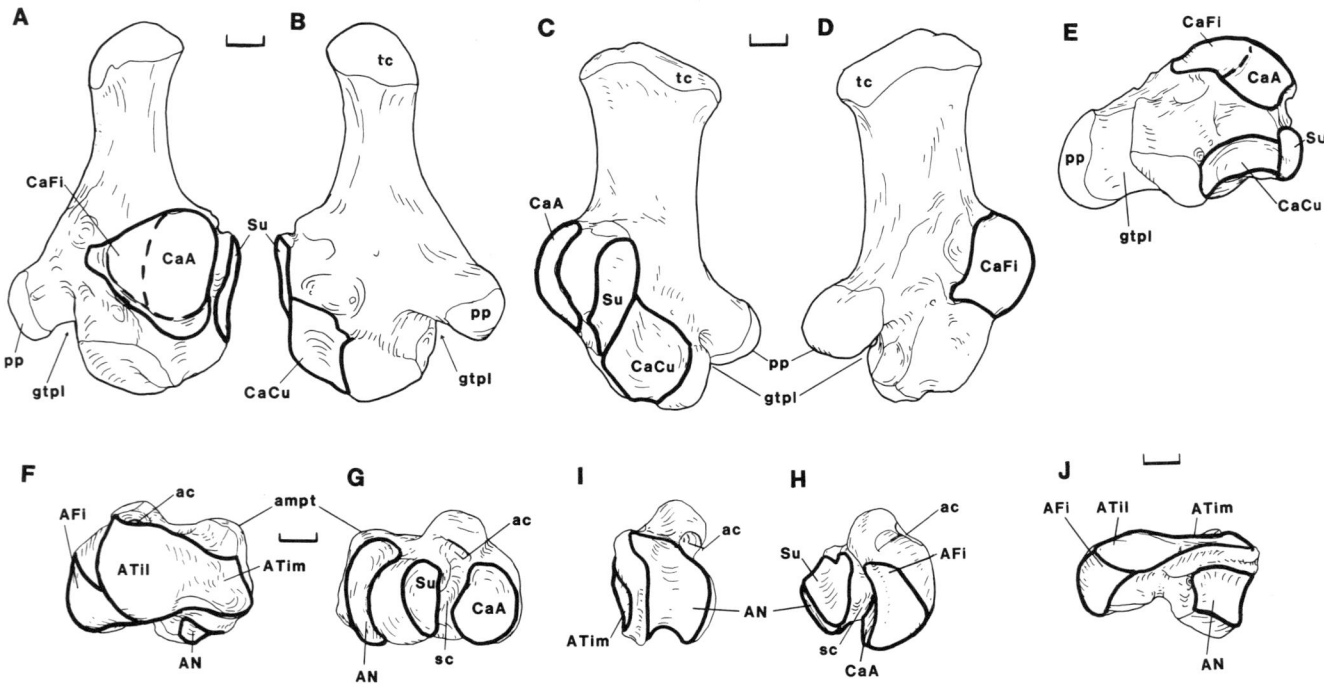

FIGURE 9.8. Ptilodontid, right calcaneus and left astragalus (reversed), Tyrrell Museum of Paleontology collections, Gryde loc., Alberta, late Cretaceous. From left to right: dorsal, plantar, medial, lateral, and distal views. For abbreviations see Table 9.1. Scales represent 1 mm.

Chapter 9. Pedal Evolution of Mesozoic Mammals

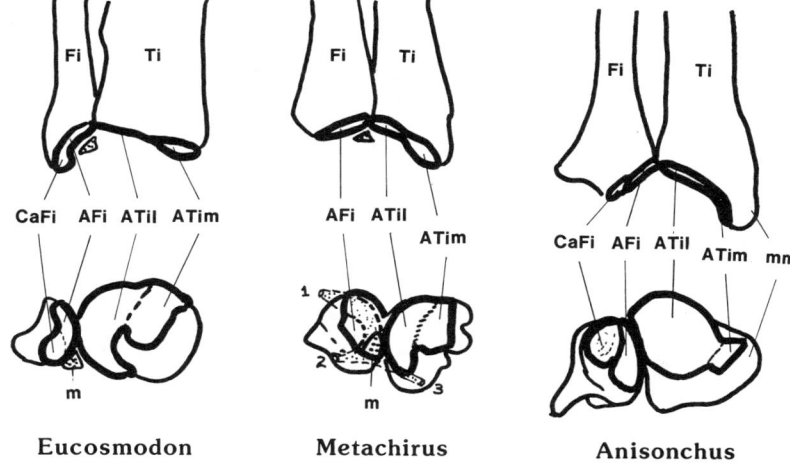

FIGURE 9.9. Comparison of the distal crus in a multituberculate (*Eucosmodon*), an opossum (*Metachirus*), and a Paleocene periptychid ("*Gillisonchus*" = *Anisonchus*). Anterior (above) and distal (below) views. For abbreviations see Table 9.1.

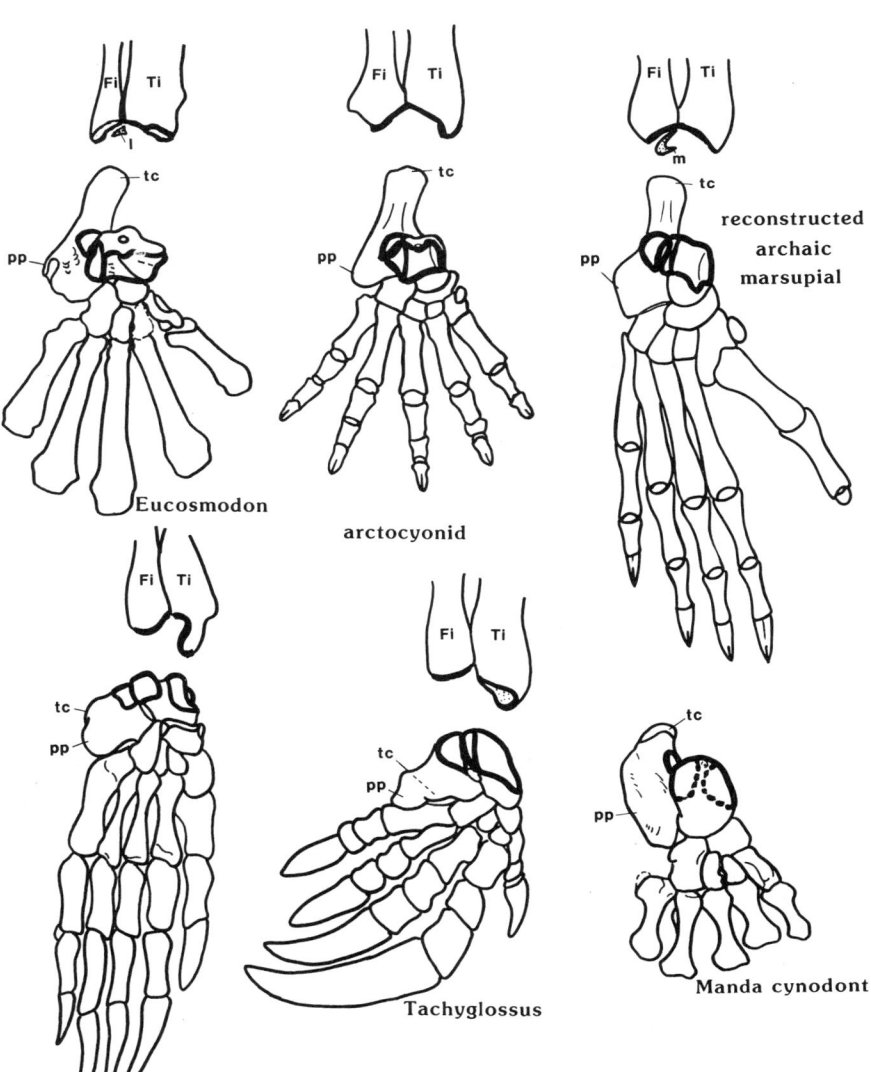

FIGURE 9.10. Comparison of the right distal crus and osseous foot in a multituberculate (*Eucosmodon*), an early Paleocene eutherian (a composite of various arctocyonids), a hypothetical primitive marsupial (based on Paleogene proximal tarsals and the remaining on living opossums), monotremes, and a Triassic cynodont. Articular areas of the **UAJ** are outlined heavily. For abbreviations see Table 9.1.

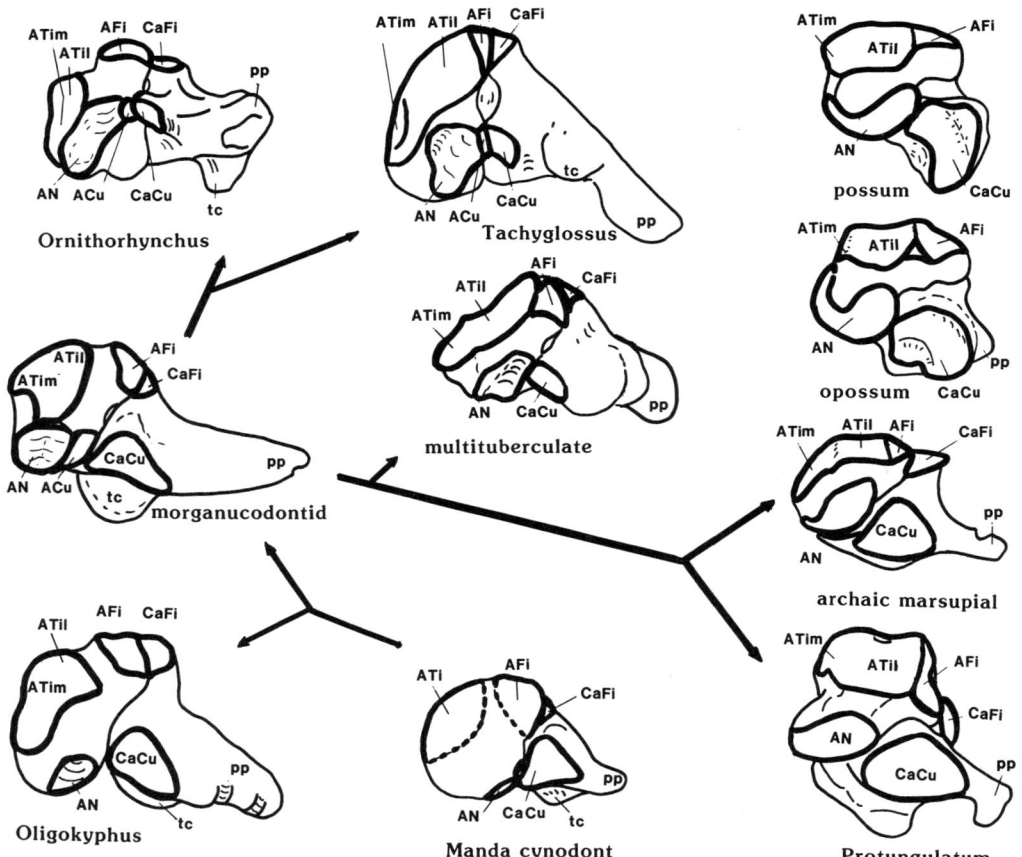

FIGURE 9.11. Summary of some aspects of tarsal transformations for selected Mesozoic, Paleogene, and extant taxa—left astragali and calcanea is distal view—to show **UAJ** and **TTJ** changes. The approximate transformation sequences are shown with arrows. For abbreviations see Table 9.1.

monotremes is the peroneal process and not the tuber. The real tuber of the calcaneus is easy enough to identify in *Ornithorhynchus*, both from the attachment of the tendon of the two-headed gastrocnemius and from the position of the process on cleaned bones. In echidnas, however, an extreme transformation has taken place. The phyletic depression of the tuber flexad resulted in its migration to become a tuberosity prominent on the edge of the extensad surface of the pes. In incomplete muscle preparations (e.g., Lewis, 1963, fig. 1) it appears that the tendon attaches laterally to the most lateral bony projection, misidentified as the tuber by Lewis (1983). In fact it loops to the flexad edge of the actual tuber, identified as the peroneal process by Lewis (1983, 1989). Even from Lewis's (1963) figure, it is evident that the tendon does not extend to the laterally and distally visible peroneal process, which is parallel with the fifth ray. Having made the (wrong) assessment of homology for the tuber on echidnas, Lewis has transferred this assumed topology to the platypus.

The necessary explanation for such seemingly outlandish morphological change, which I suggest for echidnas, resides in the fact that the arc on the medial side of the astragalus, and the ability of the tibia to travel through it, is nearly 100 degrees. To implement such rotation on a transverse axis of the **UAJ**, the mechanical advantage is greatly increased by the flexad and distal migration of the insertion of the tendon of the gastrocnemius. This unique increase in the **UAJ** arc is probably related to the original digging or shoveling adaptation of the monotreme ancestry, carried to potentially confusing extremes in echidna morphology. Concomitant with the extreme rotary ability of the foot in echidnas, on a transverse axis, is the additional supplemental motion along the same axis in the greatly stabilized **LAJ** (actually a sagittally oriented joint primitively and in monotremes). The premium on mobility as well as stability manifests itself in highly conical, pivot-like calcaneal **CaA** and astragalar **Su** facets, each of which moves in the conarticular female facets. This supplies additional flexion to the foot on the same axis as the **UAJ**. A functional and adaptational perspective is clearly indispensable in testing this homology hypothesis.

There is an occasional atavistic remnant of the astragalar canal in the nonarticular groove between the **ATil** and **AFi** facets on some astagali of the platypus, but not in echidnas. The latter have **ATil** and **AFi** facets that are closely abutted against one another, as in therians—but this is almost certainly a convergent attainment in the two groups.

The joint between the astragalus and navicular of

monotremes, particularly in the platypus is of particular significance in its great similarity in some important attributes to non-therians. No living monotremes have a neck-head distal configuration, not even one similar to *Oligokyphus* or morganucodontids. Furthermore, the **ANJ** of the platypus has an unmistakable combination of a concave lateral channel on the astragalus and a convex medial component that may be considered a "head." What is significant is that this joint is sellar, unlike the therian concave (navicular) and convex (astragalar) combination. Echidnas also retain a smaller concave astragalar component.

The presence of prehallux and a "tibiale" are well documented (Emery, 1901; see also *Ornithorhynchus* in Fig. 9.2). The outline of the bony plate (Os calcis; **Oc**) to which the horny perforated spur is attached, via a syndesmosis (Lewis, 1963), is outlined on the flexad view of the platypus's foot (Fig. 9.2).

Lewis misidentified some critical aspects of the tarsus because he did not take into account factors evident from functional and ecological morphology. The platypus appears to make little or no use of its hind legs while swimming (Howell, 1937; personal observations), and therefore most of its functionally significant pedal features may be related to its terrestrial, partly burrowing, activities. It does not, however, depend on digging and rapid "sinking" into the earth for its livelihood and safety, as echidnas probably do. The misidentification of the critical tarsal homologies led to a general inability to differentiate between the diagnostic monotreme apomorphy and shared primitive mammalian traits in the ancestor of the Monotremata, compared to the unique attributes of either the echidnas or the platypus.

MULTITUBERCULATA. The accounts of Granger and Simpson (1929) and Simpson and Elftman (1928) laid the firm foundations for subsequent understanding of multituberculate postcranial morphology, functions, and bioroles. Following McKenna's (1961) and Kielan-Jaworowska's (1979, 1989) papers, Krause and Jenkins (1983) and Jenkins and Krause (1983) have described and illustrated much new evidence and made the most significant contribution to multituberculate postcranial form and function to date. Their ecological account of multituberculate adaptations for arboreality, at least in Ptilodontoidea, appears to be exceptionally strong.

In addition to the reanalysis and comparison of the Eocene cruropedal specimen of *Eucosmodon* reported by Granger and Simpson, and illustrated in detail by Krause and Jenkins (1983), I also evaluate the Late Cretaceous proximal tarsal specimens of a ptilodontoid (Fig. 9.8) from the Gryde Locality of the Frenchman Formation, Alberta (Storer, 1990). My reconstruction of articular relationships in the tarsus is shown in Fig. 9.11, with the rays of the foot somewhat artificially spread. The mechanics, as noted, have been clearly described by Krause and Jenkins (1983) and Jenkins and Krause (1983). The combined evidence shows tarsal material to be relatively diagnostic of this diverse order.

The distal crus of *Eucosmodon* described by Granger and Simpson (1929) essentially mirrors the **UAJ** configuration (Fig. 9.9). I believe, judging from the figures of Granger and Simpson (1929, fig. 22C), that the correction by Krause and Jenkins (1983, p.224) that the former illustrated a left rather than a right fibula is wrong. The articulation of the crus, as shown in Fig. 9.9, may support a calcaneofibular contact. As in the multituberculate astragali seen by me, the medial articular condyle (i.e., the **ATim** facet, not a medial malleolus) of the tibia is a surface largely in line, without forming an angle, with the **ATil** facet. Although in the Eocene *Eucosmodon* the two tibial facets are demarcated by a sharp diagonal crest, in the Cretaceous material shown here this crest is only a faint feature of the **UAJ**. Both this lack of medial buttressing in multis and the acute angle that the astragalar **AFi** facet forms with the tibial articulation are diagnostic of the multituberculate **UAJ**. Added to this is the observation that the **AFi** facet is contiguous with the **ATil** facet only distally. This suggests to me that the wedge-shaped nonarticular area on the astragalus is the equivalent of the primitive synapsid separation of the crural articulations. Krause and Jenkins (1983) were understandably silent on the role of fibular contact with the calcaneus. They were perhaps perplexed by the deep lateral buttressing of the astragalus by a fibular malleolus, which articulates with the calcaneus. In my view, calcaneofibular contact was probably present, as the faceting of the calcaneus and the configuration of the distal fibula appear to indicate, even given a bony lunula between the fibula and astragalus posteriorly.

The unique and arched buttress around the astraglar canal (**ac**) of multis requires an eventual explanation—it is highly diagnostic. The articulation of the astragalus with the navicular is equally diagnostic, as it is an extreme and decidedly unique modification of the primitive mammalian trait of a convex, or sellar, navicular **AN** facet. This **AN** facet in multis is developed as a large, curved, and concave sulcus on the astragalus and a convex male equivalent on the navicular. There is no resemblence to the well-established therian apomorphy of a rounded, rather convex (in all directions) head. This is so in spite of the eutherian-slanted semantics that the marsupials allegedly have no astragalar head.

The relatively large peroneal process of the calcaneus of the Cretaceous sample shown is much reduced compared either to Triassic-Jurassic mammals or to known early therians. The size of the peroneal process in the Eocene *Eucosmodon* is further reduced, but the deep and confining groove for the tendon of the peroneus longus is of great stability-providing importance. The tuber at the end of a long and slender proximal extension of the calcaneus in multituberculates, as in advanced therians, is, as far as such an observation can go, strikingly similar to that of therians. This is a fact, but it does not necessarily represent a homologous taxonomic property of the two groups.

It is interesting that multis, like monotremes, have no calcaneal sustentacular shelf, present in tritylodontids, morganucodontids, and therians. This may well be an independent effect of the modification of the primitive mammalian pattern rather than any significant shared apomorphy between the former two groups.

THERIA. The few known specimens from the Cretaceous that can be unequivocally allocated to marsupials and placentals, respectively, add new insight and confirm expectations based on the inductive and temporally based evaluation of the Paleogene evidence. The major outlines of the eutherian pattern, now badly dated, have been reviewed (Szalay, 1977, 1985; Szalay and Decker, 1974; Szalay and Drawhorn, 1980), and an interim study (Szalay, 1984) has touched on the ancestral therian and marsupial traits of the crus and tarsus. I will briefly present a summary for the most probable therian, metatherian, and eutherian diagnostic traits of the crus and tarsus. The details of the evidence for this are treated in Szalay (1993).

There are no described Cretaceous or even early Paleogene crural remains of marsupials. The posterior extension of the tibial **ATil** facet (Szalay, 1984), however, which is helically crescent shaped in didelphids (and even shows this remnant in such terrestrial marsupials as caenolestids), is quite similar to its homologue in multituberculates, but is distinct from monotremes. This certainly suggests that didelphids may have retained this aspect of a primitive therian and even more ancient mammalian **UAJ** configuration and function. Eutherians, on the other hand, like the most ancient relatives of various "ungulates," appear to have developed a well-restricted **UAJ** (see Fig. 9.11), which has been dubbed a mortise-tenon variety (Szalay, 1984) for its greatly increased transverse constraint and stability and the flexad-extensad mobility. This is quite distinct from the medially restricted, but laterally virtually open (osteologically), monotreme condition. The narrow astragalar **ATil** facets of Campanian Cretaceous and Paleogene marsupials and the well-defined calcaneal **CaFi** facets of both marsupial and early eutherian fossils, in contrast to what we find in didelphids, for example, suggest the retention of a tricontact **UAJ** in the therian, metatherian (unlike in didelphids), and eutherian ancestries (contra Lewis, 1983, 1989). Secondary tricontact **UAJs** have evolved repeatedly in therians, but this does not explain calcaneofibular articulations in the early fossil record of the two groups of therians.

The **LAJ** in the therian common ancestor was unlikely one in which astragalar superposition was achieved, as it is in living didelphids or eutherians. Based on the Campanian marsupial specimens discussed in Szalay (1993), the calcaneal sustentaculum was on a supporting shelf posteriorly (a calcaneal sustentacular shelf), while its more distal extension was nearly as sharply angled with the **CaA** facet as in multituberculates. The relatively more acute angle (on the flexad side of the foot) between the **Su** facet and the peroneal process suggests that early marsupials had a less "superimposed" astragalus than the earliest known tarsals of eutherians.

The question of astragalocuboid articulation is a problematic one. In transversely more compressed tarsal complexes, as in those of rapid terrestrial scampering, "cursorial" forms (in a broad sense this can be applied to early eutherians), contact between the astragalus and cuboid is to be expected. More widely spread tarsals of arboreal climbers and graspers, however, would have these bones separated from one another. Well-established pedal transformations clearly support this, and some marsupial groups supply excellent examples for this functional-adaptive account of morphology (Szalay, 1993).

Contrary to statements in the general literature, the rounded, convex articulation of the head of the astragalus is a therian but not a mammalian character. As noted previously, the neck is less discernible in metatherians than in even the earliest eutherians. The shape and mechanics of the **ANJ**, therefore, together, is a highly informative character for evaluating pedal evolution. There is, of course, the danger of circularity in attempting to identify astragali as eutherian based on the criterion of a well-differentiated neck alone.

Character Transformations

Jenkins's (1971) monograph on the postcranial skeleton of African cynodonts discusses in great detail some aspects of the functional and adaptive evolution of the cruropedal complex from synapsid beginnings into modern mammals. Kemp (1978), in a thought-provoking functional-adaptive analysis dealing with the skeleton of a therocephalian (*Regisaurus*), used crocodilian analogy and made some implicit suggestions concerning cynodont-mammal pedal evolution. Issues raised in these contributions cannot be discussed here. A more detailed account of the evolution of mammalian feet will be pursued elsewhere.

The facts of the fossil record and analyses of function and ecological morphology of the hand and foot in the living form the boundaries within which significant characters should be proposed and the taxonomic properties of the respective taxa assessed. My restricted aim here is to pursue the phylogeny of the known Mesozoic tarsal complex through a large number of available groups.

The methodology of vertical comparisons in the boundary context of fossils, ecological accounts, and functional analyses is essentially similar to that employed somewhat less explicitly by Jenkins (1971). The specific modifications of the characters chosen, and their tested transformational hypotheses (and of all other such studies), are the accounts against which the phylogenetic hypotheses dealing with the taxa themselves must be tested. This procedure is quite distinct from both cladistic analysis, or what Lucas (1993) dubbed "stratophenetic transformationism." A functional and adaptive perspective in historical-narrative explanation (H-N E) is the underlying framework of this phylogenetic character analysis, which considers the known morphocline.

CALCANEAL TUBER. There is little doubt that the tuber is a cynodont acquisition. One of its most extreme modifications is seen in the shortening in monotremes (specifically echidnas), and its other extreme, elongation, is documented for the therians and multituberculates. The usual dual questions remain, however: Did this occur independently in the two latter groups and, if not, what other postcranially unknown mammals shared the extreme expression of this character? I suspect that calcaneal tuber elongation is an independently attained feature. The well-defined tuber, unlike this condition in echidnas, is an ancient synapsid attribute. It is, I believe, the drastic reduction of the peroneal shelf into a peroneal process that results in the slender shaft of the tuber (the "elongate, square tuber calcis" as an unaccepted diagnostic character for "Theriiformes" of Rowe, 1988). It may well be that this **effect** of peroneal shelf reduction has arisen independently several times. The large remnant of the shelf, hypothesized to be morphotypic in the Theria, is reduced independently in several lineages of both metatherians and eutherians. If that is the case, then this trait alone is a poor diagnostic feature for a taxon. As far as the shape of the cross section of the "shaft" and the end of the tuber are concerned, these are largely determined by mechanical loadings, not only by the tendon of the gastrocnemius but by the importance of plantar flexors such as a flexor digitorum brevis (in therians). This, or an equivalent muscle, is probably responsible for the great depth of the cross section of the "shaft" and the flexad extension of the calcaneal tuber (a plantar process on the tuber) in the multituberculate illustrated here (Fig. 9.8).

The persistence of the very large peroneal processes of monotremes strongly suggests that their reduction of the calcaneal tuber originated from a stage not different in many ways from the Triassic mammals, i.e., from an ancestry with a primitively short tuber and huge peroneal process, and not from a stage of elongated tuber and reduced peroneal process. While the tuber was clearly present in cynodonts and stem mammals, the great expanse of the peroneal process was available for sundry functional solutions. In the marsupial mole *Notoryctes*, with a calcaneus that was already all but devoid of a peroneal process in the protoaustralidelphian, lateral buttressing is accomplished by a hypertrophied lateral process of **Mt5** (see Szalay, this volume, chapter 15). The monotreme ancestor achieved this function, probably for somewhat overlapping bioroles, through the ligamentous binding of a relatively huge peroneal process to the lateral ray of the foot. The predominant hand use in contrast to the nearly complete inactivity of the platypus's hindfoot in swimming may be a consequence of the very solutions that emerged from the conflicting bioroles of swimming and burrowing. This issue needs detailed functional-adaptive investigation studying the living animals.

UAJ (JOINT BETWEEN CRUS AND TARSUS). The evidence is unequivocal that a tricontact cruropedal joint (i.e., contact between fibula, astragalus, and calcaneus; Szalay, 1984, 1993) is primitive for monotremes and all mammal groups that make their appearance in the Mesozoic. In spite of Lewis's (1983, 1989) statements that therians are primitively bicontact (i.e., contact between fibula and astragalus only) in the **UAJ**, the secondary redevelopment of the tricontact condition from fully superimposed bicontact ones can be documented in both metatherian and eutherian lineages (Szalay, 1977, 1993).

LAJ (JOINT BETWEEN ASTRAGALUS AND CALCANEUS) AND ASTRAGALAR SUPERPOSITION. The extreme sagittal orientation of the monotreme **LAJ** may be a consequence of the forces generated by the shortening and the flexad, forward, and lateral migration of the tuber and the lateral twist of the peroneal process. It may have started from a condition of "astragalar lean" seen in the morganucodontids. It is highly unlikely that it represents a pre-cynodont condition.

What has become evident to me in this study is that the evolutionary and directionally dynamic concept of "astragalar superposition" is a useful heuristic, but it has highly specific morphological and functional meaning and a varying reality in different groups of mammals. I do not believe that there was an initial and boundary condition, a single phylogenetic constraint at the base of mammalian lineages that we can call "mammalian superposition." Although it undoubtedly began among synapsids (see Manda cynodont), it was, I believe, achieved to different degrees. In tritylodontids, morganucodontids, Paleogene fossil marsupials, and even (to a lesser degree) didelphids, the presence of a relatively enormous astragalar medial plantar tuberosity (**ampt**) is well established. In the didelphids the relatively reduced **ampt** (compared to archaic marsupials) occurs together with the equally robust, in line, navicular tuberosity distal to the astragalar protuberance. This is, in my view, a reflection of substantial direct astragalar support by the substrate, even in such pedally advanced australidelphian marsupials as the opossums. The tuberosities, while they are also the bases of ligamentous binding, probably also reflect the load resistance by these bones at those points. These conditions, together with the helically oriented **Su** articulation (which is distally nearly sagittal), strongly support astragalar contact with the substrate at least in the metatherian common ancestor. This, for example, could not be considered as a degree of astragalar superposition equivalent to that seen in most eutherians.

As noted above, the multituberculate astragalus extends laterally almost horizontally onto the calcaneus, while the sustentacular contact is nearly 90 degrees to the **CaA** articulation along its length. The facts that the **Su** and **CaA** facets form close to a 90-degree angle with one another, and that there is no noticeable shelf for the sustentaculum on the calcaneus, could mean that astragalar superposition in multituberculates was initiated by the lateral part of the astragalar facets. Conversely, the multituberculates may well have reduced the calcaneal sustentacular shelf as noted above. Such an initially small (or secondarily reduced) condition also may have been the source for the similarly small sustentacular shelf discerned in the earliest known calcanea of marsupials. There is suggestion from archaic metatherians that the development of the shelf was initiated at the proximal (posterior) extremity of the sustentaculum. The distal segment is still nearly sagittal in orientation (at least in relation to the **CaA** facets) in these therians.

ACJ (JOINT BETWEEN ASTRAGALUS AND CUBOID). The phylogenetic significance of the morphological attributes of this joint is, I believe, limited. If this articular contact is present, then it transmits loads of a relatively transversely constricted tarsus. Unlike much of the early literature that was concerned with the alternating versus parallel tarsal arrangements, my view is that the presence or absence of this articulation, at least in the manner of minor differences seen in archaic Mesozoic and Paleogene forms, has more functional than significant phylogenetic relevance. Loading-related evolution of this contact can be documented in some groups of the diprotodontians (Szalay, 1993).

ANJ (JOINT BETWEEN ASTRAGALUS AND NAVICULAR). The clearly differentiated condition of therians, and the additional differences within the Theria, coupled with the distinctive multituberculate state, render this character very useful. The fossil evidence indicates that the sulcal (concave) astragalar "head" of cynodonts and some Mesozoic mammals is primitive, and that this condition is further, and extremely, accentuated in multituberculates and retained in a modified form in monotremes. There is no doubt about a "neck," a narrowed distal portion of the astragalus, in tritylodontids and morganucodontids. The convex distal articulation of the astragalus, the "head," is a development in therians, independent from the early appearance of the "neck." Multituberculates probably lost this pre-mammalian trait as they expanded **ANJ** articulation transversely. The "neckless" appearance of the astragali of didelphids is a reflection of the large and medially distributed articular surface of this arboreal radiation. The powerfully grasping hallux generates great loads medially in this joint. Some Paleogene lineages of marsupials either retain or, more likely, develop independently (and secondarily) narrow astragalar distal halves. Such a loading of the **ANJ** is probably indicative of nongrasping locomotion.

CCJ (JOINT BETWEEN CALCANEUS AND CUBOID). The nature of this articulation can be one of the most significant mirrors of pedal and tarsal movements in many groups of mammals. The movements and force transmission from calcaneus to cuboid tend to reflect flexion-extension, resistance to habitual abduction, and ability to rotate (pronate-supinate) the forefoot on this joint. Although the stem mammalian and therian **CCJ** condition suggests forefoot rotational abilities, the limited knowledge of multituberculates indicates great flexion-extension and abduction-adduction at the **CCJ** and **ANJ**, but shows a derived calcaneal restriction of forefoot rotation.

TRANSVERSE TARSAL ARCH. Jenkins (1971) has discussed the presence of this character in the Manda cynodont. The wedging of the cuboid and the cuneiforms is confirmed by me, and a similar condition is present in monotremes. It is almost certain that this predates the last common ancestor of morganucodontids and other mammals. Therefore it should not be regarded as a mammalian feature.

EMT1J (JOINT BETWEEN ENTOCUNEIFORM AND FIRST METATARSAL). The articular surfaces of this joint can be highly diagnostic and their detailed transformation can be followed in metatherians and some grasping eutherians (Szalay and Dagosto, 1988; Szalay, 1993). The diversity in morphology is clearly constrained and channeled by specific initial conditions in phylogeny, and therefore this is a potentially extremely useful character among some therians. Due to the distinctive articular facets between the entocuneiform and first metatarsal of modern didelphids and various groups of both fossil and extant eutherians (Szalay and Dagosto, 1988), little can be said of the diagnostic (i.e., primitive) therian condition. No fossil entocuneiforms or first metatarsals are known to me of the Cretaceous or early Paleogene marsupials. While therian graspers show distinct convergent solutions (see Szalay and Dagosto, 1988), the multituberculates have a textbook example of a sellar articulation (Granger and Simpson, 1929). It does not resemble that of the most primitive living marsupial graspers, the didelphids. This potentially important area, therefore, offers very little evidence for a particular transformation hypothesis for Mesozoic mammals. Ability to slightly abduct-adduct and rotate **Mt1** on the entocuneiform is not, by itself, adequate evidence for or against obligate arboreality or terrestriality.

OTHER TARSO-METATARSAL ARTICULATIONS. Little systematic work has been done in these areas in either therians or other mammals.

Pedal Evolution and Mesozoic Mammal Phylogeny

PEDAL EVOLUTION. This is a very brief and interim account of this complex topic. The potential number of characters that may be noted when describing parts of the anatomy of organisms is great indeed. Nonetheless, from the descriptive, functional, and ecological morphology accounts of the past that have dealt with the comparative biology of the crus and the foot, a number of characters have emerged (see "**Character Transformations**" above) that make far more biological sense than those from the often less comparative and nonfunctional descriptive accounts. It is these characters that form the bases of the various taxonomic properties against which one can potentially test hypotheses of the phylogeny of the taxonomic groups.

As established by Jenkins (1971), the pedal attributes of the Manda cynodont are "mammalian" in a retrospective, albeit not a formal and diagnostically taxonomic, sense. Not only because of its antiquity, but also because of the structural and functional attributes, the morganucodontid tarsal material is in every way potentially antecedent to that of other mammals. The tritylodontid *Oligokyphus*, however, is extremely similar to the former, but not to therians or others, as has been stated in the literature. It follows that many "mammalian" pedal attributes predate the traditional and taxonomically justifiable boundary between Mammalia and other synapsids. The calcaneus of *Gobiconodon* is quite similar to (and therefore "derivable" from) a morganucodontid one—a rather insignificant point.

There is an important question to ask at this time: when did plantigrady first evolve in mammals and what characters indicate it? Jenkins is very convincing: why " . . . the pelycosaur calcaneum and astragalus probably did not have a plantar contact during the initial phase of propulsion, but remained off the ground throughout the entire stride . . . " (1971, p. 197). The early appearance of the calcaneal tuber is, therefore, unquestionably correlated with the beginnings of mobility in the **UAJ**. The subsequent arguments, however, that the increasingly mobile **UAJ** could be stabilized only by a previous or synchronous evolution of plantigrady rest on the lack of evidence for stabilizing malleoli in cynodont crural remains. Is it likely that a gradual shift away from body support on the digits or the distal ends of the proximal tarsus was followed by a plantigrade stance? The problem, corollary to the cynodont one, is that of the ancestral therian plantigrady in the strict sense, meaning that the stride involves full contact with the skin under the calcaneal tuber. There are as yet no studies on living forms to probe and corroborate this likely primitive therian condition.

The pedal morphology of monotremes discussed above suggests derivation from a stage in which astragalar superposition was not achieved, the tibial and fibular facets of the **UAJ** were not contiguous, and the **ANJ** was sellar. The medial buttressing of the **UAJ** by the tibial malleolus is probably a hypertrophied version of a well-differentiated ancestral process. The level of this stage of evolution, or the level at which these may be considered shared attributes of monotremes with other mammalian groups, is that of the morganucodontids, or some postcranially unknown derivatives such as docodonts or pantotheres. Some of these attributes, however, are present in groups traditionally excluded from the Mammalia, such as *Oligokyphus*. From the retrospective of the Recent and extra-mammalian considerations, these are then primitive traits, the retained (and modified) apomorphies of some cynodont lineage.

The other attributes are the pedal differences of monotremes from other known mammals, and they probably represent independent evolution. None of these latter traits are shared with therians. The transformation of the tuber and the peroneal process, and the ligamentous attachment of the latter to the proximal **Mt5** (see **CaMt5l** in Fig. 9.2) are specific to monotreme ancestry. Lewis's (1983, pp. 30–31) view on " . . . how the transition from a monotreme to a therian foot architecture has, as its essence, the bending outward and backward of the primitively downwardly projecting heel . . . [and that] . . . a consequence of this realignment is that the massive flexor fibularis tendon now enters the sole undercutting the distal talar facet" is wrong. The tuber of monotremes, in order to increase the mechanical advantage for added flexion of the foot in the **UAJ**, has migrated increasingly flexad, reaching its extreme in echidnas, in which it has come to be located distolaterally on the calcaneus, while the peroneal process (Lewis's "tuber") is twisted and laterally extended into alignment with **Mt5**, to which it is bound by the broad ligament noted above. The platypus is clearly not that derived in its tuber and peroneal process morphology. In light of monotreme calcaneal homologies, Lewis's statement that "Only when this structural grade [i.e., the realignment noted by him above] is reached is it truly accurate to speak of a sustentaculum tali and sustentacular facet; attribution of these terms to the cynodont condition is not, therefore, strictly appropriate" (1983, p. 31) is without meaning. A flexor fibularis tendon undoubtedly passed medial to the tuber, since the first synapsids developed this extension before the Manda cynodont. Pedal attributes, then, do not cladistically link monotremes with any known group of mammals, but do help focus on an anagenetic stage from which the pedal characters were evolved.

The multituberculates, as expected, pose a considerable problem in interpretation. In spite of the seemingly near astragalar superposition on the calcaneus, the combined pattern of **LAJ** and **CCJ**, and **ANJ** articulations (but not the crural part of the **UAJ**; see below), while structurally derivable from a morganucodontid one, are not really similar to the apomorphies attributable to the first therian. It is at the large, "headless," concave, semicircular articulation of the **ANJ** with its extensive sellar astragalar facet (nearly the opposite of the therian condition), and at the large **CCJ**, where abduction-adduction and flexion-extension of the foot were achieved. This is a fundamental difference between multis and therians. Although the long tuber is a shared similarity with therians, its possible independent elongation from a well-established early mammalian tuber renders this a poor character. In contrast, however, the **UAJ** offers some intriguing interpretations. The nearly horizontally oriented flat or sulcal (concave) **ATim** facets in multituberculates are diagnostic of this group—as is the peculiar astragalar canal. There is virtually no arc in this **UAJ**, yet the configuration of the distal crus and the motion possible on its surface appear to be very similar to that of didelphids (and possibly of ancestral therians). The form-function of the **UAJ** of multituberculates and therians, therefore, may well have originated from an exclusively shared ancestry, whatever the level of this ancestry. Perhaps it was the more primitive morganucodontid-like condition. Needless to say, better material of crura of various Mesozoic groups may not corroborate this hypothesis.

The foot structure of the earliest reproductively eutherian mammals may have been identical to that of their reproductively marsupial sisters. It is of course also questionable whether the first dental eutherians were reproductively eutherian as yet. There is probably no hope at present to correctly correlate the disparate systems and identify the ancestral eutherians by the tarsus, or even the remainder of the skeleton.

TAXON PHYLOGENY. As new fossil material, and its significant temporal context, becomes available, and as new functional-adaptive analyses yield well-tested characters, revised taxonomic properties and new groupings of organisms emerge. The living "crown groups" should take their appropriate places in a genuinely phylogenetic perspective of carefully tested historical-narrative explanations, without any artificial special concern for them. But what does the present analysis of pedal remains contribute to the current taxic state of affairs of Mesozoic mammal evolution?

1. The tarsal evidence supports the view that the Tritylodontidae are the sistergroup of the Mammalia (Kemp, 1983; see also Crompton and Luo, this volume).

2. The monotreme taxonomic properties derived from the cruropedal evidence can be reconciled with any of the possibilities noted below, but they do not corroborate an exclusively shared common stem with either therians or multituberculates, as so clearly stated by Simpson in 1929. In fact, either of the latter possibilities are highly unlikely, based not only on cruropedal evidence but other evidence as well (see Miao, this volume). In their analysis, Kielan-Jaworowska et al. (1987) have shown that teeth of the Cretaceous *Steropodon* were unlikely to have evolved from a tribosphenic ancestor, and Archer et al. (this volume, Chapter 7) offer a discussion for close ties with mesungulatid dryolestoids, rather than with peramurids or vincelestids.

3. The general derivation of both the protopaulchoffatiid-like cheek dentition of the first multis (see Hahn, 1969, on Oxfordian, Late Jurassic forms; approximately 155 MYBP) and the much later Cretaceous and Eocene cruro-tarsal evidence is compatible with, although does not necessaily corroborate, their origins from morganucodontid-like conditions. Nevertheless, sharing of some of the noted crus-based **UAJ** similarities as homologies (but not astragalar ones), which may still turn out to be too ancient and widespread to be meaningful, from an early common stem with therians (as defined here) is not contradicted by the pedal evidence. Hahn's recent (1987) study of cranial structure of early multituberculates also suggests this derivation from morganucodontid-like stock. In spite of the suggestion of the ties with various groups, the evidence is entirely unsatisfactory, in my view, for any formal holophyly implied taxonomic grouping of multituberculates and therians.

ACKNOWLEDGMENTS. I extend my gratitude for permission to study specimens in their care to Drs. K. A. Joysey and A. Friday of the University Museum of Zoology, Cambridge University; to Drs. J. Hooker and P. Andrews of the British Museum of Natural History; Dr. John E. Storer of the Saskatchewan Museum of Natural History; Dr. Richard C. Fox of the University of Alberta; and to Drs. Guy G. Musser, Michael J. Novacek, and Richard Tedford of the American Museum of Natural History.

I especially thank Dr. Richard L. Cifelli for his careful reading of the manuscript and for his useful suggestions. Part of the costs were paid for by CUNY PSC Grants 666141 and 66731.

CORRESPONDENCE ADDRESS. Frederick S. Szalay, Departments of Anthropology and Ecology and Evolutionary Biology, Hunter College, 695 Park Avenue, New York, NY 10021, USA.

References

Archer, M., Flannery. T.F. Ritchie, A., and Molnar, R.E. 1985. First Mesozoic mammal from Australia—An early Cretaceous monotreme. *Nature* 318:363–366.
Bock, W.J. 1981. Functional-adaptive analysis in evolutionary classification. *Amer. Zool.* 21:5–20.
Bock, W.J. 1990. From Biologische Anatomie to Ecomorphology. *Nether. J. Zool.* 40:254–277.
Cartmill, M. 1981. Hypotheses testing and phylogenetic reconstruction, *Z. Zool. Syst. Evolut. Forsch.* 19:73–96.
Cracraft, J. 1981. The use of functional and adaptive criteria in phylogenetic systematics. *Amer. Zool.* 21:21–36.
Eldredge, N., Cracraft, J. 1980. *Phylogenetic patterns and the evolutionary process. Method and theory in comparative biology.* New York: Columbia University Press.
Emery, C. 1897. Beitrage zur Entwickelungsgeschichte und morphologie des Hand und Fusskelets der marsupialier. In: *Des Ganzen Werkes Leiferung 9: Zoologische Forschungensreisen in Australien* (Semon, R., ed.), pp. 663–676.
Emery, C. 1901. Hand- und Fusskelet von *Echidna hystrix*. *Semons Zoologische Forschungsreisen in Australien und dem Malayischen Archipel* 3:663–76.
Granger, W., Simpson, G.G. 1929. A revision of the Tertiary Multituberculata. *Bull. Amer. Mus. Nat. Hist.* 56:601–676.
Gregory, W.K. 1910. The orders of mammals. *Bull. Amer. Mus. Nat. Hist.* 27:3–524.
Gregory, W.K. 1947. The monotremes and the palimpsest theory. *Bull. Amer. Mus. Nat. Hist.* 88(1):1–52.
Gregory, W.K. 1951. *Evolution Emerging: A survey of changing patterns from primeval life to man.* New York: Macmillan.
Hahn, G. 1969. Beitrage zur Fauna der Grube Guimarota Nr. 3. Die Multituberculata. *Palaeontographica* 133(A):1–100.
Hahn, G. 1987. Neue Beobachtungen zum Schadel- und Gebiss-bau der Paulchoffatiidae (Multituberculata, Ober-Jura). *Palaeovertebrata* 17(4):155–196.
Howell, A.B. 1937. The swimming mechanism of the platypus. *J. Mammal.* 18(2):217–222.
Jenkins, F.A. Jr. 1970a. Cynodont postcranial anatomy and the "prototherian" level of mammalian organization. *Evolution* 24:230–252.
Jenkins, F.A. Jr. 1970b. Limb movements in a monotreme (*Tachyglossus aculeatus*): A cineradiographic analysis. *Science* 168:1473–1475.
Jenkins, F.A. Jr. 1971. The postcranial skeleton of African cynodonts. *Bull. Peabody Mus. Nat. Hist.* 36:1–216.
Jenkins. F.A. Jr., and Krause, D.W. 1983. Adaptations for climbing in North American multituberculates (Mammalia). *Science* 220(4598):712–715.
Jenkins, F.A. Jr., and McClean D. 1984. Mechanisms of hind-foot reversal in climbing mammals. *J. Morph.* 182:197–219.
Jenkins, F.A. Jr., and Parrington. F.R. 1976. The postcranial skeletons of the Triassic mammals *Eozostrodon, Megazostrodon* and *Erythrotherium*. *Phil. Trans. of the Royal Soc. of London B. Biol. Sci.* 273(926):387–431.
Jenkins. F.A. Jr., and Schaff, C.R. 1988. The early Cretaceous mammal *Gobiconodon* (Mammalia, Triconodonta) from the Cloverly Formation in Montana. *J. Vert. Paleo.* 8:1–24.
Kemp, T.S. 1983. The relationships of mammals. *Zool. J. Linn. Soc* 77:353–384.
Kemp. T.S. 1987. Stance and gait in the hindlimb of a therocephalian mammal-like reptile. *J. Zool London.* 186:143–161.
Kielan-Jaworowska, Z. 1979. Pelvic structure and nature of reproduction in Multituberculata. *Nature* 277:402–403.
Kielan-Jaworowska, Z. 1989. Postcranial skeleton of a Cretaceous multituberculate mammal. *Acta Palaeont. Polonica* 34:75–85.
Kielan-Jaworowska, Z., Crompton, A.W., and Jenkins, F.A. Jr. 1987. The origin of egg-laying mammals. *Nature* 326:871–873.
Krause. D.W., and Jenkins, F.A. Jr. 1983. The postcanial skeleton of North American multituberculates. *Bull. Mus. Comp. Zool.* 150:199–246.
Kuhne, W.G. 1956. *The liassic therapsid Oligokyphys* (Brit. Mus. Nat. Hist.). London: Adlard and Son.
Lewis, O.J. 1963. The monotreme cruro-pedal flexor musculature. *J. Anat. London* 97:55–63.
Lewis, O.J. 1964. The homologies of mammalian tarsal bones. *J. Anat. London* 98:195–208.
Lewis, O.J. 1983. The evolutionary emergence and refinement of the mammalian pattern of foot architechture. *J. Anat. London* 137:21–45.
Lewis, O.J. 1989. *Functional morphology of the evolving hand and foot.* Oxford: Oxford University Press.
Lucas, S.G. 1993. Pantodonts, Tillodonts, Uintatheres, and Pyrotheres are not ungulates. In: *Mammal phylogeny: Placentals* (Szalay, F.S., Novacek, M.J., and Mckenna, M.C., eds.). New York: Springer-Verlag.
McKenna, M.C. 1961. On the shoulder girdle of the mammalian subclass Allotheria. *Amer. Mus. Novitates* 2066:1–27.
Parker, T.J., and Haswell, W.A. 1897. *A test-book of zoology, vol. 2.* New York: Macmillan Co.
Rowe, T. 1988, Definition, diagnosis, and origin of Mammalia. *J. Vert. Paleont.* 8:241–264.
Schaeffer, B. 1941. The morphological and functional evolution of the tarsus in amphibians and reptiles. *Bull. Amer. Mus. Nat. Hist.* 78:395–472.

Simpson, G.G. 1929. The dentition of *Omithorhynchus* as evidence of its affinities. *Amer. Mus. Novitates* 390:1–15.

Simpson, G.G. and Elftman, H.O. 1928. Hind limb musculature and habits of a Paleocene multituberculate. *Amer. Mus. Novitates* 333:1–19.

Sober, E. 1989. *Reconstructing the past: Parsimony, evolution, and inference.* Cambridge, MA: MIT Press.

Szalay, F.S. 1977. Phylogenetic relationships and a classification of the eutherian Mammalia. In: *Major patterns in vertebrate evolution: Macroevolutionary trends and their implications in vertebrate phylogeny* (Hecht, M.K., Goody, P.C., and Hecht, B., eds.). New York: Plenum Press, pp. 315–374.

Szalay, F.S. 1981. Functional analysis and the practice of the phylogenetic method as reflected by some mammalian studies. *Amer. Zool.* 21:37–45.

Szalay, F.S. 1984. Arboreality: Is it homologous in metatherian and eutherian mammals? In: *Evolutionary biology, vol. 18* (Hecht, M.K., Wallace, B., and G.T. Prance, eds.). New York: Plenum Publishing Corporation, pp. 215–258.

Szalay, F.S. 1985. Rodent and lagomorph morphotype adaptations, origins, and relationships: Some postcranial attributes analyzed. In: *Evolutionary relationships among rodents—A multidisciplinary analysis* (Luckett, W.P., and Hartenberger, J.L., eds.). *NATO ASI series A, Life Sciences* 92:83–157. New York: Plenum Press.

Szalay, F.S. 1993. *Evolutionary history of the marsupials and an analysis of osteological characters.* New York: Cambridge University Press.

Szalay, F.S., and Bock, W.J. 1991. Evolutionary theory and systematics: Relationships between process and pattern. *Z. Zool. Syst. Evolut.-Forsch.* 29:1–39.

Szalay, F.S., and Dagosto, M. 1988. Evolution of hallucial grasping in the primates. *J. Human Evol.* 17:1–33.

Szalay, F.S., and Decker, R.L. 1974. Origins, evolution, and function of the tarsus in late Cretaceous eutherians and Paleocene primates. In: *Primate locomotion*, F.A. Jenkins, Jr., ed.) New York: Academic Press, pp. 223–254.

Szalay, F.S., and Drawhorn, G. 1980. Evolution and diversification of the Archonta in an arboreal milieu. In: *Comparative biology and evolutionary relationships of tree shrews* (Luckett, W.P., ed.). New York: Plenum Press, pp. 133–169.

Szalay, F.S., Rosenberger, A.L., and Dagosto, M. 1987. Diagnosis and differentiation of the order primates. *Yearbook of Physical Anthropology* 30:75–105.

CHAPTER 10

Phylogenetic Systematics and the Early History of Mammals

TIMOTHY ROWE

Overview

Numerous recent authors have used phylogenetic systematics to study mammalian evolution. As a result, there have been many fundamental changes in our view of early mammalian history compared with the view of a decade ago. However, even phylogenetic analyses have produced conflicting interpretations of this history. On closer inspection, many of the conflicts may simply reflect the different samples of taxa and characters that have been brought to bear on this issue. In a series of computer parsimony analyses, different rates of evolution in the dentition, skull, and postcranium were responsible for different tree topologies that resulted when different, restricted character samples were analyzed. When sampling artifact is removed and all available character data analyzed, a highly corroborated, stable phylogeny remains, which is largely consistent with the temporal distributions of taxa recorded in the fossil record. Several patterns dominate this phylogeny. Most transformations in the head involved elaborate repackaging of an expanded brain and special sense organs, remodeling of the masticatory system, and accelerated evolution of a highly complex dentition. Postcranial evolution involved differentiation of the vertebral column and remodeling of the limbs and girdles, associated with parasagittal gait. Another pattern involved evolutionary miniaturization similar in detail to historical patterns in other miniaturized tetrapod lineages, suggesting the existence of developmental constraints common to all tetrapods. Although some of these patterns have long been recognized, others have become evident only as the effects of rate and sampling in phylogenetic analyses have become more fully understood.

Contents

Introduction, 130
Issues in Early Mammalian History, 130
    Mammalian Monophyly, 130
    Incompleteness, 131
    Rates of Evolution and Missing Data, 131
    Phylogenetic Resolution and Stability, 132
Eucynodont Phylogeny and the Early History of Mammals, 133
    Node 1: Eucynodontia, 133
    Node 2: (Unnamed), 133
    Node 3: (Unnamed), 134
    Node 4: (Unnamed), 134
    Node 5: (Unnamed), 135
    Node 6: Mammaliamorpha, 136
    Node 7: (Unnamed), 137
    Node 8: Mammaliaformes, 137
    Node 9: Mammalia, 138
    Node 10: Theriimorpha, 139
    Node 11: Theriiformes, 140
    Node 12: (Unnamed), 141
    Node 13: Theria, 141
Constraints on the Positions of Unresolved Taxa, 141
Conclusions, 142
References, 143

Introduction

The contemporary phylogenetic perspective on early mammalian history has obvious, deep foundations in earlier research, but it is nevertheless significantly different from the dominant view of a decade ago. Phylogenetic analyses have changed our measures of the most fundamental properties of Mammalia, including its definition, diagnosis, membership, the relationships among its members, and its distribution in time and space. Below, I first review some of the conceptual transformations that have arisen under phylogenetic systematics, and I then devote the bulk of this report to summarizing what now appear to be the most strongly corroborated and stable relationships in early mammalian history.

Identifying which aspects of early mammalian phylogeny are "stable" and which are not is far from a categorical process. The details of cynodont phylogeny have long been the subject of debate, particularly the relationships among the extinct taxa most closely related to mammals and among the basal mammals themselves. Numerous phylogenetic analyses have recently pursued this problem (e.g., Crompton and Sun, 1985; Crompton and Luo, this volume, chapter 4; Gauthier et al., 1988, 1989; Kemp, 1982, 1983; Hopson and Barghusen, 1986; Novacek, 1986, 1989, 1990; Novacek and Wyss, 1986; Rowe, 1986, 1988; Rowe and Simmons, unpublished; Sues, 1985; Wible, 1987, 1990, 1990; Wible et al., 1990; Wible and Hopson, this volume chapter 5). Although general agreement on a number of points has emerged, there remain conflicting views on other aspects of relationship, on diagnoses for groups whose memberships are not disputed, and on interpretations of individual characters.

Resolution of these conflicts is complicated by the fact that published analyses have all used different samples of taxa and characters. Phylogenetic methods have themselves varied considerably among studies, further complicating the issue, but I focus only on the problem of sampling. To better understand sampling effects, several series of computer-assisted parsimony analyses have been designed to measure how different assemblages of taxa and characters affect the global tree topology (Donoghue et al., 1989; Gauthier et al., 1988, 1989; Rowe, 1988; Rowe and Simmons, unpublished). All of these studies, and particularly that by Rowe and Simmons (unpublished), endeavored to bring together a large and, more important, diverse osteological data base that sampled all parts of the skeleton and dentition, in a wide range of extant and extinct taxa. By selectively adding and deleting different subsets of taxa and characters in a large data matrix, these tests helped to identify the points in early mammalian phylogeny that were most consistently resolved and were relatively immune to change as the analytic sample was changed. At the same time, the tests identified taxa whose positions were more sensitive to sampling variations and placed constraints on their range of possible positions.

This approach to measuring phylogenetic pattern has arrived at new assessments of phylogenetic pattern in early mammalian history, while also providing strong corroboration of many points made in older literature. Not surprisingly, the new phylogenetic patterns suggest that previously unrecognized processes may have played key roles in shaping early mammalian history.

Issues in Early Mammalian History

MAMMALIAN MONOPHYLY. Prior to the advent of phylogenetic systematics, the focus of study on early mammals was to elucidate the reptile-to-mammal transition (e.g., Aulie, 1974; Crompton and Jenkins, 1973; Olson, 1959, 1962; Simpson, 1959, 1960, 1961). With a rich sequence of fossils extending from the Carboniferous to the Recent, the origin of mammals was taken as the premier example of an evolutionary transition from one Linnean class to another. Evolutionary grades played a central role in discussion of this event. Influential generalizations about issues ranging from macroevolution to natural selection to convergence were derived from study of this lineage and were later extrapolated to other metazoan lineages. The major debates waged in the literature involved whether Mammalia was polyphyletic, and which character or character assemblage most meaningfully marked a boundary between reptilian and mammalian grades (e.g., Hopson and Crompton, 1969; Olson, 1959, 1962; Reed, 1960; Simpson, 1959, 1960, 1961, 1971; Van Valen, 1960). The influence of environment via natural selection was virtually the only mechanism invoked to describe morphological patterns discovered in this history. It was argued that similar environmental demands led to the convergent evolution of "mammalian" characters in many different lineages. Implicit in many discussions is the thought that convergence was so prevalent that the true genealogy could never be known with any precision.

Phylogenetic systematics has turned our attention from many of these issues. The discovery of monophyletic taxa replaced definition of grades as the central issue in understanding early mammalian history. There has been no doubt for a century that extant mammalian species share a unique common ancestor at some point in history, and that Mammalia is monophyletic in the strict meaning of the term (e.g., Haeckel, 1897; Rowe, 1988). Whereas Simpson (1971, p. 192) and most others dismissed mammalian common ancestry as "trivial" (but see Reed, 1960), phylogenetic systematics has reorganized our analyses with this as a pivotal point (Fig. 10.1). This attitude reflects a shift in our conceptual view of Mammalia, and of taxa generally. Previous views saw taxa as classes defined by characters, while

Chapter 10. Early Mammal Phylogenetic Systematics

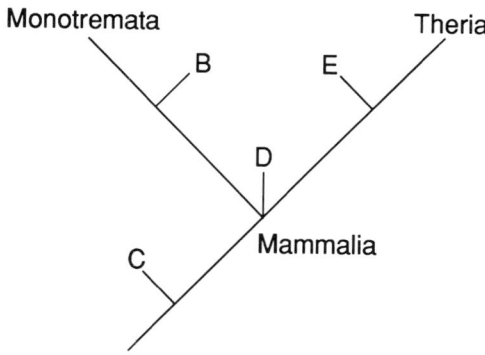

FIGURE 10.1. Potential relationships between fossils and the extant sister lineages whose common ancestor defines Mammalia. Any fossil that is not itself a monotreme or therian can have only four possible positions. It can be closest to therians, closest to monotremes, or lie outside of Mammalia. Taxa in position D are of unresolved position within Mammalia (Rowe, 1988).

contemporary phylogenetics views taxa as individuals defined by common ancestry (e.g., de Queiroz, 1988; de Queiroz and Gauthier, 1990; Gauthier, 1986; Ghiselin, 1984; Hull, 1976; Rowe, 1987, 1988). Current phylogenetic debate focuses on the content and diagnosis of Mammalia, on the relationships among its members and extinct relatives, and on interpreting the new pattern of character hierarchy and genealogy.

INCOMPLETENESS. Incompleteness of the fossil record has been, and continues to be, a major obstacle in interpreting early mammalian history. Most Mesozoic taxa relevant to this history lie within the smallest order of vertebrate size magnitudes (McMahon and Bonner, 1983). Few specimens have survived the rigors of transport, burial, and diagenesis, and their tiny size renders them very difficult to find in the field. Most that have been recovered consist of only the hardest and most resilient parts of the skeleton, and even today most Mesozoic taxa are known only from dentitions. Understandably, this led to a strong analytic bias toward dental data, with the aim of developing a taxonomy that might apply to the majority of taxa. The resulting framework was handicapped, however, in that researchers were largely unable to fit nondental data into it. Another major shortcoming of the older framework is that it treated homoplasy as an inductively recognized phenomenon. When character data from the skull or postcranial skeleton conflicted with the distributions of dental data, the dental characters were necessarily taken as evidence of relationship, while the skeletal characters were generally dismissed as convergent.

Contemporary methods choose among competing phylogenetic hypotheses based on their information content as well as on parsimony (e.g., Kluge, 1989). Consequently, researchers have recently endeavored to bring more diverse character data into their analyses, to test previous views. As a result, several hundred skeletal characters and a large body of new dental evidence have come to bear on early mammalian systematics (e.g., Novacek, 1986, 1989, 1990; Novacek and Wyss, 1986; Rowe, 1988; Rowe and Simmons, MS; Gauthier et al., 1988; Wible, 1987, 1990, 1991). This has had a very positive effect, but the shift to character-rich analyses has come at the cost of taxon richness, as most researchers have opted to omit fragmentary fossils from consideration. Although there is both theoretical and empirical justification for this (Rowe, 1988; Rowe and Simmons, in press), it is also clear that even highly incomplete taxa can preserve data of critical importance to tree topology (Gauthier et al., 1988; Donoghue et al., 1989). These and other results (Rowe and Simmons, in press) point to the sampling methods used to construct a matrix as a significant source of artifact and discrepancy between analyses.

Ironically, as more complete fossils and more encompassing analytic procedures have emerged, new sources of missing data have been discovered inherent in the data themselves. Evolution itself leads to the loss of data through the transformation and divergence of form (e.g., Doyle and Donoghue, 1987; Gauthier et al., 1988; Donoghue et al., 1989). In a real sense, characters can become extinct within a lineage, resulting in the loss of potentially critical historic information. For example, one cannot score the presence of a single versus a divided caniniform tooth root in taxa that have lost the caniniform teeth. Unlike nonpreservation, this type of information loss has a complex temporal component in that its extent is related to both the temporal scope of the problem and the evolutionary rates of the character systems analyzed.

RATES OF EVOLUTION AND MISSING DATA. The degree to which data are lost as a result of divergence can be viewed as a complex function of time. Data loss is potentially most severe in analyses that span broad temporal intervals and seek to reconstruct ancient branching patterns, and which sample small, rapidly evolving sets of characters. It is noteworthy that while nonpreservation is usually a problem only with fossils, divergence can affect both extinct and extant taxa. Moreover, by virtue of having survived to the recent, extant taxa may be especially prone to data lost through this process. The effects of divergence are such that analyses of extant taxa alone can produce mistaken estimates of character polarities, distributions, and tree topologies (Doyle and Donoghue, 1987; Donoghue et al., 1989; Gauthier et al., 1988).

Variation in evolutionary rates of different characters and character systems is the primary cause of "mosaic evolution," which has long been acknowledged as a problem in discerning phylogenetic patterns. However, only recently have efforts been directed at understanding its effects in any particular phylogenetic analysis. Minimizing the adverse effects of mosaic evolution involves discovering data that transform at an average rate appropriate to resolving the problem at hand. Characters or systems evolving too slowly may have recorded nothing of the history of interest, while systems evolving too rapidly may become too transformed to preserve any useful information.

Rates of character evolution cannot be determined *a priori*, but methods are emerging to base measurements on a data matrix and the most parsimonious phylogeny found within it when all data are considered. One technique compares average homoplasy levels for different subsets of characters within a single matrix, using a single tree topology. Higher than average homoplasy levels in one region, for example, the dentition, reflect more rapid transformation rates for that region than for the other regions sampled in the matrix. The homoplasy retention index (Farris, 1989a, 1989b) is a convenient measure for comparing average homoplasy levels of different character subsets (Cloutier, 1991; Rowe and Simmons, unpublished). Rowe and Simmons (unpublished) compared homoplasy levels in the dentition, the skull exclusive of the dentition, and the postcranium. Homplasy was found in all three regions, but it was not distributed equally in each. The skull and postcranium recorded similar levels, whereas the dentition recorded minimally a 30% higher homoplasy level. The enormous diversity of mammalian dental morphology reflects a rapid rate of evolution expressed over a 140-million-year history.

Additional tests showed that, predictably, the phylogenetic resolving power of the dentition was strongest over relatively short segments of the tree. Dental characters by themselves resolved few relationships, and omitting them altogether had little effect on the tree topology found with all data. However, this is not to say that the dentition was uninformative, or even that it was uninformative for older events. Dental characters were diagnostic at most levels on the phylogeny described below, but without the signal provided by the more slowly evolving cranium and postcranium, the distributions of most dental characters remained equivocal (Rowe and Simmons, unpublished).

PHYLOGENETIC RESOLUTION AND STABILITY. Although it may be obvious that no taxon or character is relevant to phylogenetic analyses at all levels, methods to identify the most informative sample of taxa and characters for any given problem have been less than clear. A step in this direction was recently taken by the design of series of tests to analyze some of the effects of sampling in understanding early mammalian history (Rowe and Simmons, unpublished). The tests focused on the relationship between completeness, measured from a data matrix, and phylogenetic resolution, the degree to which taxa are arrayed in a strictly dichotomous branching pattern (Mickevich and Platnick, 1989). The interactive capabilities of PAUP (Swofford, 1989) and Hennig 86 (Farris, 1986) were used to study a matrix of 151 characters for 24 of the major groups of mammals and their closest extinct relatives. By systematically adding and deleting taxa according to their degrees of completeness, it was possible to examine the relationships among some of the different kinds of information recorded in the matrix and to evaluate the relevant contributions of different taxa and character subsets to the final tree topology.

The tests found a general correspondence between taxon completeness and phylogenetic resolution. However, it was also clear that the two are not strictly coupled and that incomplete taxa offered potentially critical information to resolving phylogenetic questions. Nevertheless, ten relatively incomplete taxa remained unresolved in this analysis, in that each could be placed in several equally parsimonious phylogenetic positions and removing them had no effect on relationships among the other taxa. There was also a cumulative effect of incompleteness, which led to an exponential decrease in resolution as more and more incomplete taxa were added to the matrix. Hence, simply including all available taxa may not produce as unambiguously informative results as analyzing a select sample of available taxa. Selection is not an *a priori* process and can be made only with tests such as these.

Despite lability in the positions of some taxa, the tests consistently found a stable topology among fourteen other taxa (Fig. 10.2). This topology remained unchanged when different samples of taxa were added or removed from the analysis. Only when large sets of character data were deleted did resolution diminish.

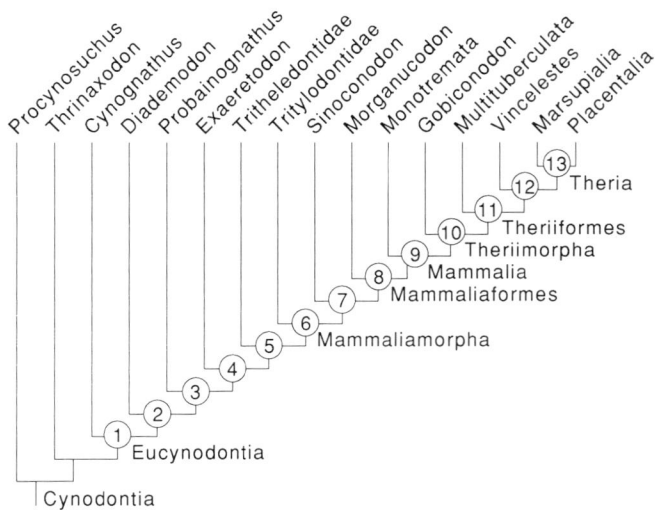

FIGURE 10.2. Eucynodont phylogeny (see text).

This effect was most marked when the dentition was analyzed alone, and at only a few nodes did resolution diminish when either the cranial or postcranial data set was deleted. Even during the rapid drop in resolution that accompanied the cumulative addition of incomplete taxa, the topology in Figure 10.2 was found in all of the equally parsimonious trees. Because it consistently summarizes the largest and most diverse osteological data sample yet analyzed, and includes the taxa most relevant to understanding early mammalian history, I take the phylogeny depicted in Figure 10.2 to be the preferred hypothesis for now. This phylogeny supports many aspects of earlier work, although it also differs from all previously published studies at least in detail.

Eucynodont Phylogeny and the Early History of Mammals

The diagnoses discussed below are based on the recent analysis by Rowe and Simmons (unpublished). Character distributions were measured using the DELTRANS option of PAUP for all character data and the sixteen taxa depicted in Figure 10.2. Readers are referred to Rowe and Simmons (unpublished) for details of methodology and for a complete listing of all character distributions and taxon diagnoses.

Geological dates for the taxa in this phylogeny are from Harland et al. (1990). As noted in that work, the time span over which the phylogenetic events of interest occurred is one of the most poorly constrained of the entire Phanerozoic. Very few radiometric dates have been obtained; only three are known for the entire Jurassic, and linear interpolations provided all of the Jurassic dates cited below. All dates have an error margin of between 10 and 30 million years, roughly 7% to 20% of the total time involved, and considerable future improvement in the numbers listed below can be expected. In addition to this imprecision, superpositional relationships are unresolved for several clusters of taxa. What is known of the temporal distribution of these taxa is consistent with the sequence of branching in this phylogeny. Temporal resolution is sufficiently poor, however, that it would be consistent with a range of other phylogenetic hypotheses as well.

Node 1: Eucynodontia. Eucynodontia offers a convenient place to begin discussion, because it is one of the most stable and widely recognized monophyletic groups within Cynodontia, and there is abundant support for using the cynodonts *Thrinaxodon* (Fourie, 1974; Jenkins, 1971) and *Procynosuchus* (Kemp, 1979, 1980) as consecutively more distant outgroups, to polarize characters transforming among eucynodonts (Gauthier et al., 1988; Hopson and Barghusen, 1986; Kemp, 1982; Rowe, 1986, 1988; Wible, unpublished). Eucynodontia is the taxon stemming from the last common ancestor that mammals share with *Cynognathus*.

Eucynodontia is diagnosed by characters of the skull and dentition. The attachment of eucynodont cheekteeth was transformed from the ancestral cynodont pattern of ankylosis with the jaws (e.g., Crompton, 1963) to a new attachment via greatly elongated tooth roots that are anchored by a periodontal ligament. There was also a major reduction in the rate of cheektooth replacement and the initiation of bilateral occlusion, though only irregular wear facets were produced in early members of this group. Ligamentous attachment, the so-called thecodont gomphosis, offers a degree of mobility of each tooth within its socket that is critical to the development and maintenance of precise, complex occlusal patterns (e.g., Johnson, 1983; Noble, 1969; Ten Cate, 1969). Slowing of tooth replacement was also crucial to maintaining occlusion. These diagnostic apomorphies of Eucynodontia mark the beginning of an acceleration in dental evolutionary rates and the onset of occlusal complexity that is so characteristic of mammals among extant gnathostomes. Additional diagnostic characters of eucynodonts include participation of the surangular in the craniomandibular joint, reduction of the quadrate ramus of the pterygoid, and the absence of incisiform teeth in the maxilla anterior to the caniniform.

The fossil record of eucynodonts begins near the end of the Scythian Stage of the Early Triassic, which extended from 245 ± 10 to 241 ± 10 million years ago (Harland et al., 1990). The oldest eucynodonts are *Cynognathus* and the more derived eucynodont, *Diademodon*, which occur together in *Cynognathus* Zone sediments of the South African Karoo (Anderson and Cruickshank, 1978; Kitching, 1977), and which provide a minimum age for Eucynodontia. Because the phylogeny (Fig. 10.2) indicates that *Diademodon* originated later than *Cynognathus*, it would seem that the earliest parts of eucynodont history occurred in unrecorded, pre-*Cynognathus* Zone times. Future fossil discoveries and increased biostratigraphic resolution are likely to push eucynodont history back toward the beginning of the Early Triassic, although just how far back the lower time limit might extend is more difficult to constrain. *Cynognathus* has not been recovered from rocks younger than Early Triassic, but its sister lineage, represented here at Node 2, has survived to produce more than 4,000 extant species and a diversity of extinct species.

Node 2: (Unnamed). This unnamed taxon comprises the last common ancestor that mammals share with *Diademodon* and all its descendants. *Diademodon* was long classified as a "gomphodont" cynodont, a group united on resemblances in the dentition that have been taken as evidence of a herbivorous diet. The gomphodonts were thought to represent an evolutionary radiation of herbivorous cynodonts that diverged early in the Triassic from a persistently predatory lineage from which mammals ultimately descended (e.g., Crompton, 1972; Hopson, 1969; Hopson and Barghusen, 1986). However, analyses of a broader sample of data, including characters from the skull and postcranium as well as the dentition, argue that the gomphodonts form a paraphyletic assemblage, because some gomphodonts, in particular *Exaeretodon* and tritylodontids, are more closely related to mammals than to *Diademodon* (see below). In this light, it appears that either the diets of cynodonts are more evolutionarily labile than has been believed and herbivory has evolved several times, or previous interpretations of diet in these taxa are mistaken.

Node 2 is diagnosed by additional transformation of the dentition. Members of this group have cheekteeth with consistent, regular sets of wear facets, which indicate consistent occlusal patterns. Some members of this group, such as *Sinoconodon* and a number of extant species, lack occlusion because the teeth are either greatly simplified or are absent altogether. Nevertheless, all of these taxa retain other features derived within this group that identify them as members of Node 2 and that indicate the absence of occlusion and wear facets to be reversals to a pre-eucynodont state, rather than retained plesiomorphies. This node is also diagnosed by contact between the palatal processes of the premaxillae, which meet behind the incisive foramen (later reversed in *Exaeretodon*).

Like Eucynodontia, the first fossil record of Node 2 appears in the early Triassic (Scythian) *Cynognathus* Zone of the South African Karoo. Unlike its sister taxon (Node 3), *Diademodon* did not survive the end of the early Triassic.

NODE 3: (UNNAMED). This unnamed taxon comprises the last common ancestor mammals share with *Probainognathus* and all taxa stemming from it. *Probainognathus* was widely regarded as being close to the direct ancestry of mammals and representing a morphotype from which mammals descended (Allin, 1986; Romer, 1969, 1970; Hopson and Kitching, 1972; Crompton and Jenkins, 1973, 1979). However, other cynodonts now appear to be more closely related to mammals than is *Probainognathus*, notably *Exaeretodon* (Gauthier et al., 1988; Rowe, 1986, 1988) and tritylodontids (Kemp, 1982, 1988a, 1988b; Rowe, 1986, 1988; Wible, 1991; Wible and Hopson, this volume, chapter 5; Gauthier et al., 1988). The latter taxa were considered only distant relatives of mammals because their highly apomorphic dentitions seemed to demand a long expanse of independent evolution (e.g., Crompton, 1972; Hopson, 1969). However, when viewed in light of all available data, the dentitions of *Exaeretodon* and tritylodontids appear to be the result of a rapid rate rather than a distant time of divergence. Although their dentitions offer little unambiguous information on higher-level relationships of these taxa, the more slowly evolving skulls and postcrania bear a host of unequivocal, unique resemblances to mammals that are not found in *Probainognathus* (see below).

Most of the characters diagnosing Node 3 manifest increased rates of development of the bones surrounding the brain. The parietals, which primitively remained paired throughout life and formed the borders of the pineal foramen, became fused, completely closing off the pineal foramen in adults. The pineal body itself is retained in most mammals, suggesting that this transformation reflects more an increase in the rate of parietal growth than any qualitative change in the underlying structure of the brain (Roth et al., 1986). In addition, the prootic and opisthotic bones fuse to form the petrosal, reflecting a similar developmental acceleration. In the postcranium, proximal expansions of the ribs known as costal plates, which extended over both the thoracic and lumbar vertebrae in eucynodonts ancestrally (Jenkins, 1971), were lost from the thoracics, and only the short lumbar ribs retained any vestige of these structures (Romer, 1970).

The oldest member of Node 3 is *Probainognathus*. It is known from the Chanares Formation of Argentina (Romer, 1970), which was deposited during the Ladinian Stage of the Middle Triassic, roughly between 239 ± 8 and 235 ± 5 million years ago (Harland et al., 1990). *Probainognathus* did not survive the end of the Triassic, unlike its sister taxon (Node 4).

NODE 4: (UNNAMED). This unnamed taxon comprises the last common ancestor that mammals share with *Exaeretodon* and all of its descendants (Rowe, 1986, 1988). The group is diagnosed by a number of characters in the vicinity of the braincase, the dentition, and the postcranium. A partial floor developed beneath the cavum epiptericum, below the presumed position of the ganglion of the facial nerve, though the cavum remained open beneath the trigeminal ganglion (Bonaparte, 1966). The prootic was elaborated with the appearance of the novel posterolateral flange, beneath which reappeared a long quadrate ramus of the pterygoid, a structure that had been reduced in eucynodonts ancestrally (Node 1). On the mandible, the retroarticular process became elongated and strongly recurved, forming the structure referred to in mammals as the manubrium mallei and continuing an ancient trend involving remodeling of the acoustic system. In the dentition, there was a reduction to only three incisiform teeth in the premaxilla. Other dental modifications may have occurred in the ancestor of this group, but the dentitions in *Exaeretodon* and currently known basal members of its sister taxon (Node 5) are too divergently specialized to permit unequivocal interpretation of ancestral morphology.

In the postcranium, costal plates were completely lost from the ribs. In the pelvis, the iliac blade was reduced in size by a deep emargination of its dorsal edge and an overall reduction in the posterior iliac spine. In retrospect, these changes may mark a step toward parasagittal locomotion and the eventual coupling of breathing tides and locomotor cycles that facilitated the eventual increase in metabolic scope and level characteristic of modern mammals (Bramble, 1989; Bramble and Jenkins, 1989). It is doubtful, however, that those physiological modifications were more than incipient at this stage.

Exaeretodon is the oldest member of Node 4, occurring in the Santa Maria Formation of Brazil (Barbarena, 1974, cited in Hopson, 1984), which was probably deposited late in the Ladinian Stage of the Middle Triassic, roughly between 239 ± 8 and 235 ± 5 million years ago (Harland et al., 1990). *Exaeretodon* extends into the Late Triassic Ischigualasto sediments of Argentina (Bonaparte, 1966; Hopson, 1984), which are somewhere between 235 ± 5 and 223 ± 10 million years old (Harland et al., 1990).

NODE 5: (UNNAMED). This group comprises the last common ancestor mammals share with tritheledontids and all of its descendants. This is one of the most distinctive and strongly diagnosed taxa within Cynodontia. Virtually all parts of the skeleton were remodeled in association with a dramatic reduction in body size. The early members of this group lie within the smallest order of vertebrate size magnitudes and exhibit many of the classic structural features induced by evolutionary miniaturization that have been documented in various teleost (Fink, 1981; Roberts, 1981), lissamphibian (Alberch and Alberch, 1981; Hanken, 1983, 1984; Wake, 1966), and squamate (Rieppel, 1984) lineages. The origin of this taxon marked the beginning of a long history—extending from the beginning of the Jurassic into the Paleocene and encompassing nearly 140 million years—that was carried out in tiny animals (Rowe, 1989).

Cranial reorganization in Node 5 involved inflation of the nasopharyngeal cavity, remodeling of the orbit, further enclosure of the brain, and further modification of the acoustic and masticatory systems. The nasal chamber expanded to such a degree that the choana was displaced backwards from its ancestral position in front of the orbit to a new position entirely behind the orbit. The roof of the choana was also modified as the pterygopalatine ridges became reoriented and separated by a series of parallel troughs. These probably mark the positions of the auditory tubes, indicating establishment of the intricate connection between the nasopharynx and middle ear cavity that is characteristic of extant mammals (Barghusen, 1986). At the same time, the orbit developed an extensive medial wall formed by a descending flange from the frontal and an ascending flange from the palatine. The borders of the orbit also changed as the prefrontal and postfrontal bones were lost, leaving the orbit confluent with the temporal fenestra. Elsewhere in the skull, the quadrate rami of the pterygoid and epipterygoid joined the prootic to form a very broadly reflected posterolateral flange. Further closure of the braincase occurred with the development of a medial wall separating the internal acoustic meatus from the cavum cranii, and a bony separation between the fenestra rotunda and the jugular canal.

The auditory and masticatory systems remained intricately coupled, and any changes in one system were probably felt to some degree in the function of the other system. The postdentary bones became reduced to a thin rod of bones lying within a deep Meckelian sulcus, signaling greater auditory sensitivity to high frequencies, and the surangular was withdrawn from the craniomandibular joint. At the front of the jaws the dentaries, which develop from separate anlage in all gnathostomes (e.g., de Beer, 1937), remained separate throughout ontogeny, instead of fusing to form an osseous symphysis as was the case in eucynodonts ancestrally. Without a fused symphysis, some capacity for longitudinal rotation of each mandible was present, though this potential was limited by the persistence of large coronoid bones and transverse processes of the pterygoids.

The postcranial skeleton of tritheledontids has not been described in detail, apart from a brief overview by Broom (1932). However, thanks to the generosity of A. W. Crompton and J. A. Hopson, I was able to score postcranial material that they are currently studying. As they point out (personal communication), the preserved portions of the tritheledontid postcranium are virtually identical to those of tritylodontids and morganucodontids. In the vertebral column, differentiation of the cervical region was marked by great foreshortening of both the centra and neural arches, indicating a highly mobile neck. Thoracic and lumbar regions also show strong differentiation, and the sacrum was reduced from a massive structure to one that comprised only two or at most three vertebrae. The iliac blade was correspondingly reduced, and the acetabulum was rotated backwards to a new position entirely posterior to the sacrum. It is at this point in history that the rod-like iliac blade with triangular cross section, characteristic of most extant mammals, appeared. All of these changes suggest that sagittal axial flexion-extension in a roughly sagittal plane was a major component in locomotion in the members of Node 5.

The limbs were also modified, especially at their proximal articulations. The glenoid became widely open ventrally as the coracoid was reduced and shifted forward, permitting a great degree of freedom at the shoulder. The femur also took on a highly characteristic form, with a strongly differentiated subspherical head flanked laterally and medially by sharply differentiated greater and lesser trochanters. As with the shoulder, this probably enhanced stability of the hip joint over an increased range of excursion. Taken as a whole, these modifications suggest that in Node 5 a great step was taken toward the coupling of locomotion cycles and breathing tides, and perhaps also elevation in metabolic scope, as is characteristic of extant mammals (Bramble, 1989; Bramble and Jenkins, 1989).

Tritheledontidae is diagnosed by a distinctly modified dentition, in which there were both unique crown morphology and a reversal to the pre-eucynodont mode of cheektooth attachment via ankylosis to the jaws (e.g., Gow, 1980; Hopson and Barghusen, 1986). Occlusal facets were still produced, but the relationship between upper and lower teeth was less intricate and regular than in Eucynodontia ancestrally. In light of the large body of data from the skull and postcranium placing tritheledontids within Node 5, there is little doubt that these dental characters are reversals rather than plesiomorphy. Despite its seemingly primitive aspect, the tritheledontid dentition continues to reflect the rapid rate of dental evolution among eucynodonts.

Node 5 is first recorded in rocks that were long considered Late Triassic but that now appear to be Early Jurassic in age (e.g., Olsen and Galton, 1984), between 208 ± 8 and 203 ± 7 million years old. Tritheledontids appear more or less simultaneously with morganucodontids, tritylodontids, and *Sinoconodon*. The geologically "simultaneous" appearance of these taxa is another instance in which resolution of the fossil record lags behind that of the fossils themselves, whose preserved characters support fully resolved, stable relationships among all of these taxa. Prior to the Early Jurassic records of these taxa, there is a gap in the eucynodont fossil record that extends to the base of the Carnian, and which could be anywhere from 10 to 40 million years long. However, even with the outside estimate of 40 million years, the stratigraphic gap alone is unable to account for the number of characters diagnosing this node under an assumption of uniform evolutionary rate during eucynodont phylogeny (Rowe, 1989). There appears to have been a true burst in evolutionary rates that occurred along the stem of Node 5. It is probably no coincidence that this increase in rate is correlated with miniaturization.

NODE 6: MAMMALIAMORPHA. Mammaliamorpha comprises the last common ancestor that mammals share with Tritylodontidae and all its descendants (Rowe, 1986, 1988). Mammaliamorpha is diagnosed by characters of the cranium, dentition, and postcranium. A degree of uncertainty exists in the distributions of some postcranial characters listed here because the tritheledontid postcranium is as yet poorly known. This leaves open the possibility that some characters described below are more widely distributed than described here. Nevertheless, there is strong, unequivocal support for monophyly of Mammaliamorpha, and it is clear that none of these character states occurs outside of Node 5. Moreover, this degree of uncertainty was insufficient to compromise phylogenetic resolution among the taxa depicted in Figure 10.2 in any of the tests aimed at understanding sampling effects (Rowe and Simmons, unpublished).

The medial wall of the mammaliamorph orbit is even more extensive than was the case in Node 5 ancestrally, with the orbitosphenoid joining the previous contributions by the frontal and palatine. Even with this addition, however, the primitive orbital fissure was not yet completely closed. At the rear of the skull, the prootic posterolateral flange became perforated by cranial vessels. In addition, the paroccipital process was elaborately modified, bifurcating distally to form separate quadrate and mastoid processes that are separated ventrally by a deep fossa, which is often interpreted (probably erroneously) to have provided origin to a hyoid muscle. The quadrate process developed direct articulation with the quadrate, whereas in eucynodonts ancestrally the two bones were separated by an intervening slip of the squamosal.

The dentition underwent further profound modification with the development of multiple roots and a consequent escalation in complexity of crown morphology and occlusal relationships. Crown diversity in the cheekteeth is so great and so widely divergent among the taxa within Mammaliamorpha, especially tritylodontids, that little can be said of precise crown structure in mammaliamorphs ancestrally. Root morphology is also variable, with three or more roots developing on the cheekteeth in a number of taxa. Some authors have argued that division of the roots is non-homologous in basal mammaliamorphs because root morphology differs among these taxa (Sues, 1985; Hopson and Barghusen, 1986). However, in light of the other data supporting mammaliamorph monophyly, it would appear that there were divided roots in the cheekteeth of mammaliamorphs ancestrally. Like crown structure, mammaliamorph root morphology diversified at a rapid rate from a common ancestral form with divided roots.

In the postcranium, the atlas-axis complex became more flexible with loss of the atlas postzygapophysis and flattening of its centrum. At the same time, a stout dens formed as a neomorphic outgrowth from the axis centrum, contributing strength and stability to the joint (Jenkins, 1969, 1971). At the opposite end of the vertebral column, the distal caudal centra became elongated with greatly reduced neural and haemal arches. Associated with vertebral differentiation was the segmentation of the sternum, to produce sternebrae. The sternum originated developmentally and functionally as a part of the shoulder girdle. However, in mammaliamorphs it appears to have become functionally linked to the vertebral column as well, by facilitating the parasagittal flexion-extension of the vertebral column. The appearance of epipubic bones, which lay embedded in the abdominal musculature, may also be linked to parasagittal locomotion, although a host of other functions have been suggested.

Tritylodontids include large animals, such as *Kayentatherium* (Sues, 1986a), as well as very small forms such as *Oligokyphus* (Kuhne, 1956). The highly derived dentition of tritylodontids offers both strong corroboration of the group's monophyly and evidence pertaining to their phylogeny. Recent phylogenies (Clark and Hopson, 1985; Sues, 1986b) indicate that tritylodontids arose from a small ancestor, and that size increase within the group was secondary.

Mammaliamorphs first appear in the Early Jurassic, between 208 ± 8 and 203 ± 7 million years ago (see above, Node 5). Tritylodontids have a record that extends to into the Middle and perhaps the Late Jurassic (e.g., Clark and Hopson, 1985).

NODE 7: (UNNAMED). This unnamed taxon comprises the last common ancestor that mammals share with *Sinoconodon* and all its descendants. The diagnosis of this taxon consists solely of features of the skull, which may simply reflect that the postcranium of *Sinoconodon* is virtually unknown. The diagnosis of this group is more extensively detailed by Crompton and Luo (this volume, chapter 4). Information used to score character states for *Sinoconodon* (Rowe and Simmons, unpublished) was generously provided by Z. Luo and A. W. Crompton (personal communication), supplementing previous literature (Crompton and Sun, 1985; Patterson and Olson, 1961).

The medial wall of the orbit became completely closed by further expansion of the orbitosphenoid, which contributed to the orbital processes of the frontal and palatine. The hindbrain was expanded, causing the parietals to bulge outward into the temporal fenestrae and the basicranium to become wider than the choana. Correspondingly, the cavum epiptericum became completely enclosed beneath the trigeminal ganglion, and the petrosal promontorium also appeared.

On the dentary is a large condyle that articulates with a distinctive glenoid fossa on the squamosal (Crompton and Sun, 1985), although the quadrate and articular remained involved in the articulation and the postdentary bones remained attached to the mandible throughout life. The dentary-squamosal articulation is a widely discussed character. Under the character-based definitions of the Linnean System, many authors regarded it as the definitive structure dividing mammals from reptiles.

This engendered debate about whether Mammalia was mono- or polyphyletic because some authors asserted that the dentary condyle evolved convergently among the various Triassic and Jurassic taxa that possess it (e.g., Barghusen and Hopson, 1970). In light of the recent phylogenetic analyses summarized here and elsewhere, it is now evident that the dentary-squamosal articulation evolved only once, but that it is more widely distributed than previously believed. It is true that all mammals possess this articulation, but it now also appears true that not all taxa with a dentary-squamosal joint are mammals under a definition based on common ancestry.

The dentition of *Sinoconodon* is highly apomorphic in that the upper and lower cheekteeth did not occlude, and there was little or no replacement of them during ontogeny (Crompton and Luo, this volume, chapter 4). The position of the cheekteeth is also unique in that the rear end of the row extends onto the lower edge of the zygoma (Crompton and Sun, 1985; Crompton and Luo, this volume, chapter 4). This is yet another manifestation of the high rate of dental evolution in eucynodonts.

As with tritylodontids and tritheledontids, the earliest members of Node 7 first appear in the Early Jurassic (see above, Node 5).

NODE 8: MAMMALIAFORMES. Mammaliaformes comprises the last common ancestor shared by Mammalia and *Morganucodon* and all its descendants (Rowe, 1986, 1988). In older works (e.g., Hopson and Crompton, 1969), morganucodontids were considered to be among the oldest "true" mammals and were believed most closely related to monotremes among extant taxa. This was properly taken as an indication that Mammalia had both originated and split into its two major daughter lineages sometime before the Late Triassic. Implicit in this view was the idea that the origin of Mammalia corresponds to the appearance of the last common ancestor of monotremes and therians, and not a more distant ancestor. This view is formalized in the contemporary phylogenetic definition of Mammalia, which is based on the last common ancestor of living taxa. However, the strength of evidence now indicates that monotremes and therians are more closely related to each other than to morganucodontids (Kemp, 1983; Rowe, 1988; Gauthier et al., 1988; Wible, 1991). Under the phylogenetic definition, morganucodontids are not mammals, a conclusion hinging on their revised phylogeny. A number of authors continue to refer to morganucodontids and *Sinoconodon* as "mammals," evidently in an effort to preserve the traditionally recognized membership of Mammalia. However, this implicitly casts the definition of mammals in terms of characters and evolutionary grade, a perspective largely abandoned by the phylogenetic system (but see Szalay, this volume, chapter 9).

Mammaliaformes is diagnosed by characters of the skull and dentition. There was further modification of the orbital floor, in which the maxilla has come to participate (Luo, personal communication). There was also further modification of the dentition, in which the cheekteeth came to occlude in a unilateral pattern (Crompton and Jenkins, 1979; Crompton, 1989), although lateral excursion of the mandibles was still greatly constrained by the transverse process of the pterygoids and the persistence of a robust coronoid bone on the medial surface of the dentary. Mammaliaform cheekteeth are also differentiated into a pattern of relatively simple premolariform teeth in front, with more complex molariform teeth behind. Whether these dental groups also possessed the replacement patterns commonly associated with the terms "premolar" and "molar" is as yet unclear in most relevant extinct taxa.

Morganucodontids first appear together with *Sinoconodon*, tritylodontids, and tritheledontids in the Early Jurassic (see above, Node 5).

NODE 9: MAMMALIA. Mammalia comprises the last common ancestor of monotremes and therians and all of its descendants (Rowe, 1986, 1987, 1988; Rowe and Simmons, unpublished; de Queiroz and Gauthier, 1990). In older literature, the mammalian boundary was ambiguous because no consensus could be reached on which character or character assemblage marked the most meaningful discontinuity between mammals and reptiles. This uncertainty is removed by a phylogenetic definition of Mammalia, whose common ancestry unequivocally sets it apart from all other taxa, living and extinct. The phylogenetic definition has also significantly altered our view of many basic properties of Mammalia, including its diagnosis, membership, and distribution in time. This is not merely a semantic transformation, because it reflects a revised genealogical hypothesis among the Mesozoic taxa that have long figured centrally in this debate, and it alters our measures of the most fundamental evolutionary properties of many of these taxa (Rowe, 1988).

Most of the skeletal characters diagnosing Mammalia, and which set the last common ancestor of monotremes and therians apart from morganucodontids, involve "repackaging" of an enlarged brain and special sense organs. In the snout, the prenasal process was lost in adults, rendering the external nares confluent. The nasal chamber expanded to such an extent posteriorly that its rearmost portion lies in a subcerebral position. An ossified cribriform plate appeared at the same time, fully separating the nasopharynx from the braincase. At the rear of the skull, the hindbrain became greatly inflated, resulting in profound remodeling of the skeletal structures in that region. The parietals bulged outward into the temporal fenestrae, and the paroccipital processes were rotated sharply downward as the brain expanded outward over the region of the trigeminal ganglion and middle ear structures, from its ancestral position lying almost entirely between them. In addition, the occipital condyles were greatly expanded, coming to enclose the entire ventral two-thirds of the foramen magnum.

The visceral arch skeleton was also affected by inflation of the hindbrain. The middle ear ossicles, which derive from the first arch, shifted, in the most famous transformation of this entire history, from their primitive attachment throughout ontogeny to the mandible, to a new position suspended beneath the adult cranium. As a result, the quadrate (now the incus) and articular (now the malleus) were removed from the craniomandibular joint, which in adult mammals is built entirely from the dentary and squamosal, with occasional secondary contributions from other bones such as the alisphenoid. The second visceral arch was also affected. The stapes was reduced and the stapedial foramen, retained in adult morganucodontids, tritylodontids, and most mammalian embryos, was lost in adult mammals. Lateral to the stapes, Reichert's cartilage became attached to the cranium but, unlike the ear ossicles, its proximal end coossified with the cranium to form the adult styloid process, while its more distal corpus became detached and continued its primitive role in the hyoid apparatus. Technically a visceral arch derivative, the transverse process of the pterygoid became reduced from a massive structure that fit closely between the coronoid bones of the mandible to a vestigial structure, the pterygoid hamulus. The modifications of the middle ear probably signal greater auditory sensitivity to high frequencies. Taken collectively, these visceral arch modifications all contributed far greater ranges of lateral movement and longitudinal rotation of the mandibles than ever before.

In the postcranium, modification of the occiput was complemented by the atlas, whose formerly separate parts fused to form a solid ring. In addition, the cervical ribs became fused to their vertebrae, enclosing the foramina transversaria, and perhaps also contributing strength and stability to the craniocervical mobility. Elsewhere in the postcranium, the limbs and girdles developed secondary ossification centers or epiphyses. In many areas these are related to the more intricate and precise sculpting of the joints, for example, in the styloid processes of the radius, tibia, and fibula. Secondary ossifications are also involved in the fibular flabellum, patella, and sesamoid bones in the flexor musculature of the hands and feet. At least a small degree of opposition of the hallux to the other toes was possible in mammals ancestrally, though the degree of movement is left ambiguous by the highly modified feet on monotremes (see Szalay, this volume, chapter 9).

Because Mammalia is defined on the basis of the common ancestor of two extant lineages, it is also possi-

Chapter 10. Early Mammal Phylogenetic Systematics

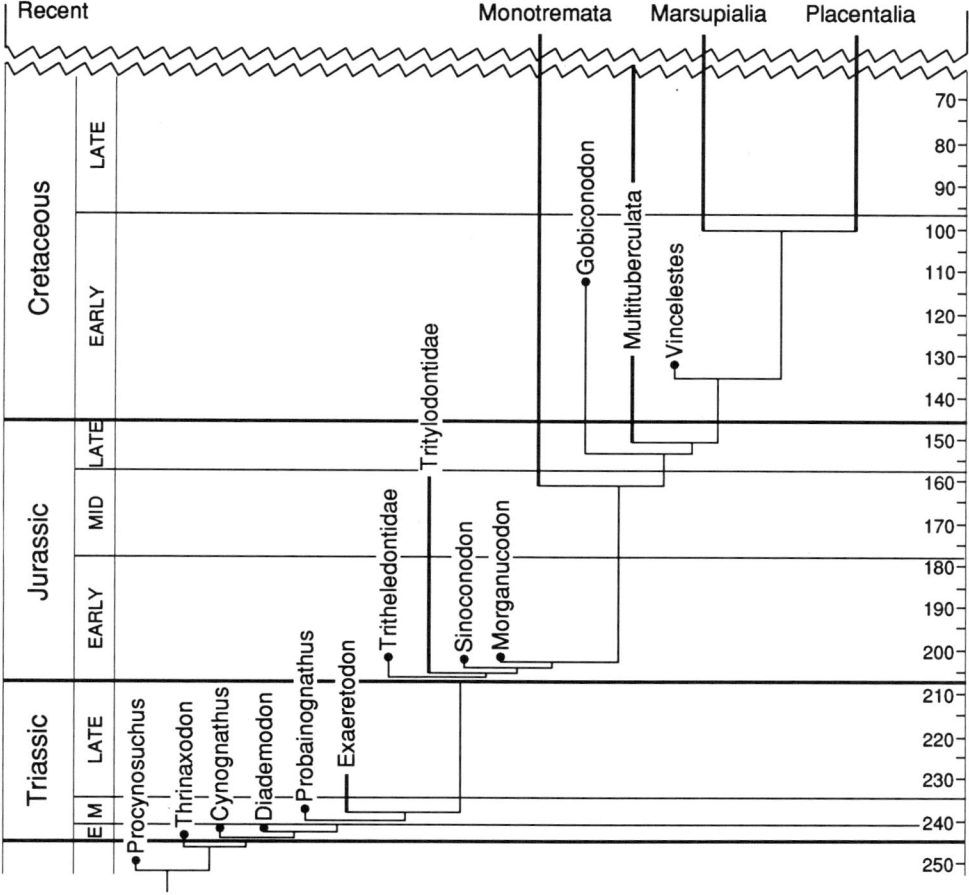

FIGURE 10.3. Cynodont phylogeny superimposed on a Mesozoic time scale, based on Harland et al. (1990). Scale is in millions of years before present. Extinct taxa known from only very limited stratigraphic ranges are represented by small circles; taxa with extended ranges are depicted with heavy line. The range for monotremes is inferred from the fossil record of its sister taxon.

ble to provide a "soft" diagnosis. As detailed elsewhere (Gauthier et al., 1988, Appendix B; Gauthier et al. 1989), extensive remodeling of "soft" tissues affected the nervous system, circulatory system, pulmonary system, digestive system, excretory system, endocrine system, and the integument and modified a number of developmental and metabolic pathways. Behavioral modifications have also been documented. Although we will probably never know the exact levels within Synapsida at which these modifications arose, these characters nevertheless distinguish mammals among extant tetrapods (Gauthier et al., 1988, 1989) and can be postulated to have been present in, if not apomorphic of, the ancestral mammal.

A long gap in the stratigraphic record separates the oldest mammal fossils from morganucodontids, the sistergroup of mammals. The oldest fossil possessing character states derived within Mammalia is *Phascolotherium bucklandi*, a member of Theriimorpha (Rowe, 1988). It is known from the Middle Jurassic (Bathonian) Stonesfield Slate of England, which is between 166 ± 8 and 161 ± 8 million years old (Harland et al., 1990). This estimate of the minimum age of Mammalia represents a reduction by nearly one-fourth over previous estimates, which treated morganucodontids as the earliest mammals.

The large number of osteological characters diagnosing Mammalia may reflect a jump in evolutionary rates, as was the case with the origin of Node 5 (above). However, with the wide margin of dating error and a major gap in the mid-Jurassic fossil record (Fig. 10.3), it remains possible that this number merely reflects average evolutionary rates during the Jurassic, and that stratigraphic incompleteness accounts for their accumulation at this node.

NODE 10: THERIIMORPHA. Theriimorpha is the only taxon discussed here that I prefer to diagnose using a stem-based definition (de Queiroz and Gauthier, 1990). The monophylies of theria and Mammalia are both firmly established; hence there is little doubt that a monophyletic taxon exists that includes therians and all extinct taxa closer to therians than to monotremes. Despite this certainty on monophyly of Theriimorpha, its diagnosis is among the weakest of any on the tree, and relationships among the basal theriimorphs are sufficiently unresolved that stable node-based definitions are not yet practical. This weakness is due in part to the high incompleteness of virtually all Mesozoic fossils representing the group, as well as to the lack of a fossil record and the highly apomorphic nature of monotremes, the sister taxon of Theriimorpha. Doubtless, as the Jurassic through Early Cretaceous segment of theriimorph history is more completely understood, the diagnoses of Theriimorpha and a number of subsequent nodes will be revised.

The basal theriimorph represented in Figure 10.2 is *Gobiconodon*, which is the most complete of the theriimorphs that consistently fell outside of Theriiformes in the analyses summarized above. There are many other taxa that cluster unresolved in this position, such as *Trioracodon ferox*, *Triconodon mordax*, *Phascolotherium bucklandi*, and the unnamed "eupantothere" briefly described by Krebs (1987). In one of the many equally parsimonious trees, *Gobiconodon* and *Triconodon* are sister taxa in what would constitute a monophyletic Triconodonta, based on tooth crown morphology. However, there are numerous other possible relationships among basal theriimorphs, and stable resolution of their relationships will probably be gained only with the discovery of more complete specimens.

The diagnosis of Theriimorpha includes the inflected angle, or pterygoideus shelf, of the dentary. In addition, the external acoustic meatus became redirected to occupy a characteristic position behind the craniomandibular joint. Both characters are equivocal because of the highly apomorphic state of monotremes. In the shoulder girdle, the coracoid was reduced, as indicated in *Gobiconodon* by a reduced articular facet on the scapula.

As described earlier (see Node 9, Mammalia) the oldest theriimorph is *Phascolotherium bucklandi*, which preserves the inflected angle, from the Middle Jurassic Stonesfield Slate (Rowe, 1988).

NODE 11: THERIIFORMES. Theriiformes comprises the last common ancestor therians share with multituberculates (*sensu* Simmons, this volume, chapter 11), and all its descendants. The position of Multituberculata has long been controversial, and no consensus can be found even in recent phylogenetic literature. However, the solution advocated here was stable in the face of the sampling tests described above and takes into account all available date (Rowe, 1988; Rowe and Simmons, unpublished; see also Sereno and McKenna, 1990).

Equivocation on the position of multituberculates appears in large part an artifact of sampling, which in the past was restricted to dental and cranial data. Although many authors have looked to the dentition for information on higher-level relationships, even the most primitive multituberculate teeth are so apomorphic that little unambiguous information relevant to the broad temporal scope of this problem remains. This dentition has been highly informative at the more restricted temporal scope of relationships within the group, however (Greenwald, 1989; Simmons, this volume, chapter 11).

Relatively few cranial characters have been found to support any of the competing hypotheses of multituberculate relationships to other taxa, and homoplasy complicates the issue in every analysis. Multituberculates were long linked with morganucodontids and monotremes, based on resemblances of the braincase that subsequent studies have shown to be plesiomorphic (Presley, 1981; Kemp, 1983; Novacek and Wyss, 1986; Rowe, 1988). More recently, however, Wible (1991) analyzed selected subsets of the cranial and dental data used by Rowe (1988) and found what appeared to be new support for sister relationships between multituberculates and monotremes. Comparing these results is complicated because the two studies analyzed different sets of taxa and characters. The most inclusive analysis (Rowe and Simmons, unpublished) found that multituberculates and therians share communication of the post-temporal fenestra with an enclosed paracranial passage rather than with the temporal fossa, as it did in mammals ancestrally (Kielan-Jaworowska et al., 1986; Wible, 1990). In addition, the incus lies postertor to the malleus, and the ectotympanic lies on a plane inclined from the horizontal (Miao and Lillegraven, 1986).

Reference to the postcranium provides a great deal of additional corroboration of this relationship. It was recognized for many years that multituberculates and therians share striking postcranial resemblances (Gidley, 1909; Simpson and Elftman, 1928; Jenkins and McClearn, 1984), but most authors have asserted the similarities to be convergent because the dentitions of multituberculates and therians are so different. When all available data are considered, however, it now appears that the multituberculate dentition diverged at a rapid rate rather than at a distant in time, and that the postcranial similarities reflect common ancestry. The most marked similarities involve the shoulder girdle and humerus, the femur, and the ankle. In the shoulder, the procoracoid became lost as a discrete adult ossification and the acromion process became strongly down-turned, projecting to a level below the roof of the glenoid. The humeral head became spherical and strongly inflected dorsally, and the teres and deltopectoral crests became reduced to narrow tubercles separated by a narrow bicipital groove. The distal end was also modified with only weakly developed entepi- and ectepicondyles. The femoral head became spherical, inflected medially, and set apart from the shaft on a constricted neck. The lesser trochanter rotated from its former position medial to the shaft to a new position on the ventral surface of shaft, while the greater trochanter became elongated and massive. In the ankle, the calcaneal tuber was elongated to a length greater than its width. In the foot, the entocuneiform developed a strongly saddle-shaped articulation (viz., fully opposable hallux), and a styloid process developed on the proximal end of metatarsal V for the insertion of the peroneal musculature.

The oldest members of Multituberculata (*sensu* Simmons, this volume, chapter 11) are from the Kimmeridgian Guimarota deposits of Portugal, which are between 155 ± 8 and 152 ± 12 million years old. Many authors have included haramiyids, a poorly known taxon based entirely on isolated and highly derived teeth, within Multituberculata, extending the range of the group into the Triassic (Hahn, 1973; Sigogneau-Russell, 1989). However, haramiyids were the most phylogenetically labile taxa studied by Rowe and Simmons (unpublished). When analyzed with the taxa depicted in Figure 10.2, there were at least six other equally parsimonious positions for haramiyids besides the possibility that they are the sister taxon to multituberculates, and several thousand possible positions when all taxa were considered. This lability leaves little justification for using haramiyids to mark the minimum time of multituberculate origin. Moreover, the broad temporal separation between haramiyids and multituberculates speaks, albeit circumstantially, against the possibility of sister relationships between them.

NODE 12: (UNNAMED). This unnamed taxon comprises the most recent common ancestor of *Vincelestes* and Theria and all its descendants. *Vincelestes* is known from several nearly complete skulls and postcranial skeletons, but to date only a preliminary description of the skull (Bonaparte and Rougier, 1987) and details of the snout (Wible et al., 1990) have been published. Nevertheless, the conclusion that *Vincelestes* is the sister taxon of Theria has been reached by all who have treated it (Bonaparte and Rougier, 1987; Rowe and Simmons, unpublished; Wible et al., 1990); further testing of this conclusion will occur as Rougier completes his ongoing analysis of the postcranium. The skull in *Vincelestes* shares with therians a broad contact between the alisphenoid and frontal, and a cochlea that is elongated and coiled 360 degrees or more. Bonaparte and Rougier (1987) also described a small participation by the squamosal in the braincase.

Vincelestes is the earliest representative of Node 12. It was collected from the La Amarga Formation of Argentina, of Early Cretaceous (Hauterivian) age, roughly between 135 ± 8 and 132 ± 8 million years old (Harland et al., 1990).

NODE 13: THERIA. Theria comprises the most recent common ancestor shared by extant marsupials and placentals and all of its descendants (but see Cifelli, this volume, chapter 14, and Szalay, this volume, chapters 9 and 15, for a different concept of Theria) Because it is defined by two extant lineages, data from virtually every anatomical system diagnose Theria (e.g., Marshall, 1979; Lillegraven et al., 1987). Although the exact point within Theriimorpha at which most soft characters arose is uncertain, these characters are nevertheless apomorphic of therians, compared to other extant taxa, and we may infer the characters to have been present in extinct therians. There are a number of poorly known Mesozoic taxa, in particular deltatheridians and various other taxa commonly referred to as "therians of the metatherian-eutherian grade," that have not yet been analyzed phylogenetically. It is uncertain whether these taxa lie within Theria or only close to it, and their eventual inclusion into future phylogenetic analyses is likely to revise the following "hard" diagnosis.

Theria is diagnosed by modifications of both the cranium and postcranium. In the vicinity of the braincase, the anterior lamina of the prootic was lost, and the floor of the cavum epiptery cum became enclosed predominantly by the alisphenoid. In addition, the pila antotica no longer formed an ossified adult structure, although atavistic remnants of the structure have been reported in rare Recent individuals (e.g., de Beer, 1937). The postcranium was marked by more subtle modifications. In the vertebral column, the inferior lamella appeared on the ventrum of the sixth cervical vertebra, and secondary ossifications developed on the faces of most or all presacral centra. Elsewhere in the postcranium, the acetabulum was remodeled by the development of an inverted, U-shaped articular tract for the femoral head.

Kuehneotherium was long viewed as the earliest therian (e.g., Hopson and Crompton, 1969; Prothero, 1981) and as such, it was properly taken as evidence of the Triassic origin of mammals because it seemed to indicate that monotremes and therians had, by the Late Triassic, already diverged from their last common ancestor. However, using ancestry-based taxon definitions, it now appears that *Kuehneotherium* is neither a therian nor a mammal. The oldest fossils preserving characters derived within Theria are teeth of the marsupial *Pariadens kirklandi*, from Late Cretaceous (Cenomanian) Dakota Formation of Utah (Cifelli and Eaton, 1987), which is between 97 ± 3 and 90 ± 3 years old.

Constraints on the Positions of Unresolved Taxa

A number of additional taxa were treated in the analyses by Rowe and Simmons (unpublished), including *Massetognathus*, *Asioryctes*, Docodonta, *Triconodon*, *Trioracodon*, *Kuehneotherium*, *Dinnetherium*, Haramiyidae, and an unnamed "eupantothere" described briefly by Krebs (1987). These taxa were less complete than those described above, and their positions could not be unambiguously resolved. However, it was found that each of these taxa ranged in position over finite segments of the tree depicted in Figure 10.2. *Massetognathus* was closer to Mammaliamorpha than either *Diademodon* and *Cynognathus*, but it was not as close as *Exaeretodon*. In a similar way, Docodonts, *Dinnetherium*, and *Kuehneotherium* each consistently varied between Nodes 7 and 9, while *Triconodon*, *Trioracodon*, and Krebs's unnamed "eupantothere" fell between Nodes 9 and 11. Haramiyidae was the least complete and the most labile taxon studied, ranging in positions between Nodes 6 and 13.

Conclusions

Methods of sampling taxa and characters for particular systematic problems can be a determining factor in the outcome of phylogenetic analyses, and they may be a major source of current conflicts in interpreting early mammalian phylogeny. Some of the most persistent systematic problems in early mammalian history have been the result of restricted character sampling, in part a result of the preservational bias toward dental data. But because different systems transform at different rates, we should not expect any particular region to be informative for phylogenetic questions at all temporal scales. Resolution of a host of long-standing systematic problems in early mammalian history, such as the position of multituberculates, has been gained by reference to a diverse data set and by filtering out artifacts attributable to sampling. Further examination of sampling and rates may be the most direct avenue to resolving many remaining conflicts.

The eucynodont phylogeny that has emerged from recent phylogenetic analyses is dominated by several patterns. One involves a great acceleration in the rate and complexity of dental evolution. The initiation of this trend can be traced to the ancestral eucynodont, which lived no later than the earliest Triassic, and in which appeared several structures critical to occlusion, most notably the periodontal ligament. From that ancestor evolved descendants with an enormous diversity of complex tooth morphologies and occlusal relationships. Because of its rapid transformation rate, the dentition in fossil and extant eucynodonts is most likely to be informative with respect to phylogenetic problems encompassing short temporal spans, such as relationships at low taxonomic levels. By reference to the more slowly evolving skull and postcranium, the broader history of the dentition can be unambiguously interpreted.

Another pattern, this one long recognized, involves a major reorganization of the skull involving repackaging of a greatly inflated brain and special sensory organs. This trend was manifested throughout eucynodont history, and during the Mesozoic it resulted in the complete and intimate osseous enclosure of the brain along with some of its associated vessels. The brain expanded to nearly one-half the length of the skull and greatly increased in width. This was associated with ontogenetic fusions among bones that primitively remained separate throughout life, a great increase in the extent of both membranous and endochondral bone, and profound remodeling of inherited structures. The nasal capsule and associated structures also expanded enormously, extending backward to a subcerebral position, displacing the choana backward as well. The orbit, lying between the inflated nasal capsule and brain, became enclosed medially and was extensively reorganized in other respects.

The postcranium was also reorganized during this time as the vertebral column became fully differentiated and the limbs correspondingly transformed. The neck became markedly differentiated, enhancing mobility of the head. Specialization of thoracic and lumbar regions was accompanied by reorientation of the principal plane of flexure of the vertebral column, indicating a shift toward parasagittal gait. An associated transformation involved segmentation of the sternum, whose role in the shoulder girdle diminished over time, becoming functionally linked instead to the vertebral column. The appendicular skeleton was correspondingly modified as the limbs became long and slender, developing intricately sculpted articulations, and the girdles were modified to permit great ranges of stable excursion. These changes probably reflect coupling of breathing tides and gait cycles, and an expanded metabolic scope.

The various character states that record these transformations are not distributed equally along the phylogeny, pointing to yet another pattern. The greatest number of changes occurred in the last common ancestor of tritheledontids and mammaliamorphs (Node 5), in association with miniaturization. The early parts of cynodont history were carried out by relatively large animals. But, beginning at Node 5, a long segment of the mammalian history involved animals that fall into the smallest order of vertebrate size magnitudes. With this reduction in size occurred a rapid remodeling of virtually all parts of the skeleton, involving a complex interplay of both paedomorphosis and peramorphosis (Rowe, 1989, 1990). Similar changes have been described in other miniaturized tetrapod lineages, suggesting that developmental constraints common to all tetrapods had profound influence on the shape of mammalian history.

This general picture bears strong resemblance to older views, and the broader features of several of these patterns have long been recognized. However, the picture has also been altered in fundamental respects as our measures of the evolutionary properties of taxa have changed under phylogenetic systematics. In addition, many new detalis of evolutionary pattern have come to light, and entirely new patterns involving both tempo and mode of evolution have been discovered through efforts to understand the effects of rate and sampling on the outcome of phylogenetic analyses. The picture will undoubtedly continue to change as we collect new data from extant and extinct forms, and as we further scrutinize the effects of our analytic methods on the patterns we seek to explain.

CORRESPONDENCE ADDRESS. Timothy Rowe, Department of Geological Sciences, and Vertebrate Paleontology Laboratory, The University of Texas at Austin, Austin, TX 78712, USA.

References

Alberch, P., and Alberch, J. 1981. Heterochronic mechanisms of morphological diversification and evolutionary change in the neotropical salamander *Bolitoglossa occidentalis* (Amphibia: Plethodontidate). *J. Morph.* 167:249–264.

Allin, E.F. 1986. The auditory apparatus of advanced mammal-like reptiles and early mammals. In *Ecology and biology of mammal-like reptiles* (Hotton N.H. III, MacLean, P.D., Roth, J.J., and Roth, E.C., eds.). Washington, D.C.: Smithsonian Institution and National Institute of Mental Health, pp. 283–294.

Anderson, H.M., and Cruickshank, A.R.I. 1978. The biostratigraphy of the Permian and Triassic. Part 5: A review of the classification and distribution of Permo-Triassic tetrapods. *Palaeontl. Afr.* 21:15–44.

Aulie, R.P. 1974. The origin of the idea of mammal-like reptile. Parts I–III. *Am. Biol. Teacher* 36:476–485; 36:545–553; 37:21–32.

Barbarena, M.C. 1974. *Contribuicao ao conhecimento dos cinodontes gonfodontes (Cynodontia, Tritylodontidae) do Brasil*. Ph.D. thesis, University Federal do Rio Grande del Sul.

Barghusen, H.R. 1986. On the evolutionary origin of the therian tensor veli palatini and tensor tympani muscles. In: *Ecology and biology of mammal-like reptiles* (Hotton N.H. III, MacLean, P.D., Roth, J.J., and Roth E.C., eds.). Washington, D.C.: Smithsonian Institution and National Institute of Mental Health, pp. 253–262.

Barghusen, H.R., and Hopson, J.A. 1970. Dentary-squamosal joint and the origin of mammals. *Science* 168:573–575.

Bonaparte, J.F. 1966. Sobre las cavidades cerebral, nasal y otras estructuras del cranio de *Exaeretodon* sp. (Cynodontia-Traversodontidae), Tucuman. *Acta Gelogica Lilloana* 8:5–31.

Bonaparte, J.F., and Rougier, G. 1987. Mamiferos del Cretacico inferior de Patagonia. *IV Congreso Latinoamericano de Paleontologia, Bolivia* 1:343–359.

Bramble, D.M. 1989. Axial-appendicular dynamics and the integration of breathing and gait in mammals. *Syst. Zool.* 29:171–186.

Bramble, D.M., and Jenkins, F.A. Jr. 1989. Structural and functional integration across the reptile-mammal boundary: The locomotor system. In: *Complex organismal functions: Integration and evolution in vertebrates* (Wake, D.B., and Roth, G., eds.). New York: John Wiley, pp. 133–146.

Broom, R. 1932. *The mammal-like reptiles of South Africa and the origin of mammals*. London: H. F. & G. Witherby.

Cifelli, R.L., and Eaton, J.G. 1987. Marsupial from the earliest late Cretaceous of western U.S. *Nature* 325:520–522.

Clark, J.M., and Hopson, J.A. 1985. Distinctive mammal-like reptile from Mexico and its bearing on the phylogeny of the Tritylodontidae. *Nature* 315:389–400.

Cloutier, R. 1991. Patterns, trends, and rates of evolution within the Actinistia. In: *Environmental biology of fishes*. Netherlands: Kluwer Academic Publishers, 32:23–58.

Cromton, A.W. 1963. Tooth replacement in the cynodont *Thrinaxodon liorhinus*. *Ann. S. Afr. Mus.* 46:479–521.

Crompton, A.W. 1972. Postcanine occlusion in cynodonts and tritylodontids. *Bull. Br. Mus. Nat. Hist. (Geol.)* 21:27–71.

Crompton, A.W. 1989. The evolution of mammalian mastication. In: *Complex organismal functions: Integration and evolution in vertebrates* (Wake, D.B., and Roth, G., eds.). New York: John Wiely, pp. 23–40.

Crompton, A.W., and Jenkins, F.A. Jr. 1973. Mammals from reptiles: A review of mammalian origins. *Annual Review of Earth and Planetary Sciences* 1:131–155.

Crompton, A.W., and Jenkins, F.A. Jr. 1979. Origin of mammals. In: *Mesozoic mammals: The first two thirds of mammalian history* (Lillegrave, J.A. Kielan-Jaworowska, Z., and Clemens, W.A., eds.). Berkeley: University of California Press, pp. 59–73.

Crompton, A.W., and Sun, A.-L. 1985. Cranial structure and relationships of the Liassic mammal *Sinoconodon*. *Zool J. Linnean Soc.* 85:99–119.

de Beer, G.R. 1937. *The development of the verterate skull*. Oxford: Oxford University Press.

de Queiroz, K. 1988. Systematics and the Darwinian revolution. *Philos. Sci.* 55:238–259.

de Queiroz, K., and Gauthier, J.A. 1990. Phylogeny as a central principle in taxonomy: Phylogenetic definitions of taxon names. *Syst. Zoo.* 39:307–322.

Donoghue, M.J., Doyle, J.A., Gauthier, J., Kluge, A.G., and Rowe, T. 1989. The importance of fossils in phylogeny reconstruction. *Ann. Rev. Ecol. Syst.* 20:431–460.

Doyle, J.A., and Donoghue, M.J. 1987. The importance of fossils in elucidating seed plant phylogeny and macroevolution. *Rev. Paleobot. Palynol.* 50:63–95.

Farris, J.S. 1986. Hennig 86, version 1.5. Computer software distributed on 5-inch diskette by J.S. Farris.

Farris, J.S. 1989a. The retention index and the rescaled consistency index. *Cladistics* 5:417–419.

Farris, J.S. 1989b. The retention index and homoplasy excess. *Syst. Zool.* 38:406–407.

Fink, W.L. 1981. Ontogeny and phylogeny of tooth attachment modes in actinopterygian fishes. *J. Morph.* 167:167–184.

Fourie, S. 1974. The cranial morphology of *Thrinaxodon liorhinus* Seeley. *Ann. S. Afr. Mus.* 65:337–400.

Gauthier, J.A. 1986. Saurischian monophyly and the origin of birds. In: The origin of birds and the evolution of flight (Padian, K., ed.). *Mem. Calif. Acad. Sci.* 8:1–55.

Gauthier, J.A., Cannatella, D., de Queiroz, K., Kluge, A.G., and Rowe T. 1989. Tetrapod phylogeny. In: *The hierarchy of life: Molecules and morphology in phylogenetic analysis* (Fernholm, B., Bremer, K., and Jornvall, H., eds.). Amsterdam: Nobel Symposium 70, Excerpta Medica, pp. 337–354.

Gauthier, J.A., Kluge, A.G., and Rowe, T. 1988. Amniote phylogeny and the importance of fossils. *Cladistics* 4:105–209.

Ghiselin, M.T. 1984. "Definition," "character," and other equivocal terms. *Syst. Zool.* 33:104–110.

Gidley, J.W. 1909. Notes on the fossil mammalian genus *Ptilodus*, with descriptions of new species. *Proc. U.S. Nat. Mus.* 36:611–626.

Gow, C.E. 1980. The dentitions of the Tritheledontidae (Therapsida: Cynodontia). *Proc. Roy. Soc. London* B208:461–481.

Greenwald, N.S. 1989. *Phylogeny and systematics of multituberculate mammals*. Ph.D. Dissertation, University of California, Berkeley.

Haeckel, E. 1897. *The evolution of man*, vols. 1 and 2. New York: Appleton and Co.

Hahn, G. 1973. Neue Zahne von Haramiyiden aus der deutschen Ober-Trias und ihre Beziehungen zu den Multituberclaten. *Palaeontographica Abt. A* 142:1–15.

Hanken, J. 1983. Miniaturization and its effects on cranial morphology in plethodontid salamanders, genus *Thorius* (Amphibia, Plethodontidate): II. The fate of the brain and sense organs and their role in skull morphogenesis and evolution. *J. Morph.* 177:255–268.

Hanken, J. 1984. Miniaturization and its effects on cranial morphology in plethodontid salamanders, genus *Thorius* (Amphibia: Plethodontidae). I. Osteological variation. *Biol. J. Linn. Soc.* 23:55–75.

Harland, W.B., Armstrong, R.L., Cox., A.V., Craig, L.E., Smith, A.G., and Smith, D.G. 1990. *A geological time scale 1989*. Cambridge: Cambridge University Press.

Hopson, J.A. 1969. The origin and adaptive radiation of mammal-like reptiles and non-therian mammals. *Ann. New York Acad. Sci.* 167:199–216.

Hopson, J.A. 1984. Late Triassic traversodont cynodonts from Nova Scotia and southern Africa. *Palaeontol. Afr.* 25:181–201.

Hopson, J.A., and Barghusen, H.R. 1986. An analysis of therapsid relationships. In: *Ecology and biology of mammal-like reptiles* (Hotton N.H. III, MacLean, P.D., Roth, J.J., and Roth, E.C., eds.). Washington, D.C.: Smithsonian Institution and National Institute of Mental Health, pp. 83–106.

Hopson, J.A., and Crompton, A.W. 1969. Origin of mammals. *Evol. Biol.* 3:15–72.

Hopson, J.A., and Kitching, J.W. 1972. A revised classification of cynodonts. *Palaeont. Afr.* 14:71–85.

Hull, D.L. 1976. Are species really individuals? *Syst. Zool.* 25:174–191.

Jenkins, F.A. Jr. 1969. The evolution and development of the dens of the mammalian axis. *Anat. Rec.* 164:173–184.

Jenkins, F.A. Jr. 1971. The postcranial skeleton of African cynodonts. *Bull. Peabody Mus. Nat. Hist.* 36:1–216.

Jenkins, F.A. Jr., and McClearn, D. 1984. Mechanisms of hind foot reversal in climbing mammals. *J. Morph.* 182:197–219.

Johnson, A. 1983. The physiology of tooth eruption. *Birth Defects: Original Article Series* 19:67–82.

Kemp, T. 1979. The primitive cynodont *Procynosuchus*: Functional anatomy of the skull and relationships. *Phil. Trans. Roy. Soc. London* B285:73–122.

Kemp, T. 1980. The primitive cynodont *Procynosuchus*: Structure, function and evolution of the postcranial skeleton. *Phil. Trans. Roy. Soc. London* B288:217–258.

Kemp, T. 1982. *Mammal-like reptiles and the Origin of mammals*. London: Academic Press.

Kemp, T., 1983. The relationships of mammals. *Zool. J. Linn. Soc. London* 77:353–384.

Kemp, T. 1988a. A note on the Mesozoic mammals, and the origin of therians. In: *The phylogeny and classification of the tetrapods, vol. 2: Mammals* (Benton, M. J., ed.). Syst. Assoc. Spec. Vol. No. 35B. Oxford: Clarendon Press, pp. 23–29.

Kemp, T. 1988b. Interrelationships of the Synapsida. In: *The phylogeny and classification of the tetrapods, vol. 2: Mammals* (Benton, M.J., ed.). Syst. Assoc. Spec. Vol., No. 35B. Oxford: Clarendon Press, pp. 1–22.

Kielan-Jaworowska, Z., Presely, R., and Poplin, C. 1986. The cranial vascular system in taeniolabidoid multituberculate mammals. *Phil Trans. Roy. Soc. London* B313:525–602.

Kitching, J.W. 1977. The distribution of the Karroo vertebrate fauna. Mem. No. 1, Bernard Price Institute for Palaeontological Research, University of Witwatersrand, pp. 1–131.

Kluge, A.G. 1989. A concern for evidence and a phylogenetic hypothesis of relationships among *Epicrates* (Boidae, Serpentes). *Syst. Zool..* 38:7–25.

Krebs, B. 1987. The skeleton of a Jurassic eupantothere and the arboreal origin of modern mammals. In: *Fourth symposium on Mesozoic terrestrial ecosystems, short papers* (Currie, P.M., and Koster, E.H., eds.). Occ. Pap. No. 3., Tyrell Museum of Paleontology, Drumheller, pp. 132–137.

Kühne, W.G. 1956. *The Liassic therapsid Oligokyphus*. London: Brit. Mus. (Nat. Hist.).

Lillegraven, J.A., Thompson, S.D., McNab, B.K., and Patton, J.L. 1987. The origin of eutherian mammals. *Biol. J. Linn. Soc.* 32:281–336.

Marshall, L.G. 1979. Evolution of metatherian and eutherian (mammalian) characters: A review based on cladistic methodology. *Zool. J. Linn. Soc. London* 66:369–410.

Marshall, L.G., and Corruccini, R.S. 1978. Variability, evolutionary rates, and allometry in dwarfing lineages. *Paleobiology* 4:101–119.

McMahon, T.A., and Bonner, J.T. 1983. *On size and life*. New York: Scientific American Books, Inc.

Miao, D., and Lillegraven, J.A. 1986. Discovery of three ear ossicles in a multituberculate mammal. *Nat. Geograph. Res.* 2:500–507.

Mickevich, M.F., and Platnick, N.I. 1989. On the information content of classifications. *Cladistics* 5:33–47.

Noble, H.W. 1969. Evolution of the mammalian periodontum. In: *Biology of the Periodontum* (Melcher, A.H., and Bowen, W.H., eds.). London, New York: Academic Press, pp. 1–26.

Novacek, M.J. 1986. The skull of leptictid insectivorans and the higher-level classification of eutherian mammals. *Bull. Am. Mus. Nat. Hist.* 183:1–112.

Novacek, M.J. 1989. Higher mammalian phylogeny: The

morphological-molecular synthesis. In: *The hierarchy of life: Molecules and morphology in phylogenetic analysis* (Fernholm, B., Bremer, K., and Jornvall, H., eds.). Amsterdam: Nobel Symposium 70, Excerpta Medica, pp. 421–435.

Novacek, M.J. 1990. Morphology, paleontology, and the higher clades of mammals. In: *Current mammalogy, vol. 2* (Genarys, H.H., ed.). New York: Plenum, pp. 507–543.

Novacek, M.J., and Wyss, A.R. 1986. Higher-level relationships of the Recent eutherian orders: Morphological evidence. *Cladistics* 2:257–287.

Olsen, P.E., and Galton, P.M. 1984. A review of the reptile and amphibian assemblages from the Stormberg of southern Africa, with special emphasis on the footprints and the age of the Stormberg. *Palaeontol. Afr.* 25:87–110.

Olson, E.C. 1959. The evolution of mammalian characters. *Evolution* 13:344–353.

Olson, E.C. 1962. Late Permian terrestrial vertebrates, U.S.A. and U.S.S.R. *Trans. Am. Phil. Soc.* 52:1–224.

Patterson, B., and Olson, E.C. 1961. A triconodontid mammal from the Triassic of Yunnan. In: *International colloquium on the evolution of lower and specialized mammals* Part 1 (Vandebroek, G., ed.). Kon. Vlaamse Acad. Wetensch. Lett. Schone Kunstten Belgie, Brussels, pp. 129–191.

Presley, R. 1981. Alisphenoid equivalents in placentals, marsupials, monotremes and fossils. *Nature* 294:668–670.

Prothero, D.R. 1981. New Jurassic mammals from Como Bluff, Wyoming, and the interrelationships of non-tribosphenic Theria. *Bull. Am. Mus. Nat. Hist.* 167:280–325.

Reed, C.A. 1960. Polyphyletic or monophyletic ancestry of mammals, or: What is a class? *Evolution* 14:314–322.

Rieppel, O. 1984. Miniaturization of the lizard skull: Its functional and evolutionary implications. *Symp. Zool. Soc. London* 52:503–520.

Roberts, T.R. 1981. Sundasalangidae, a new family of minute freshwater salmoniform fishes from southeast Asia. *Proc. Calif. Acad. Sci.* 42:295–302.

Romer, A.S. 1969. Cynodont reptile with incipient mammalian jaw articulation. *Science* 166:881–882.

Romer, A.S. 1970. The Chanares (Argentina) Triassic reptile fauna VI. A chiniquodontid cynodont with an incipient squamosal-dentary jaw articulation. *Breviora* 344:1–18.

Roth, J.J., Roth, E.C., and Hotton, N. III. 1986. The parietal foramen and eye: Their function and fate in therapsids. In: *Ecology and biology of mammal-like reptiles* (Hotton, N.H. III, MacLean, P.D., Roth, J.J., and Roth, E.C., eds.). Washington, D.C.: Smithsonian Institution and National Institute of Mental Health, pp. 173–184.

Rowe, T. 1986. *Osteological diagnosis of Mammalia, L. 1758, and its relationship to extinct Synapsida*. Ph.D. Dissertation, University of California, Berkeley.

Rowe, T. 1987. Definition and diagnosis in the phylogenetic system. *Syst. Zool.* 36:208–211.

Rowe, T. 1988. Definition, diagnosis and origin of Mammalia. *J. Vert. Paleo.* 8(3):241–246.

Rowe, T. 1989. Skeletal ontogeny and the origin of mammals. Amer. Soc. Mammal., Abstracts to 69th Ann. Meet., pp. 34.

Rowe, T. 1990. Tempo and mode in early mammalian morphogenesis. *J. Vert. Paleo.* 10 (Suppl. to No. 3):18A.

Rowe, T., and Simmons, N. Unpublished. Early mammalian phylogeny and the effects of incompleteness on phylogenetic resolution.

Sereno, P.C., and McKenna, M.C. 1990. The multituberculate clavicle and interclavicle, and the early evolution of the mammalian pectoral girdle. *J. Vert. Paleon.* 10 (Suppl. to No. 3):42A.

Sigogneau-Russell, D. 1989. Haramiyidae (Mammalia, Allotheria) en Provenance du Trias Superieur de Lorraine (France). *Palaeontographica, Abt. A.* 206:137–198.

Simpson, G.G. 1959. Mesozoic mammals and the polyphyletic origin of mammals. *Evolution* 13:405–414.

Simpson, G.G. 1960. Diagnosis of the classes Reptilia and Mammalia. *Evolution* 14:388–392.

Simpson, G.G. 1961. Evolution of Mesozoic mammals. In: *International colloquium on the evolution of lower and specialized mammals.* Part 1 (Vandebroek, G., ed.). Kon. Vlaamse Acad. Wetensch. ,Lett. Schone Kunstten Belgie, Brussels, pp. 57–95.

Simpson, G.G. 1971. Concluding remarks. In: Early mammals (Kermack, D.M., and Kermack, K.L., eds.). *Zool. J. Linn. Soc. London* 50 (Suppl. 1): 181–198.

Simpson, G.G., and Elftman, H.O. 1928. Hind limb musculature and habits of a Paleocene multituberculate. *Novitates* 333:1–19.

Sues, H.-D. 1985. The relationships of the Tritylodontidae (Synapsida). *Zool. J. Linn. Soc. London* 85:205–217.

Sues, H.-D. 1986a. The skull and dentition of two tritylodontid synapsids from the Lower Jurassic of western North America. *Bull. Mus. Comp. Zool.* 151:217–268.

Sues, H.-D. 1986b. Relationships and biostratigraphic significance of the Tritylodontidae (Synapsida) from the Kayenta Formation of northeastern Arizona. In: *The beginning of the age of dinosaurs* (Padian, K., ed.). New York: Cambridge University Press, pp. 279–284.

Swofford, D.L. 1989. Phylogenetic analysis using parsimony [PAUP], Release 3. Macintosh computer software distributed on 31/2-inch diskette by the Illinois Natural History Survey.

Ten Cate, A.R. 1969. The mechanism of tooth eruption. In: *Biology of the Periodontum* (Melcher, A.H., and Bowen, W.H., eds.). London, New York: Academic Press, pp. 91–103.

Van Valen, L. 1960. Therapsids as mammals. *Evolution* 14:304–313.

Wake, D.B. 1966. Comparative osteology and evolution of the lungless salamanders, family Plethodontidae. *Mem. S. Calif. Acad. Sci.* 4:1–111.

Wible, J.R. 1987. The eutherian stapedial artery: Character analysis and implications for superordinal relationships. *Zool. J. Linn. Soc.* 91:107–135.

Wible, J.R. 1990. Petrosals of late Cretaceous marsupials from North America, and a cladistic analysis of the petrosal in therian mammals. *J. Vert. Paleo.* 10:183–205.

Wible, J.R. 1991. Origin of Mammalia: The craniodental evidence reexamined. *J. Vert. Paleont.* 11:1–28.

Wible, J.R., Miao, D., and Hopson, J.A. 1990. The septomaxilla of fossil and recent synapsids and the problem of the septomaxilla of monotremes and armadillos. *Zool. J. Linn. Soc.* 98:203–228.

CHAPTER 11
Phylogeny of Multituberculata

NANCY B. SIMMONS

Overview

Despite the importance of multituberculates as major components of many Mesozoic and early Tertiary faunas, the evolutionary relationships among these mammals have remained poorly understood. To address this issue, a cladistic analysis of relationships among forty-nine multituberculate taxa was conducted using dental and cranial characters. This analysis resulted in a strict consensus tree with a resolution of 66%. The relatively low level of resolution in this tree seems largely to be a result of including incompletely known taxa in the analysis. Sequential addition analyses (in which taxa were added to the analysis in order of decreasing completeness) indicated that even the most incomplete taxa preserve information important for understanding multituberculate relationships. Sequentially deleting such taxa provided a method for testing the stability of observed clades.

Based on the hypothesis of relationships presented in this study, Plagiaulacoidea is recognized as a paraphyletic group composed of several lineages of primitive multituberculates. Ptilodontoidea appears to form a monophyletic group if *Boffius* and *Liotomus* are removed to Taeniolabidoidea. Although support for the group is somewhat ambiguous, Taeniolabidoidea may also be monophyletic, provided that *Eobaatar* and *Monobaatar* are removed. Members of Cimolomyidae all appear to belong to various lineages within Taeniolabidoidea. Ptilodontoidea and Taeniolabidoidea together constitute a more inclusive monophyletic group, Cimolodonta McKenna, 1975.

The hypotheses of phylogeny presented in this study provide a framework within which evolution of multituberculate morphology can be studied. Results of the current analyses indicate that while cusp numbers and tooth size often vary in direct relationship to one another, these aspects of tooth morphology were not tightly coupled in any lineage during multituberculate evolution. As suggested by Krause and Carlson (1987), "gigantoprismatic" enamel appears to have evolved only once in multituberculates, within derived "plagiaulacoids" ancestral to Cimolodonta. In contrast, small prismatic enamel apparently evolved three times—once in late Cretaceous ptilodontoids and twice in late Paleocene taeniolabidoids. Fully restricted enamel on the lower incisors, long thought to be limited to taeniolabidoids, seems to have evolved once in "plagiaulacoids" and several times in taeniolabidoids. Partial restriction of the enamel appears to be diagnostic for Taeniolabidoidea.

Contents

Introduction, 147
History of Phylogenetic Studies of Multituberculata, 147
Relationships of Multituberculata Within Mammalia, 148
Definition and Diagnosis of Multituberculata, 148
Methods and Results, 148
    Phylogenetic Analyses, 148
    Optimization of Character Transformations, 152
Discussion, 152
    Phylogenetic Relationships and Classification, 152
    Implications for Evolution of Selected Characters, 154
Conclusions, 156
Appendix 1: Taxa Included in Analysis, 157
Appendix 2: Character Descriptions, 157
Appendix 3: Taxon-Character Matrix, 159
Appendix 4: Taxonomic Diagnoses, 160
Acknowledgments, 163
References, 163

Chapter 11. Phylogeny of Multituberculata

Introduction

Multituberculata is an extinct order of non-therian mammals known from various Jurassic through Oligocene deposits in Europe, Asia, North America, and South America. Unfortunately, evolutionary relationships among multituberculates remain poorly understood. Previous studies of multituberculate systematics have been variously hampered by exclusive reliance on dental characters, use of family or higher-level groups of untested monophyly as operational taxonomic units, and application of phenetic or unspecified methodologies. The goal of this chapter is to provide a substantiated hypothesis of multituberculate phylogeny based on cladistic analysis of both cranial and dental characters at generic and specific levels. This phylogeny offers a basis for assessing the monophyly of currently recognized multituberculate families and suborders, and will provide a practical framework for future studies of multituberculate relationships, functional morphology, biogeography, and origin and extinction.

History of Phylogenetic Studies of Multituberculata

The first comprehensive systematic review of multituberculates was conducted by Simpson (1928, 1929), who recognized three groups within Multituberculata: Plagiaulacidae, Ptilodontidae, and Taeniolabididae. Plagiaulacidae was interpreted as a primitive group that eventually gave rise to the latter two taxa (Simpson, 1928, 1929). Microcleptidae (=Haramiyidae) was included in Multituberculata by Simpson (1928, 1929), but was later removed from the group and left in Mammalia *incertae sedis* (Simpson, 1945). The three families recognized by Simpson (1945) were elevated to subordinal rank in the 1960s and continue to be accepted by workers today as the major subdivisions of Multituberculata: Plagiaulacoidea, Ptilodontoidea, and Taeniolabidoidea.

Several phylogenies of multituberculates were proposed before the widespread use of cladistic methods of phylogenetic analysis. These phylogenies were not supported with explicit data, but summarized the opinions of highly knowledgeable authors. A phylogeny proposed by Hahn (1969, fig. 85) suggests that Ptilodontoidea and Taeniolabidoidea are sister taxa; Plagiaulacidae is the sistergroup to this lineage, *Paulchoffatia* in turn being the sistergroup to Plagiaulacidae + Ptilodontoidea + Taeniolabidoidea. Kielan-Jaworowska (1974, fig. 1) similarly indicated paraphyly of Plagiaulacoidea, and further suggested possible polyphyly of Taeniolabidoidea and Ptilodontoidea (the latter two groups each containing more than one lineage descendant from plagiaulacoids).

McKenna (1975) offered a strict cladistic classification of the higher-order groups within Multituberculata that implied specific relationships among presumably monophyletic taxa. Multituberculata was defined as comprising the groups Haramiyoidea and Plagiaulacoidea; Plagiaulacoidea included Plagiaulacida (presumably containing some or all of the taxa traditionally included in Plagiaulacoidea) and a new taxon, Cimolodonta. Cimolodonta was defined as including Ptilodontoidea and Taeniolabidoidea (McKenna, 1975). Under this scheme, Ptilodontoidea and Taeniolabidoidea are both contained *within* Plagiaulacoidea—a radical departure from the traditional concept of the latter group. Because this phylogeny/classification was neither discussed nor supported by data in McKenna (1975), all subsequent authors have chosen to ignore this classification in favor of more traditional schemes (e.g., Clemens and Kielan-Jaworowska, 1979; Hahn and Hahn, 1983).

Archibald (1982) conducted the first cladistic analysis of multituberculate interrelationships as part of his study of late Cretaceous/Tertiary mammals. Archibald's (1982) analysis was based on sixteen dental characters surveyed in Haramiyidae, Paulchoffatiidae, Plagiaulacidae, Ptilodontoidea, Taeniolabidoidea, Cimolomyidae, *Paracimexomys*, and *Cimexomys*. Monophyly of each of these groups was assumed for purposes of the study and was not tested. Manual (noncomputer) analysis resulted in identification of two "most probable cladistic relationships" for the taxa considered (Archibald, 1982, p. 119). Both cladograms agree that Ptilodontoidea and Taeniolabidoidea form a monophyletic group that also includes Cimolomyidae, *Paracimexomys*, and *Cimexomys*; the sistergroup to this clade is Plagiaulacidae. Paulchoffatiidae and Haramiyidae probably form the next two consecutive sister taxa (Archibald, 1982, fig. 40).

Krause and Carlson (1987) conducted a cladistic analysis of relationships among the thirty most completely known late Cretaceous/Tertiary multituberculates as part of their study of enamel ultrastructure. Plagiaulacoidea was used as the outgroup for this analysis, which was based on twenty-two dental characters. This analysis confirmed the monophyly of Ptilodontoidea and Taeniolabidoidea as defined by Hahn and Hahn (1983). Cimolomyidae was identified as the sistergroup of Ptilodontoidea; Ptilodontoidea + Cimolomyidae + Taeniolabidoidea were found to form a larger monophyletic group; and *Cimexomys* and *Paracimexomys* were found to be the consecutive sistergroups to this larger group (Krause and Carlson, 1987). Relationships of taxa within Taeniolabidoidea were not discussed.

Stevens (1988) reported in an abstract the results of cladistic analysis of forty-three dental characters in taeniolabidoid multituberculates; Plagiaulacoidea was used as the outgroup. Three groups were identified by this analysis: "a clade group consisting of Eucosmodontinae as defined by Jepsen (1940), a clade consisting of *Buginbaatar*, *Nemegtbaatar*, and the Taeniolabididae as defined by Granger and Simpson (1929), and a paraphyletic group of more primitive Late Cretaceous Mongolian genera" (Stevens, 1988, p. 26A). These results suggest that Eucosmodontidae—as recognized by Clemens and Kielan-Jaworowska (1979) and Hahn and Hahn (1983)—is a paraphyletic taxon.

Hahn et al. (1989, fig. 12) presented a phylogeny indicating that Multituberculata is composed of Paulchoffatiidae plus a lineage comprising "later multituberculates" (presumably including Plagiaulacidae, Ptilodontoidea, and Taeniolabidoidea). The sistergroup of Multituberculata was identified as Haramiyidae; Theroteinidae formed the next proximate sister taxon.

Relationships of Multituberculata Within Mammalia

The phylogenetic position of Multituberculata within Mammalia has been a subject of controversy for decades (for a summary, see Greenwald, 1989a). Most recently, a series of phylogenetic analysis of higher-level taxa within Synapsida has supported the hypothesis that Multituberculata is a sistergroup to Theria (Marsupialia + Placentalia); *Gobiconodon*, Monotremata, *Morganucodon*, *Sinoconodon*, Tritylodontidae, Tritheledontidae, and *Exaeretodon* form successive outgroups (Rowe, 1986, 1988; Rowe and Greenwald, 1987; Rowe, this volume, chapter 10). Although this phylogenetic position for multituberculates (i.e., more closely related to therian mammals than are monotremes) is still controversial, these relationships are accepted in this study for the purpose of outgroup comparisons. *Vincelestes* (and any number of "pantotheres") may eventually prove to be more closely related to Theria than is Multituberculata, but this is not expected to change the hypothesis of relative relationships described above (see discussion in Rowe, this volume, chapter 10). In any case, polarities of the characters used in this study are not changed by the inclusion or exclusion of *Vincelestes* from the formal outgroup comparisons. Haramiyidae and Theroteinidae, taxa that are known only from isolated teeth, do not represent useful outgroups to Multituberculata because they cannot yet be demonstrated to share any unequivocally derived character states with multituberculates (this assessment is contrary to the position of Hahn et al., 1989; for discussion see Rowe, this volume, chapter 10).

Definition and Diagnosis of Multituberculata

As is discussed below, those taxa traditionally placed in Ptilodontoidea and Taeniolabidoidea together form a well-diagnosed monophyletic group for which the name Cimolodonta McKenna, 1975, is most appropriate. Multituberculata can be defined as including the most recent common ancestor of *Paulchoffatia* and Cimolodonta, and all of the descendants of that ancestor (note that this is equivalent to the definition implied by Hahn et al., 1989, fig. 12). As so defined, Multituberculata includes all of the taxa traditionally assigned to Plagiaulacoidea, Taeniolabidoidea, and Ptilodontoidea (*sensu* Clemens and Kielan-Jaworowska, 1979; Hahn and Hahn, 1983). The following derived characters unequivocally diagnose Multituberculata as it is defined above: (1) presence of a single lower incisor tooth that is separated from the cheekteeth by a large diastema; (2) absence of an enlarged caniniform tooth in the lower dentition; (3) presence of blade-like lower premolars with multiple serrations; (4) rotative eruption of primary generation lower premolars; (5) maximum of three upper incisors implanted in the premaxilla; (6) second upper incisor greatly enlarged; (7) absence of an enlarged caniniform tooth in the upper dentition; (8) absence of transverse cristae between cusps on molar teeth; (9) medial offset of last upper molar; (10) posterior end of lower tooth row located medial to coronoid process of dentary; (11) jugal thin, restricted to medial side of zygoma; (12) glenoid fossa with "tear-drop" shaped outline, posterior end broadly expanded mediolaterally; (13) absence of orbital process of palatine; (14) vestibule of petrosal greatly inflated; (15) supraglenoid foramen located in raised lateral wall of lateral flange of petrosal; (16) large, deep post-temporal fossa in contact with subarcuate fossa; (17) aquaeductus vestibuli with slit-like opening on posteromedial wall of subarcuate fossa; (18) prootic canal opening in lateral side of subarcuate fossa and large prootic sinus (see Miao (1988); Chapter 6, this volume; and Hahn (1988) for further discussion of basicranial characters). One additional derived character minimally diagnoses Cimolodonta but may diagnose Multituberculata: (19) calcaneum with a ventrally projecting peroneal tubercle comprising a bulbous head set on a distinct neck.

Methods and Results

PHYLOGENETIC ANALYSES. Phylogenetic relationships among forty-nine multituberculate taxa were investigated using computer-assisted cladistic analysis. Unfortunately, many multituberculate taxa are known only from a few teeth, making comparable taxonomic units difficult to identify. The operational taxonomic units (OTUs) defined for this analysis consist of both species and genera (see summary in Appendix 1). OTUs consisting of single species were preferred (as these seem to have the greatest chance of being monophyletic), but incomplete data at the species level often forced consideration of larger groups. OTUs comprising more than ones species generally include all species referred to the genus by Hahn and Hahn (1983) or a more recent worker (see Appendix 1). Because of the broad scope of this study, monophyly of each group could not be tested; an initial assumption of this analysis is that each OTU is monophyletic.

A combined total of sixty-seven characters from the dentition and the skull were considered in this analysis (Appendix 2); postcranial remains were judged to be too rare (and too difficult to assign to recognized taxa) to be useful. Relatively continuous quantitative characters involving cusp formulae and measurements were divided into multiple states in such a way as to minimize the number of taxa found to overlap two character states. Most of the characters considered in the analysis were entered as unordered, nonadditive characters. Quantitative multistate characters (e.g., P_4 cusps, M_1 length) and some multistate characters apparently involving progressive changes in structures (e.g., position of the posterior border of the palate) were entered as ordered, additive characters (Appendix 2). Following the method of Maddison et al. (1984), the ancestral character states for Multituberculata were estimated using the outgroups discussed above. Taking into account uncertainty concerning homologies between multituberculate and therian teeth (Greenwald, 1988, 1989a), this method allowed unambiguous determination of the ancestral state for 63% of the characters considered in the analysis. The ancestral states of the remaining 37% of the characters were recorded as unknown ("9") in the taxon-character matrix.

The data in the taxon-character matrix (Appendix 3) were analyzed using Swofford's (1989) Phylogenetic Analysis Using Parsimony (PAUP) program version 3.0a. Character weights were scaled for equal character weighting regardless of the number of states (base weight = 1,000). The size of the data set required use of the Heuristic Search Procedure to identify most parsimonious trees; MAXTREES = 10,000 was set as a limit for the maximum number of trees to be saved in any analysis. Strict consensus trees were constructed by PAUP, based on all equally parsimonious trees found by the heuristic searches. MacClade 2.97.36 Test (Maddison and Maddison, in preparation) was used for input of the taxon-character matrix, character tracing, and some graphics output.

Resolution within a tree can be measured as a percentage by dividing the number of actual nodes in the tree by the maximum number of nodes possible (#taxa - 1). Using this system, a tree exhibiting a fully resolved dichotomous branching pattern would have resolution = 100%; a completely unresolved tree (with only one node) would have resolution < 1%. Analysis of the complete taxon-character matrix (including forty-nine OTUs plus a hypothetical ancestral taxon) resulted in identification of 10,000 equally parsimonious trees, from which a strict consensus tree with 66% resolution was constructed (Fig. 11.1; see discussion below). The resolution might have been even worse if the MAXTREE limit had been set higher and more trees had contributed to the final strict consensus tree.

Low levels of resolution in consensus trees that include poorly known taxa may sometimes be a function of the condition of strict consensus, which allows only those groups that are found in all the trees to be included in a final consensus tree (Rohlf, 1982). When taxa that are highly incomplete in the context of a given matrix (i.e., are coded "9" for a high percentage of matrix cells) are included in a phylogenetic analysis, large numbers of equally parsimonious trees are frequently found. These trees may differ from one another principally in the placement of incomplete taxa relative to those that are more complete. When the trees are compared using the strict consensus method, relationships that are stable in all trees may appear unresolved in the consensus tree simply because a few incomplete taxa can be placed in many variable positions.

Incomplete taxa may nevertheless preserve information that can positively affect the outcome of an analysis. Several recent workers have clearly demonstrated that completeness of a taxon and its informativeness in a phylogenetic analysis are not necessarily correlated (Gauthier et al., 1988; Donoghue et al., 1989; Novacek, 1989). Results of preliminary analyses of multituberculate phylogeny (Greenwald, 1989a, 1989b) have suggested that unique combinations of character states preserved in some poorly known (incomplete) taxa may affect tree topology significantly, altering hypothesized relationships between more complete taxa and indicating different distributions of character transformations. The disturbing implications of these results suggest that the "best" phylogeny (i.e., the phylogeny most likely to reflect the actual evolutionary relationships of the taxa in question) is one that takes into account the largest array of ingroup taxa. The trade-off seems to be that resolution is frequently very poor in cases in which many taxa are relatively incomplete.

Two different approaches seem obvious—either (a) include as many taxa as possible, and maximize information at the expense of resolution (e.g., Fig. 11.1), or (b) include only the most complete taxa, gaining resolution at the possible expense of information. In an effort to explore compromises between these two strategies, a method of sequential addition of taxa was employed. The level of completeness of each OTU in the matrix was calculated by dividing the number of matrix cells containing positive data for that taxon (not a "9") by the total number of characters (sixty-seven) to give a percentage of completeness. OTUs in this analysis ranged from 96% complete (*Ptilodus*) to only 12% complete (*Buginbaatar*; see Appendix 3). A sequence of forty-seven phylogenetic analyses were conducted

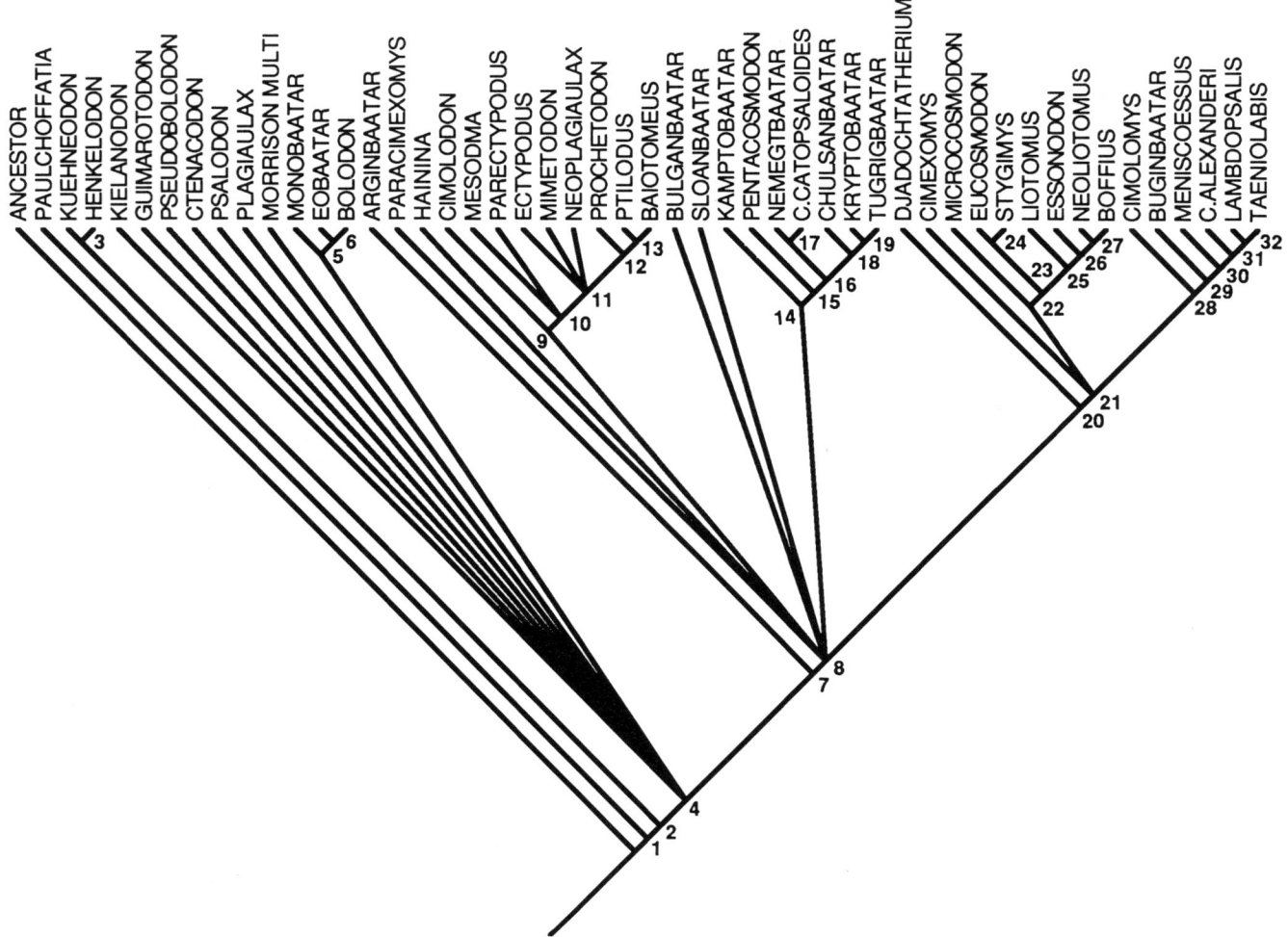

FIGURE 11.1. Strict consensus tree based on 10,000 equally parsimonious trees discovered by the heuristic search procedure of PAUP (see text). This tree is based on analysis of the complete taxon-character matrix considered in this study, which includes 49 multituberculate taxa and 67 characters (see Appendix 3). Tree length = 237 steps; consistency index (CI) = .391; retention index (RI) = .680. See Appendix 4 for diagnoses of the numbered nodes. Taxonomic names and group descriptors associated with numbered nodes (see text for explanation): Node 1: Multituberculata; Node 3: Kuehneodontinae; Node 8: Cimolodonta; Node 9: Ptilodontoidea; Node 12: Ptilodontidae; Node 14: Group X; Node 22: Group Y; Node 24: Eucosmodontidae; Node 28: Group Z; and Node 31: Taeniolabididae.

in such a way that each analysis included one more OTU than the preceding one (minimum = 3 OTUs, maximum = 49). OTUs were added to the data set in decreasing order of completeness, and the number of equally parsimonious trees, consistency index (CI) of each of these trees, and percent resolution of the strict consensus tree were recorded for each analysis (Fig. 11.2).

Inclusion of additional incomplete taxa generally (though not always) resulted in discovery of higher numbers of equally parsimonious trees with increasingly low CIs (Figs. 11.2A, 11.2B). Both results were expected, as CI is generally expected to decrease as the number of taxa is increased (Sanderson and Donoghue, 1989). Resolution varied widely within the resultant strict consensus trees (Fig. 11.2C) and was apparently not just a function of the completeness of the included taxa. Resolution dropped from over 80% to less than 60% early in the sequence of analyses, but addition of three more taxa raised the resolution to over 90%. Because of this subsequent return to high resolution, and because fewer than ten trees contributed to the poorly resolved consensus trees, it seems likely that the low resolution in this case was caused by the particular combinations of character states preserved in the included taxa, not by the level of missing data in these taxa.

Toward the end of the sequential analyses, resolution decreased (from 90% to 66%) in concert with a substantial increase in the number of equally parsimonious trees (from 45 to >10,000). It was initially hoped that a steep drop in resolution—associated with a sharp increase in the number of trees—might provide a cutoff

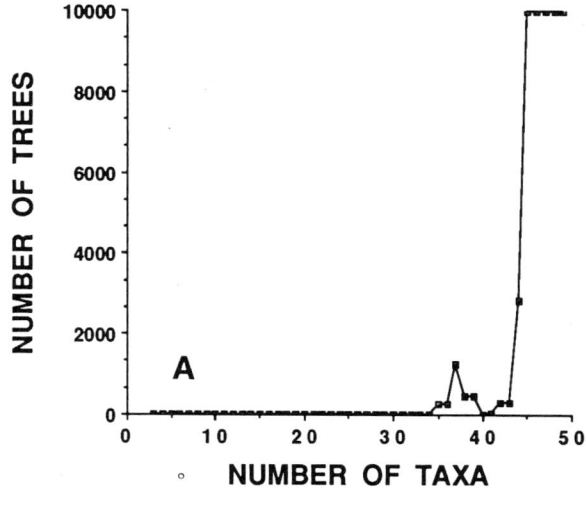
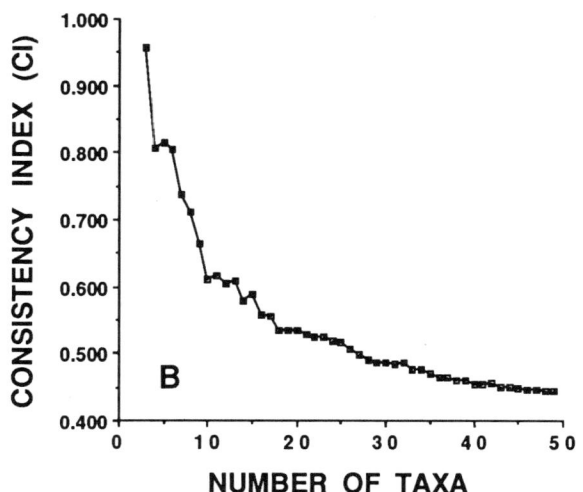
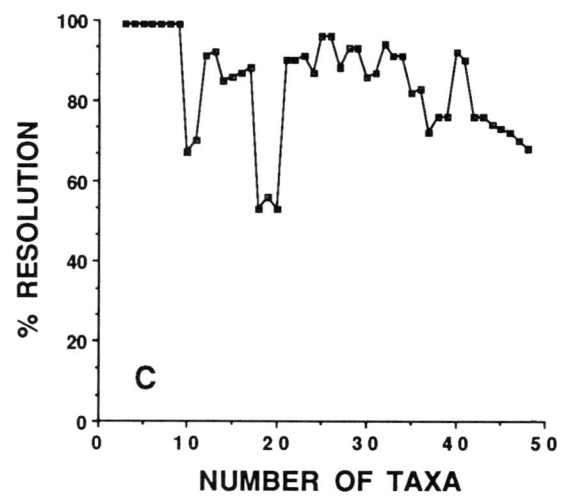

FIGURE 11.2. Results of sequential addition of increasingly incomplete taxa to the analysis. Graphs depict the observed relationships between the number of taxa included in an analysis and A: the number of equally parsimonious trees found, B: the consistency index, and C: the percent resolution (see text for description of analyses and discussion).

point that could be used to justify exclusion of additional incomplete taxa from the analysis. Compared with previous fluctuations in resolution and numbers of trees, no such point could be easily identified (see Fig. 11.2). Examination of tree topologies of the consensus trees generated in the latter parts of the "sequential addition" analyses confirmed that addition of some extremely incomplete taxa (e.g., *Hainina*, 21% complete; *Buginbaatar*, 12% complete) significantly affected the relationships among some more complete taxa. Because these incomplete taxa clearly contribute data that are important for understanding multituberculate relationships, eliminating them from the analysis cannot be justified despite the decrease in resolution associated with their inclusion. Accordingly, the tree based on the most possible information (e.g., the complete data matrix) in this case seems preferable to any better resolved tree that excludes some taxa based on their incompleteness. Figure 11.1 seems to be the best "big picture" phylogeny of multituberculates that can be constructed given the available data.

Comparisons of consensus trees generated at different stages of the sequential addition analyses provide a method for assessing the stability of the relationships implied by Figure 11.1. Interestingly, most of the instability in the tree is apparently concentrated in a few specific areas of the tree. The following nodes in Figure 11.1 were consistently supported throughout the sequential analyses: Nodes 1–8, 10–14, 17–18, 22–27, and 29–32. Although some of these nodes are not supported by many derived character states, their persistence suggests that they indicate "real" monophyletic groups. On the other hand, Nodes 9, 15–16, 19–21, and 28 were not supported in other consensus trees that lacked

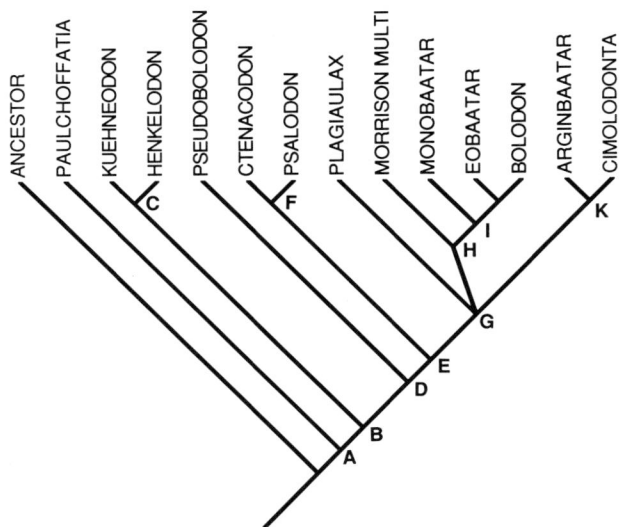

FIGURE 11.3. Hypothesis of relationships among "plagiaulacoid" multituberculates considered in this study. This tree is based on consideration of results of several sequential addition analyses (see text). See Appendix 4 for diagnoses of the various nodes. Formal taxonomic names associated with numbered nodes (see text for explanation): Node A: Multituberculata; Node C: Kuehneodontinae; Node F: Allodontidae; and Node H: Bolodontidae.

various poorly known taxa. Relationships of the taxa associated with these nodes must be considered questionable (see discussion below).

In contrast, some of the uncertainty implied by the polytomies shown in Figure 11.1 may not fairly reflect our knowledge. Node 4 (with nine branches) and Node 8 (with seven branches) appear much better resolved in most consensus trees generated during the sequential analyses. Figure 11.3 illustrates possible relationships among various "plagiaulacoid" multituberculates; these relationships were supported throughout most of the sequential analyses, becoming obscured only when *Guimarotodon* (18% complete) and *Kielanodon* (15% complete) were added to the analysis. Note that unresolved Node 4 from Figure 11.1 is resolved into five separate nodes (Nodes D–H) in Figure 11.3.

The situation with Node 8 in Figure 11.1—the node defining Cimolodonta—is considerably more complicated. Just as in the case described above, removing various incomplete taxa fully resolves this node. However, the relationships that appear in these resolved trees differ according to the taxa removed. Taxa referred to Taeniolabidoidea by Hahn and Hahn (1983) frequently form a monophyletic group, but *Cimolodon* sometimes appears to be more closely related to these taxa than to ptilodontoids. Relationships among various taeniolabidoid clades change slightly as incomplete taxa are removed, even if these taxa (such as *Guimarotodon*) clearly do not belong within Cimolodonta;

Nodes 20 and 21 in Figure 11.1 are particularly unstable. Given this uncertainty, choosing one resolved tree of cimolodont taxa over other trees cannot be justified.

OPTIMIZATION OF CHARACTER TRANSFORMATIONS. After construction of the trees illustrated in Figures 11.1 and 11.3, character optimization was done using the DELTRAN option of PAUP to identify the most parsimonious arrangement of transformations given the specified tree topologies. The results of the optimization were checked to identify transformations for which equally parsimonious interpretations are possible (e.g., convergence and reversal are equally likely). The character states resulting from such ambiguous transformations are included in the taxonomic diagnoses (Appendix 4) because they may represent synapomorphies of the taxa in question; these transformations are marked with an asterisk (*) to distinguish them from unequivocal synapomorphies.

An additional problem in assigning transformations to particular nodes occurs as a result of missing data. DELTRAN assigns unequivocal transformations to the minimal level at which they can be observed; sometimes missing data in sister taxa leaves open the possibility that a particular transformation might actually have evolved at a lower level in the tree. Such transformations are clearly unequivocal synapomorphies, but the level at which they apply remains uncertain. In the taxonomic diagnoses presented below (Appendix 4), these transformations are included at the minimal level at which they can be demonstrated, and are marked "#" in order to distinguish them from unequivocal transformations of known distribution (unmarked) and equivocal transformations (marked "*").

Discussion

PHYLOGENETIC RELATIONSHIPS AND CLASSIFICATION. Novacek et al. (1988, p. 55) cautioned that "expectations [for reflecting true evolutionary relationships] would be rather low for a morphologically based classification where more than 30% of the relevant data were missing." The matrix on which Figure 11.1 is based is only 53% complete; well over 40% of the "relevant data" are missing. Unfortunately, little short of discovering new specimens can be done to improve the completeness of this data set, because nonpreservation is the source of most of the missing data. Because Multituberculata includes no living taxa, the fossil record (incomplete as it may be) is the only source of data available, and any phylogenies or classifications must be based upon this record.

Hahn and Hahn (1983) provided a comprehensive classification of multituberculates that summarized all previous work on the group. Many of the monophyletic groups identified in the current analysis correspond roughly with higher taxa described in Hahn and Hahn (1983), while others do not. In many cases, adding or subtracting a few taxa from the groups recognized by Hahn and Hahn (1983) apparently renders these groups monophyletic. Because of the questionable stability of parts of the phylogenies presented in Figures 11.1 and 11.3, recommendations for changing the existing classification scheme (that of Hahn and Hahn, 1983) are made only in cases in which such changes are clearly supported by the analyses described above.

Plagiaulacoidea *sensu* Hahn and Hahn (1983) is clearly a paraphyletic group that consists of several lineages of primitive multituberculates. This name, if used at all, should be explicitly defined as an informal non-monophyletic group, and should be cited in quotation marks ("Plagiaulacoidea") to distinguish it from monophyletic taxa. Similarly, Paulchoffatiidae, Paulchoffatiinae, and Kuehneodontinae *sensu* Hahn and Hahn (1983) are also not monophyletic. *Paulchoffatia* is the most primitive known multituberculate; if used at all, Paulchoffatiidae (and Paulchoffatiinae) should be restricted to include only *Paulchoffatia*. Other taxa previously placed in Paulchoffatiinae (e.g., *Pseudobolodon*) appear to be more closely related to late Cretaceous/Tertiary multituberculates than to *Paulchoffatia*. Kuehneodontinae may be redefined as a monophyletic group by restricting its membership to *Kuehneodon* and *Henkelodon*; *Guimarotodon*, *Kielanodon*, and *Bolodon* apparently belong to groups more closely related to late Cretaceous/Tertiary multituberculates.

Plagiaulacidae *sensu* Hahn and Hahn (1983) does not appear to be monophyletic. *Ctenacodon* and *Psalodon* form a stable monophyletic group, but *Plagiaulax* seems to be more closely related to late Cretaceous/Tertiary multituberculates than to these taxa. Because *Plagiaulax* is a very incomplete taxon (only 21% complete), it is recommended that Plagiaulacidae be restricted to only *Plagiaulax* until more complete specimens become available for analysis. *Ctenacodon* + *Psalodon* can be referred to Allodontidae Marsh, 1889, to indicate their relationship to one another (*Allodon* Marsh is a junior synonym of *Ctenacodon*).

An interesting monophyletic assemblage identified by this analysis includes the "Morrison multituberculate" (see Appendix 1) + (*Monobaatar* + (*Eobaatar* + *Bolodon*)). *Bolodon* was previously placed in Plagiaulacoidea: Paulchoffatiidae: Kuehneodontinae (Hahn and Hahn, 1983), while *Eobaatar* (Eobaataridae) and *Monobaatar* (family *incertae sedis*) were considered to represent primitive members of Taeniolabidoidea (Kielan-Jaworowska et al., 1987). This assemblage, for which the name Bolodontidae Osborn, 1887, seems most appropriate, was supported in all of the sequential analyses except those including *Guimarotodon* (18% complete) and *Kielanodon* (15% complete). The final "plagiaulacoid" multituberculate taxon included in this analysis, *Arginbaatar* (assigned to the monotypic family Arginbaataridae by Hahn and Hahn, 1983), appears to be the sistergroup to Cimolodonta.

McKenna (1975) proposed the name Cimolodonta to refer to the monophyletic group composed of Ptilodontoidea + Taeniolabidoidea, but did not diagnose this taxon. The current analysis strongly supports monophyly of Cimolodonta, which can diagnosed by at least six and perhaps as many as seventeen synapomorphies (see Appendix 4). Cimolodonta, as defined here, includes *Ptilodus* and *Taeniolabis* and all taxa that have evolved from their most recent common ancestor. Ideally, Ptilodontoidea should be defined as including all multituberculates that are more closely related to *Ptilodus* than to *Taeniolabis*, and Taeniolabidoidea should be defined as including all multituberculates that are more closely related to *Taeniolabis* than to *Ptilodus*. Unfortunately, monophyly of Ptilodontoidea and Taeniolabidoidea (as thus defined) is not easily addressed. Both taxa appear polyphyletic if one includes all the taxa referred to them by Hahn and Hahn (1983) and Kielan-Jaworowska et al. (1987). However, reassignment of a few problem genera permits redefinition of Ptilodontoidea and Taeniolabidoidea such that these groups may be monophyletic (see below). Based on the results of the current analysis, *Boffius* and *Liotomus* must be removed from Ptilodontoidea to Taeniolabidoidea; *Monobaatar* and *Eobaatar* must be removed from Taeniolabidoidea to Bolodontidae; and *Cimexomys* and members of Cimolomyidae should be placed in Taeniolabidoidea. Following these changes, the next task is to assess the monophyly of the resulting groups.

In the case of Ptilodontoidea, which appears monophyletic in Figure 11.1, removal of various incomplete taxa from the analysis (e.g., *Plagiaulax*, *Essonodon*) produced consensus trees suggesting that *Cimolodon* may be more closely related to Taeniolabidoidea than to *Ptilodus*. The other taxa traditionally referred to Ptilodontoidea (*Mesodma*, *Parectypodus*, *Ectypodus*, *Mimetodon*, *Neoplagiaulax*, *Prochetodon*, *Ptilodus*, and *Baiotomeus*) together form a stable monophyletic group that was supported throughout the sequential addition analyses. Monophyly of Ptilodontoidea is therefore relatively certain if *Cimolodon* is excluded; monophyly is clearly possible if *Cimolodon* is included, but it is not strongly supported by the current data set. Because *Cimolodon* is traditionally included in Ptilodontoidea, and because it groups with other ptilodontoids in Figure 11.1, it is recommended that *Cimolodon* be retained in Ptilodontoidea until additional data become available (see Appendix 4 for a diagnosis of Ptilodontoidea as so defined). The status of *Cimolodon* might be resolved through consideration of *Anconodon*, a poorly known taxon placed in Ptilodontoidea: Cimolodontidae by

Hahn and Hahn (1983); unfortunately, *Anconodon* was not included in the current analysis.

The situation with Taeniolabidoidea is somewhat different. In Figure 11.1, the relationships of taeniolabidoid taxa to other cimolodonts are ambiguous—taeniolabidoids might form a monophyletic group, or they might not. The sequential analyses, however, demonstrate that removal of a few incomplete taxa from the analysis (e.g., *Hainina*, *Plagiaulax*) results in consensus trees in which Taeniolabidoidea appears as a monophyletic group that includes all those taxa listed to the right of Ptilodontoidea in Figure 11.1 (see Appendix 4 for a diagnosis of this group). One can only conclude that Taeniolabidoidea is probably monophyletic, but that the available data do not provide unambiguous support for that grouping.

The current analysis provides information useful for assessing monophyly of lower-level taxa within Taeniolabidoidea and Ptilodontoidea. As discussed above, Ptilodontoidea *sensu* Hahn and Hahn (1983) can be redefined as a monophyletic group (as described above) only if *Boffius* and *Liotomus* are removed from the group. Two clades are tentatively recognized within Ptilodontoidea: one containing only *Cimolodon*, and another clade including taxa traditionally placed in either Ptilodontidae or Neoplagiaulacidae. Within this latter group, Ptilodontidae can be defined as a monophyletic group that contains *Prochetodon*, *Ptilodus*, and *Baiotomeus*; this definition is concordant with that of Hahn and Hahn (1983). Neoplagiaulacidae, on the other hand, appears to be a paraphyletic group consisting of various taxa that are more closely related to Ptilodontidae than to *Cimolodon*. Accordingly, Neoplagiaulacidae should be removed from usage.

Taeniolabidoidea as defined above is a large, potentially monophyletic taxon that includes three smaller monophyletic groups: Group X (defined by stable Node 14 in Figure 11.1), Group Y (Node 22), and Group Z (Node 28). In addition to the taxa included in these groups, *Cimexomys* (generally left *incertae sedis*), *Bulganbaatar* (previously assigned to Eucosmodontidae, e.g., Hahn and Hahn, 1983), *Sloanbaatar* (Sloanbaataridae), and *Djadochtatherium* (Taeniolabididae) belong to Taeniolabidoidea, although their positions within the group are unclear.

Group X includes a number of taxa previously assigned to Eucosmodontidae (*Nemegtbaatar*, *Kryptobaatar*, *Tugrigbaatar*, *Chulsanbaatar*, and *Pentacosmodon*) and two previously assigned to Taeniolabididae (*Catopsalis catopsaloides* and *Kamptobaatar*). Although the relationships of these taxa appear fully resolved in Figure 11.3, reference to other cladograms from the sequential analyses indicates that these relationships are not particularly stable—adding or deleting taxa elsewhere on the tree may affect placement of various members of Group X relative to one another. Monophyly of Group X, however, was supported by all of the analyses. Two points of stability within this group involve *Nemegtbaatar* + *Catopsalis catopsaloides*, which always appear in the same clade, and *Krytobaatar* + *Tugrigbaatar* + *Chulsanbaatar*, which always form a monophyletic group relative to the former clade.

Group Y includes *Microcosmodon*, Eucosmodontidae *sensu stricto*, and Boffiidae (see below). Microcosmodontinae, originally defined to include *Microcosmodon* and *Pentacosmodon* (Hahn and Hahn, 1983), should be removed from usage because *Pentacosmodon* apparently belongs to Group X. Eucosmodontidae should be restricted to only *Eucosmodon* + *Stygimys*. The remaining monophyletic taxon includes an interesting assemblage of taxa previously placed in widely separated groups (Hahn and Hahn, 1983); *Boffius* (previously placed in Ptilodontoidea: Boffiidae), *Neoliotomus* (Taeniolabidoidea: Eucosmodontidae), *Essonodon* (?Cimolomyidae, suborder *incertae sedis*), and *Liotomus* (Ptilodontoidea: Cimolodontidae). Boffiidae Hahn and Hahn, 1983, can be expanded to include the members of this assemblage, which apparently represents the sister clade of Eucosmodontidae.

Group Z includes Taeniolabididae, *Meniscoessus*, *Buginbaatar*, and *Cimolomys*. Taeniolabididae is restricted to *Catopsalis alexanderi*, *Lambdopsalis*, and *Taeniolabis*; the other taxa traditionally referred to this group (*Kamptobaatar*, *Djadochtatherium*, *Catopsalis catopsaloides*) do not appear to be closely related to Taeniolabididae. Cimolomyidae, generally considered to include *Cimolomys*, *Meniscoessus*, and perhaps *Essonodon* (e.g., Hahn and Hahn, 1983) cannot be defined monophyletically and should therefore be removed from usage. Although Figure 11.3 indicates that Groups Y and Z may be each other's closest relatives, these relative positions—and the positions of *Cimexomys* and *Djadochtatherium*—were not consistently supported in the sequential analyses.

IMPLICATIONS FOR EVOLUTION OF SELECTED CHARACTERS. The phylogeny presented above has numerous implications for interpretation of multituberculate morphology. Relatively high levels of homoplasy within Multituberculata have clearly produced a complex pattern of mosaic evolution that will require further study. Discussion of all of the evolutionary changes implied by this study is beyond the scope of this chapter, but hypothesized changes in several sets of characters seem particularly important in view of the roles these characters have played in multituberculate systematics.

Cusp formulae and tooth size. Homoplasy is obviously present in many of the characters considered in this study, particularly dental characters involving size and cusp formulae of the cheekteeth. However, the levels of homoplasy in cusp numbers and tooth lengths (Table 11.1) are not so high that such characters are rendered useless—good news for systematists, as many

TABLE 11.1. Homoplasy in length and cusp formulae of upper and lower fourth premolars and first molars

Character	CI	RI
(9) P_4 cusps	0.200	0.625
(10) P_4 length	0.167	0.364
(13) M_1 cusps	0.250	0.786
(14) M_1 length	0.300	0.741
(35) M^4 cusps	0.667	0.933
(36) P^4 length	0.273	0.529
(38) M^1 cusps	0.250	0.667
(39) P^1 length	0.300	0.741
Mean for all 67 characters	0.553	0.694

The consistency index (CI), which provides a simple measure of homoplasy, is calculated by dividing the minimum possible number of transformations by the actual number of transformations observed. The retention index (RI) compares the amount of observed homoplasy to the amount of homoplasy possible given the data matrix and tree topology that are specified. RI is calculated as follows: (maximum possible transformations − actual transformations)/(maximum transformations − minimum possible transformations). Both indices scale such that low values indicate relatively high amounts of homoplasy. Note that while the CI suggests that these dental characters (with the exception of P^4 cusps) exhibit substantially more homoplasy than mean, the RI values for several characters approach or exceed the mean RI. All calculations are based on the tree topology indicated in Figure 11.1; character optimization was conducted using the DELTRAN option of PAUP (see text). See Appendix 2 for character descriptions.

multituberculate taxa are known only from their teeth. These results simply indicate that caution is appropriate when such characters provide the principal source of data in systematic studies. Luckily, the dental characters considered in this study seem to vary independently in most lineages (see Appendix 4), suggesting that consideration of a complete dentition will generally provide enough data to estimate the phylogenetic relationships of a taxon fairly accurately.

Working at the species level, Cole and Krause (1988) identified a positive correlation between tooth size and cusp number in taeniolabidoid multituberculates; a similar correlation was not found in ptilodontoids. The results of the current study suggest that while cusp numbers and tooth size often vary in direct relationship to one another, these aspects of tooth morphology were not tightly coupled in any lineage during multituberculate evolution. Size of a tooth could apparently increase or decrease without a concomitant change in the cusp formula, and vice versa. In many instances there seems to have been some sort of evolutionary time lag between changes in these features—a tooth increased in size prior to evolving additional cusps, or vice versa. It is possible, however, that these patterns may be an artifact of the coding system used in this study. Further study is needed to resolve the relationship between tooth size and cusp number in multituberculates.

Enamel ultrastructure. Three distinct ultrastructural patterns occur in the dental enamel of multituberculate mammals; for summary and detailed accounts of enamel types, see Carlson and Krause (1985), Fosse et al. (1985), and Krause and Carlson (1987). Two patterns of prismatic enamel are recognized: "small prismatic" enamel, which contains small circular enamel prisms with complete prism boundaries; and "giantoprismatic" enamel, which contains large arc-shaped prisms with incomplete prism boundaries. In addition, some multituberculates have "preprismatic" enamel that is structurally intermediate between nonprismatic and prismatic types of enamel.

Krause and Carlson (1987) analyzed ultrastructural types in a phylogenetic framework provided by a cladistic analysis of gross morphology of multituberculate dentitions. Results of their analysis suggested that (a) preprismatic enamel is primitive for multituberculates; (b) giantoprismatic enamel represents a derived condition relative to preprismatic enamel; (c) giantoprismatic enamel evolved only once within multituberculates, in a common ancestor of *Paracimexomys*, ptilodontoids, and taeniolabidoids; (d) small prismatic enamel represents an additional derived condition, one that presumably evolved from giantoprismatic enamel; and (e) small prismatic enamel evolved at least twice within multituberculates, once in ptilodontoids and once in the taeniolabidoid *Neoliotomus* (Krause and Carlson, 1987). These conclusions are all supported by the current analysis, in which enamel ultrastructure was considered as an unordered character.

An addendum to the results of Krause and Carlson (1987) is the apparent evolution of small prismatic enamel a third time, in *Liotomus*. Previously considered to be a derived ptilodontoid, removal of this taxon to Taeniolabidoidea (see above) requires the hypothesis that this taxon evolved its small prismatic enamel independently from ptilodontoids and *Neoliotomus*. Based on the ages of relevant taxa, small prismatic enamel appears to have evolved once in the late Cretaceous (in ptilodontoids) and twice in the late Paleocene (*Liotomus* and *Neoliotomus*).

The current analysis indicates that the origin of giantoprismatic enamel within multituberculates dates back to at least the late Jurassic. Presence of giantoprismatic enamel in *Arginbaatar* (Lower Cretaceous) (Kielan-Jaworowska et al., 1987) and the "Morrison Multituberculate" (Upper Jurassic) (Engelmann et al., 1990) demonstrates that this derived condition minimally applies to Node 4 in Figure 11.1 and Node G in Figure 11.3. A study of enamel ultrastructure in other "plagiaulacoids" (e.g., *Henkelodon*, *Pseudobolodon*, *Ctenacodon*, *Psalodon*, *Plagiaulax*, *Bolodon*) could prove instrumental in resolving the relationships among these relatively primitive multituberculate taxa.

Enamel covering on incisor teeth. Incisors with a continuous layer of uniformly thick enamel are presumably primitive for multituberculates. Restriction of the enamel on the lower incisor to the ventrolabial surface of the tooth (to produce a "gliriform" or rodent-like incisor) has long been considered a diagnostic feature of taeniolabidoid multituberculates (e.g., Sloan and Van Valen, 1965; Clemens and Kielan-Jaworowska, 1978). Most recently, Kielan-Jaworowska et al. (1987, p. 7) stated that "the limited enamel on the lower incisor . . . is a taeniolabidoid synapomorphy and does not occur in any other multituberculate group." Full restriction of the enamel, producing a tooth with no enamel covering at all on the lingual surface, is seen in many taeniolabidoid taxa (e.g., *Taeniolabis*). Other taxa referred to Taeniolabidoidea exhibit conditions of partial restriction, in which some enamel (usually a very thin or tapering layer) still remains on the lingual surface of the tooth (e.g., *Kamptobaatar*, *Sloanbaatar*, *Meniscoessus*).

The results of this study indicate that the evolution of restricted enamel is not as clear-cut as previously suggested. Lower incisors with partially restricted enamel may be a synapomorphy of Taeniolabidoidea, but fully restricted enamel seems to have evolved as many as four or five times within taeniolabidoids (once within Group X, once within Group Z, once or twice in Group Y, and once in *Djadochtatherium*). More important, fully restricted enamel has evolved at least once outside Taeniolabidoidea—in the "plagiaulacoid" group Bolodontidae. The "Morrison Multituberculate" exhibits an interesting combination of a relatively primitive cheektooth dentition (e.g., canine and five premolars in the upper jaw) and fully restricted enamel on the incisors (Engelmann et al., 1990). This pattern of mosaic evolution indicates that great care should be taken in using enamel restriction as a "key" character when assigning poorly known species to higher-level taxa.

An even more interesting picture emerges when the enamel covering of the enlarged upper incisor is considered. Most taeniolabidoids for which upper incisors are known do not show any restriction of the enamel on these teeth; uniformly thick enamel on the upper incisor is clearly primitive for Taeniolabidoidea. Incisors referred to one taxon, *Bulganbaatar*, show partially restricted enamel. Fully restricted enamel appears limited to two clades within Taeniolabidoidea: *Nemegtbaatar* + *Catopsalis catopsaloides*, and Taeniolabididae. Fully restricted enamel also occurs on the upper incisors of the "Morrison Multituberculate."

When the enlarged upper and lower incisors are considered together, a more informative pattern emerges. The primitive condition of the incisors (both upper and lower) in multituberculates is for these teeth to be covered with a relatively uniform layer of enamel. This is retained in most "plagiaulacoids" and in ptilodontoids. In the most primitive taeniolabidoids, partial enamel restriction on the lower incisor apparently evolved without concomitant change in the covering of the upper incisor. From this condition, fully restricted enamel on the lower incisor evolved perhaps as many as five times (see above). In one case (Taeniolabididae), fully restricted enamel on the upper incisor apparently evolved at the same time as did full restriction on the lower incisor. In Group Y, fully restricted enamel on the lower incisor apparently evolved first, to be followed later by full restriction of the enamel on the upper incisors in a limited subset of that clade (i.e., *Nemegtbaatar* + *Catopsalis catopsaloides*). In the "plagiaulacoid" Bolodontidae (which has fully restricted enamel on both the upper and lower incisors), there is no evidence for evolution of this state from an intermediate condition of partial restriction, although this may be an artifact of inadequate sampling of that lineage.

The current study suggests that full enamel restriction on both upper and lower incisors evolved independently at least three times: in Bolodontidae, Taeniolabididae, and *Nemegtbaatar* + *Catopsalis catopsaloides*. Partial restriction of the enamel on the lower incisor may be a synapomorphy of taeniolabidoids, but full restriction is not. In all cases, restriction of enamel on the upper incisor seems to have evolved either at the same time or later than restriction of enamel on the lower incisor. This suggests that "gliriform" lower incisors were functional in multituberculates in the absence of gliriform upper incisors, but not vice versa. Detailed future studies of the functional morphology of the incisor dentition may further elucidate the evolution of this system.

Conclusions

Analysis of the phylogenetic relationships among multituberculates is complicated by a combination of homoplasy and large amounts of missing data. As should be obvious by this time, much remains to be learned about the relationships and evolution of multituberculates. As expected, this study suggests more new questions than it answers. Addition of new data, be they from newly discovered specimens, additional taxa (e.g., *Ancondon*), or new characters (e.g., cusp structure, cranial vasculature, the postcranial skeleton), will undoubtedly change the resulting picture of multituberculate evolution. Nevertheless, the cladistic analyses described above provide reasonably well resolved hypotheses of phylogeny (Figs. 11.1 and 11.3). Encouragingly, many of the relationships appear quite stable in the face of the addition or deletion of various taxa.

Based on the hypothesis of relationships presented in this study, "Plagiaulacoidea" is recognized as a paraphyletic group composed of several lineages of primitive multituberculates. Ptilodontoidea appears to form a monophyletic group if *Boffius* and *Liotomus* are removed to Taeniolabidoidea. Although support for the group is somewhat ambiguous, Taeniolabidoidea may also be monophyletic, provided that *Eobaatar* and *Monobaatar* are removed. Members of Cimolomyidae all appear to belong to various lineages within Taeniolabidoidea. Ptilodontoidea and Taeniolabidoidea together constitute a more inclusive monophyletic group, Cimolodonta McKenna, 1975.

The hypotheses of phylogeny presented in this chapter provide a framework within which evolution of multituberculate morphology can be studied. Results of the current analyses indicate that while cusp numbers and tooth size often vary in direct relationship to one another, these aspects of tooth morphology were not tightly coupled in any lineage during multituberculate evolution. As suggested by Krause and Carlson (1987), gigantoprismatic enamel appears to have evolved only once in multituberculates, within derived "plagiaulacoids" ancestral to Cimolodonta. In contrast, small prismatic enamel apparently evolved three times—once in late Cretaceous ptilodontoids and twice in late Paleocene taeniolabidoids. Fully restricted enamel on the lower incisors, long thought to be limited to taeniolabidoids, seems to have evolved once in "plagiaulacoids" and several times in taeniolabidoids. Partial restriction of the enamel, however, appears to be diagnostic for Taeniolabidoidea.

Relatively high levels of homoplasy within Multituberculata have produced a complex evolutionary mosaic that clearly requires further study. It is hoped that the analyses and discussion presented here will stimulate future research on multituberculate mammals by pointing out conspicuous gaps in our knowledge (such as lack of data on enamel ultrastructure in most "plagiaulacoids"), by providing a testable hypothesis of relationships, and by offering a framework for interpretation of multituberculate morphology, function, and biogeography.

Appendix 1: Taxa Included in Analysis

The operational taxonomic units (OTUs) of the analysis are listed below in upper case, followed by a list of the species included in the OTU for the purposes of this study. Species marked with an asterisk (*) were included as sources of data on enamel ultrastruture only; no morphological data were collected from these species. Sources of all data (literature references, specimen numbers, etc.) are available upon request. Institutional abbreviations: DNM-NPS: Dinosaur National Monument, National Park Service, Vernal, Utah; LACM: Los Angeles County Museum of Natural History, Los Angeles.

PAULCHOFFATIA: *P. delgadoi*, ?*P. delgadoi* (Hahn, 1969), *Paulchoffatia* sp. A (Hahn, 1978), *Paulchoffatia* sp. (Fosse et al., 1985)
KUEHNEODON: *K. dietrichi*, *K. simpsoni*, *K. uniradiculatus*, *Kuehneodon* sp. (Fosse et al., 1985)
HENKELODON: *H. naias*, *H. ?guimarotensis*
GUIMAROTODON: *G. leiriensis*
KIELANDON: *K. hopsoni*
PSEUDOBOLODON: *P. oreas*, *P. ?robustus*
CTENACODON: *C. laticeps*, *C. scindens*, *C. serratus*
PSALODON: *P. fortis*, *P. ?marshi*, *P. potens*
PLAGIAULAX: *P. becklesii*
MORRISON MULTI: LACM 120453 and DNM-NPS 10822, both from the Morrison Formation of western Colorado / eastern Utah
MONOBAATAR: *M. mimicus*
EOBAATAR: *E. magnus*
BOLODON: *B. crassidens*, *B. elongatus*, *B. osborni*
ARGINBAATAR: *A. dimitrievae*
PARACIMEXOMYS: *P. magister*, *P. priscus*
HAININA: *H. belgica*, *H. godfriauxi*
CIMOLODON: *C. nitidus*
MESODMA: *M. thompsoni*
PARECTYPODUS: *P. clemensi*, *P. lunatus*
ECTYPODUS: *E. powelli*, *E. tardus*
MIMETODON: *M. churchilli*, *M. silberlingi*, *M. trovessartianus*
NEOPLAGIAULAX: *N. hunteri*
PROCHETODON: *P. cavus*, *P. foxi*, *P. taxus*
PTILODUS: *P. montanus*, *P. tsosiensis*, *P. wyomingensis*
BAIOTOMEUS: *B. douglassi*
KRYPTOBAATAR: *K. dashzevegi*
TUGRIGBAATAR: *T. saichansensis*
PENTACOSMODON: *P. pronus*
CHULSANBAATAR: *C. vulgaris*
SLOANBAATAR: *S. mirabilis*
KAMPTOBAATAR: *K. kuczynskii*
BULGANBAATAR: *B. nemegtbaataroides*
NEMEGTBAATAR: *N. gobiensis*
C. CATOPSALOIDES: *Catopsalis catopsaloides*
DJADOCHTATHERIUM: *D. matthewi*
CIMEXOMYS: *C. judithae*, *C. minor*, *C. antiquus*, *C. hausoi*
MICROCOSMODON: *M. conus*, *M. rosei*, *M. woodi*
EUCOSMODON: *E. americanus*, *E. molestus*, *E. primus*
STYGIMYS: *S. kuszmauli*
LIOTOMUS: *L. marshi*
NEOLIOTOMUS: *N. conventus*, *N. ultimus*
BOFFIUS: *B. splendidus*
ESSONODON: *E. browni*
CIMOLOMYS: *C. clarki*, *C. gracilis*
BUGINBAATAR: *B. transaltaiensis*
MENISCOESSUS: *M. robustus*
C. ALEXANDERI: *Catopsalis alexanderi*
LAMBDOPSALIS: *L. bulla*
TAENIOLABIS: *T. lamberti*, *T. taoensis*

Appendix 2: Character Descriptions

DENTAL CHARACTERS. The dental terminology used for this study assumes that the primitive multituberculate dentition includes eleven upper teeth and seven lower teeth. These tooth loci are designated I^{1-3}, C, P^{0-4}, and M^{1-2} in the upper dentition, and I, P^{1-4}, and M^{1-2} in the lower dentition. Note that the tooth designated "C" is not an enlarged caniniform in terms of its morphology; it is a single-rooted premolariform that occupies a position similar to that of the "true" canine of placental mammals (see discussion in Hahn, 1977).

1. Enamel ultrastructure preprismatic (0), small prismatic (1), or gigantoprismatic (2). Unordered.
2. Root of lower incisor does not extend posteriorly under P_3 and P_4 (0), or does extend under P_3 and P_4 (1).
3. Enamel covering of lower incisor of uniform thickness (0), thicker on labial surface than on lingual surface (1), or completely restricted to labial surface of tooth (2). Ordered.
4. Lower incisor not laterally compressed (0), or laterally compressed (1).
5. P_1 present (0), or absent (1).
6. P_2 present (0), or absent (1).
7. P_3 present (0), or absent (1).
8. P_3 cusp count 3–5 (0), or 1–2 (1).
9. P_4 cusp count 3–4 (0), 5–7 (1), 8–10 (2), or 11–15 (3). Ordered.
10. P_4 length 1.0–2.4 mm (0), 2.5–5.2 mm (1), 5.3–7.4 mm (2), or 7.5–15.0 mm (3). Ordered.
11. P_4 in lateral view has straight anterior (leading) edge set at sharp angle to crown (0), or has arcuate leading edge blending into arcuate crown (1).
12. P_4 diphyodont (0), or monophyodont (1).
13. M_1 cusp formula 4:3 or lower (0), 4:4 (1), 5:4 (2), or 7:4 or higher (3). Ordered.
14. M_1 length 1.0–2.9 mm (0), 3.0–5.4 mm (1), 5.5–6.4 mm (2), 6.5–10.9 mm (3), 11.0–22.0 mm (4). Ordered.
15. M_2 cusp formula 0:1 (0), or 2:2 or more (1).
16. M_2 length 1.0–2.9 mm (0), 3.0–4.4 mm (1), 4.5–9.9 mm (2), 10.0–17.0 mm (3). Ordered.
17. M_2 with central basin present (0), or absent (1).
18. I^1 present (0), or absent (1).
19. I^2 cusp count 4 (0), or 3 or fewer (1).
20. Enamel covering of I^2 of uniform thickness (0), thicker on labial surface than on lingual surface (1), or completely restricted to labial surface of tooth (2). Ordered.
21. I^3 with multiple cusps (0), or single cusp (1).
22. I^3 located on margin of palate (0), or in medial position near sagittal midline (1).
23. Canine present in upper dentition (0), or absent (1).
24. Canine cusp count 4 (0), or 3 or fewer (1).
25. P^0 present (0), or absent (1).
26. P^0 cusp count 4 (0), or 2–3 (1).
27. P^1 present (0), or absent (1).
28. P^1 cusp count 4 (0), or 3 (1).
29. P^2 present (0), or absent (1).
30. P^2 cusp count 5–6 (0), 4 (1), or 3 (2). Ordered.
31. P^2 double-rooted (0), or single-rooted (1).
32. P^3 present (0), or absent (1).
33. P^3 cusp formula 1:2 (0), 2:2 (1), 2:4 (2), or 3:4 (3). Ordered.
34. P^3 double-rooted (0), or single-rooted (1).
35. P^4 cusp formula 0–5:1–4:0–5 (0), 0–5:5–10:0–5 (1), or 0–5:5–10:8–12 (2). Ordered.
36. P^4 length 0.5–3.4 mm (0), 3.5–5.4 mm (1), 5.5–8.4 mm (2), or 8.5–11.0 mm (3). Ordered.
37. P^4 double-rooted (0), or single-rooted (1).
38. M^1 cusp formula 2–3:2–4:0 (0), 4–5:4–5:0–1 (1), 5–7:5–8:2–5 (2), or 5–11:7–10:6–11 (3). Ordered.
39. M^1 length 1.0–3.9 mm (0), 4.0–5.9 mm (1), 6.0–7.9 mm (2), 8.0–11.9 mm (3), or 12.0–25.0 mm (4). Ordered.
40. M^2 lingual cusp row absent (0), or present (1).
41. M^2 length 0.5–3.9 mm (0), 4.0–5.4 mm (1), 5.5–8.9 mm (2), or 9.0–17.0 mm (3). Ordered.

CRANIAL CHARACTERS

42. Premaxilla with facial process height > anteroposterior length (0), or height ≤ length (1).
43. Number of infraorbital foramina 2 (0), or 1 (1).
44. Primary infraorbital foramen (F_1 sensu Hahn, 1985) opens anteriorly and laterally on face, not visible in ventral view (0), or opens ventrally (1).
45. Primary infraorbital foramen (Fi_1 sensu Hahn, 1985) positioned dorsal to P^1 (0), dorsal to P^2 (1), or dorsal to P^3 or P^4 (2). Ordered.
46. Maxilla with facial process length (measured from posterior edge of base of zygomatic arch) > height (measured from same point) (0), or length < height (1).
47. Bony roof over anterior orbital space absent (0), or present (1).
48. Flare present in zygomatic arch, snout not confluent with zygomatic arch (0), or flare absent, snout confluent with zygomatic arch (1).
49. Base of zygomatic arch (as marked by posterior edge) directly dorsal or anterior to P^3/P^4 embrasure (0), dorsal to P^4 (1), or dorsal or posterior to P^4/M^1 embrasure (2). Ordered.
50. Zygomatic ridge absent or very faint (0), or strongly developed (1).
51. Postorbital process absent or very small (0), or large (1).
52. Snout length <45% of total skull length (0), or >50% of skull length (1).
53. Frontal-parietal suture V-shaped (0), or U-shaped (1).
54. Contact between nasals and parietals absent (0), or present (1).
55. Lacrimal large, extends anteriorly as far as frontals do (0), small, does not extend as far as frontals (1), or absent from face (2). Unordered.
56. Thickenings in palatal process of premaxilla absent (0), or present (1).

57. Incisive foramina small, round to oval (0), medium size, elongate (1), or very large with palatal fossae (2). Unordered.
58. Palatal vacuities absent (0), or present (1).
59. Posterior border of palate medial or posterior to posterior edge of M^2 (0), medial to middle of M^2 (1), or media to M^1/M^2 embrasure (2). Ordered.
60. Ethmoid foramen present (0), or absent (1).
61. Foramen masticatorium (or subdivisions of same) lateral and slightly anterior to foramen ovale inferium (0), or anterior and slightly medial to foramen ovale inferium (1).
62. Jugular fossa small and shallow (0), or large and deep (1).
63. Sulcus medialis present (0), or absent (1).
64. Internal carotid foramen pierces basisphenoid, carotid canal enters pituitary fossa from below (0), or foramen opens at junction of basisphenoid, petrosal, alisphenoid, and pterygoid; carotid canal enters pituitary fossa laterally (1).
65. Condylar foramina present (0), or absent (1).
66. Condylar formaina paired (0), or single (1).
67. Angle of coronoid process relative to tooth row steep, >55 degrees (0), or low, <45 degrees (1).

Appendix 3: Taxon-Character Matrix

Taxon										
ANCESTOR	90000 00990	00909	09090 90090	90909	00909	00909	09009 00090 00000 00900	90000 90	67% complete
PAULCHOFFATIA	00000 00000	09000	00090 00000	00099	00100	00000	00000 00000 99910 99099	09990 20	79% complete
KUEHNEODON	01009 00000	00000	00900 19001	90000	09900	00009	09000 10000 99999 09099	09099 90	70% complete
HENKELODON	91001 09990	90909	09900 10191	90000	00100	00100	00999 19999 99999 99999	99999 90	49% complete
KIELANODON	99999 99999	99999	99999 99999	99999	00200	00999	99999 99000 99999 99999	99999 99	15% complete
GUIMAROTODON	91000 00010	09009	99999 99999	99999	99999	99999	99999 99999 99999 99999	99999 99	18% complete
PSEUDOBOLODON	91000 00(01)90	99909	09000 90010	10002	00200	00999	99000 09010 99999 99999	09009 99	58% complete
CTENACODON	91000 00010	09009	91999 99190	10102	00300	00100	09001 99009 99999 99999	99999 90	55% complete
PSALODON	91000 00911	09009	99000 00190	10102	00300	00999	99999 99999 99999 99999	99999 99	45% complete
PLAGIAULAX	91001 00021	09009	91999 99999	99999	99999	99999	99999 99999 99999 99999	99999 90	21% complete
MORRISON MULTI	11210 00010	09001	01012 10010	10102	00200	00101	00000 11020 90000 000(01)9	99999 90	86% complete
MONOBAATAR	99999 99999	99999	99999 99990	10102	00300	09991	09101 99000 99999 99999	99999 99	30% complete
EOBAATAR	11201 00021	09001	01999 99999	10101	00300	00001	09999 99999 99999 99999	99999 90	51% complete
BOLODON	99999 99999	99999	99099 19011	90101	00300	00009	90191 10990 99999 99999	99999 99	36% complete
ARGINBAATAR	11001 00131	09009	91999 990(01)0	10102	00000	00990	09001 99010 99999 99999	99999 99	55% complete
PARACIMEXOMYS	19909 99921	19(01)01	01999 99999	99999	99999	99101	09999 99999 99999 99999	99999 99	23% complete
HAININA	19999 99921	19001	01999 99999	90101	00101	00(23)01	09999 99999 99999 99999	99999 99	21% complete
CIMOLODON	11001 10132	19211	01999 99191	90199	00101	10211	09102 99010 99999 99919	99999 99	59% complete
MESODMA	21001 10131	19(23)01	01999 99999	99999	99991	00201	09999 99099 99999 99999	99999 99	36% complete
PARECTYPODUS	21001 10131	19(23)01	01999 99191	90102	00101	00201	09102 09019 99999 99999	99999 90	61% complete
ECTYPODUS	21001 10131	19301	01910 99191	90101	00101	00301	09102 00010 00002 92121	10110 10	88% complete
MIMETODON	29991 101(23)1	19301	01999 99191	99999	90001	00301	09102 09019 99999 99999	99999 90	48% complete
NEOPLAGIAULAX	21001 10131	11301	01999 99191	90102	00101	003(01)1	09102 00010 99999 99999	99999 90	64% complete
PROCHETODON	21001 10133	19309	99999 99191	9010(12)	00202	10211	09102 99999 99999 99999	99999 99	49% complete
PTILODUS	21001 10132	19211	01110 00191	90101	00202	10301	00102 00010 00002 02121	00110 10	96% complete
BAIOTOMEUS	99999 90932	19(23)11	01999 99999	90199	90202	10999	99999 99999 99999 99999	99999 99	25% complete
CIMEXOMYS	11101 10121	19211	01999 99191	90102	00101	00201	01102 19999 99999 99999	99999 90	59% complete
DJADOCHTATHER	91201 10101	19119	91199 01191	90102	00109	90999	91191 10111 99012 00199	99999 90	63% complete
KRYPTOBAATAR	11211 10121	19001	01110 01191	90102	0010(01)	00101	01111 11111 01099 90109	11199 90	85% complete
TUGRIGBAATAR	91211 10121	19001	01190 01191	90102	00109	00101	01999 91110 01901 00019	91199 91	76% complete
PENTACOSMODON	11111 11910	19001	01999 99999	99999	99999	99999	99999 99999 99999 99999	99999 99	29% complete
CHULSANBAATAR	91211 10110	19001	01190 01191	90102	00101	00101	01111 11111 11901 10019	11111 90	88% complete
SLOANBAATAR	91101 19921	19001	01110 01191	90102	00101	00101	01111 11010 01102 00100	01199 90	87% complete
KAMPTOBAATAR	11101 10110	19009	99110 01191	90102	00101	00101	01191 11011 01102 00000	91111 91	88% complete
BULGANBAATAR	99999 99999	19999	99191 99191	90102	00101	00201	01111 11011 99999 00129	99999 99	51% complete
C. CATOPSALOIDES	11201 10101	19121	11192 01191	90119	90101	10221	11111 11120 11100 00019	99199 90	82% complete
NEMEGTBAATAR	11211 10111	19211	01112 01191	90102	00101	00211	01111 11111 01100 10110	01111 91	94% complete
MICROCOSMODON	91111 10119	19301	01900 99999	99999	9999(01)	00209	99999 99999 99999 99999	99999 90	33% complete
EUCOSMODON	11211 119(23)3	19219	91900 99999	91902	10911	20211	09102 09011 99999 99999	99999 91	54% complete
STYGIMYS	11211 11931	19311	01900 01191	90102	10111	10211	09101 19011 99999 11199	99999 91	70% complete
LIOTOMUS	29999 99933	19311	01999 99999	90102	00101	19311	09999 99999 99999 99999	99999 99	34% complete
NEOLIOTOMUS	21211 10933	19(23)39	99999 99999	99999	90991	30329	99999 99999 99999 99999	99999 99	25% complete
BOFFIUS	11119 99999	99999	99999 99999	99999	99991	30349	99999 99999 99999 99999	99999 99	14% complete
ESSONODON	19999 99999	99331	01999 99999	99999	90909	00320	09999 99999 99999 99199	99999 99	22% complete
CIMOLOMYS	19999 99921	19311	11999 99999	99999	99991	00211	09999 99999 99999 99999	99999 99	24% complete
BUGINBAATAR	99999 99999	99999	99999 99999	99999	99991	10331	19999 99099 99999 99999	99999 99	12% complete
MENISCOESSUS	11101 10122	19231	21100 00191	90102	0090(01)	10331	21001 19029 99999 01199	99999 91	72% complete
C. ALEXANDERI	91201 11901	99231	21992 90191	91919	91991	11331	29999 99999 99999 99999	99999 90	42% complete
LAMBDOPSALIS	91201 11900	91231	21192 01191	91919	91990	01331	21109 10020 01012 01020	91110 10	81% complete
TAENIOLABIS	11201 11902	91331	21902 90191	91919	91990	21341	20109 10020 01012 01009	99999 90	72% complete

Appendix 4: Taxonomic Diagnoses

The following diagnoses apply to monophyletic groups discussed in the text and identified in Figures 11.1 and 11.3. Derived character transformations are described as follows: (character number) character abbreviation, state number→state number. For example, "(1) enamel ultrastructure, 0→1" indicates a change in enamel ultrastructure (character 1) from state 0 to state 1. Unequivocal transformations of known distribution are unmarked; unequivocal transformations that may apply at a lower level (the level of the transformation cannot be established due to missing data) are marked "#"; and equivocal character transformations are indicated by "*". See text for a discussion of optimization procedures and types of transformations, and Appendix 2 for character-state descriptions.

DIAGNOSIS OF CIMOLODONTA. (This diagnosis assumes monophyly of both Ptilodontoidea and Taeniolabidoidea.) (6) P_2 presence, 0→1; (11) P_4 shape, 0→1; #(12) P_4 replacement, 0→1; #(18) I^1 presence, 0→1; #(19) I^2 cusps, 0→1; (23) C presence, 0→1; (25) P^0 presence, 0→1; (35) P^4 cusps, 0→1; *(40) M^2 cusps, 0→1; (43) number of infraorbital foramina, 0→1; *(44) orientation of infraorbital foramen, 0→1; *(47) roof over anterior orbit, 0→1; #(55) lacrimal, 0→2; (58) palatal vacuities, 0→1; #(63) sulcus medialis, 0→1; #(64) internal carotid foramen, 0→1; *(66) condylar foramina, 0→1.

DIAGNOSIS OF PTILODONTOIDEA. (9) P_4 cusps, 2→3; (13) M_1 cusps, 0→2; (38) M^1 cusps, 1→2; (45) position of infraorbital foramen, 1→2; *(47) roof over anterior orbit, 1→0; #(57) incisive foramina, 0→2; (59) posterior border of palate, 0→1.

DIAGNOSIS OF TAENIOLABIDOIDEA. (3) enamel covering on lower I, 0→1; (22) I^3 position, 0→1; (42) premaxilla shape, 0→1; (46) maxilla shape, 0→1; (50) zygomatic ridge, 0→1; (52) snout length, 0→1; (62) jugular fossa, 0→1.

NODE/OTU DIAGNOSES FOR FIGURE 11.1

NODE 1 (Multituberculata): See diagnosis in text.
PAULCHOFFATIA: *(17) M_2 basin, 1→0; (54) nasal-parietal contact, 0→1; *(66) condylar foramina, 1→0.
NODE 2: (2) lower I root, 0→1; *(38) M^1 cusps, 0→1.
NODE 3 (Kuehneodontinae): (21) I^3 cusps, 0→1; (25) P^0 presence, 0→1; (46) maxilla shape, 0→1.
KUEHNEODON: *(38) M^1 cusps, 1→0.
HENKELODON: #(5) P_1 presence, 0→1; (23) C presence, 0→1; *(38) M^1 cusps, 0→1.
NODE 4: (1) enamel ultrastructure, 0→1; (9) P_4 cusps, 0→1; #(12) P_4 replacement, 0→1; (15) M_2 cusps, 0→1; *(17) M_2 basin, 0→1; (24) C cusps, 0→1; #(26) P^0 cusps, 0→1; (28) P^1 cusps, 0→1; (30) P^2 cusps, 0→2; (33) P^3 cusps, 1→2; *(38) M^1 cusps, 0→1; *(40) M^2 cusps, 0→1; *(45) position of primary infraorbital foramen, 0→1.
PSEUDOBOLODON: (28) P^2 cusps, 1→0; (46) maxilla shape, 1→0; *(45) position of primary infraorbital foramen, 1→0; (49) position of base of zygoma, 0→1.
CTENACODON: (23) C presence, 0→1; (33) P^3 cusps, 2→3; *(40) M^2 cusps, 1→0; *(45) position of primary infraorbital foramen, 0→1.
PSALODON: (10) P_4 length, 0→1; (23) C presence, 0→1; (33) P^3 cusps, 2→3.
PLAGIAULAX: (5) P_1 presence, 0→1; (9) P_4 cusps, 1→2; (10) P_4 length, 0→1.
MORRISON MULTITUBERCULATE: (3) enamel covering on lower I, 0→2; (4) lateral compression of lower I, 0→1; (19) I^2 cusps, 0→1; (20) enamel covering on I^2, 0→2; (21) I^3 cusps, 0→1; *(40) M^2 cusps, 0→1; *(45) position of primary infraorbital foramen, 1→0; (47) roof over anterior orbit, 0→1.
NODE 5: (33) P^3 cusps, 2→3; *(40) M^2 cusps, 0→1; (43) number of infraorbital foramina, 0→1; *(45) position of primary infraorbital foramen, 0→1.
NODE 6: (30) P^2 cusps, 2→1; (38) M^1 cusps, 1→0.
EOBAATAR: (3) enamel covering on lower I, 0→2; (5) P_1 presence, 0→1; (9) P_4 cusps 1→2; (10) P_4 length, 0→1.
BOLODON: #(21) I^3 cusps, 0→1; (25) P^0 presence, 0→1.
NODE 7: (5) P_1 presence, 0→1; (8) P_3 cusps, 0→1; (9) P_4 cusps, 1→2; (10) P_4 length, 0→1; (33) P^3 cusps, 2→1; *(45) position of primary infraorbital foramen, 0→1; (49) position of base of zygoma, 0→1.
ARGINBAATAR: (9) P_4 cusps, 2→3; (33) P^3 cusps, 1→0; *(40) M^2 cusps, 1→0.
NODE 8 (Cimolodonta): (3) enamel covering on lower I, 0→1; (6) P_2 presence, 0→1; (11) P_4 shape, 0→1; #(12) P_4 replacement, 0→1; #(18) I^1 presence, 0→1; #(19) I^2 cusps, 0→1; (22) I^3 position, 0→1; (23) C presence, 0→1; (25) P^0 presence, 0→1; (35) P^4 cusps, 0→1; *(40) M^2 cusps, 0→1; (43) number of infraorbital foramina, 0→1; *(44) orientation of infraorbital foramen, 0→1; *(47) roof over anterior orbit, 0→1; *(50) zygomatic ridge, 0→1; (52) snout length, 0→1; #(55) lacrimal, 0→2; (58) palatal vacuities, 0→1; (62) jugular fossa, 0→1; #(63) sulcus medialis, 0→1; #(64) internal carotid foramen, 0→1; *(66) condylar foramina, 0→1.
HAININA: (30) P^2 cusps, 2→1; (38) M^1 cusps, 1→2.

Chapter 11. Phylogeny of Multituberculata

NODE 9 (Ptilodontoidea): (3) enamel covering on lower I, $1 \to 0$; (9) P_4 cusps, $2 \to 3$; (13) M_1 cusps, $0 \to 2$; (38) M^1 cusps, $1 \to 2$; (45) position of infraorbital foramen, $1 \to 2$; *(47) roof over anterior orbit, $1 \to 0$; *(50) zygomatic ridge, $1 \to 0$; #(57) incisive foramina, $0 \to 2$; (59) posterior border of palate, $0 \to 1$.

CIMOLODON: (10) P_4 length, $1 \to 2$; (14) M_1 length, $0 \to 1$; (36) P^4 length, $0 \to 1$; (39) M^1 length, $0 \to 1$.

NODE 10: (1) enamel ultrastructure, $1 \to 2$; *(13) M_1 cusps, $2 \to 3$; (46) maxilla shape, $1 \to 0$.

NODE 11: *(13) M_1 cusps, $2 \to 3$; *(30) P^2 cusps, $2 \to 1$; (38) M^1 cusps, $2 \to 3$; (52) snout length, $1 \to 0$; #(57) incisive foramina, $0 \to 2$; #(59) posterior border of palate, $1 \to 2$; #(60) ethmoid foramen, $0 \to 1$; #(62) jugular fossa, $1 \to 0$.

ECTYPODUS: *(30) P^2 cusps, $2 \to 1$; (61) foramen masticatorium, $0 \to 1$.

MIMETODON: (33) P^3 cusps, $1 \to 0$.

NEOPLAGIAULAX: *(30) P^2 cusps, $1 \to 2$.

NODE 12 (Ptilodontidae): (10) P_4 cusps, $1 \to 2$; (33) P^3 cusps, $1 \to 2$; (35) P^4 cusps, $1 \to 2$; (36) P^4 length, $0 \to 1$.

PROCHETODON: (10) P_4 length, $2 \to 3$; (38) M^1 cusps, $3 \to 2$; (39) M^1 length, $0 \to 1$.

NODE 13: *(13) M_1 cusps, $3 \to 2$; (14) M_1 length, $0 \to 1$.

PTILODUS: *(13) M_1 cusps, $3 \to 2$; (22) I^3 position, $1 \to 0$; *(30) P^2 cusps, $2 \to 1$; (42) premaxilla shape, $1 \to 0$.

BULGANBAATAR: (20) I^2 enamel, $0 \to 1$; (38) M^1 cusps, $1 \to 2$; *(44) orientation of infraorbital foramen, $0 \to 1$; *(47) roof over anterior orbit, $0 \to 1$; *(50) zygomatic ridge, $0 \to 1$; (59) position of posterior border of palate, $0 \to 2$.

SLOANBAATAR: *(44) orientation of infraorbital foramen, $0 \to 1$; *(47) roof over anterior orbit, $0 \to 1$; *(50) zygomatic ridge, $1 \to 0$; (53) frontal-parietal suture, $0 \to 1$.

NODE 14 (Group X): (9) P_4 cusps, $2 \to 1$; *(10) P_4 length, $1 \to 0$; *(44) orientation of infraorbital foramen, $1 \to 0$; *(47) roof over anterior orbit, $0 \to 1$; *(50) zygomatic ridge, $0 \to 1$; *(53) frontal-parietal suture, $0 \to 1$; (58) palatal vacuities, $1 \to 0$; (65) condylar foramina, $0 \to 1$.

KAMPTOBAATAR: *(10) P_4 length, $1 \to 0$; *(53) frontal-parietal suture, $0 \to 1$; (67) coronoid process, $0 \to 1$.

NODE 15: (4) lateral compression of lower I, $0 \to 1$.

PENTACOSMODON: (7) P_3 presence, $0 \to 1$; *(10) P_4 length, $1 \to 0$.

NODE 16: (3) enamel on lower I, $1 \to 2$; (44) orientation of infraorbital foramen, $0 \to 1$; #(48) zygomatic flare, $0 \to 1$; #(59) posterior border of palate, $0 \to 1$.

NODE 17: *(10) P_4 length, $0 \to 1$; (13) M_1 cusps, $0 \to 1$; (14) M_1 length, $0 \to 1$; (20) enamel covering on I^2, $1 \to 2$; (38) M^1 cusps, $1 \to 2$; (39) M^1 length, $0 \to 1$; *(53) frontal-parietal suture, $0 \to 1$; (55) lacrimal, $2 \to 0$.

NEMEGTBAATAR: (13) M_1 cusps, $1 \to 2$; (56) palatal thickenings, $0 \to 1$; (58) palatal vacuities, $0 \to 1$; (67) coronoid process, $0 \to 1$.

CATOPSALIS CATOPSALOIDES: (4) lateral compression of lower I, $1 \to 0$; (9) P_4 cusps, $1 \to 0$; (14) M_1 length, $1 \to 2$; (16) M_2 length, $0 \to 1$; (29) P^2 presence, $0 \to 1$; (36) P^4 length, $0 \to 1$; (39) M^1 length, $1 \to 2$; (41) M^2 length, $0 \to 1$; (49) position of base of zygoma, $1 \to 2$; (50) zygomatic ridge, $1 \to 0$; (51) postorbital process, $0 \to 1$.

NODE 18: *(53) frontal-parietal suture, $1 \to 0$; (55) lacrimal, $2 \to 1$; (61) foramen masticatorium, $0 \to 1$.

CHULSANBAATAR: *(10) P_4 length, $1 \to 0$; (51) postorbital process, $0 \to 1$; (56) palatal thickenings, $0 \to 1$.

NODE 19: (9) P_4 cusps, $1 \to 2$; *(10) P_4 length, $0 \to 1$.

KRYPTOBAATAR: (58) palatal vacuities, $0 \to 1$; (59) posterior border of palate, $1 \to 0$.

TUGRIGBAATAR: (50) zygomatic ridge, $1 \to 0$; (67) coronoid process, $0 \to 1$.

NODE 20: (13) M_1 cusps, $0 \to 1$; (14) M_1 length, $0 \to 1$; *(47) roof over anterior orbit, $1 \to 0$; *(50) zygomatic ridge, $0 \to 1$; (54) nasal-parietal contact, $0 \to 1$.

DJADOCHTATHERIUM: (3) enamel covering on lower I, $1 \to 2$; (9) P_4 cusps, $2 \to 0$; (48) zygomatic flare, $0 \to 1$; *(50) zygomatic ridge, $0 \to 1$.

NODE 21: (13) M_1 cusps, $1 \to 2$; (19) I^2 cusps, $1 \to 0$; (38) M^1 cusps, $1 \to 2$; *(44) orientation of infraorbital foramen, $1 \to 0$; (57) incisive foramina, $0 \to 1$.

CIMEXOMYS: (45) position of infraorbital foramen, $1 \to 2$.

NODE 22 (Group Y): (4) lateral compression of lower I, $0 \to 1$; (13) M_1 cusps, $2 \to 3$; #(56) palatal thickenings, $0 \to 1$.

MICROCOSMODON: (9) P_4 cusps, $2 \to 1$; (14) M_1 length, $1 \to 0$.

NODE 23: *(3) enamel covering on lower I, $1 \to 2$; (9) P_4 cusps, $2 \to 3$; *(10) P_4 length, $1 \to 3$; (36) P^4 length, $0 \to 1$; (39) M^1 length, $0 \to 1$.

NODE 24 (Eucosmodontidae): *(3) enamel covering on lower I, $1 \to 2$; (7) P_3 presence, $0 \to 1$; (31) P^2 roots, $0 \to 1$; (34) P^3 roots, $0 \to 1$; *(50) zygomatic ridge, $0 \to 1$; #(67) coronoid process, $0 \to 1$.

EUCOSMODON: *(10) P_4 length, $1 \to 3$; (13) M_1 cusps, $3 \to 2$; (27) P^1 presence, $0 \to 1$; (36) P^4 length, $1 \to 2$; (45) position of infraorbital foramen, $1 \to 2$; (46) maxilla shape, $1 \to 0$.

STYGIMYS: *(10) P_4 length, $3 \to 1$; #(56) palatal thickenings, $0 \to 1$.

NODE 25: *(10) P_4 length, $1 \to 3$; (38) M^1 cusps, $2 \to 3$.

LIOTOMUS: (1) enamel ultrastructure, $1 \to 2$.

NODE 26: (14) M_1 length, $1 \to 3$; (39) M^1 length, $1 \to 2$.

ESSONODON: (36) P_4 length, $1\to 0$; (40) M^2 cusps, $1\to 0$.
NODE 27: (36) P_4 length, $1\to 3$.
NEOLIOTOMUS: (1) enamel ultrastructure, $1\to 2$; *(3) enamel covering on lower I, $1\to 2$; (14) M_1 length, $1\to 3$.
BOFFIUS: *(3) enamel covering on lower I, $1\to 2$; (39) M^1 length, $2\to 4$.
NODE 28 (Group Z): (16) M_2 length, $0\to 1$; (39) M^1 length, $0\to 1$.
CIMOLOMYS: (13) M_1 cusps, $2\to 3$.
NODE 29: #(16) M_2 length, $1\to 2$; (36) P^4 length, $0\to 1$; (38) M^1 cusps, $2\to 3$; (39) M^1 length, $1\to 3$; (41) M^2 length, $1\to 2$.
NODE 30: *(10) P_4 length, $1\to 2$; (14) M_1 length, $1\to 3$; #(16) M_2 length, $1\to 2$; #(22) I^3 position, $1\to 0$; (41) M^2 length, $1\to 2$; #(49) position of base of zygoma, $1\to 2$.
MENISCOESSUS: *(10) P_4 length, $1\to 2$; (43) number of infraorbital foramina, $1\to 0$; (67) coronoid process, $0\to 1$.
NODE 31 (Taeniolabididae): (3) enamel covering on lower I, $1\to 2$; (7) P_3 presence, $0\to 1$; (9) P_4 cusps, $2\to 0$; (20) enamel covering on I^2, $0\to 2$; (27) P^1 presence, $0\to 1$; (29) P^2 presence, $0\to 1$; (32) P^3 presence, $0\to 1$; (37) P^4 roots, $0\to 1$.
CATOPSALIS ALEXANDERI: *(10) P_4 length, $2\to 1$.
NODE 32: (35) P_4 cusps, $1\to 0$; *(50) zygomatic ridge, $1\to 0$; #(58) palatal vacuities, $1\to 0$.
LAMBDOPSALIS: (10) P_4 length, $1\to 0$; (36) P^4 length, $1\to 0$; (59) posterior palate, $0\to 2$.
TAENIOLABIS: *(10) P_4 length, $1\to 2$; (13) M_1 cusps, $2\to 3$; (36) P^4 length, $1\to 2$; (39) M^1 length, $3\to 4$; (42) premaxilla shape, $1\to 0$.

Node/OTU Diagnose for Figure 11.3

NODE A (Multituberculata): See diagnosis in text.
PAULCHOFFATIA: *(17) M_2 basin, $1\to 0$; (54) nasal-parietal contact, $0\to 1$; *(66) condylar foramina, $1\to 0$.
NODE B: (2) lower I root, $0\to 1$; *(46) maxilla shape, $0\to 1$.
NODE C (Kuehneodontinae): (21) I^3 cusps, $0\to 1$; (25) P^0 presence, $0\to 1$; *(46) maxilla shape, $0\to 1$.
HENKELODON: #(5) P_1 presence, $0\to 1$; (23) C presence, $0\to 1$; (38) M^1 cusps, $0\to 1$.
NODE D: (24) C cusps, $0\to 1$; #(26) P^0 cusps, $0\to 1$; (30) P^2 cusps, $0\to 2$; (33) P^3 cusps, $1\to 2$.
PSEUDOBOLODON: (46) maxilla shape, $1\to 0$; (49) position of base of zygoma, $0\to 1$.
NODE E: #(9) P_4 cusps, $0\to 1$; *(10) P_4 length, $0\to 1$; *(17) M_2 basin, $0\to 1$; (28) P^1 cusps, $0\to 1$; *(38) M^1 cusps $0\to 1$; (45) position of primary infraorbital foramen, $0\to 1$.
NODE F: (23) C presence, $0\to 1$; (33) P^3 cusps, $2\to 3$.
CTENACODON: *(10) P_4 length, $1\to 0$.
PSALODON: *(10) P_4 length, $0\to 1$.
NODE G: #(1) enamel ultrastructure, $0\to 1$; *(5) P_1 presence, $0\to 1$; (9) P_4 cusps, $1\to 2$; *(10) P_4 length, $0\to 1$; #(15) M_2 cusps, $0\to 1$; (19) I^2 cusps, $0\to 1$; *(40) M^2 cusps, $0\to 1$.
PLAGIAULAX: *(5) P_1 presence, $0\to 1$.
NODE H (Bolodontidae): (3) enamel covering on lower I, $0\to 2$; (21) I^3 cusps, $0\to 1$; *(40) M^2 cusps, $0\to 1$; *(46) maxilla shape, $0\to 1$.
MORRISON MULTITUBERCULATE: (4) lateral compression of lower I, $0\to 1$; *(5) P_1 presence, $1\to 0$; (9) P_4 cusps, $2\to 1$; #(20) enamel covering on I^2, $0\to 2$; (45) position of primary infraorbital foramen, $1\to 0$; (47) roof over anterior orbit, $0\to 1$.
NODE I: (33) P3 cusps, $2\to 3$; (43) number of infraorbital foramina, $0\to 1$.
NODE J: (30) P^2 cusps, $2\to 1$; *(38) M^1 cusps, $1\to 0$.
EOBAATAR: *(5) P_1 presence, $0\to 1$; (9) P_4 cusps $1\to 2$; (10) P_4 length, $0\to 1$.
BOLODON: (25) P^0 presence, $0\to 1$.
NODE K: *(5) P_1 presence, $0\to 1$; (8) P_3 cusps, $0\to 1$; (9) P_4 cusps, $1\to 2$; (10) P_4 length, $0\to 1$; (33) P^3 cusps, $2\to 1$; (49) position of base of zygoma, $0\to 1$.
ARGINBAATAR: (9) P_4 cusps, $2\to 3$; (33) P^3 cusps, $1\to 0$; *(40) M^2 cusps, $1\to 0$.
CIMOLODONTA: (6) P_2 presence, $0\to 1$; (11) P_4 shape, $0\to 1$; #(12) P_4 replacement, $0\to 1$; #(18) I^1 presence, $0\to 1$; (23) C presence, $0\to 1$; (25) P^0 presence, $0\to 1$; (35) P^4 cusps, $0\to 1$; *(40) M^2 cusps, $0\to 1$; (43) number of infraorbital foramina, $0\to 1$; *(46) maxilla shape, $0\to 1$; #(55) lacrimal, $0\to 2$; *(58) palatal vacuities, $0\to 1$; *(63) sulcus medialis, $0\to 1$; #(64) internal carotid foramen, $0\to 1$; *(66) condylar foramina, $0\to 1$.

Acknowledgments. This chapter represents a study I began as part of my doctoral dissertation, which I completed in the Department of Paleontology at the University of California, Berkeley. Special thanks go to the members of my dissertation committee, W. Clemens, K. Padian, J. Patton, and M. Wake, for their helpful comments and continuing support. Thanks also to T. Rowe, G. Engelmann, and F. Szalay for reading various drafts of this manuscript. This study would not have been possible without the marvelous descriptions of multituberculates published by G. Hahn, D. Miao, Z. Kielan-Jaworowska, D. Krauser, and others. G. Callison, G. Engelmann, J. Hopson, and R. Sloan were kind enough to share unpublished drawings of several key specimens, for which I am grateful. Thanks also to the following people for allowing me to study specimens in their care: W. Clemens and H. Hutchison (UCMP), M. Novacek and M. McKenna (AMNH), J. Horner

(MOR), D. Chure (DNM), L. Martin and R. W. Wilson (KU), M. Lambert (CCM), and D. Whistler (LACM). Finally, special thanks to W. Maddison for providing me with a useful test copy of MacClade. This work was supported in part by a National Science Foundation Graduate Fellowship (1983–1986); final data analyses and manuscript preparation were completed with support from an American Museum of Natural History Kalbfleish Postdoctoral Research Fellowship (1989–1990).

CORRESPONDENCE ADDRESS. Nancy B. Simmons, Department of Mammalogy, American Museum of Natural History, Central Park West at 79th Street, New York, NY 10024, USA.

References

Archibald, J.D. 1982. A study of Mammalia and geology across the Cretaceous-Tertiary boundary in Garfield County, Montana. *Univ. California Publ., Geol. Sciences* 122:1–286.

Carlson, S.J., and Krause, D.W. 1985. Enamel ultrastructure of multituberculate mammals: An investigation of variability. *Contrib. Mus. Paleo., Univ. Michigan* 27:1–50.

Clemens, W.A., and Kielan-Jaworowska, Z. 1979. Multituberculata. In: *Mesozoic mammals: The first two-thirds of mammalian history* (Lillegraven, J.A., Kielan-Jaworowska, Z., and Clemens, W.A., eds.). Berkeley: University of California Press, pp. 99–149.

Cole, T.M., and Krause, D.W. 1988. Interspecific relationships between tooth size and cusp numbers in the Multituberculata (Mammalia). *J. Vert. Paleo.* 8(3):12A.

Donoghue, M.J., Doyle, J.A., Gauthier, J., Kluge, A.G., and Rowe, T. 1989. The importance of fossils in phylogeny reconstruction. *Ann. Rev. Ecol. Syst.* 20:431–446.

Engelmann, G.F., Greenwald, N.S., Callison, G., and Chure, D.J. 1990. Cranial and dental morphology of a late Jurassic multituberculate mammal from the Morrison Formation. *J. Vert. Paleo.* 10(3):22A.

Fosse, G., Kielan-Jaworowska, Z., and Skaale, S.G. 1985. The microstructure of tooth enamel in multituberculate mammals. *Palaeont.* 28:435–449.

Gauthier, J., Kluge, A.G., and Rowe, T. 1988. Amniote phylogeny and the importance of fossils. *Cladistics*, 4:105–209.

Granger, W., and Simpson, G.G. 1929. A revision of the Tertiary Multituberculata. *Bull. Amer. Mus. Nat. Hist.* 56:601–676.

Greenwald, N.S. 1988. Patterns of tooth eruption and replacement in multituberculate mammals. *J. Vert. Paleo* 8:265–277

Greenwald, N.S. 1989a. *Phylogeny and systematics of multituberculate mammals*. Ph.D. Dissertation, University of California, Berkeley.

Greenwald, N.S. 1989b, Effects of missing data and homoplasy on estimates of multituberculate phylogeny. *J. Vert. Paleo.* 9(3):24A.

Hahn, G. 1969. Beitrage zur Fauna der Grube Guimarota Nr. 3, die Multituberculata. *Palaeontographica, Abt. A* 133:1–100.

Hahn, G. 1977. Neue Schadel-Reste von Multituberculaten aus dem Malm Portugals. *Geol. et Palaeo.* 11:161–186.

Hahn, G. 1978. Neue Unterkiefer von Multituberculaten aus dem Malm Portugals. *Geol. et Palaeo.* 12:177–212.

Hahn, G. 1988. Die Ohr-region der Paulchoffatiidae (Multituberculata, Ober-Jura). *Palaeovert.* 18:155–185.

Hahn, G., and Hahn, R. 1983. Multituberculata. *Fossilium Catalogus 1: Animalia* 127:1–409.

Hahn, G., Sigogneau-Russell, D., and Wouters, G. 1989. New data on Theroteinidae—Their relations with Paulchoffatiidae and Haramiyidae. *Geol. et Palaeont.* 23:205–215.

Jepsen, G.L. 1940. Paleocene faunas of the Polecat Bench Formation, Park County, Wyoming. *Proc. Amer. Phil. Soc.* 83:217–340.

Kielan-Jaworowska, Z. 1974. Migrations of the Multituberculata and the late Cretaceous connections between Asia and North America. *Ann. S. Afr. Mus.* 64:231–243.

Kielan-Jaworowska, Z., Dashzeveg, D., and Trofimov, B. 1987. Early Cretaceous multituberculates from Mongolia and a comparison with late Jurassic forms. *Acta Palaeont. Polonica* 32:3–47.

Krause, D.W., and Carlson, S.J. 1987. Prismatic enamel in multituberculate mammals: Tests of homology and polarity. *J. Mammalogy* 68:755–765.

Maddison, W.P., Donoghue, M.J., and Maddison, D.R. 1984. Outgroup analysis and parsimony. *Syst. Zool.* 33:83–103.

Marsh, O.T. 1889. Discovery of Cretaceous mammals. Part II. *Amer. J. Sci.*, Ser. 3, 38:177–180.

McKenna, M.C. 1975. Toward a phylogenetic classification of the Mammalia. In: *Phylogeny of the primates* (Luckett, W.P., and Szalay, F.S., eds.) New York: Plenum Press, pp. 21–46.

Miao, D. 1988. Skull morphology of *Lambdopsalis bulla* (Mammalia, Multituberculata) and its implications to mammalian evolution. *Cont. to Geol., Univ. Wyoming, Special Paper* 4:1–104.

Novacek, M.J. 1989. Higher mammal phylogeny: The morphological-molecular synthesis. In: *The hierarchy of life* (Fernholm, B., Bremer, K., and Jornwall, H., eds.). New York: Elsevier Science Publications, pp. 421–435.

Novacek, M.J., Wyss, A.R., and McKenna, M.C. 1988. The major groups of eutherian mammals. In: *The phylogeny and classification of the tetrapods, vol. 2: The mammals* (Benton, M.J., ed.), Systematics Association Special Volume 35B, pp. 31–71. Oxford: Clarendon Press.

Osborn, H.F. 1887. The structure and classification of the Mesozoic Mammalia. *J. Acad. Nat. Sci. Philadelphia* 9:186–265.

Rohlf, F.J. 1982. Consensus indices for comparing classifications. *Mathematical Biosciences* 59:131–144.

Rowe, T. 1986. *Osteological diagnosis of Mammalia, L. 1758, and its relationship to extinct Synapsida*. Ph.D. Dissertation, University of California, Berkeley.

Rowe, T. 1988. Definition, diagnosis and origin of Mammalia. *J. Vert. Paleo.* 8:241–264.

Rowe, T., and Greenwald, N.S. 1987. The phylogenetic position and origin of Multituberculata. *J. Vert. Paleo.* 7:24A–25A.

Sanderson, M.J., and Donoghue M.J. 1989. Patterns of variation in levels of homoplasy. *Evolution* 43:1781–1795.

Simpson, G.G. 1928. *A catalogue of the Mesozoic Mammalia in the geological department of the British Museum.* London: Oxford University Press.

Simpson, G.G. 1929. American Mesozoic Mammalia. *Peabody Mus. (Yale Univ.) Memoirs* 3:1–171.

Simpson, G.G. 1945. The principles of classification and a classification of mammals. *Bull. Amer. Mus. Nat. Hist.* 85:1–350.

Sloan, R.E., and Van Valen, L. 1965. Cretaceous mammals from Montana. *Science* 148:220–227.

Stevens, W.P. 1988. Phylogeny of taeniolabidoid multituberculates. *J. Vert. Paleo.* 8:26A.

Swofford, D.L. 1989. Phylogenetic analysis using parsimony. Release 3. Illinois Natural History Survey (computer program distributed on 3 1/2-inch disk).

CHAPTER 12

Cranial Morphology of the Therian Common Ancestor, as Suggested by the Adaptations of Neonate Marsupials

WOLFGANG MAIER

Overview

Skull structures are properly understood only if their ontogenetic development is considered. It has been argued by a number of authors that the reproductive biology of marsupials may be reminiscent of a primitive ovo-viviparous phase of therian evolution. Therefore, it seemed rewarding to test the hypothesis that the skull structures of neonatal didelphids may be adapted to such a primitive mode of development as well. It is assumed that the head skeleton of the altricially born didelphid is constructed mainly to sustain mechanical strain caused by sucking at the teat, where it is also suspended. It is investigated which structures of the skull may be functionally correlated with the "lactation-complex." It is mainly the strong chondrocranium that has to provide sufficient stability at first, because exocranial bone development is retarded. It is further argued that the ala temporalis, which ossifies as the alisphenoid, lends support to the sidewall of the braincase; the alisphenoid is considered to be fully homologous to the reptilian epipterygoid. The ontogenetic development of the didelphid middle ear is also reminiscent of a primitive evolutionary state. The postdentary elements lie at first close to the medial side of the dentary; in a complicated process of growth they become fixed to the basicranium and then detached from the mandible. The "horizontal anlage" of the tympanic in monotremes and eutherians is not considered to be primitive (postulate of Van Kampen) but secondarily derived two times independently. The nasal capsule of all extant mammals shows a common "bauplan," which is presented schematically; the importance of macrosmatic adaptations in mammals is pointed out. Most of the derived features of the eutherian skull seem to be linked with its prolonged intrauterine development, which sets different functional conditions on the growing structures; therefore, phenomena of heterochrony play an important role for understanding the eutherian conditions. Whereas early skull development in marsupials is characterized by a number of narrow functional and biological constraints resulting in a certain structural uniformity, the eutherian mode of development seems to provide more freedom for fetal differentiation.

Contents

Introduction, 166
Principles of Cranial Morphology, 167
Cranial Morphology of Neonate Didelphids, 169
    Feeding apparatus, 169
    Alisphenoid problem, 171
    Therian middle ear, 174
    Therian nasal capsule, 175
Discussion and Conclusions, 177
Acknowledgments, 179
References, 179

Introduction

This chapter covers only a few aspects of the skull morphology of the inferred therian common ancestor. It is hoped that some of those structural complexes chosen are of particular relevance for understanding the evolutionary differentiation of the therian skull. Skull morphology can be fully appreciated only if its ontogenetic development (craniogenesis) is duly considered (Gaupp, 1906; Stadtmüller, 1936; de Beer, 1937; Starck, 1967). Comparative craniogenetic research has accumulated a huge bulk of information, which also needs to be further analyzed in terms of phylogenetic systematics. On the other hand, many higher mammalian taxa are known so far by only one ontogenetic stage of one species, if at all. This lack of information severely limits comparative analysis, and much further work on craniogenesis needs to be done.

In the past, many of the embryogenetic investigations were performed with reference to the "biogenetical principles" of Haeckel (Gould, 1977). Although many examples of "recapitulation" from different organ systems are known (e.g., Luckett and Maier, 1982; Maier and Schrenk, 1989), I do not believe that early ontogenetic stages are, *a priori*, more revealing of evolutionary history than the adult ones. Selection forces are acting on all phases of development; this position is inherently in accordance with a statement of de Beer (1937) that "the chondrocranium is not a relic, but a functional structure progressing in evolution along its own lines" (p. 449). What is needed is a better knowledge of whole life histories as a basis for comparison and phylogenetic reconstruction (Fig. 12.1).

Phylogeny may ultimately be understood as a long sequence of ontogenies ("hologeny" of Zimmermann, 1943). Therefore phylogeny largely determines the present ontogenetic processes, but past ontogenies have deeply influenced phylogeny as well. This mutual or dialectical interrelationship of ontogeny and phylogeny should be more thoroughly considered in modern evolutionary theory (Bonner, 1982; Wake and Roth, 1989). Hennig (by profession a systematist of insects, in which group "larval life" plays an important role) repeatedly pointed out the need for studying different ontogenetic character states ("semaphoronts"); to him, a comprehensive "holomorphic" and biological approach was an important prerequisite for further progress in phylogenetic systematics. Thus, due consideration of all ontogenetic stages should be part of a careful character analysis (Hennig, 1948; Bock, 1988).

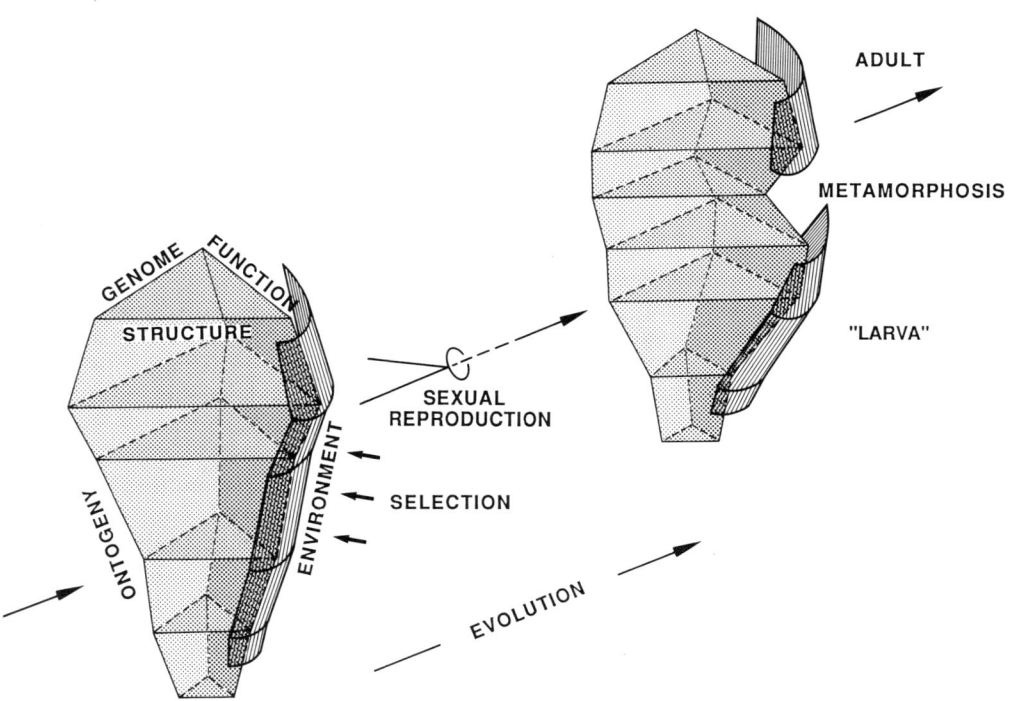

FIGURE 12.1. Diagrammatic illustration of an organism concept considering both the ontogenetic and the evolutionary dimensions. Any individual organism is a complex four-dimensional unit ("raum-zeit-gestalt"), which is exposed to selection at any phase of ontogeny. Therefore, any life phase is "adapted" to its specific requirements. The life history of any organism is a continuous process; for practical reasons, a limited number of ontogenetic stages is to be studied. At any level the specific synthesis of genetic, structural, and functional factors with defined environmental parameters should be analyzed. Individual organisms are members of a reproducing population, and as such they are members of the evolutionary chain. Each evolutionary line is determined by the interaction of phylogenetic and ontogenetic factors.

The following study concentrates on perinatal stages of skull development. These stages are of special importance, because many structures and functions seem to be correlated with certain aspects of reproductive biology; in the reverse, adaptations of skull structures to perinatal functions may be used to develop hypotheses on the presumable reproductive biology of the therian common ancestor.

It is here assumed, of course, that marsupials and eutherians are sistergroups that separated at or after the Jurassic-Cretaceous boundary (Lillegraven, 1976; Marshall, 1979; Kielan-Jaworowska et al., 1979; Clemens, 1979). This phylogenetic frame appears to be so well established that it need not be confirmed by further formal phylogenetic-systematic reasoning; there is rather a need to further analyze the supposed evolutionary scenario by further functional and biological arguments.

The only available living mammalian outgroup for craniogenetic studies are the monotremes, which are very well known (Kuhn, 1971; Zeller, 1989); only a few comparisons will be made with fossil therapsids and early mammals (see also Wible & Hopson, Zhexei & Crompton, Archer et al., Miao, Presley, this volume; chapters 3–7).

Principles of Cranial Morphology

Comparative craniogenetic studies are usually performed with plate reconstructions on the basis of histological serial sections, which also provide detailed information about cartilages, membranes, and other tissue and organ structures. Microscopic anatomical studies also yield a lot of information about the functionally determined textures of the connective and supporting tissues. These data provide some insight into the distribution of mechanical strains (Pauwels, 1960; Kummer, 1972; and thus provide a basis for the constructional morphology of the head skeleton.

For didactic reasons, I have compiled a "schematized model" of the skull structure of a primitive stem reptile (Fig. 12.2): I have taken the outlines of the adult exocranium of *Captorhinus* and added a hypothetical fetal endocranium as derived from studies of extant lepidosaurs; this "monster" is primarily meant to illustrate the more important components of the endocranium: the neural endocranium, the occipital cranium, and the visceral cranium. The palatoquadrate of the mobile mandibular arch is articulating with the neurocranium (autosynstyly), and the primary mandibular joint is supported by a strong hyomandibula (a sort of amphistylic suspension of the jaw) (de Beer, 1937; Hofer, 1945). The hyal and branchial arches are only vaguely indicated. The circle marks the future position of the tympanic membrane in therapsids and mammals (Maier, 1990).

Whereas the exocranial bones are directly formed by intramembraneous ossification, the endocranial structures of the head skeleton are ontogenetically mostly preformed by cartilage; therefore, their morphology can be evaluated only by ontogenetic studies. Many parts of the "chondrocranium" become secondarily ossified perichondrally and enchondrally ("replacement bone"); the "replacement bones," with which we are mostly concerned in the present study, are the epipterygoid (=alisphenoid), quadrate (=incus), articulare (=malleus), hyomandibulare (=stapes), and ethmoid. Other elements of the endocranium become resorbed in later ontogenetic stages, and few persist into adult stages (nasal cartilages, for example). Sometimes, membranes derived from the perichondrium or periosteum may be the last remnants of ontogenetically or phylogenetically reduced skeletal structures (temporal fascia, tympanic membrane, etc.).; on the other hand, membranes may be guiding structures for newly outgrowing cartilages and bones (i.e., Membrana sphenoobturatoria). Some portions of the endocranium are directly formed as "appositional bone" ("Zuwachsknochen"; Starck, 1989). In the adult skull, endocra-

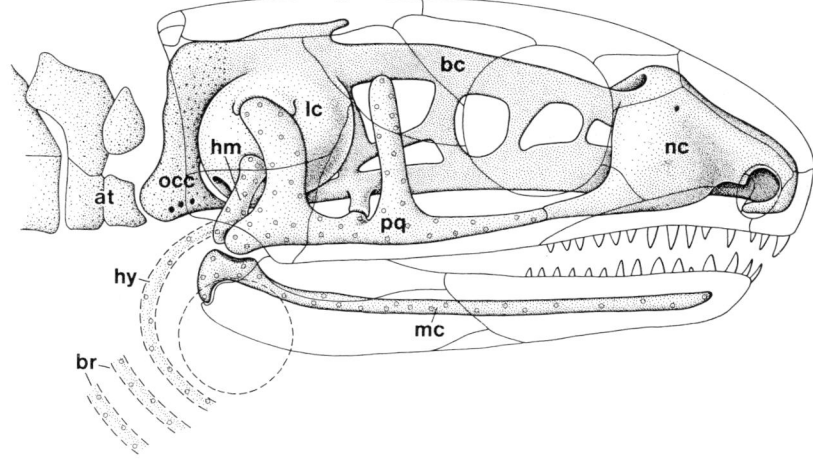

FIGURE 12.2. Diagrammatic "bauplan" of the primitive amniote cranium. The outlines of the exocranium are taken from *Captorhinus*, whereas the hypothetical endocranial components are derived from embryos of extant lepidosaurs. (The mixture of embryonic and adult structures merely serves didactical purposes.) The endocranium is composed of neurocranial and viscerocranial elements; the whole cranium is the syncranium. Note the mobile, amphistylic fixation of the mandibular arch. The future position of the mammalian tympanic membrane is indicated by a stippled circle. For abbreviations, see Table 12.1.

TABLE 12.1. Abbreviations

ac	carotid artery
aci	a. carotis interna
ah	hypochiasmatic wing
ahy	hypochiasmatic wing
al	ascending lamina
ali	alisphenoid
ane	external nasal aperture
ang-ty	angular-tympanic
aor	ala orbitalis
at (Figs. 12.7, 12.8, 12.9)	ala temporalis
at (Figs. 12.2, 12.3)	atlas
at (Fig. 12.11)	atrioturbinale
ax	axis
az	zygomatic arch
azyg	zygomatic arch
bc	braincase
br	branchial arches
bs	basisphenoid
cac	alicochlear commissure
cd	chorda dorsalis
cda	cartilage of the nasopalatine duct
cdn	(12.11)
cep	cavum epitericum
cm	Meckel's cartilage
cna	anterior nasal cupola
cnp	posterior nasal cupola
con	(12.9) orbitonasal commissure
cop	orbitoparietal commissure
cpa (Fig. 12.11)	palatine cartilage
cpa (Fig. 12.5)	paraseptal cartilage
cps	paraseptal cartilage
cs	semicircular crest
csa	(12.11) semicircular crest
cup	(12.11) cupola posterior
d	dentary
de	dentale
dien	diencephalon
dnl	nasolacrimal duct
dnp	nasopharyngeal duct
dvn	vomeronasal duct
ep	epipterygoid
et I	ethmoturbinale I
et II	ethmoturbinale II
fc	foramen caroticum
fhy	hypophyseal foramen
fi	foramen incisivum
fl	fused lip
fo	foramen opticum
fol	foramen olfactorium
fon	orbitonasal fissure
fr	frontale
ft (Fig. 12.11)	frontoturbinalia
ft (Fig. 12.7)	temporal fascia
ft (Fig. 12.6)	temporal fenestra
gg	geniculate ganglion
go	(12.10) goniale (=prearticulare)
gs	trigeminal ganglion
gtr	trigeminal ganglion
hm	hyomandibula (12.2)
hy	hyoid
hyf	hypophyseal fossa
II	optic nerve
inc	incus
jo	Jacobson's organ
ju	jugale
la (Fig. 12.11)	lacrimale

TABLE 12.1. Continued

la (Fig. 12.9)	lamina ascendans
lac	lacrimale
lar	larynx
lc	labyrinth capsule
lh	lamina horizontalis (frontomaxillary septum)
li	tongue
lo	(12.10) lamina obturans
lorb	lamina orbitalis of ethmoid
lta	anterior transversal lamina
ltp	lamina transversalis posterior (terminal lamina)
ltr (Fig. 12.9)	lamina trabecularis
ltr (Fig. 12.7)	trabecular plate
ma	masseter muscle
mall	malleus
mam	adductor mandibulae muscle
mc	Meckel's cartilage
meth	mesethmoid
mm	masseter muscle
mpm	medial pterygoid muscle
mpmx	medial process of premaxillary
mptl	pterygoideus lateralis muscle
mptm	(12.6) medial pterygoid muscle
mt (Fig. 12.11)	marginoturbinale
mt (Figs. 12.6, 12.7)	temporalis muscle
mx	maxillary
mxt	maxilloturbinale
N.V	(12.9) trigeminal nerve
nc	nasal capsule
nd	nasal duct
ne	nasal entrance
np	superficial petrosal nerve
ns	nasal septum
nt	nasoturbinale
oc (Figs. 12.2, 12.3)	occipital region
oc (Figs. 12.8, 12.9)	otic capsule
orsph	orbitosphenoid
pa	ascending process
paa	ascending process of palatine
pal	palatine
par	parietale
pas	superior alar process
peth	ethmoidal process of orbitosphenoid
pmx	premaxillary
pn	sidewall of nasal capsule
pp	pterygoid cartilage
ppp	posterior paraseptal process
ppt	pterygoid process of the ala temporalis
pq	palatoquadrate
pra	alar process
psph (Fig. 12.6)	parasphenoid
psph (Fig. 12.11)	presphenoid
pt	pterygoid
retht	ethmoturbinal recessus
rl	recessus lateralis
sn (Figs. 12.9, 12.11)	nasal septum
sn (Fig. 12.5)	septal cartilage
sq	squamosal
st	stapes
syp	symphyseal plate
tel	telencephalon
tn	roof of nasal capsule
ty	tympanic
$V_{1,2,3}$	trigeminal branches
vo	vomer
x (Fig. 12.6)	"restknorpel"
x (Fig. 12.9)	basipterygoid joint
y	"restknorpel"

nial and exocranial bones are forming a complex "syncranium," and only careful morphological analysis can help to distinguish the heterogeneous components, which may even be completely fused in adult stages.

The exocranial bones are usually well known to the paleontologist. Endocranial bones are often quite delicate and complicated, and they are mostly hidden underneath the exocranium; therefore, their morphology is not too well understood by paleontologists and systematists. On the other hand, these bony structures are of special morphological importance because they are—for ontogenetic reasons—closely related with the most important nerves and vessels of the head region.

In this chapter I concentrate on the following four parts of the cranium: (1) the mouth cavity and its importance for sucking and swallowing, (2) the middle portion of the palatoquadrate (ascending process and epipterygoid) and its relationship with the primary wall of the neurocranium, (3) the transformation of the primary jaw articulation into the therian middle ear and the formation of a secondary mandibular articulation, and (4) the structure of the expanded therian nasal capsule. Only those aspects of these complicated morphological problems that are related to the hypothetical reconstruction of the "grundplan" of the therian common ancestor will be addressed.

Cranial Morphology of Neonate Didelphids

I have been able to study the postnatal skull development of *Monodelphis domestica* and of a considerable number of other marsupials (Maier, 1987a, 1987b, 1988, 1989a), but I do not intend to present here a comprehensive analysis of the chondrocranium with its complicated terminology. Instead I concentrate on a few characteristic structural adaptations to the functional requirements of neonatal life; it will be argued later why the skull structure of neonatal didelphids may be especially helpful for elucidating some basic adaptations of the therian skull.

It is well known that the newborns of marsupials are delivered in a very altricial state after a relatively short intrauterine phase of development. Only a few organ systems, which are indispensable for survival, are differentiaed: specialized shoulder girdle and upper limbs, by which the newborns actively climb to the teat (Klima, 1987; Hughes and Hall, 1988); strongly developed skeleton of the head and the upper vertebral column, by which the body of the young is fixed and suspended at the nipple; well-developed feeding apparatus, by which the milk is actively sucked from the milk ducts of the teat and swallowed; huge heart and well-differentiated peripheral circulatory system; lungs that have not yet differentiated definitive alveolae (Krause and Leeson, 1975); and completed but weak diaphragma (Hughes and Hall, 1988).

FEEDING APPARATUS. The feeding system consists of most parts of the facial skull, the intrinsic and extrinsic muscles of the tongue and the mouth floor as well as the adductor musculature. This system has the task to fix the mouth to the teat and to extract and swallow the secretions of the milk glands (Sharman, 1973). The mouth is sealed by a peridermal fusion of the lips; the lips and cheeks have only begun to be invaded by the facial muscles (m. orbicularis oris and m. buccinator are hardly differentiated), and there does not even exist a vestibulum oris.

Unfortunately, we do not yet have exact data on the milking mechanism of pouch young of marsupials, but we know a bit more about the feeding behavior of newborn eutherians (Herring and Scapino, 1973; Lakars and Herring, 1980). Herring (1985) described the feeding of young piglets as follows:

> Suckling is accomplished by the cheeks, tongue and oral floor, with a minimal contribution from the adductor muscles. . . . In bottle suckling the the piglet curls its protruded tongue around the sides of the nipple and holds the nipple against the anterior end of the palate. Contractions of the oral floor the are accompanied by small orthal jaw movements. EMG recordings show simple but sharp simultaneous bursts of activity in the adductor muscles, alternating with longer bursts in the oral floor. (p. 213).

It is likely that the muscle activities occur on both sides simultaneously. Hiiemae and Crompton (1985), Crompton (1989), and others have repeatedly stated that the various mechanisms of ingesting, handling, transporting, and swallowing different sorts of food are based on a strikingly similar pattern of muscle activities; pending further information, we consider these interpretations to be more or less true for the sucking, transporting, and swallowing of milk in young marsupials.

It would be of great interest to the functional morphologist not only to know more exactly how suckling is actually performed by the tongue as well as by other structures of the mouth and the pharynx, but also to better understand how the involved mechanical forces are distributed and thus influence the structures of the head skeleton. Since no detailed functional analysis is available, we make a few tentative deductions from the structures themselves, i.e., we are relying on technomorphic analogies.

First of all, we realize that in the neonate all those structures that are likely to be involved in sucking are precociously well developed. The cranium is built by massive cartilage; exocranial bone elements play an important mechanical role only at the upper and lower jaws (Fig. 12.3). The lower jaw is still suspended by the primary jaw joint and by the incus (which is the homologue of the quadrate) at the otic capsule. The hyoid and the laryngeal cartilages (which are derived from the branchial arches) are well developed.

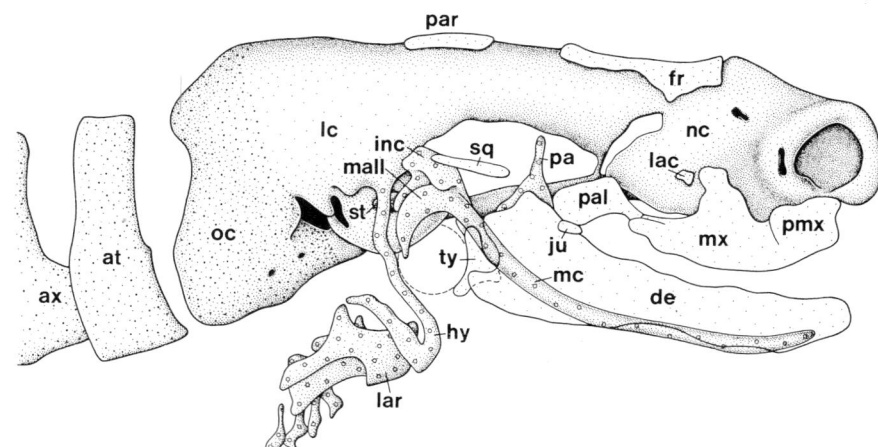

FIGURE 12.3. Cranium of a neonate of *Monodelphis domestica* (Didelphidae; Marsupialia), based on a plate reconstruction. Absolute skull length about 3.5 mm. Mainly the exocranial elements of the upper and lower jaws are developed. The cartilaginous endocranium (chondrocranium) is the main skeletal component. The primary jaw articulation is not yet replaced by the secondary squamoso-dental joint. The palatoquadrate is represented by the separated incus (=pars quadrata) and the ascending process of the ala temporalis (=pars palatina). The "anlage" of the tympanic membrane is outlined by a stippled circle. For abbreviations, see Table 12.1.

The tongue, with its intrinsic and extrinsic musculature, is very well developed, although the nuclei of the cross-striated mucle fibers are still in a central position, being surrounded by a smaller number of contractile fibrils. The intrinsic tongue muscles show the typical arrangements of longitudinal, and vertical muscle fibers (Fig. 12.4.) The vertical fibers converge toward the central groove containing the teat (Merchant and Sharman, 1966); by their contraction, the surface of the tongue must become depressed, and the teat with its milk ducts dilatated. The horizontal fibers bring the tongue back into the original position. Swallowing needs an elevation and backward shifting of the posterior parts of the tongue against the hard and soft palate (Crompton, 1989); here the longitudinal as well as the extrinsic muscles play a major role. The soft palate seals the oropharynx against the nasopharynx.

The medial pterygoid muscle runs from the prominent pterygoid process of the ala temporalis to the adjacent cartilage of Meckel. At its posterior portion, the pterygoid muscle gives rise to the tensor muscle of the soft palate as well as to the tensor tympani muscle. The lateral pterygoid muscle, which got carried posteriorly by the condylar process of the dentary, originates from its upper and posterior end. All of these muscles get innervated by the mandibular branch of the trigeminal nerve. A levator muscle of the soft palate, which is derived from the constrictor muscles of the pharynx, is not yet discernible. The nasopharyngeal ducts get closed from behind by two plugs of connective tissue, which seem to be derived from the torus of the Eustachian tube. An epiglottis of considerable size extends up into the nasopharynx between the palatopharyngeal folds; an epiglottic cartilage is already developed.

Cross sections more clearly illustrate the architecture of the facial skull (Fig. 12.5): The septum and the sidewalls of the nasal capsule consist of thick cartilage plates and pillars; the premaxillary, maxillary, and palatine bones are still very thin, and they rest on the ventral edges of these cartilages. The secondary palate and the tongue spare out the space for the teat. The lateral margins of the tongue are inserted into a longitudinal groove of the maxillary. The premaxillary bones of both sides are medially fused to form a concave shell, which underlies the ventral side of the nasal floor cartilages; the lateral margins of the premaxillary and maxillary are partially fused as well in the neonate. Therefore we recognize caenogenetic specializations, whose mechanical tasks seem to be obvious. The sutures of the hard palate bones develop somewhat later.

The maxillary shows an ascending process, which spares out the passage for the nasolacrimal duct; both this process and the body of the maxillary show specially designed areas of contact with the nasal capsule, at which forces may be transmitted. The palatine processes of the maxillary do not yet meet in the midline, but they are bent upward to get supported from the paraseptal cartilages. In contrast to sauropsids, in mammals the paraseptal cartilages surround the vomeronasal organ. The lateral margins of the maxillary are turned downward, but they are not equivalent to the alveolar processes; instead, they should be considered as strengthening frames of the palatal shell construction.

The palatine bones form bony plates whose lateral walls are bent dorsally. Their medial palatine processes form the posterior ends of the hard palate, whereas the dorsal edge rests on the cartilages of the posterior nasal cupola, the hypochiasmatic wing, and the pterygoid process of the ala temporalis. The posterior portion of the hard palate has no firm medial support, and the two nasal cavities communicate underneath the free lower end of the nasal septum or trabecular plate, respectively. The pressure of the tongue can push the mobile medial part of the hard palate upward against the central stem, but the two laterally situated nasopharyngeal ducts will always be kept open (Fig 12.7A). Therefore,

Chapter 12. Cranial Morphology of Therian Common Ancestor

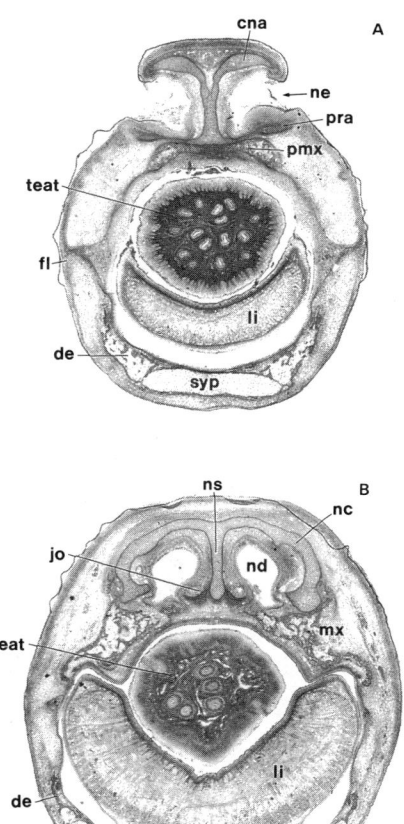

FIGURE 12.4. Histological cross sections of the head of a pouch young of *Petrogale penicillata* (CRL 19.5 mm; Hubrecht Collection). **A**: Section across the nasal entrance and the tip of the tongue. The lips are fused by periderm. The mouth cavity is filled with the teat. **B**: Section across the midth of the nasal capsule showing the open nasal ducts. The vomeronasal organ is differentiated. For abbreviations, see Table 12.1.

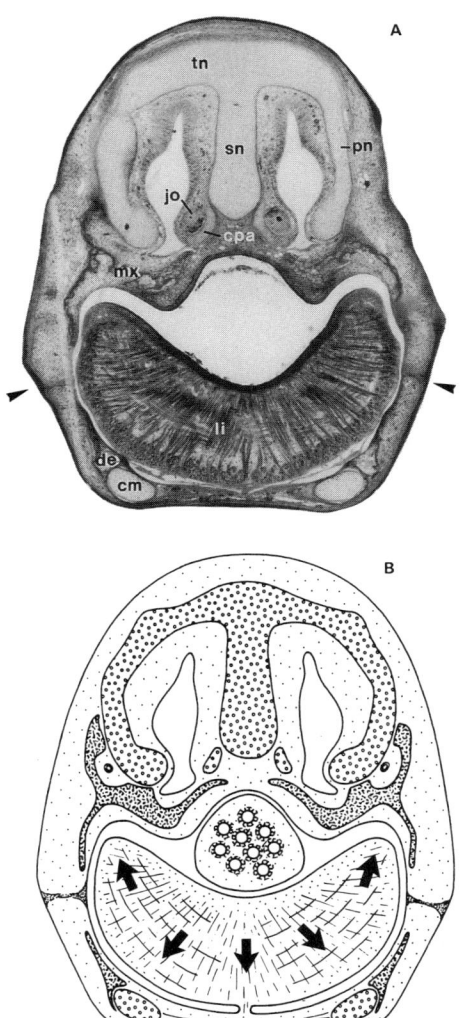

FIGURE 12.5. **A**: Histological cross section of the snout of a neonate of *Monodelphis domestica* (CRL 10 mm). **B**: Semidiagrammatic drawing of the most relevant elements of constructional morphology. For abbreviations, see Table 12.1.

the palatine bones serve as elastic frames for these ducts, and they convey pressure forces to the base of the skull.

The dentary bones consist of thin bony lamellae, which anteriorly rest upon the dorsal and lateral sides of the strong cartilages of Meckel; posteriorly, Meckel's cartilage moves more to the medial side of the dentary. The thin layer of genuine muscle of the mouth floor, the mylohyoid muscle, spans between the cartilages of Meckel. Again, the upper rim of the dentary is not equivalent to the alveolar process, because the "anlagen" of the teeth are situated at its medial side.

ALISPHENOID PROBLEM. The homology of the mammalian alisphenoid is a problem that is still debated. Maier (1987b, 1989a) has reviewed the different opinions and has proposed hypotheses that are largely based on evidence derived from marsupials. It is argued that the cartilaginous ala temporalis of marsupials is completely equivalent to the basal and ascending processes of the palatoquadrate in reptiles (cf. Fig. 12.2); it is ontogenetically transformed into the alisphenoid bone by "Zuwachsknochen" (appositional bone) invading the sphenoobturator membrane from the perichondrium of the ala as well as by enchondral ossification of the ala itself. The ascending process of the ala temporalis shows varying relationships with the trigeminal branches in different taxa of marsupials, but in most cases maintains a direct contact with the lower border of the primary sidewall of the braincase, i.e., the orbitoparietal commissure (Maier, 1989a). This connection has been interpreted as a mechanical support of the sidewall against

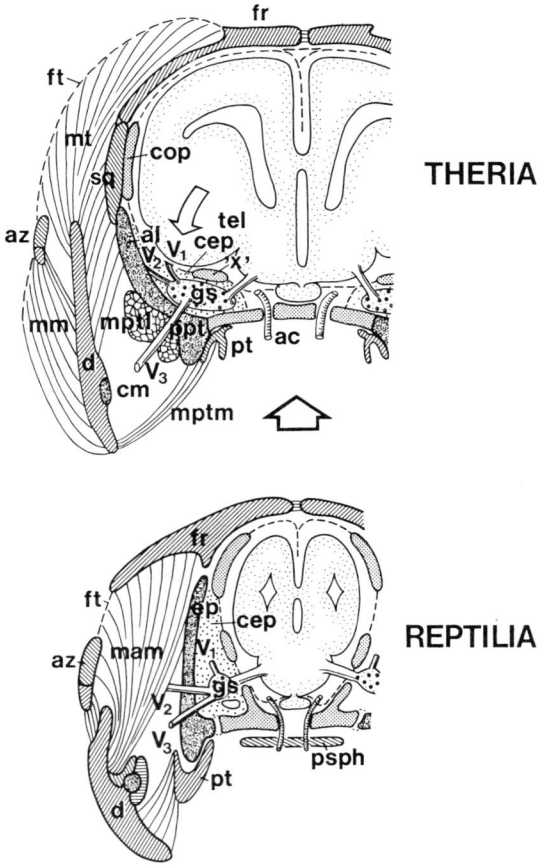

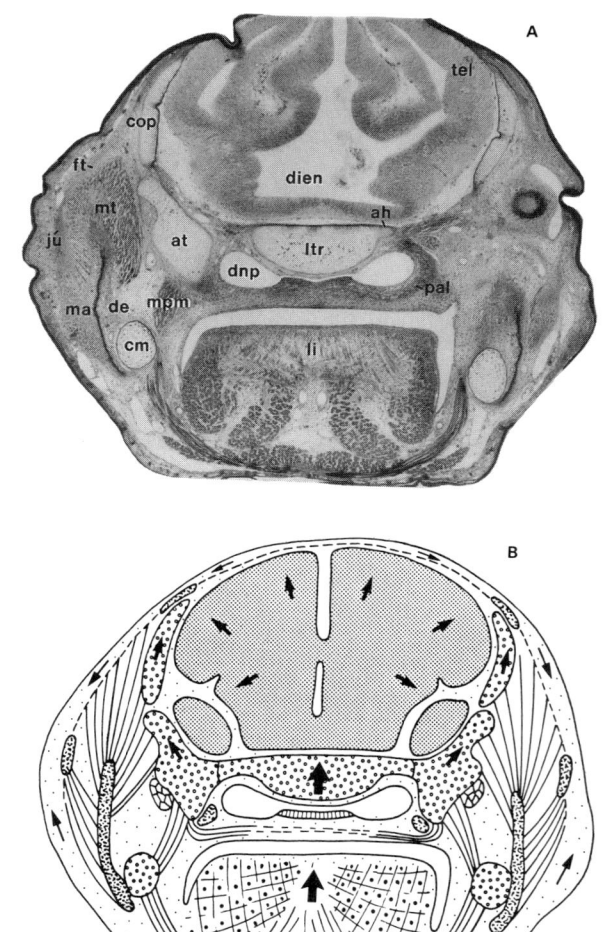

FIGURE 12.6. Diagrammatic cross sections of the temporal region of fetal stages of lepidosauran reptiles and therian mammals. In the reptile, the endocranium (stippled) and the exocranium (hatched) are well separated. The small brain is enclosed by the endocranium; the cavum epitericum with the trigeminal ganglion lies outside the braincase; it is laterally bordered by the epipterygoid (ossified ascending process of the palatoquadrate) and the adjacent sphenoobturatory membrane. Epipterygoid and skull base are linked by the basipterygoid joint. In therians, a secondary sidewall of the braincase is formed by deep lamellae of exocranial elements (frontal, parietal, and squamosal) as well as by the alisphenoid, which is a broadened epipterygoid. The cavum epitericum and the trigeminal ganglion are enclosed in the braincase by the secondary sidewall, but in fact they always remain outside the dura mater (cavum trigeminale) and other equivalents of the primary wall of the braincase. For abbreviations, see Table 12.1.

FIGURE 12.7. Histological cross section through the orbitotemporal region of a neonate *Monodelphis domestica* (CRL 10 mm). The section is running somewhat obliquely; at the left side, the ascending process of the ala temporalis and the soft palate, at the more rostral right side, the palatine and the eye are met. **A**: A somewhat more posterior section shows the big size of the trigeminal ganglion, which bulges into the medial fossa of the brain cavity; the temporal lobes are not yet well developed at this stage. The basal plate is pierced by the hypophyseal canal. **B**: A semidiagrammatic compilation of the structures of the temporal region with tentative distribution of forces caused by upward pressure of the tongue base. For abbreviations, see Table 12.1.

the pull of the adductor muscles during sucking activities (Fig. 12.6–7).

The "anlage" of the ala temporalis is separated from the basitrabecular process of the hypophyseal plate by a cleft, which most probably is reminiscent of the ancient basipterygoid articulation. Later on, there remains a synchondrosis between the alisphenoid and the basisphenoid, which probably is equivalent to the basipterygoid joint as well. This synchondrosis is a functional joint and a center of growth at the same time (in German, "Wachstumsgelenk"). This important suture does not even have an official anatomical name; Gaupp (1910) has called it "junctura basipalatina," which signifies it as a joint between the basal plate and the palatal process of the palatoquadrate.

Semidiagrammatic cross sections show how the secondary sidewall of the therian braincase is composed (Fig. 12.6). The alisphenoid, which is here interpreted as an expanded epipterygoid, is the most important element of the new sidewall; dorsally, it is completed by

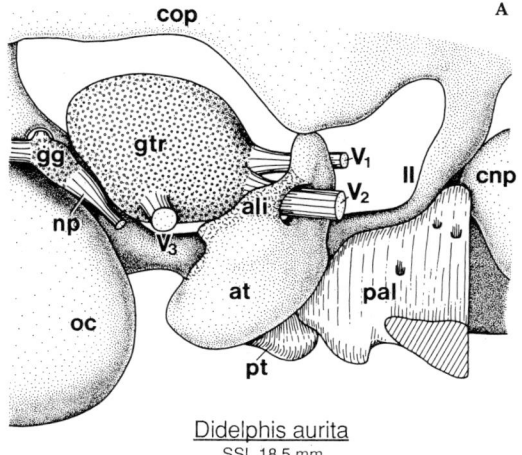

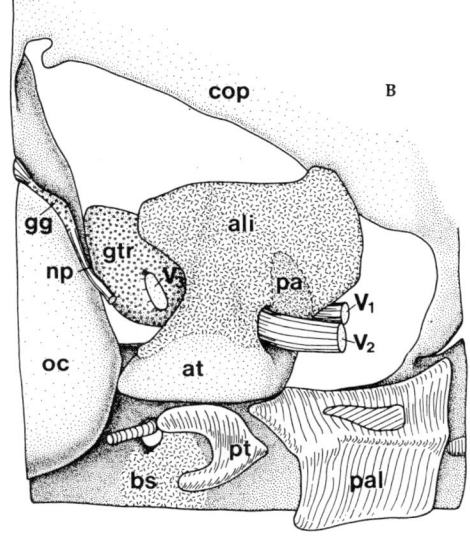

FIGURE 12.8. Lateral aspects of the orbitotemporal region of two pouch young marsupials, based on plate reconstructions. **A**: *Didelphis aurita* (CRL 18.5 mm; Hubrecht collection). Note that the cartilaginous ascending process of the ala temporalis reaches the orbitoparietal commissure; the formation of the "zuwachsknochen" of the alisphenoid begins as an ossification of the sphenoobturate membrane surrounding the maxillary branch. The ascending process lies between the first two trigeminal branches as in reptiles. **B**: Somewhat later stage of *Metachirops opossum* (CRL 17.5 mm), in which the alisphenoid is much expanded dorsally and posteriorly. For abbreviations, see Table 12.1.

limitating membrane; the primary exit of the trigeminal nerve is the prootic foramen (Gaupp, 1911). In mammalian embryos, the primary sidewall is represented by the orbitoparietal commissure and by the dura mater; occasionally, we find pieces of "restknorpel" ("x") embedded in the dura above the trigeminal porus, which are commonly interpreted as atavistic relics of the primary sidewall of the braincase. The trigeminal cave is the morphological equivalent of the cavum epitericum.

Figure 12.7 shows that in *Monodelphis domestica* the ascending process of the ala temporalis almost reaches the lower border of the orbitoparietal commissure. In a two-day-old young of the same species, the ascending process is broadly fused with the commissure. Somewhat later, "zuwachsknochen" of the alispenoid provides the connection between the cartilages of the ala and the commissure (Maier, 1987b). Figure 12.8 presents two stages of the alisphenoid formation in two different taxa of didelphids, showing the gradual inclusion of the huge trigeminal ganglion.

It has been mostly argued that it was the increasing size of the forebrain that resulted in the incorporation of extracranial space into the braincase (Gaupp, 1902). Maier (1989a) has shown that it is more likely that in the neonate marsupials the enormously enlarged trigeminal ganglion is pushed medially, and is then protected by the outgrowing alisphenoid against friction by the chewing musculature. The precocious development of the sensible trigeminal system as well as the early action of the adductors are closely correlated with the sucking activity.

In eutherian mammals, the ala temporalis has usually lost its supporting functions for the primary sidewall; due to prolonged intrauterine life, and due to the precocious expansion of the brain, which brings the ventral parts of the sidewall into an almost horizontal position, the constructional role of the ascending process may have become superfluous. It has been argued elsewhere that the postoptical pillar, which possibly is a secondary neomorph in eutherians, takes some of its tasks (Maier, 1987b). The ventral process of the ala temporalis gives origin to the pterygoid muscles and lends support to the pterygoid remains, namely the hamulus, which is a hypomochlion for the tensor muscle of the soft palate (Barghusen, 1986). The ascending process of the ala temporalis usually lends support to the trigeminal ganglion and its branches as well as for the maxillary artery. In a few eutherians, the first two branches of the trigeminal nerve may be more or less completely separated by a thin ascending process (Fig. 12.9), thus representing a transitional stage of alisphenoid differentiation (Maier, 1987b, 1989a). In my view, there is no need to make a clear morphological distinction between an ascending process and an ascending lamina of the ala temporalis (Presley and Steel, 1976).

deep processes of the frontal, parietal, and squamosal. The new sidewall of the braincase encloses the cavum epitericum and its contents (trigeminal ganglion, etc.) into the braincase. In primitive tetrapods, the cavum epitericum is situated outside the primary sidewall structures, which consist of cartilaginous bars and the

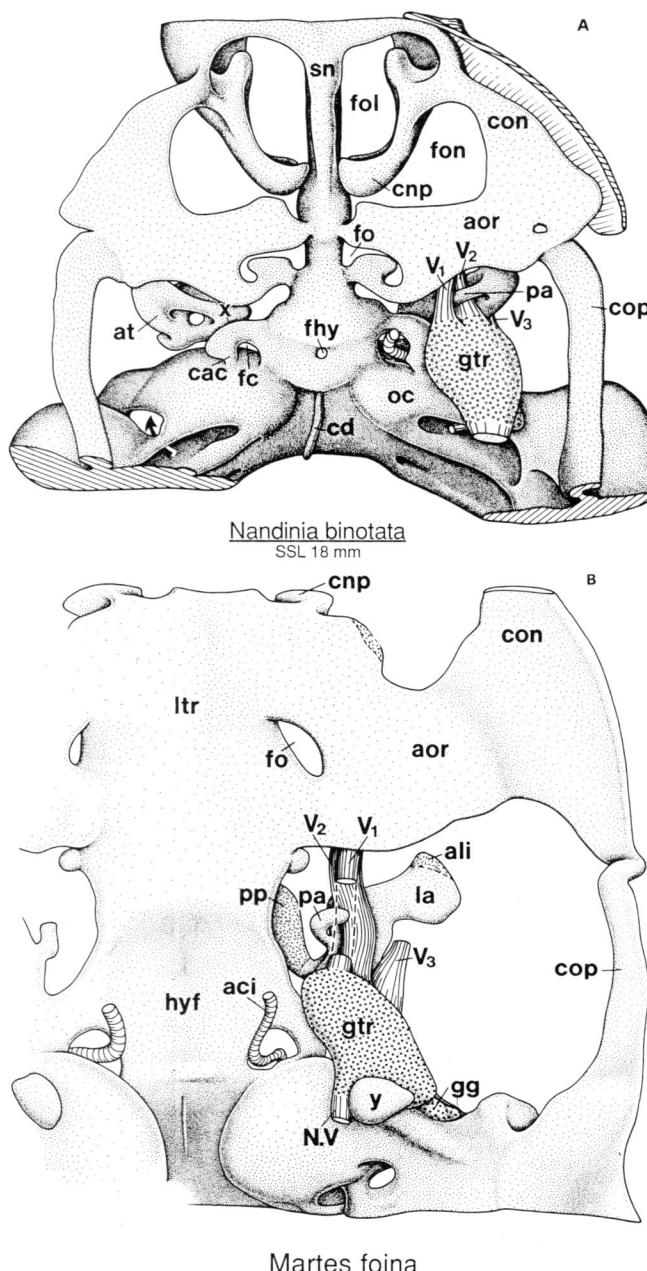

FIGURE 12.9. **A**: Dorsal view of immature chondrocranium of *Nandinia binotata* (according to a plate reconstruction). The isolated "anlage" of the ala temporalis articulates with the alar process of the hypophyseal plate at the equivalent of the basi-pterygoid joint (x); the ala temporalis shows a slender ascending process medial to the "ascending lamina", which encloses the maxillary branch of the trigeminus; the lamina ascendens reaches the double-rooted ala orbitalis. **B**: Dorsal view of an older chondrocranium of *Martes foina*. The ala temporalis is relatively small, and it shows a tiny alisphenoid ossification; the ascending process is a small hook-like structure, which partially separates the first two trigeminal branches (V_1 partially stippled); the exocranial pterygoid is (as in many eutherians) developed as a secondary "pterygoid cartilage"; the trigeminal porus is overlain by a "restknorpel" (y), which may be an atavistic relic of the primary sidewall. For abbreviations, see Table 12.1.

THERIAN MIDDLE EAR. Van Kampen (1905) has postulated that the horizontal "anlage" of the tympanic ring and of the tympanic membrane is primitive for mammals; this hypothesis, which has been accepted by most authors (Gaupp, 1913; Van der Klaauw, 1924, 1931; Starck, 1967, 1979), was based on the evidence of the monotremes and eutherian fetuses. On the other hand, Van Kampen and Gaupp together with Palmer (1913) were the first authors to suggest the homology of the tympanic and the angulare; this homology has subsequently been accepted by most authors (Goodrich, 1930; Romer, 1956). However, the angulare of amphibians and reptiles is an exocranial element of the lower jaw, whereas the mammalian tympanic appears to be fixed to the basicranium from the beginning in most cases. This problem of homology and evolutionary translocation of a skeletal element has not yet been properly addressed; it was assumed that it had occurred in pre-mammalian evolution.

Allin (1975), Crompton and Jenkins (1979), Kemp (1982), and others have presented the paleontological evidence for the gradual evolutionary transformation of the primary quadrato-articular jaw joint into the sound-transmitting middle ear ossicles as well as the concomitant formation of the secondary squamoso-dental articulation. Kermack et al. (1981) have shown that in the early mammal *Morganucodon* from the Triassic-Jurassic boundary the angulare or tympanic was still fixed to the angle of the lower jaw in the adult. Therefore, we are faced with the situation that the positional change of the tympanic must have taken place during the evolutionary differentiation of early mammals (Carroll, 1988). Of course, we are also confronted with the problem of explaining how and why this transformation took place.

Maier (1987a, 1990) has argued that postnatal skull development in marsupials may shed some light on these questions. In early postnatal stages of *Monodelphis*, the developing tympanic/angulare and goniale/prearticulare, together with the cartilage of Meckel, closely fit into the medial trough of the dentary; the whole complex of these skeletal elements is at first in a relatively vertical position. As mentioned before, the still functioning primary jaw joint becomes replaced by the secondary squamoso-dentary joint only during the second week of postnatal life and is thus released from mandibular mechanics. Because of the postnatal expansion of the forebrain and of the cochlear part of the otic capsule, the glenoid area of the squamosum becomes shifted laterally, and the dentale together with the adjacent postdentary elements as well as the "anlage" of the tympanic membrane are gradually brought into an almost horizontal position (Maier, 1987a, fig. 20; Maier, 1990).

During the third week or so, the upper portion of the ramus of the dentary gradually moves back into a vertical position again, the angular process remaining be-

FIGURE 12.10. Evolution of the middle ear structure in mammals. In *Morganucodon*, the angulare/tympanicum—and probably the tympanic membrane—remained situated at the angle of the lower jaw; in extant marsupials, this primitive position is maintained during several weeks of postnatal life before it becomes fixed to the basicranium (see text). In both monotremes and eutherians, the angulare/tympanic develops in a horizontal position underneath the basicranium; it is well separated from the dentale from the beginning. This derived mode of development probably occurred two times independently in the two taxa, whereas marsupials have retained a largely plesiomorphic condition. (The drawing of *Morganucodon* is modified after Kermack et al., 1981; *Ornithorhynchus* is taken from Zeller, 1989). For abbreviations, see Table 12.1.

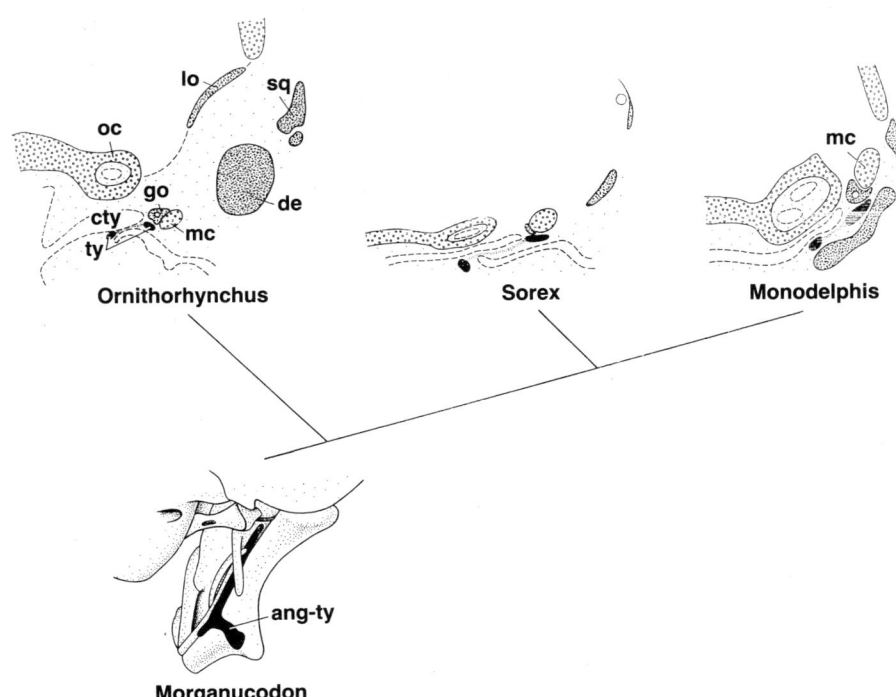

hind in its typical inflected state. The backshift of the mandibular ramus occurs at about the time when the lip fusion is dissolved and when the first teeth begin to erupt. Meanwhile, the tympanic ring has become fixed to the tympanic processes of the alisphenoid and petrosal by ligaments; therefore, it stays behind at the basicranium when the dentary takes its definitive form and position. This decoupling may possibly be directly caused by the increasingly mobile dentary, because some destruction of the connective tissues between the angular process and the tympanic ring can be observed at this time. Shortly after the separation from the tympanic, the dentary begins to grow rapidly, whereas malleus, incus, and tympanic retain about their initial size; Fürbringer (1904) had already suggested that unequal mechanical stimulation may be responsible for this differential growth of the primary and secondary jaw elements, which eventually results in a complete separation of both subsystems.

Thus we can still observe during postnatal development of marsupials how the tympanic system is moving from the lower jaw to the basicranium. The more difficult question to answer is what this ontogenetic process tells us about the reasons for the past evolutionary transformation. It is here again postulated that ontogenetic adaptations to specific mammalian breeding biology also influenced this aspect of mammalian evolution. Gegenbaur (1898) and Fürbringer (1904) suggested that fusion of the lips and immobilization of the jaw as well as lateral expansion of the braincase were important factors in allowing the exchange of the primary and secondary jaw articulations to occur. Precocious ontogenetic expansion of the rhombencephalon must have been an important factor as well.

It is here assumed that the mode of ontogenetic differentiation of the middle ear in marsupials represents a primitive feature, and that it provides a model for explaining the evolutionary development of the mammalian middle ear in general. It is concluded that this developmental mode reflects the mode of the common mammalian ancestor as well as that of the common therian ancestor. Obviously, this assumption is at variance with the "Van Kampen postulate": It is suggested that the horizontal "anlage" of the tympanic in monotremes and eutherians is a derived feature that developed two times independently from a more primitive upright condition. Marsupial ontogeny is thought to continue to represent such a primitive condition, one that may be directly linked with the adult situation of morganucodontids (Fig. 12.10).

THERIAN NASAL CAPSULE. It is well established that primitive mammals display pronounced macrosmatic adaptations; together with their mechanosensitivity (rhinarium, vibrissae), this is considered to be closely connected with their specific nocturnal, terrestrial, and social way of life (Starck, 1978; Crompton and Jenkins, 1979; Eisenberg, 1981; Klima and Maier, 1988). Macrosmatic olfaction needs a big and complicated nasal cavity providing sufficient surface for an expanded olfactory epithelium with a large number of chemoreceptors. The olfactory bulbs and the olfactory centers of the telencephalon are affected by this specialization as well. The vomeronasal organ (Jacobson's organ) is enlarged and elongated in a posterior direction; this sense organ, possibly specialized for intraspecific olfactory communication (Maier, 1980), possesses distinct central nervous pathways as well (Schilling, 1970). A cribriform plate is a synapomorphy of mammals, including monotremes (Zeller, 1988). The production of pheromones by skin glands is a corollary of the advanced olfactory system (Zeller et al., 1988).

The morphology of the nasal cavity of mammals is not well known. It can be properly understood only if its ontogenetic development is considered. The classical study of Paulli (1900) still has to serve as a major source of information, but Paulli did not have sufficient embryogenetic evidence either. Although Hershkovitz (1977) carefully described the bony structures of the nasal skeleton in many primates, he considered neither craniogenesis nor soft tissues. Kermack et al. (1981) have made a reasonable attempt to reconstruct the missing ethmoidal structures in *Morganucodon*. I do not intend to present a comprehensive account of the mammalian or even therian nasal skeleton here, but rather to provide some hints and incentives for its further study.

Ontogenetically, the epithelial nasal sac is almost completely encased in a cartilaginous nasal capsule. It shows anterior openings for the nares, and at its ventral side it is widely open to the mouth cavity (primary choanae). During early fetal development, the palatine processes form a secondary palate to separate the nasal from the oral cavity; the nasopalatine ducts (Stenon's ducts) are remnants of the primary choanae. The secondary choanae open into the nasopharynx. The nasal sac produces a specific pattern of conchae, which become secondarily supported by cartilaginous turbinals. Comparative craniogenetic research has worked out a "bauplan" of the fetal nasal capsule; Figure 12.11 may help to provide an idea of the major features.

In newborn marsupials, the internal relief of the nasal cavities is very simple, showing only the major outlines of the "bauplan" presented above. This "empty" state of the nasal cavities may be interpreted as a specific adaptiation to the predominance of breathing functions in early postnatal stages. The rounded external nasal apertures are stiffened by the strong cartilaginous frame of the anterior nasal cupola. The lower breathing ducts are kept open by the well-developed hyoidal, laryngeal, and tracheal skeleton (cf. Fig. 12.3). Further postnatal development results in a highly elaborated conchal system of the nasal cavity, the greater parts of which are covered with olfactory epithelium. In these stages, some parts of the cartilaginous nasal capsule and the turbinals are developing into the ethmoidal skeleton by enchondral ossification; other parts become resorbed, and still others remodeled by "zuwachsknochen" (appositional bone). Still further complications are caused by the highly variable processes of pneumatization of ad-

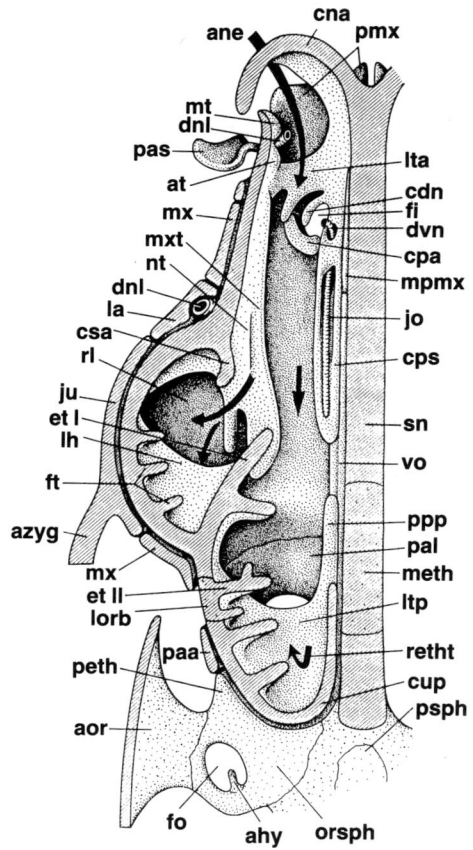

FIGURE 12.11. Schematic horizontal section through the nasal capsule of a therian mammal of fetal age; dorsal view. The tripartite composition of the mammalian nasal capsule is evident. Variable pneumatization of exocranial bones adjacent to the nasal capsule further complicates the internal structure of the nasal cavity in later stages. For abbreviations, see Table 12.1.

jacent exocranial bones (paranasal cavities). However, juvenile and adult nasal cavities of marsupials need much further study.

Due to their prolonged intrauterine development, all eutherian neonates are born with a much further elaborated nasal cavity than are marsupials. But the differentiation of the turbinal system and formation of paranasal cavities continue in them long into postnatal life (Starck, 1967). As far as we can see at the present stage of knowledge, both the marsupial and the eutherian nasal systems are probably derived from the same "bauplan," such as the one outlined in Figure 12.11. Monotremes are more difficult to interpret, because both subgroups are specialized in different directions: The Tachyglossidae are extremely macrosmatic, whereas the Ornithorhynchidae, like many other aquatic mammals, show distinct reductions; but from the available data, their nasal skeleton appears to agree with the therian pattern (Kuhn, 1971; Zeller, 1989). Of special morphological interest are evolutionary lines with regressive differentiation of the olfactory system, such as in primates (Maier, 1980, 1983, in preparation).

Discussion and Conclusions

This chapter makes an attempt to reconstruct certain aspects of the skull morphology of the supposed common therian ancestor. Acknowledging the fact that skull morphology needs a complete ontogenetic analysis, I started with a comparative craniogenetic approach. For certain reasons, which shall be explained below, I used the cranium of the neonate didelphid as a starting point. The head skeleton of neonate marsupials shows structural adaptations that seem to be closely connected with the reproductive biology of this group; a number of features of the skull appear to be functionally related mainly to the sucking activities of these altrically born animals.

I confine may considerations therefore largely to the hypothetical "grundplan" of the neonate of the common therian ancestor. It is assumed that these neonatal adaptations must have had consequences for the evolutionary development of the adult skull as well; the mutual relationship of ontogenetic and phylogenetic processes is important for understanding skulls of both living and fossil mammals.

This approach leads one to considerations of the role of reproductive biology in mammalian evolution in general, and in therians in particular. In recent years, a number of attempts have been undertaken to devise a scenario of early mammalian evolution on the basis of comprehensive biological premises, including reproductive strategies (Hopson, 1973; Pond, 1977; Case, 1978; Crompton et al., 1978; Guillette and Hotton, 1986; Lillegraven, 1986; Renfree, this volume chapter 2). Recently Tyndale-Biscoe and Renfree (1987) have carefully reviewed the questions related to the specific adaptations of marsupial reproduction. They come to the conclusion "that the basic mode of mammalian reproduction evolved simultaneously with the origin of mammals in the Triassic and remained almost unchanged until the Tertiary because it was the most appropriate mode for small nocturnal insectivorous mammals" (p. 397). We have to ask what evidence can possibly be derived from the growing skull of extant primitive mammals such as the didelphids for elucidating this problem.

According to Hopson, Pond, Tyndale-Biscoe and Renfree, and others it is not primarily the difference between oviparity and viviparity that has been most important for the evolution of mammalian reproduction, but rather the way of nourishing the hatchling or neonate. It is typical for mammals that their young feed themselves by actively sucking milk secretions produced by the mother. It seems to be widely assumed that feeding of the young with skin secretions and hatching them at an incubation patch are prerequisites for understanding the evolution of the reproductive biology of small endothermic mammals (Hopson, 1973).

Assuming lactation as a diagnostic mammalian adaptation (Hopson, 1973; Pond, 1977), we can ask for other correlated structural differentiations that do not necessarily show themselves in the fossil record. On the side of the adult mother, we can postulate a very early evolutionary differentiation of the typical mammalian integument (Starck, 1981), for which hairs and glands are higly distinctive. Hairs are indicative of a homeothermic metabolism, whereas skin glands may have served primarily for producing substances that are used as chemical signals (pheromones) in social communication. As mentioned before, the progressive differentiation of the olfactory system of mammals is closely correlated with the development of skin glands as well. On the side of the fetus, we find the early differentiation of the chewing muscles, of tongue muscles (n. hypoglossus), and of cheeks and lips provided with facial muscles (n. facialis). The secondary palate, larynx, and epiglottis, as well as the massive chondrocranium, may also be functionally related to active sucking, swallowing, and breathing by an altricial neonate. The soft palate is provided with the tensor veli palatini muscle (n. trigeminus), which is derived from the medial pterygoid muscle (Barghusen, 1986).

It is not unlikely that these apparent adaptations to milk sucking that are present in both monotremes and therians (Kuhn, 1971; Griffiths 1978; Zeller, 1989) preceded the evolution of ectental shearing occlusion in adult mammals. Fossil teeth indicate that shearing developed in early mammals during the Jurassic (Crompton, 1972; Maier, 1978;), but the "lactation complex" may be significantly older. This leads to the hypothesis that structural adaptations to infantile sucking may have been preadaptive for the evolution of chewing—as they are still causally connected today in any individual ontogenesis.

How are the differences of the sidewall of the braincase, which exist between monotremes and therians, to be explained in the light of these premises? Kuhn and Zeller (1987) have shown that the position of the sidewall structures is not fully equivalent in monotremes and therians. The different skeletal elements are well known: In *Ornithorhynchus*, which is the more primitive monotreme in this respect, there is a tiny alisphenoid, whereas the sphenoparietal fenestra is closed by an isolated ossification of the sphenoobturate membrane, the lamina obturans. Later, this lamina fuses with the periotic to form a "lamina anterior periotici" (Zeller, 1989; this volume, chapter 8). In *Tachyglossus* the closure of the lateral sidewall is much more complicated (Kuhn, 1971). The chondrocrania of both monotremes have retained the pila antotica as a plesiomorphic feature; this element of the primary sidewall can provide some support to the primary sidewall of the fetal braincase. In juvenile stages, the pila antotica becomes mostly resorbed; only its posterior basis at the transversal crest becomes ossified as medial clinoidal process, bounding the hypophyseal fossa posteriorly (cf. Zeller, 1989, figs. 5 and 6).

Marsupials have a different way of closing the secondary sidewall at the orbitotemporal region, as has been described above (Maier, 1987b, 1989b). The "zuwachsknochen" of the alisphenoid, which starts to grow from the perichondrium of the ascending process of the ala temporalis, invades the sphenoobturate membrane until it reaches the otic capsule. The posterior epitympanic process of the alisphenoid covers the tympanic cavity, and the ventral tympanic process provides a substantial portion of its anterior and ventral walls (Maier, 1989b). A tegmen tympani is generally missing in marsupials (an incipient tegmen is present in a pouch young of *Thylogale*). In primitive eutherians, the alisphenoid extends far posteriorly as well, but the tympanic roof becomes formed mostly by a tegmen tympani, which is a neomorphic outgrowth of the cartilaginous otic capsule into the sphenoobturate membrane.

Eutherians mostly show a reduced "anlage" of the ala temporalis having no direct contact with the primary sidewall structures; it seems mainly to support the trigeminal ganglion and to provide origin for the pterygoid muscles. It also lends support to the posterior margin of the palatinum and to the pterygoid (which is mostly reduced to a hamulus). During the prolonged intrauterine development, the alisphenoid of eutherians prenatally enlarges by "zuwachsknochen," and the secondary sidewall of the braincase is completed at parturition by bony elements.

We know from the fossil record that late cynodonts have had a more or less expanded anterior lamina of the periotic. We do not understand the real morphological quality of this lamina, because we do not know its ontogeny; however, it seems to lie at the level of the sphenoobturate membrane, and it encloses the cavum epiptericum with its contents laterally (Crompton and Jenkins, 1979). The epipterygoid shows a tendency to become a broad plate as well (Gow, 1985; Kemp, 1982). *Morganucodon* has essentially preserved this cynodont condition (Kermack et al., 1981, fig. 96). The broadened epipterygoid is called the alisphenoid by these authors, indicating that they accept a full homology between both structures. There still exist wide gaps anteriorly and posteriorly to the alisphenoid, which were probably spanned by the sphenoobturate membrane.

The situation of marsupials can easily be derived from that of *Morganucodon* by assuming a gradual posterior expansion of the alisphenoid and a concomitant retreat of the anterior lamina of the periotic. In eutherians there exists a secondary retreat of the epitympanic process, and its complementary replacement by the neomorphic tegmen tympani. According to Kermack et al. (1981), *Morganucodon* shows a small dorsum sellae, which is not otherwise found in primitive mammals. It may be suspected that this structure is the remnant of a small fetal pila antotica resting on the transverse crest (Crompton and Jenkins, 1979). A dorsum sellae has developed several times independently within eutherians, but this dorsum is certainly not homologous to that of *Morgancodon*.

It is postulated that the fetal skull of the last common mammalian ancestor possessed both an ascending process of the palatoquadrate and an antotic pillar. The evolutionary line leading to monotremes would then have reduced the ala temporalis and retained the pila; the sphenoobturate membrane became ossified as the lamina obturans. The therian line, however, did reduce the pila and elaborated the ala temporalis with its subsequent epipterygoid/alisphenoid ossifications. Fetal cynodonts and morganucodonts possibly retained both elements; in the known adult stages, the alisphenoid plate is the predominant skeletal element, whereas the pila has almost disappeared (see also Presley, Wible and Hopson, this volume chapters 3 and 5).

It also seems not unlikely that the transformations of the mandibular joint and middle ear region have been influenced by the existing adaptation of lactation, as was already suggested by Gegenbaur (1898). Gaupp (1913) and Allin (1975) have suspected that sound reception by the postdentary elements of the lower jaw may have occurred very early in the evolution of the synapsids (Maier, 1990). Kermack et al. (1973, 1981) have been able to demonstrate that in *Morganucodon* these elements were still loosely attached to the posteromedial side of the enlarged dentary, which provided the secondary mandibular joint with the squamosal. Postnatal craniogenesis of marsupials seems to provide a model for explaining the translocation of the middle ear structures from the lower jaw to the basicranium; adaptations to sucking (fused lips; expansion of the posterior processes of the alisphenoid) play an important role in explaining this important structural change, but other factors were important as well (brain size; increased size of the cochlea). Although selection pressure may have primarily acted upon the functional improvement of the mature sense organ, the embryogenetic conditions constrained and canalized the necessary changes.

We are confronted with the fact that monotremes and many eutherians develop their tympanic ring and their tympanic membrane far away from the dentary underneath the otic capsule (Van Kampen, 1905). Given comparative anatomy and paleontology, this arrangement has to be interpreted as a highly derived mode of development. We have to conclude that the "horizontal anlage" of the tympanic system developed twice independently in mammals (Fig. 12.10). Although the structural elements are essentially homologous in all three groups, their ontogenetic detachment from the lower jaw must have occurred in convergence. The eutherian condition was probably derived from a marsupial state (as represented by didelphids) by an "ontogenetic abbreviation."

At the time of differentiation of the very small middle ear structures, the braincase of living monotremes became very broad due to its enlarged brain, and the tympanic anlage was displaced to the underside of the skull (Kuhn, 1971; Starck, 1978; Zeller, 1989, fig. 46). It should also be remembered that monotremes still show a separate ectopterygoid in front of the tympanicum; this element is not clearly distinguishable in *Morganucodon* anymore. In neonate marsupials, the forebrain is little developed, and the quadrato-articular joint provides at this age the only functioning jaw articulation. This extremely altricial state of the newborn marsupial may be responsible for the retention of a middle ear and jaw structure in young marsupials that is so similar to the primitive phylogenetic condition. The subsequent ontogenetic changes in the didelphid middle ear have been described above and elsewhere (Maier, 1987a, 1990).

Eutherians always have a distinctly prolonged intrauterine development. Although the time of pregnancy may be very short, even extremely nidicolous newborns are distinctly further developed than any marsupial. In primitive nidicolous eutherian neonates, the middle ear structures are differentiating at almost their definite adult size (Werner, 1960); this means that they look relatively large in young animals. In neonate carnivorans, for example, the tympanic ring is still very thin and fragile. It would certainly be endangered during the birth process if it were not "hidden" underneath the basicranium; therefore, I have suggested that the "horizontal anlage" of the tympanic ring in nidicolous eutherians should be interpreted as a derived protective position (Maier, 1987a); it is possibly derived from a marsupial developmental mode by a process of abbreviation. At birth, nidifugous eutherians are at a much higher state of structural and functional differentiation, and their middle ear is basically identical with that of the adults, although the tympanic cavity is not yet fully inflated.

The nasal capsule of neonate marsupials and of monotreme hatchlings is constructed of thick cartilaginous plates, indicating the biomechanical importance of this part of the chondrocranium for the development of the facial skull as well. The internal surface of the nasal cavity is not complicated at first, although the typical therian "bauplan" is already distinguishable; this indicates the predominant importance of the nasal cavity as an air passage. Eutherians, on the other hand, show a retarded development of the nasal capsule during early intrauterine development, but at the time of birth, it is structurally advanced and well equipped with turbinals. The biomechanically relevant construction of the facial skull is provided by the exocranial elements.

ACKNOWLEDGMENTS. I am grateful to the organizers for inviting me to a stimulating meeting. I thank Mrs. Margret Roser, who has made the drawings with her usual skill.

CORRESPONDENCE ADDRESS. Wolfgang Maier, Department of Systematic Zoology, Eberhard Karls-University, Auf der Morgenstelle 28, D-7400 Tübingen, Germany.

References

Allin, E.F. 1975. Evolution of the mammalian middle ear. *J. Morph.* 147:403–438.

Barghusen, H.R. 1986. On the evolutionary origin of the therian tensor veli palatini and tensor tympani muscles. In: *The ecology and biology of mammal-like reptiles* (Hotton, N., et al., eds.). Washington, D.C.: Smithsonian Institution Press, pp. 253–262.

Bock, W. 1988. The nature of explanations in morphology. *Amer. Zool.* 28:205–215.

Bonner, J.T. 1982. *Evolution and development.* Berlin: Springer.

Carroll. R.L. 1988. *Vertebrate paleontology and evolution.* New York: Freeman.

Case, T.J. 1978. Endothermy and parental care in the terrestrial vertebrates. *Amer. Naturalist* 112:861–874.

Clemens, W.A. 1979. Marsupialia. In: *Mesozoic mammals* (Lillegraven, J.A., et al., eds.). Berkeley: University of California Press, pp. 192–220.

Crompton, A.W. 1972. Postcanine occlusion in cynodonts and tritylodontids. *Bull. Brit. Mus. Nat. Hist. (Geol.)* 24:399–437.

Crompton, A.W. 1989. The evolution of mammalian mastication. In: *Complex organismal functions: Integration and evolution in vertebrates* (Wake, D.B., and Roth, G., eds.). Chichester: Wiley, pp. 23–40.

Crompton, A.W., and Jenkins, F.A. 1979. Origin of mammals. In: *Mesozoic mammals* (Lillegraven, J.A., et al., eds.) Berkeley: University of California Press, pp. 59–73.

Crompton, A.W., Taylor, C.R., and Jagger, J.A. 1978. Evolution of homeothermy in mammals. *Nature* 272:333–336.

de Beer, G. 1937. *The development of the vertebrate skull.* Oxford University Press.

Eisenberg, J.F. 1981. *The mammalian radiations. An analysis of trends in evolution, adaptation, and behaviour.* London: The Athlone Press.

Fürbringer, M. 1904. Zur Frage der Abstammung der Säugetiere. In: *Festschrift zum 70. Geburtstag von E. Haeckel.* Teil I. Jena: Fischer.

Gaupp, E. 1902. Über die Ala temporalis des Säugerschädels und die Regio orbitalis einiger anderer Wirbeltierschädel. *Anat. Hefte* 19:155–230.

Gaupp, E. 1906. Die Entwicklung des Kopfskelettes. In: *Handb.d. vergl. u. experiment. Entwickelungslehre d. Wirbeltiere* vol. III/2 (Hertwig, O., ed.). Jena: Fischer, pp. 573–874.

Gaupp, E. 1910. Säugerpterygoid und Echidnapterygoid nebst Bemerkungen über das Säuger-Palatinum und den Processus basipterygoideus. *Anat. Hefte* 42:311–431.

Gaupp, E. 1911. Über den N. trochlearis der Urodelen und über die Austrittsstellen der Gehirnnerven aus dem Schädelraum im Allgemeinen. *Anat. Anz.* 38:401–444.

Gaupp, E. 1913. Die Reichertsche Theorie (Hammer-, Amboss- und Kieferfrage). *Arch. Anat. Entwickl. gesch.* 1912 (Supple.):1–416.

Gegenbaur, C. 1898. *Vergleichende Anatomie derWirbelthiere*, vol. I. Leipzig: Engelmann.

Goodrich, E. 1930. *Studies on the structure and development of vertebrates*. London: MacMillan.

Gould, S.J. 1977. *Ontogeny and phylogeny*. Cambridge: Belknap.

Gow, C.E. 1985. The side wall of the braincase in cynodont therapsids, and a note on the homology of the mammalian promontorium. *S. Afr. J. Zool.* 21:136–148.

Griffiths, M. 1978. *The biology of the monotremes*. New York:

Guillette, L.J., and Hotton, N. 1986. The evolution of mammalian reproductive characteristics in therapsid reptiles. In: *The ecology and biology of mammal-like reptiles* (Hotton, N., et al., eds.) Washington, D. C.: Smithsonian Institution Press, pp. 239–250.

Hennig, W. 1948. *Die Larvenformen der Dipteren*. Berlin: Academie-Verlag.

Hennig, W. 1966. *Phylogenetic systematics*. Chicago: University of Illinois Press.

Herring, S.W. 1985. Postnatal development of masticatory muscle function. In: *Vertebrate morphology* (Duncker, H.-R., and Fleischer, G., eds.). Stuttgart: Fischer, pp. 213–215.

Herring, S.W., and Scapino, R.P. 1973. Physiology of feeding in miniature pigs. *J. Morph.* 141:427–460.

Hershkovitz, P. 1977. *Living new world monkeys (Platyrrhini)*, vol. I. Chicago: University of Chicago Press.

Hiiemae, K.M., and Crompton, A.W. 1985. Mastication, food transport, and swallowing. In: *Functional vertebrate morphology* (Hildebrand, M., et al., eds.). Cambridge: Belknap, pp. 262–290.

Hofer, H. 1945. Zur Kenntnis der Suspensionsformen des Kieferbogens und deren Zusammenhänge mit dem Bau des knöchernen Gaumens und mit der Kinetik des Schädels bei Knochenfischen. *Zool. Jahrb. (Abt. Anat.)* 69:321–404.

Hopson, J.A. 1973. Endothermy, small size, and the origin of mammalian reproduction. *Amer. Naturalist* 107:446–452.

Hughes, R.L., and Hall, L.S. 1988. Structural adaptations of the newborn marsupial. In: *The developing marsupial* (Tyndale-Biscoe, C.H., and Janssens, P.A., eds.). Berlin: Springer, pp. 8–27.

Kemp, T.S. 1982. *Mammal-like reptiles and the origin of mammals*. London: Academic Press.

Kermack, K.A., Mussett, F., and Rigney, H.W. 1973. The lower jaw of *Morganucodon*. *J. Linn. Soc. (Zool.)* 53:87–175.

Kermack, K.A., Mussett, F., and Rigney, H.W. 1981. The skull of *Morganucodon*. *Zool. J. Linn. Soc.* 71:1–158.

Kielan-Jaworowska, Z., Eaton, J.G., and Bown, T.M. 1979. Theria of Metatherian-Eutherian grade. In: *Mesozoic mammals* (Lillegraven, J.A., et al., eds.). Berkeley: University of California Press, pp. 182–191.

Klima, M. 1987. Early development of the shoulder girdle and sternum in marsupials (Mammalia: Metatheria). *Advances Embryol. Cell Biol.* 109:1–91.

Klima, M., and Maier, W. 1988. Körperbau (Anatomie). In: *Grzimeks Enzyklopädie Saugetiere* vol. I (Grzimek, B., ed.). München: Kindler, pp. 58–85.

Krause, W.J., and Leeson, C.R. 1975. Postnatal development of the respiratory system of *D. virginiana*. II. Electron microscopy of the epithelium and the pleura. *Acta Anat.* 92:28–44.

Kuhn, H.-J. 1971. Die Entwicklung und Morphologie des Schädels von *Tachyglossus aculeatus*. *Abh. Senckenberg. Naturforsch. Ges.* 528:1–224.

Kuhn, H.-J., and Zeller, U. 1987. The cavum epiptericum in monotreme and therian mammals. In: *Morphogenesis of the mammalian skull* (Kuhn, H.-J., and Zeller, U., eds.). Hamburg: Parey, pp. 51–70.

Kummer, B. 1972. Biomechanics of bone: Mechanical properties, functional structure, functional adaptation. In: *Biomechanics, its foundations and objectives* (Fung et al., eds.) Englewood Cliffs: Prentice Hall, pp. 237–271.

Lakars, T., and Herring, S.W. 1980. Ontogeny of oral functions in hamsters (*Mesocricetus auratus*). *J. Morph.* 165:237–254.

Lillegraven, J.A. 1969. Latest Cretaceous mammals of upper part of Edmonton Formation of Alberta, Canada, and review of marsupial-placental dichotomy in mammalian evolution. *Univ. Kansas Paleont. Contrib.* 50:1–122.

Lillegraven, J.A. 1976. Biological considerations of the marsupial-placental dichotomy in mammalian evolution. *Evolution* 29:707–722.

Lillegraven, J.A. 1979. Reproduction in Mesozoic mammals. In: *Mesozoic mammals* (Lillegraven, J.A., et al., eds.). Berkeley: University of California Press, pp. 259–274.

Lillegraven, J.A. 1986. Reproduction in Mesozoic mammals. In: *The ecology and biology of mammal-like reptiles* (Hotton, N., et al., eds.) Washington, D.C.: Smithsonian Institution Press, pp. 259–276.

Luckett, W.P., and Maier, W. 1982. Development of deciduous and permanent dentition in *Tarsius* and its phylogenetic significance. *Folia Primatol.* 37:1–36.

Maier, W. 1978. Die Evolution der tribosphenischen Säugetiermolaren. *Sonderb. Naturw. Ver. Hamburg* 3:41–60.

Maier, W. 1980. Nasal structures in Old and New World monkeys. In: *Evolutionary biology of the New World monkeys and continental drift* (Ciochon, R. L., and Chiarelli, A.B., eds.). New York: Plenum, pp. 219–241.

Maier, W. 1983. Morphology of the interorbital region of *Saimiri sciureus*. *Folia Primatol.* 41:277–303.

Maier, W. 1987a, Der Processus angularis bei *Monodelphis domestica* (Didelphidae; Marsupialia) und seine Beziehungen zum Mittelohr: Eine ontogenetische und evolutionsmorphologische Untersuchung. *Morph. Jahrb.* 133:123–161.

Maier, W. 1987b, The ontogenetic development of the orbitotemporal region in the skull of *Monodelphis domestica* (Didelphidae, Marsupialia) and the problem of the mammalian alisphenoid. In: *Morphogenesis of the mammalian skull* (Kuhn, H.-J., and Zeller, U., eds.). Hamburg: Parey, pp. 71–90.

Maier, W. 1988. Dar Säugercranium als natürliche konstruktion. *H MiH. SFB 230* 3:219–226.

Maier, W. 1989a. Ala temporalis and alisphenoid in therian mammals. In: *Trends in vertebrate morphology* (Splechtna, H., and Hilgers, H., eds.). Stuttgart: Fischer, pp. 396–400.

Maier, W. 1989b. Morphologische Untersuchungen am Mittelohr der Marsupialia. *Z. zool. Syst. Evolut.-forsch.* 27:149–168.

Maier, W. 1990. Phylogeny and ontogeny of the mammalian middle ear structures. *Netherl. J. Zool.* 40:55–74.

Maier, W., and Schrenk, F. 1989. The hystricomorphy of the Bathyergidae, as determined from ontogenetic evidence. *Z.f.Säugetierk.* 52:156–164.

Marshall, L.G. 1979. Evolution of metatherian and eutherian (mammalian) characters: A review based on cladistic methodology. *Zool. J. Linn. Soc.* 66:369–410.

Merchant, J.C., and Sharman, G.B. 1966. Observations on the attachment of marsupial pouch young to the teats and on the rearing of pouch young by foster-mothers of the same or different species. *Austral. J. Zool.* 14:593–609.

Palmer, R.W. 1913. Note on the lower jaw and ear ossicles of a foetal *Perameles. Anat. Anz.* 43:510–515.

Paulli, S. 1900. Über die Pneumaticität des Schädels bei den Säugetieren. Teile I-III. *Morph. Jahrb.* 28:147–251, 483–564.

Pauwels, F. 1960. Eine neue Theorie uber den Einflussmechanischer Reize auf die Differenzierung der Stützgewebe. *Z. Anat. Entwickl. Gesch.* 121:478–515.

Pond, C.M. 1977. The significance of lactation in the evolution of mammals. *Evolution* 31:177–199.

Presley, R., and Steel, F.L.D. 1976. On the homology of the alisphenoid. *J. Anat.* 121:441–459.

Romer, A.S. 1956. *Osteology of the reptiles.* Chicago: University of Chicago Press.

Schilling, A. 1970. L'organe de Jacobson du lémurien malgache *Microcebus murinus* (Miller, 1777). *Mém. Mus. Nat. Hist. Nat., N.S.* A61:203–280.

Sharman, G.B. 1973. Adaptations of marsupial pouch young for extra-uterine existence. In: *The mammalian fetus in vitro* (Austin, C.R., ed.). London: Chapman & Hall, pp. 67–89.

Stadtmüller, F. 1936. Kranium und Visceralskelett der Säugetiere. In: *Handb. d. vergl. Anatomie d. Wirbeltiere* vol. IV (Bolk, L., et al., eds.). Berlin: Urban & Schwarzenberg, pp. 839–1016.

Starck, D. 1967. Le crâne des mammifères. In: *Traité de Zoologie* vol. 16 (Grassé, P.P., ed.). Paris: Masson, pp. 405–549.

Starck, D. 1975. *Embryologie*, 3d ed. Berlin: Thieme.

Starck, D. 1978, 1979, 1981. *Vergleichende Anatomie der Wirbeltiere auf evolutionsbiologischer Grundlage*, vols. I–III. Berlin: Springer.

Starck, D. 1989. Considerations on the nature of skeletal elements in the vertebrate skull, especially in mammals. In: *Trends in vertebrate* morphology (Splechtna, H., and Hilgers, H., eds.). Stuttgart: pp. 375–385.

Tyndale-Biscoe, H., and Renfree, M. 1987. *Reproductive physiology of marsupials.* Cambridge:

Van der Klaauw, C.J. 1924. Bau und Entwickelung der Gehörknöchelchen. *Ergeb. Anat. Entwickl. Gesch.* 25:565–622.

Van der Klaauw, C.J. 1931. The auditory bulla in some fossil mammals. With a general introduction to this region of the skull. *Bull. Amer. Mus. Nat Hist.* 62:1–352.

Van Kampen, P.N. 1905. Die Tympanalgegend des Säugetierschädels. *Morph. Jahrb.* 34:321–722.

Wake, D.B., and Roth, G., eds. 1989. *Complex organismal functions: Integration and evolution in vertebrates.* Chichester: Wiley.

Werner, C.F. 1960. *Das Gehörorgan der Wirbeltiere und des Menschen.* Leipzig: Thieme.

Zeller, U. 1988. The lamina cribrosa of *Ornithorhynchus* (Monotremata, Mammalia). *Anat. Embryol.* 178:513–519.

Zeller, U. 1989. Die Entwicklung und Morphologie des Schädels von *Ornithorhynchus anatinus* (Mammalia: Prototheria: Monotremata). *Abh. Senckenberg. Naturforsch. Ges.* 545:1–188.

Zeller, U., Epple, G., Küderling, I., and Kuhn, H.J. 1988. The anatomy of the circumgenital scent gland of *Saguinus fuscicollis* (Callitrichidae, Primates). *J. Zool. London* 214:141–156.

Zimmermann, W. 1943. Die Methoden der Phylogenetik. In: *Die Evolution der Organismen* (Heberer, G., ed.). Jena: Fischer, pp. 20–56.

CHAPTER 13

An Ontogenetic Assessment of Dental Homologies in Therian Mammals

W. Patrick Luckett

Overview

Ontogenetic analysis of the dentition in a wide range of marsupials and eutherians provides valuable criteria for assessing tooth homologies among extant therian mammals. In addition, these developmental data offer evidence for the loss of teeth during mammalian evolution. Such studies can distinguish the deciduous or successional nature of individual teeth, even when replacement does not occur postnatally. Ontogenetic analyses also provide evidence for the formation of vestigial deciduous teeth that do not erupt; these rudiments serve as valuable clues to the homologies of some tooth positions that have been lost during mammalian phylogeny. The continued accumulation of developmental data on the pattern of early budding and differentiation of deciduous tooth germs offers useful criteria for identifying tooth class homologies of highly modified teeth, in addition to their bony and occlusal relationships.

An important finding of this analysis is that epithelial connections among the deciduous tooth, primary dental lamina, oral epithelium, successional lamina, and successor tooth furnish the most useful criteria for assessing the ontogenetic relationships between a deciduous tooth and its successor. This developmental pattern was used to test recent hypotheses of premolar-molar homologies in marsupials, where it was claimed that no true postcanine tooth replacement occurs. Epithelial connections, however, support the traditional hypothesis that the posterior premolar position is occupied by typical deciduous and successional teeth during ontogeny; this falsifies the hypothesis that five molars occurred primitively in marsupials. Finally, ontogenetic data are considered for testing the hypothesis of serial homology between the postcanine dentition of eutherians and marsupials, and for the occurrence of five premolars primitively in eutherians.

Contents

Introduction, 183
Pattern of Dental Development in Eutheria, 185
   Early Development of the Dental Lamina and Deciduous Teeth, 185
   Development of Molars, 186
   Development of the Secondary or Successional Dentition, 186
   Vestigial Deciduous Teeth, 187
   Ontogenetic Criteria for Distinguishing Tooth Homologies, 188
Pattern of Dental Development in Metatheria, 189
   Ontogeny of the Anterior Dentition in Marsupials, 190
   Development of the Anterior Premolars, 191
   Development of Deciduous and Successional Posterior Premolars, 191
   Development of the Molars, 193
   Ancestral Pattern of Marsupial Dental Development, 194
Discussion, 195
   Distinguishing Deciduous versus Successional Teeth, 195
   Posterior Premolar Homologies in Marsupials, 196
   Ontogenetic Analysis of Dental Homologies in Fossil Therians, 197
   Primitive Number of Premolars in Eutheria, 197
   Tooth Homologies among Therians, 198
   Role of Heterochrony in Therian Dental Evolution, 199
Concluding Comments, 200
Acknowledgments, 201
References, 201

Chapter 13. Ontogenetic Assessment of Dental Homologies

Introduction

Bony position, occlusal relationships, and tooth replacement have been the criteria most commonly used for assessing mammalian dental homologies, since first proposed by Owen (1868). More recently, tooth shape, rather than bony or occlusal relationships, has been considered to play a major role in identifying antemolar teeth as incisors, canines, or premolars, especially in those therians that have greatly modified their dentitions (Osborn, 1978; Schwartz, 1980, 1982). Although the latter authors acknowledged that early budding sequences from the dental lamina during development can provide valuable evidence for assessing tooth class homologies, little attempt has been made to use such early ontogenetic data for testing hypotheses of mammalian dental homologies. Studies on both fossil and extant mammals have raised serious doubts concerning the serial homology of the therian dentition, in particular, the number and replacement pattern of the postcanine teeth in both marsupials and eutherians (McKenna, 1975; Archer, 1978, 1984; Novacek, 1986).

This chapter discusses the manner in which developmental data can be interpreted to analyze possible homologies of the mammalian dentition. The following questions will serve as focal points: (1) Are there ontogenetic criteria for assessing the tooth class homologies of highly modified teeth, in addition to their osteological and occlusal relationships? (2) How can deciduous and successional teeth that belong to the same tooth family be recognized, especially when the deciduous tooth is not "displaced" by an erupting successor? (3) How can one determine the deciduous or successional nature of an antemolar tooth, when there is no evidence of postnatal tooth change at that position? (4) Are there developmental clues regarding the serial homology of tooth positions that have been lost during mammalian phylogeny? (5) Is there developmental evidence for more than two generations of teeth (diphyodonty) in therian mammals? Before addressing these questions, it is necessary to present an overview of the major stages of dental ontogeny in therian mammals and to discuss the differences that exist between

TABLE 13.1. Abbreviations

AB	alveolar bone
Am	ameloblasts
BK	buccal, condensed knot-like portion
C	canines
d	deciduous
D	dentin
DC	dentinal cap
DK	dentinal knot
DL	dental lamina
DP	dental papilla
E	enamel
I	incisors
IEE	inner enamel epithelium
M	molars
Od	odontoblasts
OE	oral epithelium
OEE	outer enamel epithelium
P	premolars
PA	paracone elevation
SL	successional lamina
SR	stellate reticulum
TB	tooth bud

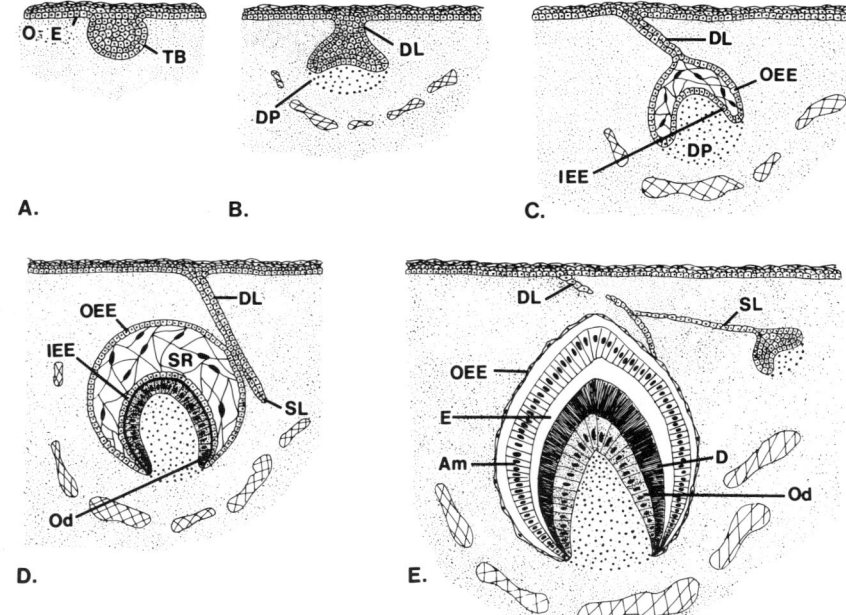

FIGURE 13.1. Overview of early dental development in Eutheria. **A**: Tooth bud forms as a thickening of the dental lamina. **B**: Mesenchyme of dental papilla indents tooth germ to form early cap stage. **C**: The enamel organ partly envelops the papilla at the middle bell stage. **D**: Odontoblasts and a lingual successional lamina form in the late bell stage. **E**: Dentin is secreted by odontoblasts, and ameloblasts secrete enamel. A successor tooth germ differentiates at the free end of the successional lamina. For abbreviations, see Table 13.1. (Modified from Luckett, 1985.)

TABLE 13.2. Sequence of dental development in the upper jaw of Macroscelididae

Stage	dI1	dI2	dI3	dC
Elephantulus 11.5 mm CR	Early-middle bud	—	—	Early bud
Macroscelides 13 mm CR	Late bud	Early bud	—	Early bud
Elephantulus 9.5 mm CR	No data	No data	No data	Early-middle cap
Elephantulus 19 mm CR	Late bell; Mod. dev. dentin	Middle bell	Middle-late bell	Late bell, no odont.
Elephantulus 31 mm CR	Mod. dev. dentin & thin enamel	Middle-late bell	Late bell; thin dentin	Late bell; thin dentin
Elephantulus 25 mm CR	Thick dentin & mod. thick enamel	Late bell; thin dentin	Mod. dev. dentin & thin enamel	Mod. dev. dentin & thin enamel
Macroscelides 37–40 mm CR	Thick dentin & mod. thick enamel	Mod. thick dentin & thin enamel	Mod. thick dentin & mod. dev. enamel	Mod. thick dentin & enamel
Elephantulus 43 mm CR	Thick dentin & enamel	Thick dentin & mod. thick enamel	Partly resorbed; middle bud I3	Thick dentin & enamel

these developmental patterns in eutherians and marsupials.

Leche (1892, 1895) partitioned the continuous process of early dental ontogeny into *bud*, *cap*, and *bell* stages (Fig. 13.1); more recently, these stages have been further subdivided into early, middle, and late phases (Luckett, in preparation), in order to compare more precisely the relative developmental rates of individual tooth loci within the jaws (Table 13.2). Such comparisons can aid in the identification of heterochronies during dental development, and subsequent analysis can provide insight into the manner in which heterochronies may have contributed to evolutionary changes in the mammalian dentition.

This chapter is part of an ongoing histological analysis of the early (pre-eruption) phases of dental ontogeny in a wide range of therians (see Table 13.3). It has been demonstrated that the dental lamina and early tooth buds differentiate at the same relative stage of ontogeny in eutherians, marsupials, and monotremes, when compared to surrounding structures of the developing head (Luckett, 1988). This facilitates the subsequent analysis of postbudding phases of dental ontogeny for possible homologies between eutherians and marsupials.

In the following discussion, the serial homology of incisors (I), canines (C), premolars (P), and molars (M) corresponds to traditionally accepted ancestral dental formulae proposed for therians: 3-1-4-3/3-1-4-3 for eutherians; and 5-1-3-4/4-1-3-4 for marsupials. Use of these traditional homologies makes it easier to compare previously published data from extant and fossil therians with the ontogenetic findings of this chapter. However, it is recognized that these homologies may prove to be incorrect, in light of paleontological evidence for the possible occurrence of a greater number of incisors and premolars in some early eutherians (McKenna, 1975; Kielan-Jaworowska, 1981).

TABLE 13.3. Therian orders and families examined during this study

MARSUPIALIA
  Families Didelphidae, Peramelidae, Dasyuridae, Phalangeridae
EUTHERIA
  Order Lipotyphla
    Families Sorcidae, Talpidae, Erinaceidae, Tenrecidae
  Order Macroscelidea
  Order Scandentia
  Order Chiroptera
    Families Pteropodidae, Phyllostomidae, Vespertilionidae, Rhinolophidae
  Order Dermoptera
  Order Primates
    Suborder Strepsirhini
    Suborder Haplorhini
  Order Rodentia
    Families Sciuridae, Muridae, Caviidae, Bathyergidae, Ctenodactylidae
  Order Lagomorpha
  Order Hyracoidea
  Order Carnivora
    Families Canidae, Felidae
  Order Xenarthra
    Families Bradypodidae, Dasypodidae
  Order Cetacea
    Suborder Odontoceti
  Order Artiodactyla
    Family Suidae

Differentiating teeth are characterized as deciduous (=primary) or successional (=secondary), on the basis of their relationships to the dental lamina (see below). The prefix "d" indicates deciduous teeth, and molars are considered to be unreplaced primary teeth. The presumed serial homologies of individual teeth are designated by a numerical superscript or subscript; thus, dP^2 indicates the second upper deciduous premolar. When referring to the same tooth locus in both jaws, the number is placed on the same line as the tooth abbreviation; i.e., P3 designates both the upper and lower successional third premolars. In considering questions of dental homology, terms such as "permanent" or "adult" teeth

TABLE 13.2. continued

dP1	dP2	dP3	dP4	M1
—	—	Middle bud	—	—
—	—	Middle cap	Early cap	—
—	Early bud	Early bell	Early bell	—
Middle cap	Late bell	Late bell; mod. dev. dentin	Late bell, odont.	Early bud
Middle bell	Thin to mod. dev. dentin	Mod. thick dentin & enamel	Mod. thick dent. & mod. dev. enamel	Early bell
Late bell	Mod. dev. dentin & enamel	Mod. thick dent. & mod. dev. enamel	Mod. thick dent. & mod. dev. enamel	Early bell
Late bell, odont.	Mod. thick dentin & enamel	Mod. thick dentin & enamel	Mod. thick dentin & enamel	Middle bell
Mod. thick dentin & enamel	Thick dentin & enamel	Thick dentin & enamel	Thick dentin & enamel	Late bell

should be avoided (as emphasized by Leche, 1892), because elements of the "permanent" dentition in the same taxon can belong to the deciduous (the unreplaced dP1 of *Elephantulus*) or the successional series (other antemolar teeth of the same genus).

Pattern of Dental Development in Eutheria

EARLY DEVELOPMENT OF THE DENTAL LAMINA AND DECIDUOUS TEETH. The dental laminae of therians are initiated as bilateral ridge-like thickenings of the oral epithelium, before the onset of ossification in both jaws. Subsequently, larger localized thickenings occur along the length of the lamina to initiate early development of discrete tooth buds (Fig. 13.2). These buds do not appear in a strict anterior-posterior sequence; instead, one differentiates at the rostral end of the dental lamina (the future dI1), another at the caudal end of the lamina (the future dP3 or dP4, depending on the species), and a third toward the middle of the dental lamina, representing the site of the future dC. The most consistent features of this pattern in mammals with a relatively complete dentition are the formation of early dC and terminal dP buds. In the incisor region, it is common that one or two buds differentiate almost simultaneously, so that it is very difficult to decide which bud has appeared first, without access to a closely graded ontogenetic series (see Table 13.2).

The reasons for this initial budding pattern are unclear. Experimental studies on the mouse embryo suggest that neural crest cells that migrate into the jaws are equipotential, and that it is the oral epithelium that provides the specification for spatial patterning and initiation of tooth germs (Lumsden and Buchanan, 1986; Lumsden, 1987). These tissue interactions are reciprocal, however, and there is evidence that neural crest cells

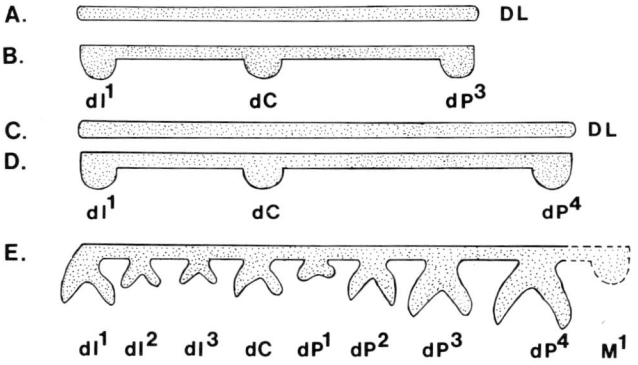

FIGURE 13.2. Diagram of the dental lamina and early tooth germs in Eutheria. **A**: Longitudinal view of early dental lamina. **B**: Initial differentiation of tooth buds along dental lamina, in those taxa (including marsupials) with dP3 as the terminal bud. **C** and **D**: Early development of dental lamina and tooth buds in those eutherians with dP4 as the terminal bud. **E**: Later formation of additional tooth germs along the dental lamina, and the secondary extension of the lamina distally to initiate the M1 bud. For abbreviations, see Table 13.1.

determine the precise form of the crown surface during late bell stages (Kollar and Baird, 1969; Lumsden, 1987). Paradoxically, the mouse embryo, the model of choice for virtually all experimental studies on mammalian odontogenesis, lacks canine and terminal premolar loci during ontogeny. Thus, it is unclear whether experimental findings on development of the highly derived mouse dentition will be generally applicable to understanding the early differentiation of spatial patterning during mammalian odontogenesis. Further consideration of these experimental studies is beyond the scope of this chapter.

As emphasized by Butler (1978a), the initial differentiation of dP3 or dP4 in different eutherian groups seems to depend mainly on the length of the dental lamina at the time of initiation of the terminal dP bud (cf. Figs. 13.2B, 13.2D). In macroscelidids (Table 13.2), carnivorans, and *Tupaia*, dP3 is the first postcanine tooth to be initiated. However, the dental lamina soon extends distally to give rise to the dP4 bud. In *Tupaia*, dP3 has attained only the late bud–early cap stage when dP4 buds are first detected (Luckett and Kuhn, in preparation). The later differentiating dP4 buds undergo a relatively accelerated rate of development in *Tupaia* and *Elephantulus*, so that both dP3 and dP4 have reached the same developmental state by the early-middle bell stage (Table 13.2). In other eutherians, including some lipotyphlans (*Sorex*, *Erinaceus*), the chiropteran *Myotis*, and the hyracoid *Procavia*, dP4 is the earliest postcanine to appear, and it remains developmentally advanced over other deciduous premolars in later stages.

Following the initial budding of an incisor, canine, and terminal premolar locus, other deciduous tooth germs develop (Fig. 13.2E), most commonly in an anterior-posterior direction in the incisor region, and in a posterior-anterior sequence for the premolars, especially when dP4 is the initial premolar bud. When present, dP1 is always the last deciduous tooth to differentiate in eutherians, as in *Elephantulus* (Kindahl, 1957a). The relative developmental retardation of dP1 (see Table 13.2) is probably related causally to the fact that a successor for dP1 is normally absent in most genera. Unfortunately, few ontogenetic data are available for the perissodactyl genus *Tapirus*, the only extant eutherian for which there is clear evidence for normal replacement of dP^1 by P^1 (Owen, 1840–45; Ziegler, 1971; Luckett, unpublished). In a 100 mm GL fetus, dP^1 is the least differentiated of the deciduous premolars, although it is clearly advanced over M^1 (Adloff, 1928). This pattern is continued postnatally, where dP^1 erupts earlier than M^1 (Ameghino, 1909; Luckett, unpublished).

DEVELOPMENT OF MOLARS. Following the early differentiation of most deciduous teeth, the dental lamina in both jaws undergoes secondary extension distal to dP4 and initiates development of a bud for M1 at its caudal extent. In most eutherians the M1 bud is not detected until after dP4 has attained the early-late bell stage (Fig. 13.2E; Table 13.2). In a similar manner, the dental lamina extends distal to M1 after it reaches the middle-late bell stage, in order to initiate the M2 bud; the same pattern is followed for M3. The relative rates of development for eutherian molars vary considerably, depending in part on the length of gestation, body size, and the animal's postnatal longevity. The pattern of dentin and enamel formation for molars, as well as for antemolar deciduous teeth and their successors (Fig. 13.1), is well documented and follows a common pathway in all therians (Butler, 1956; Ooë, 1981; Thesleff and Hurmerinta, 1981).

The ontogenetic sequence of cusp differentiation and calcification on molariform teeth usually follows the phylogenetic appearance of these cusps during mammalian evolution, at least for the major cusps. Thus, the paracone and protoconid are generally the first cusps to develop and calcify, although some exceptions have been noted (Woodward, 1896a; Butler, 1956; Marshall and Butler, 1966; Luckett, unpublished). During the present study, the developmental sequence: paracone → metacone → protocone was found for the molars of strepsirhine and haplorhine primates, *Tupaia*, *Elephantulus*, *Erinaceus*, *Talpa*, *Myotis*, and *Procavia*. When present, the hypocone is always the last major cusp to differentiate; this reflects its relatively late appearance (late Cretaceous) during evolution of the tribosphenic molar (see Butler, 1977).

DEVELOPMENT OF THE SECONDARY OR SUCCESSIONAL DENTITION. A lingual successional lamina differentiates in continuity with the outer enamel epithelium of antemolar teeth, usually during middle-late bell stages (Fig. 13.1D). This may first appear as a thickened lingual ridge; subsequently, it elongates as a flattened, ribbon-like structure, as the deciduous tooth becomes more displaced buccally from its primary dental lamina connection. The three-dimensional relationships of the primary dental lamina, deciduous tooth, and successional lamina were described and elegantly illustrated by Ooë (1981) for the developing human dentition, and the same pattern appears to hold true for other eutherians. Each successor tooth develops at the free end of the lingual successional lamina and passes through the same developmental stages (bud, cap, bell) as does its deciduous predecessor (Figs. 13.1E, 13.3). A successional tooth can be distinguished from a deciduous tooth by its epithelial lamina connections to the latter, and by the fact that it begins to differentiate later than its deciduous predecessor. In later stages, however, both the primary and successional laminae become disrupted and subsequently disappear (Fig. 13.4); this can lead to difficulties in assessing tooth homologies when only late developmental stages are available. The time of onset of differentiation of the successional tooth varies considerably in different eutherian taxa, when compared to the developmental stage of its deciduous predecessor. Generally, the successor does not begin to differentiate until dentin and enamel have been initiated on the deciduous tooth (Fig. 13.3).

In some eutherians, the developing lingual successional tooth may extend anterior or posterior to its deciduous predecessor. Subsequent eruption can result in the temporary failure of the successors to displace the deciduous teeth that they will "replace." This is a com-

Chapter 13. Ontogenetic Assessment of Dental Homologies

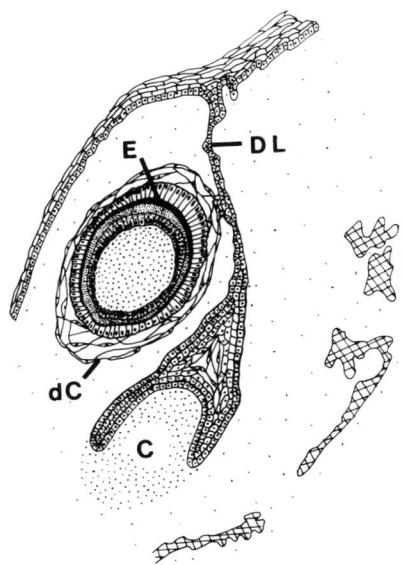

FIGURE 13.3. *Mustela erminea* 45 mm CR fetus, showing epithelial relationships among the primary dental lamina, oral epithelium, upper dC, and the lingual successional lamina leading to a successor in the middle bell stage. In this camera lucida drawing, the large dC is cut in cross section and bears both dentin and enamel. For abbreviations, see Table 13.1.

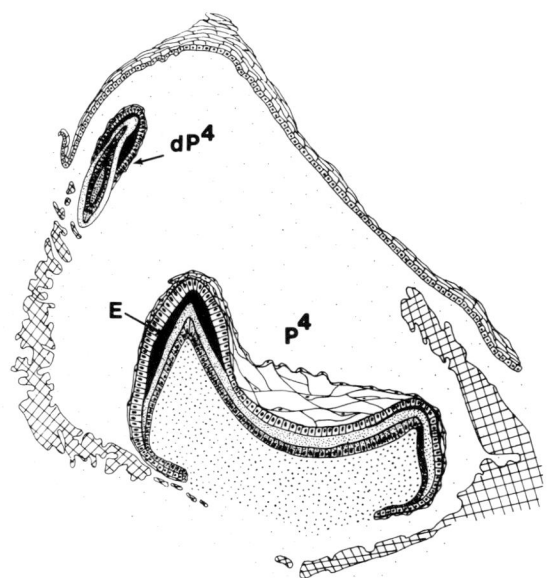

FIGURE 13.4. *Pteropus edulus* 32 mm HL (head length) fetus, showing the elongate dP^4 beneath the oral epithelium and its larger successional P^4, extending deeper into the bony alveolus. Both teeth possess dentin and enamel on their elevated "cusps," and the primary and successional laminae have disappeared at this level. Camera lucida drawing of transverse section. For abbreviations, see Table 13.1.

mon condition in both mega- and microchiropterans (Fig. 13.5), as shown by Leche (1877) and Dorst (1957). Ontogenetic analysis of the dentition in fetal chiropterans (*Rousettus, Pteropus, Myotis, Rhinolophus*) demonstrates that the successional lamina and tooth germ differentiate lingual to the deciduous tooth initially, and subsequently extend anterolingual or distolingual to it during later stages of ontogeny (Fig. 13.6). This developmental pattern results in the temporary coexistence of erupted deciduous and successional teeth in an anterior-posterior sequence in subadult chiropteran jaws. The significance of these findings for interpretation of tooth homologies in fossil eutherians will be discussed below.

VESTIGIAL DECIDUOUS TEETH. The deciduous teeth in some eutherians undergo a vestigial pattern of development at specific tooth positions. These vestigial teeth do not pass through all typical stages of differentiation, although an abnormal knot of irregular dentin is commonly formed (Fig. 13.7). Characteristic features include their relatively superficial position and the absence of stellate reticulum (Moss-Salentijn, 1978; Luckett and Maier, 1982). The latter feature is apparently correlated with the lack of enamel formation in these teeth. The distribution of vestigial teeth in eutherians, including lipotyphlans, pinnipeds, cetaceans, microchiropterans, rodents, and *Tarsius* has been summarized

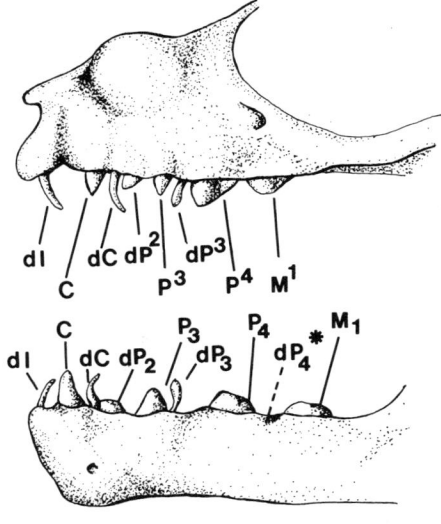

FIGURE 13.5. Juvenile skull of *Pteropus*, showing the erupted deciduous and successional teeth occurring simultaneously in the jaws. The asterisk (*) marks the alveolus for dP$_4$, which was lost in this specimen (modified from Leche, 1877). For abbreviations, see Table 13.1.

elsewhere (Moss-Salentijn, 1978; Luckett and Maier, 1982).

Vestigial deciduous teeth lack typical roots and normally do not erupt; they are either resorbed within the

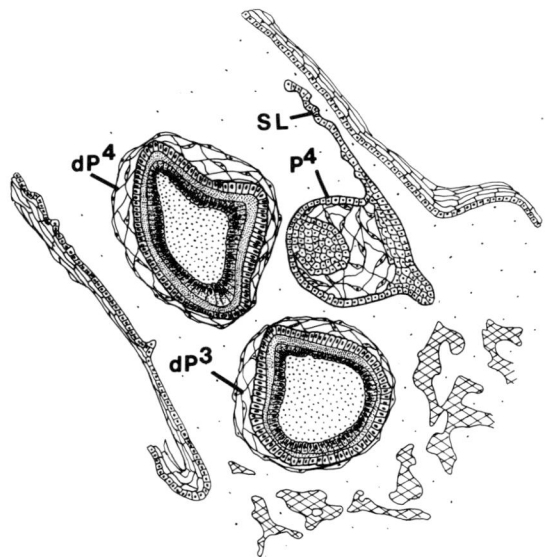

FIGURE 13.6. *Myotis myotis* 18 mm CR fetus, sectioned in a somewhat horizontal plane, to show the spatial relationships among dP³, dP⁴, and P⁴. The primary dental laminae are not evident in this section, but a portion of the successional lamina for P⁴ extends toward dP⁴. Without following serial sections, it would be difficult to determine whether P⁴ was actually the successor of dP³ or dP⁴. Camera lucida drawing. For abbreviations, see Table 13.1.

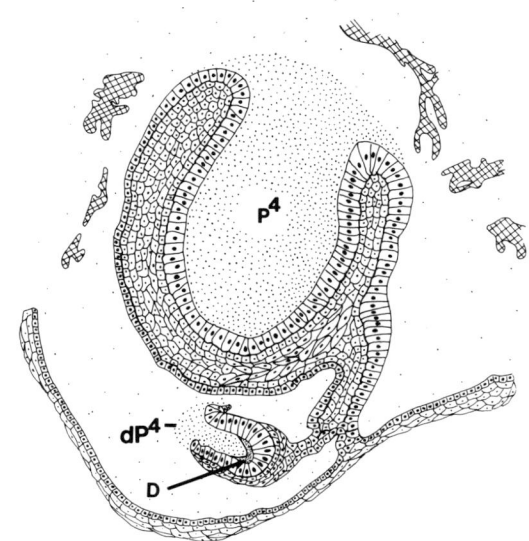

FIGURE 13.7. *Sorex araneus* 12 mm GL fetus, showing the tiny, vestigial dP⁴ beneath the oral epithelium, and covered by a thin dentinal cap. Note the epithelial connections with the larger, normal P⁴, extending lingually and deeper into the jaw. Camera lucida drawing of transverse section. For abbreviations, see Table 13.1.

jaws or, more rarely, shed into the oral cavity during the perinatal period. In most cases, a successional lamina forms lingual to the vestigial tooth, and this gives rise to a normal successor (Fig. 13.7). Commonly, the successor is accelerated in its development when its deciduous predecessor is vestigial. The combination of noneruption of the vestigial tooth and accelerated differentiation of its normal successor can result in erroneous homologies in the postnatal dentition, when early ontogenetic stages are unavailable. This was the case for P2 in the haplorhine primate *Tarsius*, in which only a single tooth generation erupts at this position postnatally. This was considered to be a retained deciduous tooth in the adult dentition by some authors (i.e., Schwartz, 1980), who were unaware of the developmental evidence for vestigial deciduous predecessors in the fetus (Leche, 1897; Luckett and Maier, 1982).

The rudimentary deciduous teeth in some eutherians lack successors; consequently, such tooth positions are undetected when only postnatal specimens are examined. This condition characterizes the entire dentition of mysticete whales (Kükenthal, 1893; Karlsen, 1962). In other eutherians, only individual tooth loci are represented by vestigial deciduous teeth, lacking successors. The transitory occurrence of such dental rudiments helps to clarify the homologies of the persisting dentition. For example, all extant rodents and lagomorphs studied possess a rudimentary deciduous incisor, immediately in front of the large, ever-growing incisors (Moss-Salentijn, 1978; Ooë, 1980; Luckett, 1985; Luckett et al., 1989). In lagomorphs and sciurid rodents, a third incisor locus develops distal to the large functional incisor. Ontogenetic occurrence of three incisor tooth families in lagomorphs and rodents thus helps to homologize their large, gliriform incisors as unreplaced dI2 (Luckett, 1985).

ONTOGENETIC CRITERIA FOR DISTINGUISHING TOOTH HOMOLOGIES. Reliance on shape as a major factor for determining tooth homologies leads to disagreements as to what is meant by "incisiform" or "caniniform," even in the same genus. Schwartz (1980, 1982) argued that homologies should be based on the "idealized morphology" of the tooth. He believed that the incisor tooth class has been lost in the haplorhine primate *Tarsius*, and that the upper jaw contains an anterior canine in the premaxilla, followed by five premolar loci and three molars. On the basis of gradient differences in shape, however, Osborn (1978, p. 194) interpreted the same teeth as two incisors, two canines, two premolars, and three molars. A detailed analysis of early dental ontogeny in *Tarsius* has been presented elsewhere by Luckett and Maier (1982), who demonstrated that the traditionally interpreted tooth homologies were correct.

In spite of their reliance on shape as the major factor for recognizing therian dental homologies, both Osborn (1978) and Schwartz (1980, 1982) acknowledged that

Chapter 13. Ontogenetic Assessment of Dental Homologies

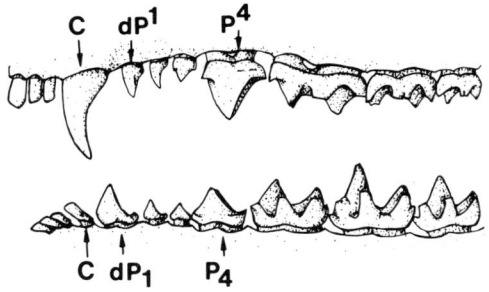

FIGURE 13.8. Adult dentition of *Talpa europaea*, showing the "typical" shapes of the upper teeth and the modified lower teeth. Note that the caniniform dP$_1$ occludes posterior to the upper canine (modified from Thenius, 1988). For abbreviations, see Table 13.1.

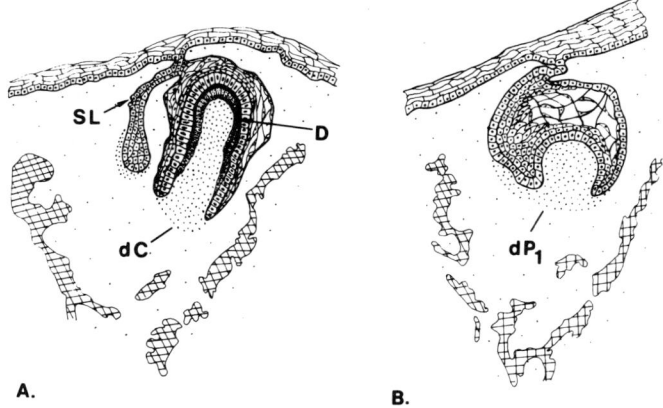

FIGURE 13.9. *Talpa europaea* 30 mm GL fetus, comparing the developmental state of dC and dP$_1$ in the lower jaw. **A**: A thin layer of dentin and a lingual successional lamina are formed on the small dC. **B**: The dP$_1$ has attained only the middle bell stage and lacks dentin and a successional lamina. Camera lucida drawings of transverse sections, at the same magnification. For abbreviations, see Table 13.1.

the early budding sequence can provide valuable evidence for assessing tooth class homologies, i.e., in distinguishing whether a tooth is a canine or a premolar. When early ontogenetic evidence is available, some suggested homologies proposed on the basis of tooth shape are clearly untenable. As an example, Osborn (1978) noted that the lower jaw of the talpid *Talpa europaea* contains four incisiform teeth (including the modified tooth traditionally interpreted as the canine), followed by a caniniform tooth that is usually identified as a modified dP$_1$ (Fig. 13.8). Based on shape gradients, Osborn homologized these teeth as four incisors and a canine. As acknowledged by Osborn, his clone model is predicated on the fact that dC is one of the earliest tooth germs to differentiate in eutherians. When ontogenetic stages of *Talpa* are analyzed, however, the "fourth" lower incisiform tooth is one of the earliest tooth germs to be initiated, whereas the larger caniniform tooth that follows distally is the last deciduous tooth to develop (Sicher, 1916; Kindahl, 1957b). At all stages of its early ontogeny (Fig. 13.9), the fourth lower incisiform tooth is advanced over the teeth immediately preceding and following it (dI$_3$ and dP$_1$, respectively), as expected if this tooth is a modified deciduous canine. The developmental sequence of dP$_1$ and dC in the lower jaw corroborates the homologies of these modified teeth with their "normal" appearing occlusal partners in the upper jaw (Fig. 13.8).

Pattern of Dental Development in Metatheria

Following differentiation of the dental lamina in late intrauterine stages, the initial budding pattern of marsupials is fundamentally similar to that of eutherians. A notable difference is that dP3 is always the first postcanine tooth to differentiate in all marsupials studied (Table 13.4; Fig. 13.2B) (see also Wilson and Hill, 1897; Berkovitz, 1972a; Archer, 1974a). As a consequence of birth during the early "fetal" phase of ontogeny, most of the process of dental development occurs while the marsupial young is suckling during the prolonged period of lactation. In general, dental onto-

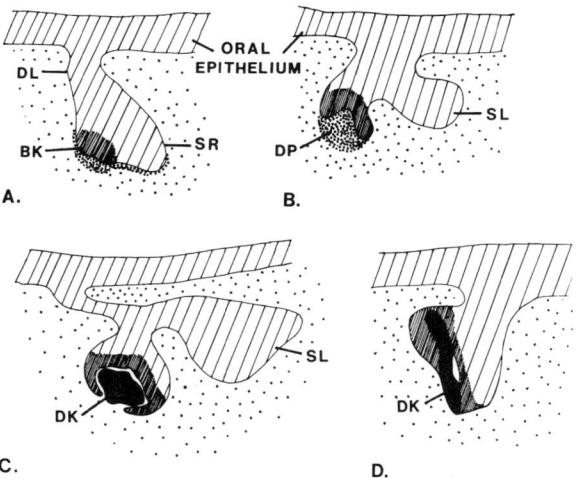

FIGURE 13.10. Diagram of four developmental stages (in transverse section) for vestigial deciduous incisors and canines of marsupials. **A**: Late bud stage, with buccal, condensed knot-like portion and more thickened, lingual successional ridge (SR). **B**: The buccal portion is now invaginated to form a cap-like structure, with a rudimentary dental papilla. The lingual ridge is more elongate and extends distally as the successional lamina. **C**: Later, a dentinal knot may develop within the buccal cap, forming a distinct, vestigial decidual tooth. A lingual successor tooth germ has formed, but it extends distal to this level. **D**: In later stages, epithelial connections are disrupted, and a thin remnant of the dentinal knot may lie beneath the oral epithelium. For abbreviations, see Table 13.1.

TABLE 13.4. Summary of dental development in the upper jaw of the marsupial Perameles

	dI1	dI2	dI3	dI4	dI5	dC
12.25 mm GL Embryo	Early bud	Early bud	Early bud	Early bud	—	Early bud
14 mm GL Neonate	Early bud	Early bud	Early bud	Not distinct	—	Early bud
15.5 mm GL (6 mm HL)	Early bud	Early bud	Middle bud	Early bud	Early bud	Early bud
15 mm GL (6.5 mm HL)	Early bud	Nodular bud	Buccal knot, with lingual projection	Early bud	Early bud	Early bud
15 mm GL (7 mm HL)	Buccal knot; lingual bud	Buccal knot; lingual bud	Buccal cap; lingual bud	Buccal knot; lingual bud	Buccal knot; lingual bud	Buccal knot; lingual bud
16 mm GL (7 mm HL)	Buccal knot; lingual bud	Buccal knot; lingual bud	Buccal cap; lingual bud	Buccal knot; lingual bud	Indistinct	Buccal cap; lingual bud
23 mm GL (11 mm HL)	Buccal cap; lingual middle cap I1	No buccal knot; early cap I2	Buccal knot; middle-late bud I3	Buccal knot; middle-late bud I4	No buccal knot; early cap I5	Buccal knot; late cap C
28 mm CR (13 mm HL)	No buccal knot; early bell I1	No buccal knot; early bell I2	No buccal knot; early-middle cap I3	No buccal knot; middle cap I4	No buccal knot; middle cap I5	Buccal knot; early-middle bell C
31 mm CR (14 mm HL)	No buccal knot; middle bell I1	No buccal knot; middle bell I2	Buccal knot; early bell I3	Buccal knot; early-middle cap I4	No buccal knot; early-middle cap I5	Buccal knot; middle bell C
35 mm GL (16 mm HL)	No buccal knot; late bell I1; thin dentin	No buccal knot; late bell I2; mod. dev. dentin	No buccal knot; late bell I3; possible thin dentin	No buccal knot; middle bell I4	No buccal knot; middle bell I5	Buccal knot; late bell C; thin dentin
45 mm GL (26 mm HL)	No buccal knot; thick dentin & enamel on I1	No buccal knot; thick dentin & enamel on I2	Buccal knot; thick dentin & enamel on I3	No buccal knot; mod. thick dentin & enam. on I4	Buccal knot; Mod. thick dentin & enamel on I5	No buccal knot; thick dentin & enamel on C

geny in marsupials is more stereotyped than in eutherians, even in those marsupials that have lost several incisor and premolar loci. This developmental pattern involves four major events: (1) extreme modification of the anterior dentition during early ontogeny, (2) relative developmental retardation at the two anterior premolar loci, (3) relatively normal deciduous and successional tooth formation at the posterior premolar position, and (4) a typical therian pattern for molar differentiation. Each of these events has been associated with controversy concerning the homologies of individual tooth positions, and the following discussion focuses on the developmental basis for these disagreements.

ONTOGENY OF THE ANTERIOR DENTITION IN MARSUPIALS. The earliest indication of differentiation in the anterior tooth germs is found in later bud stages. Viewed in transverse section, the buccal part of the tooth bud becomes organized into an aggregation of closely packed cells, whereas the lingual region consists of more normal-appearing cells (Fig. 13.10A). In subsequent stages, the buccal cells form a more distinct, spherical nodule or cap-like structure, and the lingual portion further enlarges and elongates to establish a prominent distolingual projection (Fig. 13.10B). By following the developmental fate of these two regions, it appears that the buccal nodular portion is homologous with a rudimentary deciduous tooth, and that the more normal, distolingual ridge is homologous with the successional lamina, and with the later developing successor tooth. The deciduous rudiment exhibits little further histodifferentiation in later stages; it does not form a typical bell stage and lacks stellate reticulum. However, an irregular knot of dentin may form within the rudimentary cap (Fig. 13.10C, 13.10D); this provides the strongest evidence for homologizing these buccal vestiges as rudimentary teeth. Subsequent differentiation of the successors follows the typical therian pattern and will not be described further (see Table 13.4).

This highly modified, vestigial deciduous tooth, accompanied by normal development of its lingual successor, characterizes each incisor and canine tooth family in all marsupials that have been studied carefully. These relationships were first described by Wilson and Hill (1897) in *Perameles*, and they have been further analyzed in a closely graded series of *Macropus* (Kirkpatrick, 1969) and *Dasyurus* (Luckett, 1989, in preparation) pouch young. Most subsequent authors have supported this hypothesis of anterior tooth homologies (see the excellent discussions in Berkovitz, 1966, 1968a). Paradoxically, vestigial deciduous incisors and canines

TABLE 13.4. continued

dP1	dP2	dP3	P3	M1	M2	M3
—	—	Early bud	—	—	—	—
—	—	Early cap	—	—	—	—
—	—	Early cap	—	—	—	—
—	—	Early cap	—	Possible early bud	—	—
—	—	Early bell	—	Middle-late bud	—	—
Possible early bud	Possible early bud	Early bell	—	Middle-late cap	—	—
Early cap	Early-middle cap	Middle-late bell	Thickened lingual succ. lam.	Middle-late bell	Middle cap	—
Middle cap	Early bell	Late bell, odont.	Thickened lingual succ. lam.	Late bell	Early-middle bell	—
Early-middle bell	Middle bell	Late bell, odont.	Thickened lingual succ. lam.	Late bell, odont.	Middle bell	—
Late bell, possible odont.	Late bell; thin dentin	Mod. dev. dentin	Lingual early bud	Mod. thick dentin & mod. dev. enamel	Late bell; thin dentin	Early-middle bud
Thick dentin & mod. thick enamel	Thick dentin & mod. thick enamel	Mod. thick dentin & mod. dev. enamel	Lingual early cap	Thick dentin & enamel	Mod. thick dentin & enamel	Early-middle bud

were not identified in previous studies of dental development in *Didelphis* (Kükenthal, 1891; Rose, 1892a, 1892b; Leche, 1895; Berkovitz, 1978), and this doubtlessly contributed to the controversy over whether the rudimentary toothlets found in the anterior dentitions of other marsupial genera represented deciduous or "predeciduous" teeth.

DEVELOPMENT OF THE ANTERIOR PREMOLARS. The initiation of buds for the anterior two premolars is relatively retarded in marsupials, when compared to the time of onset of other deciduous teeth (Table 13.4). They are the last deciduous tooth germs to develop (when present) in didelphids, peramelids, dasyurids, phalangerids, and macropodids. In contrast to the incisor and canine loci, there is no evidence for development of rudimentary predecessors in the anterior premolar region. A "successional" lamina develops lingual to dP1 and dP2 in later stages, and it may exhibit a globular terminal thickening in some genera (*Perameles*, *Didelphis*, *Sminthopsis*). In no marsupial, however, is there evidence for the further differentiation of a successional tooth lingual to the anterior premolars. Therefore, the anterior premolars that erupt in marsupials are primary teeth (dP1, dP2) that are not replaced; this resembles the condition for dP1 in many eutherians.

DEVELOPMENT OF DECIDUOUS AND SUCCESSIONAL POSTERIOR PREMOLARS. In contrast to the specializations that occur during development of the more anterior dentition, differentiation of the marsupial dP3 proceeds in a typical fashion through the bud, cap, and bell stages (Table 13.4). Following its initiation, dP3 is always one of the most advanced teeth developmentally, regardless of its ultimate size or fate. A successional lamina forms lingual to dP3 during late bell stages, continuous with its outer enamel epithelium and the overlying primary dental lamina. Controversy exists concerning the fate of this successional lamina in marsupials, and whether it plays a role in formation of the successor premolar, as it does in eutherians.

Several authors (Woodward, 1893, 1896b; Archer, 1974a, 1978, 1984) have suggested that marsupials are characterized by origin of P3 from the primary dental lamina anterior to dP3, rather than from a successional lamina lingual to dP3. Others (Leche, 1895; Wilson and Hill, 1897; Luckett and Hong, 1989) have insisted that P3 develops from the successional lamina of dP3 in marsupials, as in eutherians. A third alternative is that both developmental patterns occur in different groups of marsupials (Dependorf, 1898; Berkovitz, 1966, 1968b, 1972a, 1978).

A careful analysis of developmental relationships among dP3, its lingual successional lamina, and P3 was

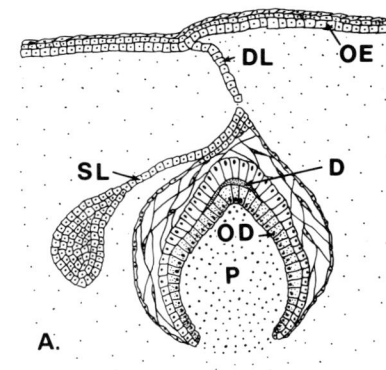

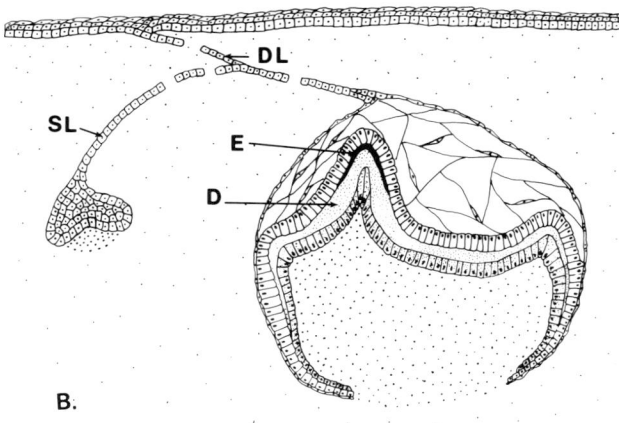

FIGURE 13.11. Diagrams of *Perameles nasuta* pouch young, showing epithelial relations between dP³ and P³. **A**: 35 mm GL young, with early rupture of dental lamina, and early bud P³ lying lingual to dP³. **B**: 45 mm GL young, with increased disruption of both primary and successional laminae. The early cap P³ lies lingual to the paracone of dP³, now covered by dentin and enamel. For abbreviations, see Table 13.1.

provided by Wilson and Hill (1897) for *Perameles*, and their observations on the epithelial connections among these structures were corroborated during the present study (see also Luckett and Hong, 1989). In 23–31 mm pouch young of *Perameles*, the late bell dP³ bears a lingual successional lamina that is slightly thickened at the level of the anterior end of the tooth (Table 13.4). In contrast, dP² in these young has only attained the cap-middle bell stage, and a lingual successional lamina has not yet developed. In a more mature 35 mm young, the successional lamina is thickened to form an early bud P³, lingual to the mesial end of dP³ (Fig. 13.11A). Both the primary dental lamina stalk and successional lamina exhibit localized regions of disruption, but examination of serial sections confirms their epithelial relationships. P³ has attained the early cap stage in a 45 mm young, and it still lies lingual to the anterior half of dP³, at the level of the paracone (Fig. 13.11B). Later ontogenetic stages of the dP3-P3 relationship in *Perameles* were described and illustrated by Wilson and Hill (1897).

A similar pattern of differentiation for the successor P3 lingual to the large dP3 was documented by Bolk (1929) for the phalangerid *Trichosurus vulpecula*. Recently, we have reexamined most of Bolk's specimens of *Trichosurus* at the Hubrecht Laboratory, and our observations indicate that the epithelial relationships among the primary dental lamina, dP3, and lingual successor P3 are identical at all stages to those described above for *Perameles*. The question remains, however, whether the developmental pattern proposed for didelphids, dasyurids, and macropodids (see Archer, 1978, 1984), in which P3 originates from the dental lamina *in front of* dP3 as a separate tooth family, differs from that of peramelids and phalangerids. In order to resolve this question, it is essential to reexamine the temporal and epithelial relationships that exist during dental ontogeny in these (and other) marsupial families. Such an analysis is currently in progress, and only a few results can be discussed here.

Examination of good ontogenetic series in the didelphid *Didelphis* and the dasyurid *Dasyurus* reveals that the developmental relationships between dP3 and P3 are fundamentally similiar to those in *Perameles* and *Trichosurus* (Luckett, 1989; Luckett and Hong, 1989; Luckett et al., 1991). In both genera, the lingual successional lamina of dP3 extends somewhat anterior to the latter, so that the developing early bud P3 lies anterior to dP3. The P3 bud, however, exhibits no primary dental lamina connection to the overlying oral epithelium. Instead, the slender epithelial stalk for P3 extends distally, parallel to the oral epithelium, and then becomes continuous with the lingual successional lamina of dP3 (Fig. 13.12). In later stages, the epithelial connections of the dental laminae become fragmented, but incomplete traces of the former relationship between dP3 and its successor are still evident (Fig. 13.13).

Dasyurus has lost one premolar position, generally considered to be P3 (Thomas, 1887; Archer, 1976), although a rudimentary dP3 develops in both jaws (Woodward, 1896b). Analysis of a closely graded series of pouch young reveals that dP3 is the earliest cheek-tooth to develop, as in other marsupials (Luckett, 1989). The successor lamina extends anterior to dP3 in 15–20 mm GL young, and P3 has already attained the cap stage in the more mature of these young (Fig. 13.12; Table 13.5). In addition, there is a diastema along the primary dental lamina of both jaws between dP1 and dP3, indicating dP2 is the missing premolar in *Dasyurus*. The tooth traditionally interpreted as P2 in *Dasyurus* (e.g., Archer, 1976) is actually the anteriorly displaced P3 (Luckett, 1989).

In later developmental stages, the small calcified dP3 of *Dasyurus* becomes vestigial, and it is destined to be

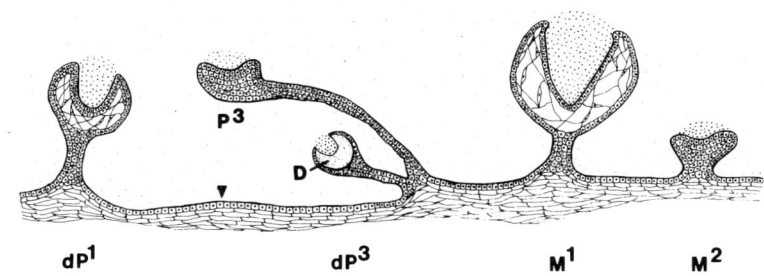

FIGURE 13.12. Diagrammatic reconstruction of epithelial relationships among dP¹-M², oral epithelium, and dental laminae, in a sagittally sectioned 20 mm GL pouch young of *Dasyurus viverrinus*. All primary teeth retain their dental lamina connections to the oral epithelium, whereas P³ arises from a lingual successional lamina in common with the vestigial dP³. P³ extends anterior to dP³ and occupies the position vacated by the lost dP² locus (indicated by arrowhead). For abbreviations, see Table 13.1.

resorbed without eruption. Associated with this, the successional lamina and P3 are accelerated in their differentiation. Corroboration for this hypothesis of accelerated development is obtained by comparing the relative rate of development among dP3, P3, M1, and M2 in other marsupials, including three-premolared dasyurids (Table 13.5). Because molars always differentiate in an anterior-posterior fashion, following attainment of the late bell stage by the preceding tooth (see Fig. 13.2 E), their developmental state serves as a useful reference point for assessing relative growth rates of the antemolar teeth. In most marsupials studied, including two genera of three-premolared dasyurids (*Sminthopsis, Antechinus*) (also see Archer, 1974a), P3 differentiates later than M2. In contrast, P3 of *Dasyurus* is relatively accelerated in its development when compared to M2, perhaps concomitant with the vestigial, nonerupting nature of dP3. This acceleration continues throughout ontogeny; P3 also erupts earlier than M2 in *Dasyurus* (Merchant et al., 1984; note that they identified P3 as P2, as do most authors).

It is unfortunate that later stages of dental development in macropodids could not be examined during this study, because they have played a central role in the controversy over the site of origin of P3. Recent studies (Berkovitz, 1966, 1972a; Kirkpatrick, 1969, 1978) have concluded that P3 arises separately from the dental lamina anterior to dP3 in macropodids, as proposed originally by Woodward (1893). Although these studies show that P3 develops anterior to dP3 in macropodids, it remains unclear whether P3 originates by a separate primary dental lamina stalk from the oral epithelium. Most authors have not presented any evidence for this, and it has received surprisingly little attention. Kirkpatrick (1969, p. 59) emphasized that P3 develops in continuity with the successional lamina of dP2 in *Macropus*, but he failed to mention that this successional lamina is also continuous distally with that of dP3. Indeed, in his Figure 17, the successor P³ bud seems to exhibit a more intimate epithelial connection with the successional lamina of dP³ than it does with dP². It remains to be demonstrated whether macropodids agree with didelphids (and other marsupials) in the pattern of epithelial origin for P3, or whether it originates in a nonhomologous fashion as a separate primary tooth.

DEVELOPMENT OF THE MOLARS. In most marsupials, as in eutherians, a distinct bud for M1 is not initiated until the posterior deciduous premolar reaches the middle-late bell stage; this is true for didelphids, phalangerids, and macropodids, and probably also for phascolarctids and vombatids. In *Dasyurus* and *Perameles*, however, M1 becomes evident somewhat earlier, when dP3 is in the cap-early bell stage (Table 13.4). Because of the limited data available, the significance of this difference is unclear. It may be correlated with the fact that in these taxa dP3 is greatly reduced in size (*Perameles*) or becomes vestigial (*Dasyurus*), in contrast to its larger molariform nature in other families studied. The subsequent differentiation of additional molars distal to M1 follows a pattern identical to that of eutherians.

The paracone and protoconid are the earliest cusps to differentiate and calcify on most marsupial molars (Woodward, 1896b; Berkovitz, 1968b); this was also confirmed for *Perameles* and *Trichosurus* during the present study. In a few genera (*Didelphis, Antechinus*),

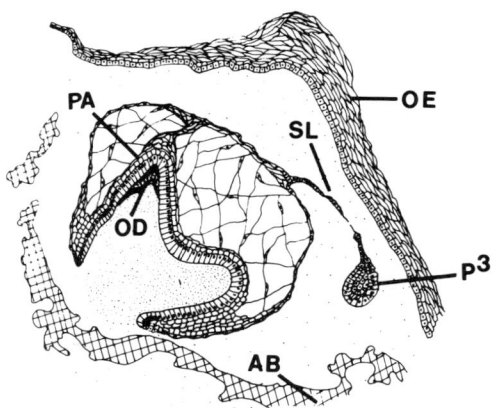

FIGURE 13.13. *Didelphis marsupialis* 20 mm HL (head length) pouch young, transverse section of dP³, at level of paracone elevation. The primary dental lamina is completely disrupted, but the lingual successional lamina remains nearly continuous with dP³. The distal end of the P³ bud is also evident, although it is more enlarged anteriorly. Camera lucida drawing. For abbreviations, see Table 13.1.

TABLE 13.5. Comparison of relative developmental stages among dP3–M2 and P3 in the upper jaw of marsupials

	dP3	P3	M1	M2
Dasyuridae				
Dasyurus 20 mm GL	Mod. thick dentin; thin enamel	Middle cap	Late bell, possible odont.	Early cap
Dasyurus 23 mm GL	Thick dentin; thin enamel	Late bell, thin dentin	Mod. thick dentin; thin enamel	Late bell
Sminthopsis 14.5 mm CR	Mod. thick dentin & enamel	Middle bud	Mod. thick dentin & enamel	Late bell
Antechinus 30 mm GL (Fosse, 1969)	Thick dentin; mod. thick enamel	Early cap	Dentin & enamel	Mod. dev. dentin
Peramelidae				
Perameles 35 mm GL	Mod. dev. dentin	Early bud	Mod. thick dentin; mod. dev. enamel	Late bell, thin dentin
Perameles 45 mm GL	Mod. thick dentin; mod. dev. enamel	Early cap	Thick dentin & enamel	Mod. thick dentin & enamel
Didelphidae				
Didelphis 49 mm CR	Thick dentin; mod. thick enamel	Middle bud	Late bell, thin dentin	Early bud
Monodelphis 34 mm CR	Thick dentin & enamel	Middle bud	Mod. thick dentin & enamel	Mod. dev. dentin
Phalangeridae				
Trichosurus 25 mm HL	Thick dentin; mod. thick enamel	Early bud	Thick dentin; mod. thick enamel	Late bell
Trichosurus 27 mm HL	Thick dentin & enamel	Early cap	Thick dentin & enamel	Mod. dev. dentin & enamel

however, the metacone is larger than the paracone and is also the first cusp to calcify on the upper molars (Rose, 1892a, 1892b; Archer, 1974b; Berkovitz, 1978). Such a developmental pattern was also recorded for *Monodelphis*, *Dasyurus*, and *Sminthopsis* during this study. In *Didelphis*, at least, there is developmental evidence for secondary reduction of the molar paracone during ontogeny, and probably also during phylogeny (Luckett and Hong, in preparation), but sufficient data are not yet available to determine whether this is true for other didelphids and dasyurids with a small paracone.

ANCESTRAL PATTERN OF MARSUPIAL DENTAL DEVELOPMENT. Despite differences in tooth size, shape, and numbers between polyprotodont and diprotodont marsupials, the basic pattern of dental development is fundamentally similar in all marsupial families studied. This makes it relatively easy to reconstruct the morphotypic pattern of dental development in extant marsupials and to speculate on the condition that probably characterized ancestral marsupials.

Both deciduous and successor tooth germs differentiate at each incisor and canine locus primitively; the deciduous teeth are rudimentary and never erupt, while the successors develop normally and erupt as the "functional" teeth of the adult. Doubtless there were five incisor loci in the upper jaw and four in the lower, as occurs in extant didelphids and most late Cretaceous–early Tertiary marsupials. Dentin probably developed in each vestigial deciduous incisor and canine, as it does to varying degrees in most marsupials studied.

Three premolar loci occur in didelphids, peramelids, many dasyurids, and late Cretaceous–early Tertiary marsupials; this doubtlessly represents the primitive marsupial condition. The suggestion (Thomas, 1887) that an additional premolar locus occurred in ancestral marsupials is not supported by careful ontogenetic analysis (Kükenthal, 1891; Leche, 1895; Wilson and Hill, 1897; Berkovitz, 1966; Archer, 1974a) or by the available fossil evidence (Clemens, 1966, 1979). Only a single tooth generation develops at the two anterior premolar positions in extant forms, and the limited fossil remains are consistent with this. These anterior teeth are members of an unreplaced primary dentition. In contrast, both deciduous and successional teeth develop and erupt at the third premolar locus. The premolariform P3 may erupt beneath the molariform dP3, replacing it immediately (as in *Trichosurus*), or it may erupt anterior to dP3 and coexist with it in the jaws for a variable period of time. This has been demonstrated for *Perameles* (Fig. 13.14) by Lyne (1982), and it

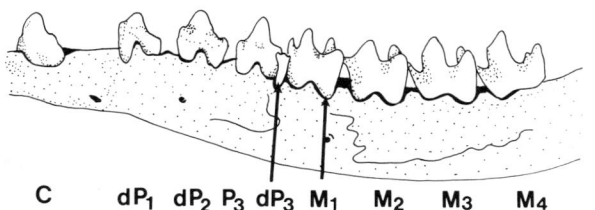

FIGURE 13.14. Part of mandible of subadult (275 days old) *Perameles nasuta*, showing dP$_3$ retained temporarily in the jaw with its anteriorly placed successor P$_3$ (drawn from photograph in Lyne, 1982). For abbreviations, see Table 13.1.

may account for the presence of an additional molariform tooth in *Myrmecobius*, as suggested by Winge (1882, 1941). The four molars of marsupials are unreplaced primary teeth.

As a consequence of the unique (autapomorphic) developmental pattern of the marsupial dentition, only a single tooth generation erupts at each position anterior to dP3. This fact was early recognized by Flower (1867), and it led to great controversy in attempts to clarify the homologies of this single tooth generation, in comparison with the diphyodonty of antemolar teeth in most eutherians. Both functional and developmental criteria were used to support hypotheses that these anterior teeth of marsupials are homologous with the secondary (Flower, 1867; Thomas, 1887; Wilson and Hill, 1897; Berkovitz, 1968a, 1972a) or the primary dentition of eutherians (Kükenthal, 1891; Leche, 1892, 1895; Woodward, 1896b; Dependorf, 1898). However, two different processes have contributed to this apparent "monophyodonty": (1) vestigial, nonerupting ontogeny of the deciduous incisors and canines, and (2) developmental retardation, but normal eruption and nonreplacement, of dP1 and dP2.

Extant marsupials can be diagnosed as possessing a highly modified "diphyodonty" of the anterior dentition, in that both deciduous and successional tooth germs develop at each incisor and canine locus, even though only the successional teeth erupt. It has been suspected for a long time (Leche, 1895; Wilson and Hill, 1897) that this modified anterior dentition is correlated with the prolonged lactation period of marsupials, especially with the "period of fixation" (Hill and O'Donoghue, 1913), during which the suckling young is continuously "attached" to the nipple for a lengthy period of time. The well-developed tongue and enlarged nipple fill the oral cavity during this period, and the continued pressure exerted by these structures probably has an effect on the developing tooth germs underlying the oral epithelium. The limited fossil evidence also suggests that the modified dental ontogenetic pattern of extant marsupials occurred in their late Cretaceous relatives. If true, this could be used as indirect evidence that some specialized features of marsupial reproduction, development, and lactation were already established in the late Cretaceous (for discussion, see Luckett, 1977; Tyndale-Biscoe and Renfree, 1987).

Discussion

Ontogenetic studies can clearly aid in resolving problems of dental homologies among extant therians, especially in assessing the deciduous or successional nature of individual teeth, even when replacement does not occur postnatally. Such studies also furnish evidence for the presence of rudimentary teeth that do not erupt; these help to clarify which tooth positions have been lost during phylogeny, and which persist in the functioning dentition. Thus, ontogenetic analyses demonstrate that dI2 is the ever-growing incisor in both jaws of rodents and lagomorphs (Luckett, 1985), and that dP2 is the missing premolar locus in the phalangerid *Trichosurus* (Bolk, 1929; Berkovitz, 1968b). The sequential pattern of budding and early differentiation can also provide valuable clues, in addition to occlusal and bony relationships, for homologies of teeth with an ambiguous phenetic appearance, such as the caniniform dP$_1$ of *Talpa*.

DISTINGUISHING DECIDUOUS VERSUS SUCCESSIONAL TEETH. Many disagreements concerning tooth homologies in therians relate directly to the imprecise criteria that have been used to distinguish between deciduous and successional teeth during different phases of ontogeny. Initially, the major or only criterion for identifying a successional tooth was that it replaced a deciduous predecessor (Owen, 1868; Flower, 1869). Later, histological studies of the developing therian dentition (Leche, 1895) indicated that successional teeth develop lingual to their deciduous predecessors, and that epithelial connections occur between the two dental generations. More recently, Berkovitz (1966, 1968a) and Moss-Salentijn (1978) have discussed additional criteria used for recognizing deciduous or successional tooth homologies.

The following developmental events appear to characterize all therian deciduous teeth. The primary dentition differentiates from buds that develop at the free end of the primary dental lamina. The deciduous tooth becomes secondarily displaced buccally during the middle-late bell stage, concomitant with differentiation of its successional lamina. This successional lamina develops in continuity with the outer enamel epithelium of its deciduous predecessor, as noted previously by Ööe (1965, 1981) and Moss-Salentijn (1978). Thus, the successional lamina differs from the primary dental lamina both temporally and positionally. A successional tooth can develop *only* from the lingual successional lamina of its deciduous predecessor. Therefore, a deciduous tooth normally differentiates earlier during ontogeny than does its successor. This is modified somewhat when the deciduous tooth becomes vestigial. However, dentin (when present) always develops earlier in vestigial deciduous teeth than it does in their successors; this is true even in the highly modified marsupial anterior dentition.

Berkovitz (1966, 1968a) believed that Kindahl's (1959) study of dental development in *Sorex* indicated that some successional teeth can form earlier than their rudimentary predecessors, contrary to the general pattern in therians. However, a careful reading of Kindahl's paper, as well as reexamination of her specimens of *Sorex* (Luckett and Bangma, in preparation), reveal that the developmental relationships between rudimentary deciduous and normal successor teeth in this genus (Fig. 13.7) are strikingly similar to those in marsupials. In this regard, it is difficult to comprehend Schwartz's (1982) logic in reversing the identification of deciduous and successor teeth in *Sorex*, based on his interpretation of Kindahl's (1959) data. This was not founded on any original observations, but rather on his belief (Schwartz, 1982, p. 133) that mammalian tooth loss is more likely to occur by "inhibiting the successor tooth," rather than by vestigial development and noneruption of the deciduous teeth. Consequently, he claimed that the large antemolar teeth of *Sorex* are deciduous, and the small, vestigial teeth that develop "from the side" of the larger teeth are rudimentary successors. In making this unfounded proposal, Schwartz seemed to be unaware of the long-standing observation that deciduous teeth develop buccal to their lingual successors, and not the reverse.

Developmental analysis indicates that it is impossible to have a secondary or successional tooth without a deciduous predecessor. Nevertheless, many authors have not been consistent in establishing the homologies of unreplaced primary teeth. For example, it is common (Berkovitz, 1966; Kirkpatrick, 1969,1978; Archer, 1974a) to designate the anterior premolars of marsupials as P1 and P2, even though they differentiate from the primary dental lamina in a manner identical to dP3. Berkovitz (1966, p. 1381) acknowledged that development of P2 "appears to be in series with the deciduous premolar", but he noted that its morphology differs from that of dP3 and more closely resembles P3. He then suggested that P2 "may be a member of the permanent series", whose development is accelerated in the absence of a deciduous predecessor. A similar problem is encountered for the unreplaced dP1 of eutherians. This has been considered as a "permanent" or successional tooth in *Canis*, in part because it lacks a rudimentary successor (Tims, 1896; Williams and Evans, 1978).

Following its initiation, the successional lamina (and its successor tooth bud) may remain lingual to its deciduous predecessor, as in most eutherians, or it may extend either anterior or posterior to this tooth, as in chiropterans and many marsupials. These latter conditions have resulted in considerable disagreement concerning the homologies of the "successor" teeth and the posterior deciduous premolars in therians. This controversy was initiated by Woodward (1893, 1896a, 1896b), who believed (as did Thomas, 1887) that the posterior premolar locus (dP4, P4) was homologous in the two therian groups. He claimed that the so-called successor of dP4 in macropodids, dasyurids, erinaceids, and centetids develops anterior to dP4, rather than lingual to it. Consequently, Woodward (1896a, p. 581) proposed that the "successional" posterior premolar of therians is a late-developing primary tooth, and that the molariform posterior deciduous premolar is a "true molar accelerated in its development," even though this tooth is later displaced by the erupting posterior premolar.

More recently, Berkovitz (1972b, 1973) demonstrated that P^4 develops somewhat anterior to dP^4 in *Cavia* and *Mustela*, and he suggested that this may represent an "alternation of dentitions." However, in the lower jaw of these genera, the successional P_4 forms in a "normal" fashion lingual to its deciduous predecessor. Although Berkovitz considered several alternatives for the homologies of these teeth, he continued to identify them as dP4 and P4. As noted above, it is the **epithelial relationships** between the posterior deciduous premolar and its successor, rather than the site of development and eruption of the successor tooth, that provide the most precise evidence for their developmental homologies.

POSTERIOR PREMOLAR HOMOLOGIES IN MARSUPIALS. Following histological analysis of dental development in an extensive series of pouch young of the dasyurid *Antechinus flavipes*, Archer (1974a) proposed that no true postcanine tooth replacement occurs in dasyurids. This was based primarily on his belief that P3 (his P4) develops from the dental lamina anterior to dP3 in *Antechinus*, rather than lingual to it, and that this relationship is identical to that previously described for didelphids and macropodids. Most of Archer's data were presented in tabular form to show the overall dental developmental pattern, and only a few sections and schematic reconstructions were provided to illustrate the critical dP3-P3 relationships in a forty-day-old pouch young. A single histological section of dP_3 shows that the bud for P_3 was connected by a slender lamina stalk to the outer enamel epithelium of dP_3 at its anterior end, and that there were no primary lamina connections of dP_3, P_3, or dP_2 to the oral epithelium at this late stage. These are the identical epithelial relationships seen for dP3 and P3 in *Didelphis* during the present investigation (cf. Archer's fig. 3b with Fig. 13.13 in this chapter).

Subsequently, Archer (1978, 1984) reinterpreted the molariform dP3 of marsupials as M1, and he proposed this as the first in a series of five molars that characterized primitive marsupials. This hypothesis was based in part on his belief that the anterior molariform tooth lacks a true "successor," even though this tooth is normally displaced during ontogeny by the late-erupting P3. In addition, Archer emphasized the morphological distinctness between the three premolariform teeth and the five molariforms, with each tooth representing a separate tooth family. This hypothesis was not new; as noted above, Woodward (1893, 1896b) interpreted dP3 of macropodids and other marsupials as a "true" molar, and Dependorf (1898, p. 295) claimed that five molars occurred primitively in diprotodont marsupials, for basically the same reasons proposed by Archer (1978).

A major difficulty with Archer's (1978) hypothesis for five molars in marsupials is his failure to account for the difference between published accounts of "normal" tooth development and replacement at the dP3 locus in peramelids and phalangerids and the dissimilar ontogenetic pattern that he believed to occur in dasyurids, didelphids, and macropodids. Archer speculated that the peramelid dental pattern may be derived from a condition similar to that he described for dasyurids, in which P3 originates from the dental lamina anterior to dP3, and then secondarily comes to lie lingual to dP3, as a result of differential growth. This was similar to Berkovitz's (1968b) suggestion to explain the appearance of P3 lingual to dP3 in the phalangerid *Trichosurus*. However, published studies on extensive developmental series from both *Perameles* and *Trichosurus* do not support these hypotheses for a secondarily lingual relationship of P3 to dP3 (Wilson and Hill, 1897; Dependorf, 1898; Bolk, 1917, 1929; Müller, 1929; Fosse and Risnes, 1972). In addition, examination of ontogenetic series from both *Perameles* and *Trichosurus* during the present analysis has provided further corroboration for the hypothesis that P3 is the true lingual successor of dP3 in these genera.

These investigations provide no support for Archer's (1978, 1984) proposal that dP3 and P3 belong to separate tooth families in marsupials, nor his corollary suggestion that dP3 should be identified instead as M1. Given the tenuous nature of the evidence on which this hypothesis is based, it is difficult to understand why some paleontologists (Marshall, 1987; Reig et al., 1987; Marshall and de Muizon, 1988) have so readily adopted Archer's (1978) reinterpretation of dP3 as M1, with little or no discussion of the merits for such a proposal. Others (Clemens, 1979; Clemens and Lillegraven, 1986) have acknowledged the conflicting ontogenetic evidence and have resisted any change in interpretation of marsupial dental homologies, without further analysis of these contradictions. Assessment of the available data provides strong corroboration for the long-standing hypothesis (Flower, 1867) that P3 is the true successor of dP3 in marsupials, and that this is the only postcanine tooth position to undergo replacement in extant and fossil marsupials.

ONTOGENETIC ANALYSIS OF DENTAL HOMOLOGIES IN FOSSIL THERIANS. Information on dental ontogeny in fossil therians is limited to the occurrence, eruption sequence, and replacement patterns of deciduous teeth, and to the sequence of eruption for the successional teeth and molars. When considered in conjunction with the known pattern of dental development in extant therians, a number of existing hypotheses on tooth homologies in fossil therians can be tested. It seems probable that the pattern of dental development and eruption in late Cretaceous–early Tertiary marsupials did not differ from the reconstructed morphotype of extant forms. Only dP3 is known to be replaced postnatally in fossil taxa (Clemens, 1966, 1979; Marshall and de Muizon, 1988). In contrast, tooth replacement probably occurred at all antemolar positions in late Cretaceous eutherians, as would be expected from a consideration of the extant eutherian morphotype. There is evidence for replacement of at least some deciduous incisors, canines, and premolars in the late Cretaceous *Kennalestes* from Asia (Kielan-Jaworowska, 1981). This pattern of antemolar diphyodonty is probably a primitive mammalian trait; it is known in some degree for many "non-therian" mammals, including morganucodontids (Parrington, 1971) and multituberculates (Greenwald, 1988).

Unfortunately, direct evidence for the occurrence and replacement of deciduous teeth in intact jaws is lacking for most Cretaceous therians. More commonly, isolated deciduous teeth are found associated with identifiable remains of fossil genera, and these are attributed on the basis of their relative size and shared resemblances with early Tertiary relatives that possess known deciduous dentitions. In juvenile jaws of *Asioryctes* and *Kennalestes* from the late Cretaceous of Mongolia, P_3 erupted and replaced dP_3 before the eruption of M3, and while dP4 still remained functional; the same relationship also occurred in the upper jaw of *Kennalestes* (Kielan-Jaworowska, 1981). In known Cretaceous eutherians, including *Asioryctes*, *Kennnalestes*, *Gypsonictops*, and *Cimolestes*, dP4 is molariform. A molariform dP4, replaced by a semimolariform P4, appears to represent the primitive eutherian condition, as suggested by Butler (1977).

PRIMITIVE NUMBER OF PREMOLARS IN EUTHERIA. Traditionally, the primitive eutherian dental formula has been considered to include four premolars and three molars in each jaw quadrant, based on an assessment of both fossil and extant taxa (i.e., Bown and Kraus, 1979). The available developmental evidence from extant eutherians is consistent with this hypothesis. However, recent fossil evidence has suggested that the presence of five premolars may characterize the ancestral eutherian pattern (McKenna, 1975; Novacek, 1986). This hypothesis is supported by (1) the identification of five lower premolars in most remains of the late Cretaceous *Gypsonictops*, including jaws with three erupted molars (Lillegraven, 1969; Clemens, 1973; Fox, 1979); (2) the presence of five premolar loci in a juvenile upper jaw of *Kennalestes*, although only four premolars occurred in adult jaws (Kielan-Jaworowska, 1981); (3) the occurrence of alveoli or teeth representing five premolars in Eocene sirenians (Domning et al., 1982); and (4) the reported presence of five premolars in the adult dentition of the early Cretaceous eutherian *Prokennalestes* (McKenna, 1975; Kielan-Jaworowska and Dashzeveg, 1989), although evidence for this has not yet been published. A detailed analysis of the strengths and weaknesses of these data is beyond the scope of this chapter. However, some ontogenetic findings in extant therians can provide alternative hypotheses for the apparent occurrence of five premolars in some of these taxa.

In both *Gypsonictops* and *Kennalestes*, the available fossil evidence has been interpreted to indicate that five premolar loci are in the process of being reduced to four, by the reduction or loss of a tooth in the middle of the premolar series. Loss at this position would represent a great departure from the pattern of tooth reduction or loss that characterizes most known extant and fossil eutherians. In these, premolars appear to have been lost in an anterior-posterior sequence (Kindahl, 1967; Ziegler, 1971; Fox, 1983). As intimated previously (Luckett and Maier, 1982), it is possible that the small central premolar in *Kennalestes* and *Gypsonictops* is a retained deciduous premolar. Further analysis (Luckett, in preparation) suggests the probability that this tooth may be a reduced dP2 in both genera, retained for a variable length of time contemporaneously with its successor P2, similar to the condition for some premolar loci in chiropterans. Whether the available fossil evidence is consistent with a similar hypothesis for a retained deciduous tooth in Eocene sirenians is unclear at present. It should be noted, however, that such a possibility was suggested previously (Abel, 1912). The discovery of intact juvenile jaws containing deciduous teeth in these groups, as well as in the recently described early Cretaceous *Prokennalestes* (Kielan-Jaworowska and Dashzeveg, 1989), might facilitate further assessment of the hypothesis for five premolar loci in eutherians.

TOOTH HOMOLOGIES AMONG THERIANS. It has long been suggested (Owen, 1868; Lydekker, 1899; Van Valen, 1974) that serial homology might exist between the seven postcanine teeth of marsupials and eutherians, despite the differences in premolar and molar count between the two taxa. These authors proposed that M1 of marsupials is homologous with dP4 of eutherians, and that the nonreplacement of this tooth in marsupials resulted in its identification as M1. This hypothesis is attractive because of its economy and simplicity, and it is supported by the fact that dP4 was molariform in the vast majority of known early eutherians.

Recent suggestions that five, rather than four, premolars represent the primitive eutherian condition (McKenna, 1975; Novacek, 1986), and that five, rather than four, molars characterized the marsupial morphotype (Archer, 1978, 1984), if accepted as valid, would render the Owen-Lydekker hypothesis less plausible. Of course, it could be argued that ancestral therians (or tribosphenids) possessed eight postcanine teeth. This is the condition believed to occur in the earliest suggested eutherian, *Prokennalestes* (Kielan-Jaworowska and Dashzeveg, 1989), and in *Peramus*, a possible sister taxon of the Tribosphenida (McKenna, 1975). Both *Peramus* and *Prokennalestes* are thought to have five premolars and three molars. If this represented the primitive tribosphenid condition, then one would have to propose that the last two posterior premolars of this ancestral group possessed molariform deciduous predecessors, in order to account for the first two molars of ancestral marsupials in Archer's scenario. Unfortunately, deciduous teeth are unknown for both *Prokennalestes* and *Peramus*. Information on cheekteeth of early Cretaceous therians from Texas is based primarily on isolated teeth (see Butler, 1978b), and the number of premolars and molars cannot be determined for any of the described genera.

As discussed above, it is more probable that the ancestral marsupial dental formula included three premolars and four molars. It also appears more likely to me that the last common ancestor of unquestioned eutherians possessed four premolars and three molars, at least until more convincing evidence is marshalled to corroborate the hypothesis that *Gypsonictops* and Eocene sirenians possessed five distinct premolar tooth families. Although *Prokennalestes* may have had five premolar loci, it is not clear whether this early Cretaceous genus shares the derived features of Eutheria. It has become increasingly difficult to distinguish between Cretaceous eutherians and marsupials on the basis of dental morphology. Consequently, Kielan-Jaworowska and Dashzeveg (1989) reported that the main reasons for eutherian allocation of *Prokennalestes* are that (1) it has five premolars and three molars, and (2) the posterior upper premolar is submolariform, without a metacone, and there is a gradual morphological transition between the premolars and molars. Although these features differ from the known morphotype of marsupials, they may represent the primitive tribosphenid condition, because both character complexes are also found in the non-tribosphenic *Peramus*, according to McKenna (1975).

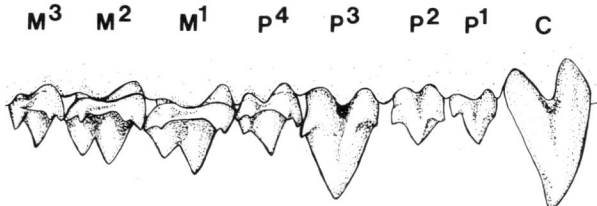

FIGURE 13.15. Upper canine and postcanine dentition of the late Cretaceous eutherian *Kennalestes gobiensis*. Note the trenchant P³ and semimolariform P⁴ (redrawn from Kielan-Jaworowska, 1969). For abbreviations, see Table 13.1.

Of great phylogenetic significance was McKenna's (1975) hypothesis that the enlarged, penultimate upper premolar of *Peramus* is homologous with the penultimate upper premolar of *Kennalestes* and *Zalambdalestes*, and with P³ of early marsupials. In addition, this trenchant appearance also characterizes the penultimate upper premolar of *Asioryctes* and *Prokennalestes*, as well as the P³ of deltatheridiids (Kielan-Jaworowska et al., 1979; Kielan-Jaworowska and Dashzeveg, 1989). In *Peramus* and in all Cretaceous tribosphenids for which the complete postcanine dentition is known, the trenchant premolar is followed by four cheekteeth primitively, regardless of the molariform or semimolariform nature of these teeth (Fig. 13.15). It is suggested here that the posterior five cheektooth positions in all these mammals are serially homologous, as proposed by McKenna (1975). Disagreement continues about the number and homologies of premolars anterior to the trenchant premolar, and McKenna (1975) acknowledged the uncertainty about which anterior premolar loci would have been lost in marsupials and deltatheridiids from his postulated eight postcanine tooth families of ancestral tribosphenids.

Ontogenetic findings from extant therians suggest a working hypothesis to account for the proposed serial homology of the five posterior cheekteeth in tribosphenids. These tooth positions have been considered by McKenna (1975) to primitively represent P4, P5, M1, M2, and M3, following his hypothesis of five premolar tooth families in tribosphenids. For convenience in discussing these tooth positions in extant therians (and consistent with my own beliefs on their homologies in extant and fossil tribosphenids), they will continue to be designated as P3, P4, M1, M2, and M3. It is proposed here that dP3 was the first postcanine tooth to differentiate in ancestral tribosphenids, as it is in all extant marsupials and in some eutherian groups, including tupaiids, macroscelidids, and carnivorans. A dichotomy subsequently occurs during later ontogeny in extant therians, and this is reflected in the differing homologies assigned to tooth positions posterior to dP3. If the fourth postcanine tooth differentiates relatively early, following the initiation of dP3, then this will develop as a deciduous tooth (dP4) and subsequently be replaced by a successor P4. Primitively, this dP4 was molariform, as in Cretaceous eutherians and in extant tupaiids and macroscelidids, and it was replaced by a semimolariform P4. This developmental pattern probably characterized the eutherian morphotype, but it may be a primitive retention of a condition inherited from pre-tribosphenic mammals, as suggested by phenetic resemblances to the cheekteeth of *Peramus*. This hypothesis could be tested by the recovery of deciduous posterior premolars from late Jurassic peramurids.

When the fourth postcanine tooth is relatively retarded in development in extant therians, so that it is not initiated until dP3 attains the bell stage, the resulting molariform primary tooth fails to develop a lingual successor. As a result, this molariform tooth persists in the "adult" dentition, without replacement, and it has been designated as M1 in marsupials. This derived developmental pattern characterizes all extant marsupials studied and probably represents the marsupial morphotype. This working hypothesis provides a plausible explanation for Owen's (1868) suggestion that the marsupial M1 is serially homologous with dP4 of eutherians, and it should be tested by the further accumulation of data on dental ontogeny in extant therians.

ROLE OF HETEROCHRONY IN THERIAN DENTAL EVOLUTION. There are three fundamental aspects of developmental processes that are subject to the temporal changes of heterochrony: initiation, termination, and rate of development (Alberch et al., 1979; Raff and Wray, 1989). The present study suggests that both acceleration and retardation in the initiation of tooth bud differentiation have played a major role in the reduction and loss of teeth during mammalian phylogeny. Raff and Wray (1989) emphasized that heterochronies that occur during early development should be measured relative to other events during that period, rather than on later ontogenetic events, such as the onset of sexual maturity. Thus, for assessing heterochronies during dental ontogeny, the developmental state of each tooth can be compared to others in the same jaws, as well as to the presumably homologous developing teeth of related taxa.

Three ontogenetic patterns provide evidence for the evolutionary loss of mammalian teeth. First, the developing deciduous tooth may become vestigial and fail to erupt, being either resorbed or shed during the perinatal period. Concomitant with this, its lingual successor is accelerated in development and subsequent eruption. As a consequence, the successor tooth may erupt at about the same time as the remaining deciduous teeth. In the absence of ontogenetic evidence, the accelerated successional tooth may be misidentified as a retained deciduous tooth, as was done by Schwartz (1980) for P2 in *Tarsius*. Such a developmental pattern, entailing loss of the deciduous tooth but accelerated development of its successor, characterizes the marsupial incisors and canines, as well as most antemolar teeth of soricids. The functional or selective advantages of this modified pathway are unknown. It seems probable that "developmental constraints" may account for the well-defined pattern of vestigial deciduous tooth reduction in unrelated therians, similar to the congruence in sequence of reduction and loss of limb bones in tetrapods (Smith et al., 1985).

When the process of vestigial tooth formation is more extreme, so that a successor also fails to develop, then neither the deciduous nor successional tooth is encountered in the erupted dentition. In such cases, evidence for the existence of a "missing" tooth position can be obtained only by careful study of early ontogenetic stages. Thus, dI1 of rodents and lagomorphs can be identified histologically in fetuses, whereas it is undetected in fossils or in adult skulls.

In other cases, there is a delay in the onset of bud initiation for a primary tooth. When this happens, the primary tooth may develop and erupt more slowly than neighboring teeth, and it fails to form a lingual successor. Such a developmental pattern characterizes the unreplaced dP1 of marsupials and some eutherians (*Talpa, Elephantulus, Sus*), and it is also the basis for homologizing therian molars as unreplaced primary teeth. As suggested above, a similar heterochrony may account for the nonreplacement of the fourth postcanine tooth in primitive marsupials. There is evidence in the triconodont *Gobiconodon* from the early Cretaceous that dental replacement occurred for at least some molar loci (Jenkins and Schaff, 1988). If true, it is likely that a delay in the onset of "molar" primary bud differentiation did not take place in this non-therian taxon, in contrast to the condition in therians.

Concluding Comments

It is obvious that much additional study remains to be done on early ontogeny of the mammalian dentition. This is a necessary prerequisite for proposing hypotheses on the controlling mechanisms of dental development and on the manner in which developmental patterns have been modified during mammalian phylogeny. In recent years, several authors have presented hypotheses of mammalian dental homologies, based primarily on theoretical considerations, or on the study of reptilian dental ontogeny, with extrapolation to the mammalian condition. As noted earlier, Osborn's (1978) reliance on gradients of shape for determining tooth class homologies is not supported by the present analysis. Westergaard (1980, 1983) suggested that seven premolar tooth families occurred primitively in therians, and that in some cases deciduous premolars are succeeded by teeth belonging to different tooth families. This proposal appears to have been based primarily on his comparison between dental development in the reptile *Lacerta* and published studies on the mammalian dentition, rather than on direct observations. There is no ontogenetic evidence to support either of these suggestions for extant therians. In addition, Westergaard's (1983, p. 76) proposal that the eutherian dental developmental pattern may have evolved "from a marsupial-like ontogeny," with eutherians having "regained" the ability to develop successors for most deciduous premolars, is highly improbable, and it finds no support in the present analysis. A more recent suggestion (Westergaard, 1989) that "predeciduous dental placodes" occur in some marsupials and eutherians, before the formation of the primary dental lamina, also seems unlikely, but further assessment of this hypothesis must await publication of the histological evidence for these placodes.

A final comment is required concerning the hypotheses proposed by Schwartz (1980, 1982), regarding patterns of dental development in mammals and his reinterpretation of many traditional tooth homologies. Although his studies have rekindled interest in the ontogenetic analysis of mammalian dental homologies, they have been based almost entirely on examination of tooth eruption and replacement in postnatal stages. Despite this lack of firsthand knowledge, Schwartz has proposed several novel reinterpretations of tooth homologies, based on his evaluation of published studies on prenatal dental ontogeny. Unfortunately, many of Schwartz's (1980, 1982) unorthodox proposals are founded on erroneous interpretations of other authors' data, as well as on a basic lack of understanding of dental developmental phenomena. Most of these misinterpretations have been discussed elsewhere (Luckett and Maier, 1982; Luckett, 1982) and will not be repeated here.

It is evident that many paleontologists are eager to incorporate ontogenetic data on the dentition into their assessments of the mammalian fossil record, as demonstrated by the recent discussions of Clemens and Lillegraven (1986) and Greenwald (1988), and by the more extensive analyses of Butler (1956, 1978a). In light of this, recent attempts (Osborn, 1978; Schwartz, 1980, 1982) to falsify traditional hypotheses of mammalian dental homologies, established more than a century ago on bony and occlusal relationships, as well as on replacement patterns, have attracted considerable attention from evolutionary biologists. Such hypotheses tend to be presented as being more progressive and consistent with recent advances in scientific thought, compared to their dated predecessors. Instead, they are often based on hastily drawn proposals that are not readily supported by a careful analysis of the available evidence. It is more time-consuming to examine the primary data, especially, as in the case of the dentition, when this requires the study of thousands of serial sections of developing jaws, than it is to postulate novel hypotheses that may appear to be supported by a few carefully selected examples. The Osborn-Schwartz hypothesis of tooth shape as an important arbiter of tooth homologies is not corroborated by the present developmental assessment of the therian dentition.

Tooth initation patterns discussed in this chapter further corroborate the traditional hypothesis (Owen, 1868) that bony position and occlusal relationships provide better clues to homologies in modified mammalian dentitions than do shape or size. Developmental studies are also consistent with the belief (i.e., Butler, 1978a) that shape differences between tooth classes, and between serially homologous teeth within the same class, have become modified during the course of mammalian evolution. This has resulted in the molarization of premolars, premolarization of canines, and other modifications that can be distinguished among genera belonging to the same family, or between earlier fossils and their presumed, more recent descendants (or sister groups). Such broad agreements among the findings of ontogeny, comparative anatomy, and paleontology for the homologies of modified mammalian teeth should provide further stimulus for the interaction of biologists from all three disciplines, in order to evaluate the developmental and functional specializations and constraints that have led to the known diversity of the dentition during mammalian phylogeny.

ACKNOWLEDGMENTS. This investigation was made possible by the cooperation and courtesy of a number of institutions and individuals, who allowed me access to their sectioned series of mammalian fetuses and juveniles. The majority of the specimens were examined at the Hubrecht Laboratory, Utrecht, the Netherlands, with the long-standing encouragement and support of Drs. Pieter Nieuwkoop, Else Boterenbrood, Siegfried de Laat, and Gesineke Bangma. Colleagues in Germany, who have contributed greatly to the number of taxa examined, include Drs. H.-J. Kuhn and U. Zeller, Göttingen; W. Maier, Tübingen; D. Starck and M. Klima, Frankfurt; and F. Schrenk, Darmstadt. Additional specimens were studied at the Carnegie Laboratories of Embryology, Davis, California; the American Museum of Natural History, New York, New York; and the Naturhistoriska Riksmuseet, Stockholm, Sweden. A number of paleontologists have devoted considerable time to discussing the problems of mammalian dental homologies with me, and their patience is gratefully appreciated (even when we disagree). These include Michael Archer, Percy Butler, Jean-Louis Hartenberger, Zofia Kielan-Jaworowska, René Lavocat, Malcolm McKenna, Michael Novacek, Gerhard Storch, Fred Szalay, and Pascal Tassy. Finally, special thanks are due to Dr. Nancy Hong for preparing the camera lucida drawings and other illustrations for this chapter, as well as for her continued support and encouragement during all phases of the study.

Financial support for this investigation was provided in part by a U.S. Senior Scientist Award from the Alexander von Humboldt Foundation, Bonn, Germany; by NATO Collaborative Research Grant No. 890377; and by RCMI Grant RR-03051 from the University of Puerto Rico Medical Sciences Campus.

CORRESPONDENCE ADDRESS. W. Patrick Luckett, Department of Anatomy, University of Puerto Rico, Medical Science Campus, P.O. Box 365067, San Juan, Puerto Rico 00936-5067, USA.

References

Abel, O. 1912. Die eocänen Sirenen der Mittelmeerregion. Erster Teil: Der Schädel von *Eotherium aegyptiacum*. *Palaeontographica* 59:289–360.

Adloff, P. 1928. Uber das Gebiss von *Tapirus americanus*. *Anat, Anz.* 66:132–136.

Alberch, P., Gould, S.J., Oster, G.F., and Wake, D.B. 1979. Size and shape in ontogeny and phylogeny. *Paleobiology* 5:296–317.

Ameghino, F. 1909. L'avant-première dentition dans le Tapir. *An. Mus. Nac. Buenos Aires* (ser. 3) 13:1–30.

Archer, M. 1974a. The development of the cheek-teeth in *Antechinus flavipes* (Marsupialia, Dasyuridae). *J. Roy. Soc. West. Aust.* 57:54–63.

Archer, M. 1974b. The development of premolar and molar crowns of *Antechinus flavipes* (Marsupialia, Dasyuridae) and the significance of cusp ontogeny in mammalian teeth. *J. Roy. Soc. West. Aust.* 57:118–125.

Archer, M. 1976. The dasyurid dentition and its relationships to that of didelphids, thylacinids, borhyaenids (Marsupicarnivora) and peramelids (Peramelina: Marsupialia). *Aust. J. Zool. Suppl.* 39:1–34.

Archer, M. 1978. The nature of the molar-premolar boundary in marsupials and a reinterpretation of the homology of marsupial cheekteeth. *Mem. Queensland Mus.* 18:157–164.

Archer, M. 1984. Origins and early radiations of marsupials. In: *Vertebrate zoogeography and evolution in Australasia* (Archer, M., and Clayton, G., eds.). Carlisle, Western Australia: Hesperian Press, pp. 585–625.

Berkovitz, B.K.B. 1966. The homology of the premolar teeth in *Setonix brachyurus* (Macropodidae: Marsupialia). *Archs. Oral Biol.* 11:1371–1384.

Berkovitz, B.K.B. 1968a. The early development of the incisor teeth of *Setonix brachyurus* (Macropodidae: Marsupialia) with special reference to the prelacteal teeth. *Archs. Oral Biol.* 13:171–190.

Berkovitz, B.K.B. 1968b. Some stages in the early development of the post-incisor dentition of *Trichosurus vulpecula* (Phalangeroidea: Marsupialia). *J. Zool.* 154:403–414.

Berkovitz, B.K.B. 1972a. Tooth development in *Protemnodon eugenii*. *J. Dent. Res.* 51:1467–1473.

Berkovitz, B.K.B. 1972b. Ontogeny of tooth replacement in

the guinea pig (*Cavia cobya*). *Archs. Oral Biol.* 17:711–718.

Berkovitz, B.K.B. 1973. Tooth development in the albino ferret (*Mustela putorius*) with special reference to the permanent carnassial. *Archs. Oral Biol.* 18:465–471.

Berkovitz, B.K.B. 1978. Tooth ontogeny in *Didelphis virginiana (Marsupialia: Didelphidae)*. *Aust. J. Zool.* 26:61–68.

Bolk, L. 1917. Die Beziehung zwischen Reptilien-, Beutler- und Plazentaliergebiss. *Z. Morph. Anthrop.* 20:259–338.

Bolk, L. 1929. Die Gebissentwicklung von *Trichosurus*. *Morph. Jahrb.* 62:58–178.

Bown, T.M., and Kraus, M.J. 1979. Origin of the tribosphenic molar and metatherian and eutherian dental formulae. In: *Mesozoic mammals: The first two-thirds of mammalian history* (Lillegraven, J.A., Kielan-Jaworowska, Z., and Clemens, W.A., eds.). Berkeley: University of California Press, pp. 172–181.

Butler, P.M. 1956. The ontogeny of molar pattern. *Biol. Rev.* 31:30–70.

Butler, P.M. 1977. Evolutionary radiation of the cheekteeth of Cretaceous placentals. *Acta Palaeont. Pol.* 22:241–271.

Butler, P.M. 1978a. The ontogeny of mammalian heterodonty. *J. Biol. Buccale* 6:217–228.

Butler, P.M. 1978b. A new interpretation of the mammalian teeth of tribosphenic pattern from the Albian of Texas. *Brevoria Mus. Comp. Zool.* 446:1–27.

Clemens, W.A. Jr. 1966. Fossil mammals of the Type Lance Formation Wyoming. Part II. Marsupialia. *Univ. Cal. Publ. Geol. Sci.* 62:1–122.

Clemens, W.A. Jr. 1973. Fossil mammals of the Type Lance Formation Wyoming. Part III. Eutheria and summary. *Univ. Calif. Publ. Geol. Sci.* 94:1–102.

Clemens, W.A. 1979. Marsupialia. In: *Mesozoic mammals: The first two-thirds of mammalian history* (Lillegraven, J.A., Kielan-Jaworowska, Z., and Clemens, W.A., eds.). Berkeley: University of California Press, pp. 192–220.

Clemens, W.A., and Lillegraven, J.A. 1986. New late Cretaceous, North American advanced theian mammals that fit neither the marsupial nor eutherian molds. *Contrib. Geol., Univ. Wyoming, Spec. Pap.* 3:55–85.

Dependorf, T. 1898. Zur Entwickelungsgeschichte des Zahnsystems der Marsupialier. *Denksch. med.-naturwiss. Ges. Jena* 6:243–402.

Domning, D.P., Morgan, G.S., and Ray, C.E. 1982. North American Eocene sea cows (Mammalia: Sirenia). *Smith. Contrib. Paleobiol.* 52:1–69.

Dorst, J. 1957. Note sur la dentition lactéale de *Tonatia amblyotis* (Chiroptères, Phyllostomidés). *Mammalia* 21:302–304.

Flower, W.H. 1867. On the development and succession of the teeth in the Marsupialia. *Phil. Trans. Roy. Soc. London* B157:631–641.

Flower, W.H. 1869. Remarks on the homologies and notation of the teeth of the Mammalia. *J. Anat. Physiol.* 3:262–278.

Fosse, G. 1969. Development of the teeth in a pouch-young specimen of *Antechinus stuartii* and a pouch-young specimen of *Sminthopsis crassicaudata*. Dasyuridae: Marsupialia. *Archs. Oral Biol.* 14:207–218.

Fosse, G., and Risnes, S. 1972. Development of the teeth in a pouch-young specimen of *Isoodon obesulus* and one of *Perameles gunnii* (Peramelidae: Marsupialia). *Archs. Oral Biol.* 17:829–838.

Fox, R.C. 1979. Mammals from the Upper Cretaceous Oldman Formation, Alberta. III. Eutheria. *Can. J. Earth Sci.* 16:114–125.

Fox, R.C. 1983. New evidence on the relationships of the Tertiary insectivoran *Ankylodon* (Mammalia). *Can. J. Earth Sci.* 20:968–977.

Greenwald, N:S. 1988. Patterns of tooth eruption and replacement in multituberculate mammals. *J. Vert. Paleont.* 8:265–277.

Hill, J.P., and O'Donoghue, C.H. 1913. The reproductive cycle in the marsupial *Dasyurus viverrinus*. *Quart. J. Micr. Sci.* 59:133–174.

Jenkins, F.A. Jr., and Schaff, C.R. 1988. The early Cretaceous mammal *Gobiconodon* (Mammalia, Triconodonta) from the Cloverly Formation in Montana. *J. Vert. Paleont.* 8:1–24.

Karlsen, K. 1962. Development of tooth germs and adjacent structures in the whalebone whale (*Balaenoptera physalus* L.). *Det Norske Vidensk-Akad., Oslo* 45: 1–56.

Kielan-Jaworowska, Z. 1969. Preliminary data on the Upper Cretaceous eutherian mammals from Bayn Dzak, Gobi Desert. *Palaeont. Pol.* 19:171–191.

Kielan-Jaworowska, Z. 1981. Evolution of the therian mammals in the late Cretaceous of Asia. Part IV. Skull structure in *Kennalestes* and *Asioryctes*, *Palaeont. Pol.* 42:25–78.

Kielan-Jaworowska, Z., Bown, T.M., and Lillegraven, J.A. 1979. Eutheria. In: *Mesozoic mammals: The first two-thirds of mammalian history* (Lillegraven, J.A., Kielan-Jaworowska, Z., and Clemens, W.A., eds.). Berkeley: University of California Press, pp. 221–258.

Kielan-Jaworowska, Z., and Dashzeveg, D. 1989. Eutherian mammals from the early Cretaceous of Mongolia. *Zool. Scripta* 18:347–355.

Kindahl, M. 1957a. Some observations on the development of the tooth in *Elephantulus myurus jamesoni*. *Ark. Zool.* 11:21–29.

Kindahl, M. 1957b. Notes on the tooth development in *Talpa europaea*. *Ark. Zool.* 11:187–191.

Kindahl, M. 1959. Some aspects of the tooth development in Soricidae. *Acta Odont. Scand.* 17:203–237.

Kindahl, M. 1967. Some comparative aspects of the reduction of the premolars in the Insectivora. *J. Dent. Res.* 46:805–808.

Kirkpatrick, T.H. 1969. *The dentition of the marsupial family Macropodidae with particular reference to tooth development in the grey kangaroo* Macropus giganteus Shaw. Unpublished Ph. D. Thesis, University of Queensland, Brisbane.

Kirkpatrick, T.H. 1978. The development of the dentition of *Macropus giganteus* (Shaw). *Aust. Mammal.* 2:29–35.

Kollar, E.J., and Baird, G.R. 1969. The influence of the dental papilla on the development of tooth shape in embryonic mouse tooth germs. *J. Embryol. Exp. Morph.* 21:131–148.

Kükenthal, W. 1891. Das Gebiss von *Didelphys*. *Anat. Anz.* 6:658–666.

Kükenthal, W. 1893. Vergleichend-anatomische und entwickelungsgeschichtliche Untersuchungen an Waltieren. *Denksch. med.-naturwiss. Ges. Jena* 3:1–448.

Leche, W. 1877. Zur Kenntniss des Milchgebisses und der Zahnhomologien bei Chiroptera. II. Theil. *Lunds Univ. Årsskrift.* 14:1–37.

Leche, W. 1892. Studien über die Entwicklung des Zahnsystems bei den Säugethieren. *Morph. Jahrb.* 19:502–547.

Leche, W. 1895. Zür Entwicklungsgeschichte des Zahnsystems der Säugethiere. *Bibl. Zoologica* 17:1–160.

Leche, W. 1897. Untersuchungen über das Zahnsystem lebender und fossiler Halbaffen. *Festschrift zum 70. Geburtstag von Carl Gegenbaur* 3:125–166.

Lillegraven, J.A. 1969. Latest Cretaceous mammals of upper part of Edmonton Formation of Alberta, Canada, and review of marsupial-placental dichotomy in mammalian evolution. *Univ. Kansas Paleont. Contrib.* Article 50 (Vertebrata 12): 1–122.

Luckett, W.P. 1977. Ontogeny of amniote fetal membranes and their application to phylogeny. In: *Major patterns in vertebrate evolution* (Hecht, M.K., Goody, P.C., and Hecht, B.M., eds.). New York: Plenum Press, pp. 439–516.

Luckett, W.P. 1982. The uses and limitations of embryological data in assessing the phylogenetic relationships of *Tarsius* (Primates, Haplorhini). *Geobios, Mém. Spéc.* 6:289–304.

Luckett, W.P. 1985. Superordinal and intraordinal affinities of rodents: Developmental evidence from the dentition and placentation. In: *Evolutionary relationships among rodents* (Luckett, W.P., and Hartenberger, J.-L., eds.). New York: Plenum Press, pp. 227–276.

Luckett, W.P. 1988. Early development and homology of the dental lamina and tooth buds in eutherian, metatherian, and prototherian mammals. *Anat. Rec.* 220:60A.

Luckett, W.P. 1989. Developmental evidence for dental homologies in the marsupial family Dasyuridae. *Anat. Rec.* 223:70A.

Luckett, W.P., Bangma, G., and Hong, N. 1991. Ontogenetic evidence for premolar homologies in marsupials. *Anat. Rec.* 229:55A–56A.

Luckett, W.P., and Hong, N. 1989. Ontogenetic evidence for tooth replacement in marsupials (Mammalia). *Cell Differ. Dev.* 27 (Suppl.):S45.

Luckett, W.P., and Maier, W. 1982. Development of deciduous and permanent dentition in *Tarsius* and its phylogenetic significance. *Folia Primatol.* 37:1–36.

Luckett, W.P., Schrenk, F., and Maier, W. 1989. On the occurrence of abnormal deciduous incisors during prenatal life in African "hystricomorphous" rodents. *Z. Säugetierk.* 54:296–303.

Lumsden, A.G.S. 1987. The neural crest contribution to tooth development in the mammalian embryo. In: *Developmental and evolutionary aspects of the neural crest* (Maderson, P.F.A., ed.). New York: John Wiley and Sons, pp. 261–300.

Lumsden, A.G.S., and Buchanan, J.A.G. 1986. An experimental study of timing and topography of early tooth development in the mouse embryo with an analysis of the role of innervation. *Archs. Oral Biol.* 31:301–311.

Lydekker, R. 1899. The dental formula of the marsupial and placental Carnivora. *Proc. Zool. Soc. London.* 1899:922–928.

Lyne, A.G. 1982. Observations on skull growth and eruption of teeth in the marsupial bandicoot *Perameles nasuta* (Marsupialia: Peramelidae). *Aust. Mamm.* 5:113–126.

Marshall, L.G. 1987. Systematics of Itaboraian (Middle Paleocene) age "opossum-like" marsupials from the limestone quarry at Sao José de Itaboraí, Brazil. In: *Possums and opossums: Studies in evolution* (Archer, M., ed.). Sydney: Surrey Beatty and Sons, pp. 91–160.

Marshall, L.G., and de Muizon, C. 1988. The dawn of the age of mammals in South America. *Nat. Geog. Res.* 4:23–55.

Marshall, P.M., and Butler, P.M. 1966. Molar cusp development in the bat, *Hipposideros beatus*, with reference to the ontogenetic basis of occlusion. *Archs. Oral Biol.* 11:949–965.

McKenna, M.C. 1975. Toward a phylogenetic classification of the Mammalia. In: *Phylogeny of the primates* (Luckett, W.P., and Szalay, F.S., eds.). New York: Plenum Press, pp. 21–46.

Merchant, J.C., Newgrain, K., and Green, B. 1984. Growth of the Eastern quoll, *Dasyurus viverrinus* (Shaw), (Marsupialia) in captivity. *Aust. Wildl. Res.* 11:21–29.

Moss-Salentijn, L. 1978. Vestigial teeth in the rabbit, rat and mouse; their relationship to the problem of lacteal dentitions. In: *Development, function and evolution of teeth* (Butler, P.M., and Joysey, K.A., eds.). London: Academic Press, pp. 13–29.

Müller, K. 1929. Uber die Zahnentwicklung bei *Perameles*. *Morph. Jahrb.* 61:457–488.

Novacek, M.J. 1986. The primitive eutherian dental formula. *J. Vert. Paleont.* 6:191–196.

Ooë, T. 1965. A study of the ontogenetic origin of human permanent tooth germs. *Okajimas Folia Anat. Japan* 40:429–437.

Ooë, T. 1980. Développement embryonnaire des incisives chez le lapin (*Oryctolagus cuniculus* L.). Interprétation de la formule dentaire. *Mammalia* 44:259–269.

Ooë, T. 1981. *Human tooth and dental arch development.* Tokyo: Ishiyaku Publishers, Inc.

Osborn, J.W. 1978. Morphogenetic gradients: Fields versus clones. In: *Development, function and evolution of teeth* (Butler, P.M., and Joysey, K.A., eds.). London: Academic Press, pp. 171–201.

Owen, R. 1840–45. *Odontography, vol. I.* London: Hippolyte Bailliere.

Owen, R. 1868. *On the anatomy of vertebrates. vol. III. Mammals.* London: Longmans, Green and Co.

Parrington, F.R. 1971. On the Upper Triassic mammals. *Phil. Trans. Roy. Soc.* B 261:231–272.

Raff, R.A., and Wray, G.A. 1989. Heterochrony: Development mechanisms and evolutionary results. *J. Evol. Biol.* 2:409–434.

Reig, O.A., Kirsch, J.A.W., and Marshall, L.G. 1987. Systematic relationships of the living and Neocenozoic American "opossum-like" marsupials (suborder Didelphimorphia), with comments on the classification of these and of the Cretaceous and Paleogene New World and European metatherians. In: *Possums and opossums: Studies in evolution* (Archer, M., ed.). Sydney: Surrey Beatty and Sons, pp. 1–89.

Rose, C. 1892a. Uber die Zahnentwickelung der Beuteltiere. *Anat. Anz.* 7:639–650.

Rose, C. 1892b. Uber die Zahnentwickelung der Beuteltiere. *Anat. Anz.* 7:693–707.

Schwartz, J.H. 1980. A discussion of dental homology with

reference to Primates. *Am. J. Phys. Anthrop.* 52:463–480.

Schwartz, J.H. 1982. Morphological approach to heterodonty and homology. In: *Teeth: Form, function and evolution* (Kurtén, B., ed.). New York: Columbia University Press, pp. 123–144.

Sicher, H. 1916. Die Entwickelung des Gebisses von *Talpa europaea*. *Anat. Hefte* 162:31–112.

Smith, J.M., Burian, R., Kauffman, S., Alberch, P., Campbell, J., Goodwin, B., Lande, R., Raup, D., and Wolpert, L. 1985. Developmental constraints and evolution. *Quart. Rev. Biol.* 60:265–287.

Thenius, E. 1988. *Zähne und Gebiss der Säugetiere*. Berlin: Walter de Gruyter & Co.

Thesleff, I., and Hurmerinta, K. 1981. Tissue interactions in tooth development. *Differentiation* 18:75–88.

Thomas, O. 1887. On the homologies and succession of the teeth in the Dasyuridae, with an attempt to trace the history of the evolution of mammalian teeth in general. *Phil. Trans. Roy. Soc. London* B1887:443–462.

Tims, H.W.M. 1896. Notes on the dentition of the dog. *Anat. Anz.* 11:537–546.

Tyndale-Biscoe, H., and Renfree, M. 1987. *Reproductive physiology of marsupials*. Cambridge: Cambridge University Press.

Van Valen, L. 1974. *Deltatheridium* and marsupials. *Nature* 248:165–166.

Westergaard, B. 1980. Evolution of the mammalian dentition. *Mém. Soc. Géol. Fr., N. S.* 139:191–200.

Westergaard, B. 1983. A new detailed model for mammalian dentitional evolution. *Z. Zool. Syst. Evolut. -forsch.* 21:68–78.

Westergaard, B. 1989. Prediciduous dental placodes in mammals, equivalent to embryonic denticles in reptiles. *Abst. VIII Int. Symp. Dental Morphology, Jerusalem, Israel*, pp. 99–100.

Williams, R.C., and Evans, H.E. 1978. Prenatal dental development in the dog, *Canis familiaris*: Chronology of tooth germ formation and calcification of deciduous teeth. *Zbl. Vet. Med. C. Anat. Histol. Embryol.* 7:152–163.

Wilson, J.T., and Hill, J.P. 1897. Observations upon the development and succession of the teeth in *Perameles*; together with a contribution to the discussion of the homologies of the teeth in marsupial animals. *Quart. J. Micr. Sci.* 39:427–588.

Winge, H. 1882. Om pattedyrenes tandskifte isaer med hensyn til taendernes former. *Vidensk. Meddel. Naturhist. Forening Kjobenhavn* 1882:15–67.

Winge, H. 1941. *The interrelationships of the mammalian genera, vol. I. Monotremata, Marsupialia, Insectivora, Chiroptera, Edentata*. Kobenhavn: C. A. Reitzels-Forlag.

Woodward, M.F. 1893. Contributions to the study of mammalian dentition. Part I. On the development of the teeth of the Macropodidae. *Proc. Zool. Soc. London* 1893:450–473.

Woodward, M.F. 1896a. Contributions to the study of mammalian dentition. Part II. On the teeth of certain Insectivora. *Proc. Zool. Soc. London* 1896:557–594.

Woodward, M.F. 1896b. On the teeth of the Marsupialia, with especial reference to the premilk dentition. *Anat. Anz.* 12:281–291.

Ziegler, A.C. 1971. A theory of the evolution of therian dental formulas and replacement patterns. *Quart. Rev. Biol.* 46:226–249.

CHAPTER 14

Theria of Metatherian-Eutherian Grade and the Origin of Marsupials

RICHARD L. CIFELLI

Overview

Although a growing number of metatherian-eutherian grade and primitive marsupial taxa are becoming known, the early radiations of the Tribosphenida and the origin of Marsupialia remain poorly understood, largely because the nature of the fossil record severely restricts the data available. Phylogenetic analysis suggests that the early history of tribosphenic mammals was complex. One group of archaic Tribosphenida, including *Picopsis* and presumed allies, is tentatively recognized; another, Deltatheridiidae, is more adequately supported by synapomorphy. *Deltatheridium* and allied taxa lack dental specializations seen in marsupials, eutherians, and various "higher" tribotheres, and for this reason do not appear to constitute a likely sister taxon to Marsupialia. Several metatherian-eutherian grade taxa, together with Eutheria and Marsupialia, can be derived from a morphotype that is advanced with respect to the condition seen in Pappotheriidae. Although no diagnostic dental specialization for Marsupialia is known, a complex of characters associated with upper molar protocone enlargement and modification of lower molar embrasures best distinguishes Cretaceous members of the group with respect to non-marsupial contemporaries. Stylar cusps D and C, in that order, probably appeared later in marsupial evolution. The hypothesized relationships among Cretaceous marsupials, together with meager negative evidence from the fossil record of South America, support a North American origin for the group.

Contents

Introduction, 206
Theria of Metatherian–Eutherian Grade and Early Marsupials, 206
   Geographic and temporal distribution, 206
   Phylogeny, 207
Discussion and Conclusions, 209
Appendix: Character Distribution, 213
Acknowledgments, 214
References, 214

Introduction

Tribosphenic mammals, which include marsupials and eutherians, are defined on the basis of dental characteristics, including an upper molar protocone and a fully basined, multicusped talonid on lower molars (Simpson, 1936). These features, which form a functional complex for crushing and shearing food items (Crompton, 1971), are lacking in monotremes and various Mesozoic taxa, and are thus presumed to be adaptations integral to the evolutionary success of marsupials and eutherians, which together constitute the bulk of the Recent mammal fauna. The history of tribosphenic mammals extends back to the Early Cretaceous; in the past four decades, discovery of Cretaceous taxa insufficiently specialized to permit reference to either major group of higher mammals has led to the recognition of a loosely knit assemblage termed Theria of metatherian-eutherian grade (Patterson, 1956), a group defined on the basis of what its constituents lack rather than what they possess. Thus, although a close relationship of Marsupialia to Eutheria is clear enough in the context of non-tribosphenic taxa such as monotremes, multituberculates, symmetrodonts, and triconodonts, consideration of metatherian-eutherian grade or tribothere (Butler, 1978) taxa complicates understanding of the phylogeny of higher mammals. Although morphologically diverse, Theria of metatherian-eutherian grade are not well-represented as fossils, and few attempts have been made to assess their relationships, despite their obvious relevance to the origin of Eutheria and Marsupialia. The discovery, in recent years, of diverse new metatherian-eutherian grade and marsupial-like mammals, coupled with investigations on the distribution, significance, and polarity of dental characters among tribosphenic mammals (e.g., Clemens and Lillegraven, 1986), provides the basis for the present review of tribothere relationships and the origin of marsupials.

Because almost all of the taxa considered here are known only by dentulous jaw fragments or isolated teeth, comparisons are, perforce, based on dental morphology. The distributions of dental characters were gathered from descriptions, published illustrations, and personal observations. Character polarities were determined by outgroup comparison; in most cases, previous studies on character distribution and functional significance (e.g., Prothero, 1981; Clemens and Lillegraven, 1986; Fox, 1975) have established polarities and appropriate outgroups with some degree of confidence. A matrix consisting of twenty-two characters (eighteen binary, four multistate) and twenty-eight taxa was compiled (Appendix 1) and analyses performed on subsets of the matrix with PAUP (Swofford, 1989). Numerous trials were run, varying the taxa and characters included and the assumptions of character transformation. The object of these trials was to determine (1) which taxon groupings are relatively stable (regardless of how characters are treated), which groupings are not, and what assumptions or character/taxon combinations effect differences in the groupings; and (2) what assumptions, if any, are necessary to replicate preexisting views on relationships, and how well-founded those assumptions are. The results of these analyses were then synthesized into a more inclusive hypothesis of relationships for Theria of metatherian-eutherian grade, marsupial-like mammals, and selected early marsupial taxa.

Theria of Metatherian-Eutherian Grade and Early Marsupials

GEOGRAPHIC AND TEMPORAL DISTRIBUTION. The distribution of early Tribosphenida, exclusive of undoubted Eutheria and Marsupialia, is given in Table 14.1. The most primitive and earliest species, *Aegialodon dawsoni*, is known by a single lower molar from the Valanginian of England (Kermack et al., 1965; Clemens et al., 1979). *Kielantherium gobiensis*, known by much of the lower dentition, may be closely similar (Dashzeveg and Kielan-Jaworowska, 1984; see also Fox, 1976, 1980), although limited knowledge of the Wealden species leaves this open to question (see "**Phylogeny**," below). *Kielantherium* is from the Early Cretaceous of Mongolia; an undescribed taxon, possibly related to deltatheridiids, has been reported from the same beds (Kielan-Jaworowska and Nessov, 1990). Important Early and Late Cretaceous mammals have recently been described from Soviet central Asia (see, e.g., Nessov, 1985a, 1985b). Of non-eutherian taxa, only Coniacian *Sulestes karakshi* has received detailed treatment. The Late Cretaceous Old World record of metatherian-eutherian grade mammals is restricted to Mongolia, where four or more genera are known (Kielan-Jaworowska, 1975; Kielan-Jaworowska and Nessov, 1990). In the North American Early Cretaceous, six genera of tribosphenic mammals are known from northern Texas; of these, *Holoclemensia* was originally described as a marsupial (Slaughter, 1968a), but its relationships remain problematic (Butler, 1978; Jacobs et al., 1989; Cifelli, 1990a). In the Late Cretaceous, two unidentified taxa are represented in the Turonian of Utah (Cifelli, 1990b), four tribotheres are known from the Aquilan of Utah and Alberta, two from the Judithian, and two from the Lancian. Because sampling for the Judithian and Lancian is good, these records almost certainly reflect a decline in diversity prior to extinction before the beginning of the Cenozoic.

Undoubted Cretaceous marsupials are restricted to North America, although mammalian faunas of that age are now known from South America (e.g., Bonaparte and Pascual, 1987). The earliest undoubted marsupial, *Pariadens*, is from the Cenomanian of Utah (Cifelli and

TABLE 14.1. Age and distribution of metatherian-eutherian grade and marsupial-like mammals

Age of stage	Taxon	Distribution	Reference(s)
Late Cretaceous (Lancian)	gen. & sp. indet. aff. *Deltatheroides*	Scollard Fm., Alberta	Fox, 1974
Late Cretaceous (Lancian)	gen. & sp. indet. aff. *Picopsis*	Lance Fm., Wyoming	Cifelli, 1990b
Late Cretaceous (Judithian)	gen. & sp. indet. aff. *Deltatheroides*	Oldman Fm., Alberta	Fox, 1974
Late Cretaceous (Judithian)	*Falepetrus barwini*	"Mesaverde" Fm., Wyoming	Clemens & Lillegraven, 1986
Late Cretaceous	*Deltatheridium pretrituberculare*	Djadoktha, Barun Goyot fms., Mongolia	Gregory & Simpson, 1926; Kielan-Jaworowska, 1975
Late Cretaceous	*Deltatheroides cretacicus*	Djadoktha Fm., Mongolia	Gregory & Simpson, 1926; Kielan-Jaworowska, 1975
Late Cretaceous	*Hyotheridium dobsoni*	Djadoktha Fm., Mongolia	Gregory & Simpson, 1926
Late Cretaceous	gen. & sp. indet.	?Nemegt Fm., Mongolia	Kielan-Jaworowska & Nessov, 1990
Late Cretaceous (Aquilan)	*Potamotelses aquilensis*	Milk River Fm., Alberta	Fox, 1972
Late Cretaceous (Aquilan)	*Picopsis pattersoni*	Milk River Fm., Alberta	Fox, 1980
Late Cretaceous (Aquilan)	*Picopsis* sp.	Milk River Fm., Alberta	Fox, 1980
Late Cretaceous (Aquilan)	gen. & sp. indet.	Milk River Fm., Alberta	Fox, 1982
Late Cretaceous (Aquilan)	*Zygiocuspis golidingi*	Wahweap Fm., Utah	Cifelli, 1990d
Late Cretaceous (Aquilan)	*Anchistodelphys archidaldi*	Wahweap Fm., Utah	Cifelli, 1990c
Late Cretaceous (Aquilan)	*Iugomortiferum thoringtoni*	Wahweap Fm., Utah	Cifelli, 1990b
Late Cretaceous (Aquilan)	*Iugomortiferum* sp. indet.	Wahweap Fm., Utah	Cifelli, 1990c
Late Cretaceous (Coniacian)	*Sulestes karakshi* (incl. *Sulestes* sp.)	Taikarshin Beds, Uzbekistan	Nessov, 1985b; Kielan-Jaworowska & Nessov, 1990
Late Cretaceous	*Kumlestes olzha*	Soviet central Asia	Nessov, 1985a
Late Cretaceous (Turonian)	gen. & sp. indet.	Straight Cliffs Fm., Utah	Cifelli, 1990b
Late Cretaceous (Turonian)	gen. & sp. indet.	Straight Cliffs Fm., Utah	Cifelli, 1990b
Late Cretaceous (Turonian)	?*Anchistodelphys delicatus*	Straight Cliffs Fm., Utah	Cifelli, 1990b
Late Cretaceous (Turonian)	gen. & sp. indet.	Straight Cliffs Fm., Utah	Cifelli, 1990b
Early Cretaceous (Albian)	*Holoclemensia texana*	Paluxy Fm., Texas	Slaughter, 1968a
Early Cretaceous (Albian)	*Pappotherium pattersoni*	Paluxy Fm., Texas	Slaugher, 1968b
Early Cretaceous (Albian)	*Kermackia texana*	Paluxy Fm., Texas	Slaughter, 1971
Early Cretaceous (Albian)	*Trinititherium slaughteri*	Paluxy Fm., Texas	Butler, 1978
Early Cretaceous (Albian)	*Slaughteria eruptens*	Paluxy Fm., Texas	Butler, 1978
Early Cretaceous (Albian)	*Comanchea hilli*	Paluxy Fm., Texas	Jacobs et al., 1989
Early Cretaceous (?Aptian or Albian)	*Kielantherium gobiensis*	Khovboor beds, Mongolia	Dashzeveg & Kielan-Jaworowska, 1984
Early Cretaceous (?Aptian or Albian)	"*Prodelttheridium kalandadzei*"	Khovboor beds, Mongolia	Reshetev & Trofimov, 1984; Kielan-Jaworowska & Nessov, 1990
Early Cretaceous (Valanginian)	*Aegialodon dawsoni*	Wealden beds, England	Kermack et al., 1965

Eaton, 1987); additional undescribed marsupials are apparently known from the same unit (Eaton, 1990). By contrast, eutherians were apparently diverse by the Albian (Kielan-Jaworowska and Dashzeveg, 1989). The record first becomes reasonably good in the Aquilan, where at least nine species are known (Fox, 1971, 1987a; Cifelli, 1990c); however, a number of "marsupial-like" taxa, perhaps remnants of an early marsupial or pre-marsupial (depending on definition) radiation, are known from the Turonian and Aquilan (Cifelli, 1990b, 1990c). The situation is further complicated by placement of *Deltatheridium* and presumed allies as a sister taxon to marsupials (Kielan-Jaworowska and Nessov, 1990). Thus, although there seems to be no doubt about the referral of latest Cretaceous North American taxa to the Marsupialia, the discovery of possibly transitional forms has tended to blur the distinction between marsupials and more primitive Tribosphenida.

PHYLOGENY. A dental morphotype for Tribosphenida has been extensively discussed (e.g., Fox, 1972, 1975; Crompton, 1971; Clemens and Lillegraven, 1986). The lower molar crown pattern probably was close to that of *Aegialodon*; the morphology of the upper molars would most closely approximate that of *Potamotelses* among known taxa. Analyses including *Potamotelses* consistently placed it as a sister taxon to all remaining tribosphenic mammals (Fig. 14.1A, B). Analyses including ten genera, sampling a broad spectrum of Tribosphenida, consistently yielded virtually identical, single most parsimonious trees, regardless of character assumptions (Fig. 14.1A, B). In these and subsequent analyses, several further points of stability emerge. Deltather-

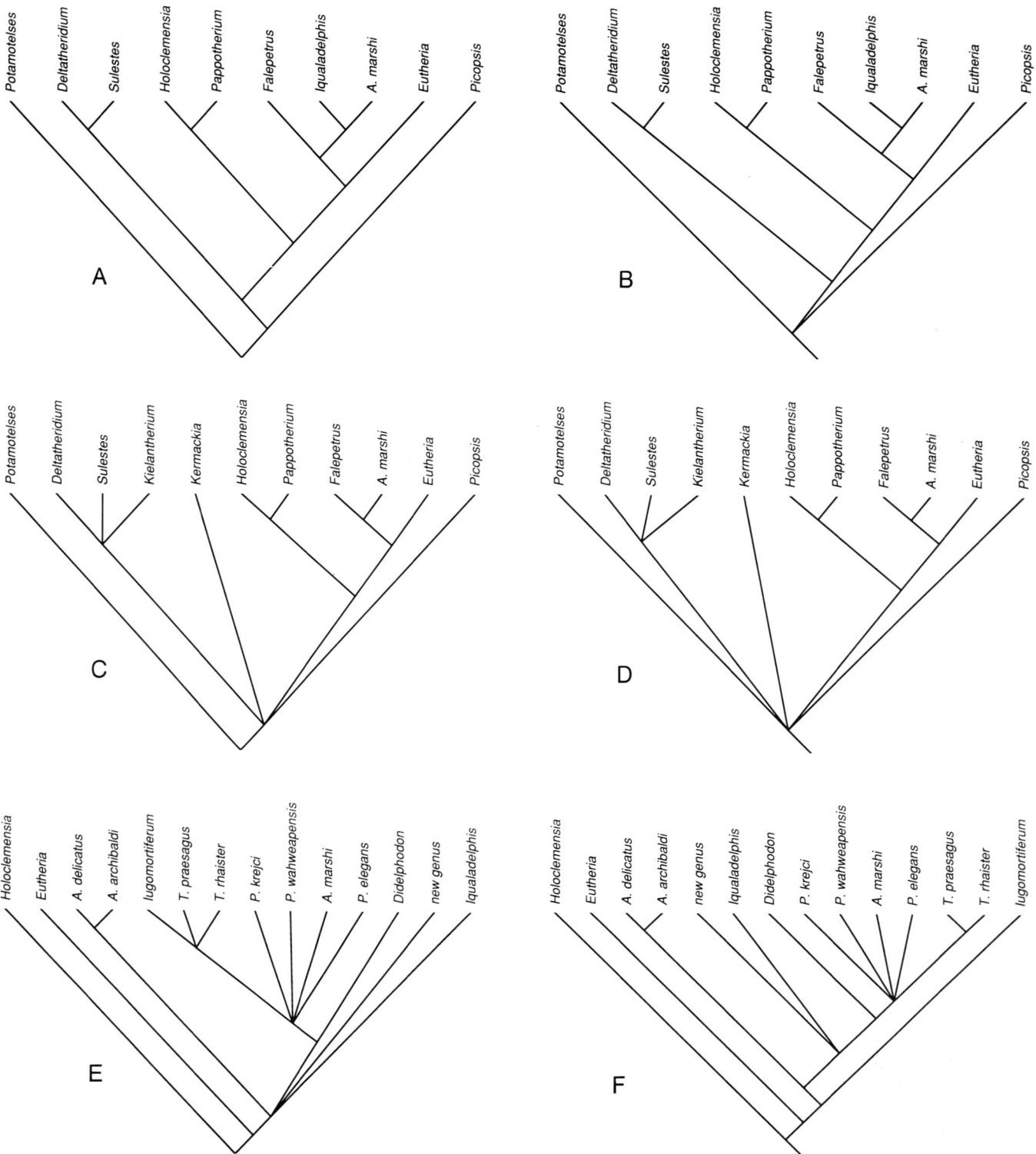

FIGURE 14.1. PAUP-generated trees of three combinations (A,B; C,D; E,F) of taxa including therians of metatherian-eutherian grade, marsupial-like mammals, and Cretaceous marsupials, based on data given in Appendix 1. Trees A, C, and E were generated with characters treated simply as ordered; ancestral states were defined for trees B, D, and F. **A**: single most parsimonious tree (L = 26; CI = .739); **B**: single most parsimonious tree (L = 29; CI = .731); **C**: consensus (Rohlf's CI = .543) of 15 trees (L = 27, CI = .741); **D**: consensus (Rohlf's CI = .586) of 18 trees (L = 29, CI = .720); **E**: consensus (Rohlf's CI = .549) of 130 trees (L = 25, CI = .684); **F**: consensus (Rohlf's CI = .771) of 6 trees (L = 35, CI = .724). *A. delicatus* and *A. archibaldi* = species of *Anchistodelphys*; *A. marshi* = *Alphadon marshi*; *P. krejcii* and *P. elegans* = species of *Pediomys*; *T. praesagus* and *T. rhaister* = species of *Turgidodon*; *P. wahweapensis* = *Protalphadon wahweapensis*.

idiidae was consistently upheld as a monophyletic group; where *Kielantherium* was included (Fig. 14.1C, D), that genus regularly clustered with deltatheridiids. Relationships of most remaining metatherian-eutherian grade mammals are difficult to assess: PAUP analyses offered no resolution unless the number of taxa was greatly restricted. A number of genera (such as *Kermackia* and *Picopsis*), although distinctive in their own right, eluded placement other than as unresolved basal branches of Tribosphenida. Other taxa, however, consistently were grouped with Marsupialia + Eutheria. Of these, the Trinity therians *Holoclemensia* and *Pappotherium* are (a) sistergroup(s) to remaining taxa; Late Cretaceous North American genera such as *Falepetrus* and *Zygiocuspis* lie closer to, or fall within, a group otherwise consisting of marsupials + eutherians.

Among marsupials and marsupial-like taxa, poor consistency resulted from trials using different character assumptions (Fig. 14.1E, F); this, and the poor resolution among some of the taxa, is probably a result of a high degree of homoplasy in some of the distinguishing characters (such as presence of certain stylar cusps). *Anchistodelphys* was consistently excluded from the group including unquestioned marsupials; another marsupial-like genus, *Iugomortiferum*, either occupied a similar place (if ancestral states were defined for characters) or was grouped with advanced species of *Turgidodon* (using strict Wagner parsimony). Early Campanian *Iqualadelphis*, an undescribed Turonian genus, and the stagodontid *Didelphodon* varied in position among the analyses, but generally fell outside the cluster including pediomyids and peradectids or didelphids (see Reig et al., 1987). Among these, the monophyly of the two included species of *Turgidodon* was consistently upheld, but no consensus tree grouped the two pediomyids together or resolved their relationships with *Alphadon* or *Protalphadon*.

Discussion and Conclusions

The results of the preceding analyses are synthesized into a hypothesis of relationships for most therians of metatherian-eutherian grade and selected early marsupials in Figure 14.2. Most of the hypothesized character changes in molar pattern are shown in Figure 14.3. The basal radiation of Tribosphenida, based on present knowledge, is not well-resolved, consisting of a multichotomy of variably autapomorphous but generally primitive taxa. One of these, *Picopsis*, from the early Campanian of Alberta, is characterized by a reduced anterior stylar shelf. Several other generally primitive taxa also have this specialization and, in the face of no other evidence as to the affinities of any of them, they are tentatively grouped together (recalling that this character appears homoplastically elsewhere among mammals). *Slaughteria eruptens* has been variously considered as a eutherian (Fox, 1980) or as a tribothere (Butler, 1978); I tentatively retain it with primitive taxa here because it appears to have a distal metacristid (Fox, 1975), although wear on the type and only specimen leaves this open to debate. *Deltatheridium* and allied taxa, mostly Old World in distribution, can be more confidently recognized as a monophyletic group, characterized by reduction to loss of the last molars, emphasis on postvallum/prevallid shearing (with the paraconid large and the metaconid small on lower molars, particularly those occupying more posterior loci; and a salient postmetacrista on upper molars), and, except for *Kielantherium*, reduction to three premolars in known taxa. Although the primitiveness of *Kielantherium* within Tribosphenida has been emphasized, especially regarding similarity to *Aegialodon* (Fox, 1980; Dashzeveg and Kielan-Jaworowska, 1984), workers have also noted a resemblance to *Deltatheridium* (e.g., Butler, 1978). *Deltatheridium* and allies were referred to a separate order, Deltatheroida, by Kielan-Jaworowska (1982). Most of the characters cited in the diagnosis of the order (Kielan-Jaworowska and Nessov, 1990) would appear to be primitive for Tribosphenida (rather than Deltatheroida). In view of this and the considerable uncertainty surrounding relationships of tribotheres, I have refrained from placing these or any other Theria of metatherian-eutherian grade in suprafamilial taxa. Deltatheridiids share with advanced Tribosphenida a labially extended preprotocrista, which provides for double-rank prevallum/postvallid shearing. Because this feature is lacking in presumably more primitive tribosphenic taxa such as *Potamotelses* and *Picopsis* (Fox, 1975, 1980), it is hypothesized to be a synapomorphy within Tribosphenida (Figs. 14.2, 14.3). However, a deltatheridiid morphotype lacks many other dental characters seen in more advanced Tribosphenida, as discussed in more detail below. If the Deltatheridiidae is correctly interpreted as a monophyletic group of primitive tribotheres, then some advanced taxa within it independently acquired similarity to more advanced mammals. In Coniacian *Sulestes*, for instance, the upper molar protocone is broadened and conules are well developed; in an undescribed taxon (represented by the "Gurlin Cav skull"), these are "winged," with conular cristae developed (Kielan-Jaworowska and Nessov, 1990).

In remaining taxa, lower molar trigonid cusps form a more acute angle, the distal metacristid of the lower molars is lost, and the cristid obliqua attaches to the back of the trigonid; these derived features (Fox, 1975) provide tentative evidence to group them into an unnamed unit. Of these, the Trinity therians *Holoclemensia* and *Pappotherium* were once placed within the Marsupialia and Eutheria, respectively (Slaughter, 1971). More recent studies (e.g., Turnbull, 1971; Butler, 1978; Kielan-Jaworowska et al., 1979; Jacobs et al., 1989) have considered the genera in question simply as

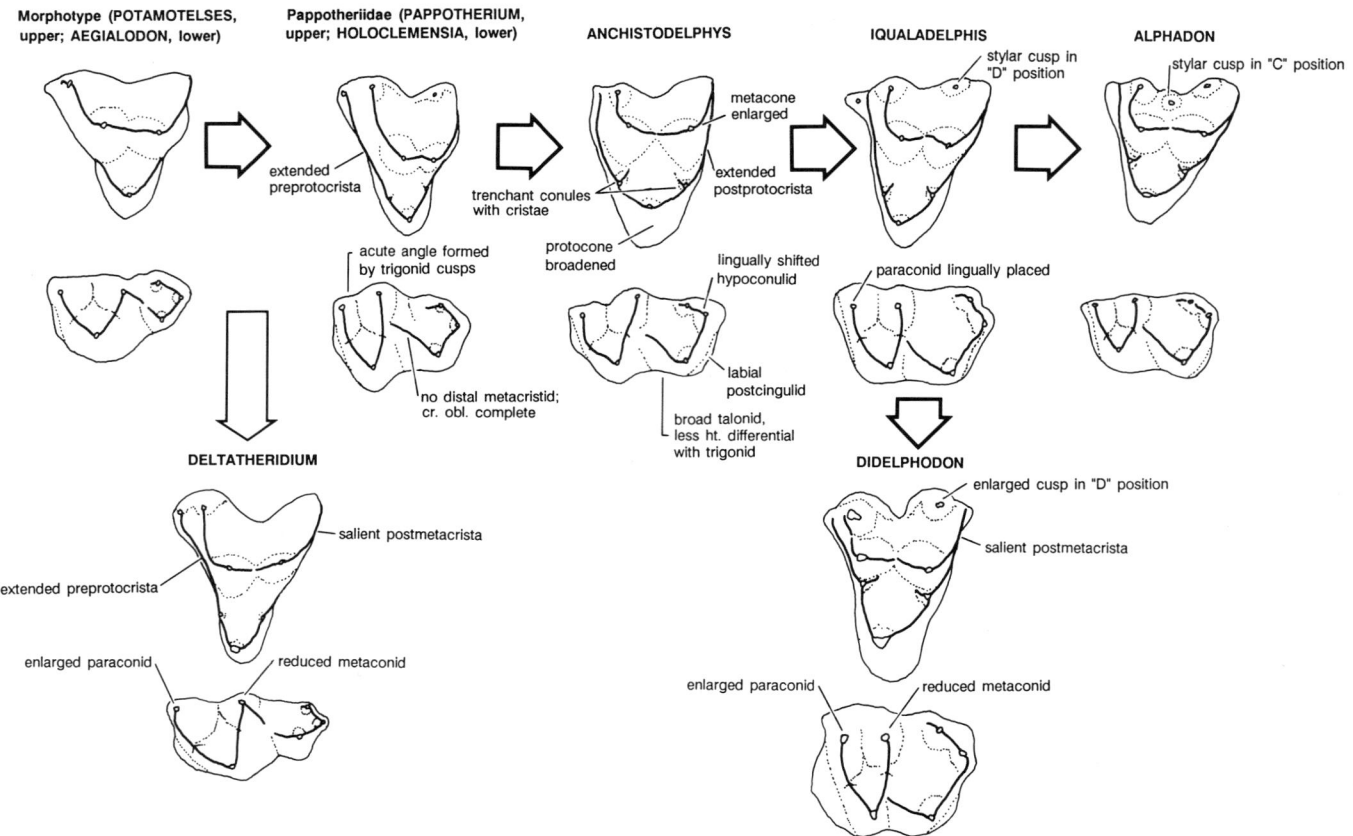

FIGURE 14.2. Hypothesized stages in the evolution of the marsupial molar pattern. Note the apparent independent acquisition of similar characters in Deltatheriidae (represented by *Deltatheridium*) and Stagodontidae (represented by *Didelphodon*).

therians of metatherian-eutherian grade, in the absence of unquestioned synapomorphies uniting them with either group of higher mammals. *Holoclemensia* and *Pappotherium* are herein excluded from the group including marsupials, eutherians, and several genera of uncertain affinities because they lack the following synapomorphies of that group: postprotocrista extending labially past the base of the metacone, providing for double-rank postvallum/prevallid shearing (Fox, 1975); and conules well-developed, labially placed, and bearing pre- and postcristae. Within this group, at least three Campanian taxa cannot be confidently grouped with either marsupials or eutherians. *Falepetrus*, from the Judithian (Clemens and Lillegraven, 1986), may be allied to an unnamed taxon described from the Aquilan (Fox, 1982), as both have the anterior part of the stylar shelf reduced, although this character appears elsewhere among the Tribosphenida (Cifelli, 1990d). Eutherians, which will not be dealt with here, are apparently known from as early as the ?Aptian or Albian of both Mongolia and Soviet central Asia (Kielan-Jaworowska and Dashzeveg, 1989), while unequivocal marsupials do not appear until the Cenomanian (Cifelli and Eaton, 1987; Eaton, 1990).

Dental criteria for defining the Marsupialia have been discussed extensively (e.g., Clemens, 1979; Fox, 1987a; Clemens and Lillegraven, 1986; Cifelli, 1990b, 1990c). Clemens and Lillegraven (1986) suggested that the primary specializations of marsupial molar teeth may involve a functional complex associated with protocone and metacone enlargement: On the upper molars, the metacone is enlarged and the protocone anteroposteriorly expanded; on the lowers, the paraconid and hypoconulid assume ligual positions, and a posterolabial cingulum is developed (see also Clemens, 1966). As with most mammalian dental characters, none of these is individually unique: several are present in some taxa that on other grounds cannot be reliably referred to the Marsupialia or Eutheria (*Zygiocuspis*, for instance, has an enlarged metacone and an anteroposteriorly developed protocone), and all are expressed, in one or more major taxa, among the Eutheria. Nonetheless, as an associated complex at least, these characters are restricted to marsupials in the context of Cretaceous Mammalia. Most undoubted Cretaceous marsupials also have stylar cusps in addition to the stylocone; these have generally been considered to be diagnostic of the group (Clemens, 1979). The PAUP analyses show that

Chapter 14. Theria of Metatherian-Eutherian Grade

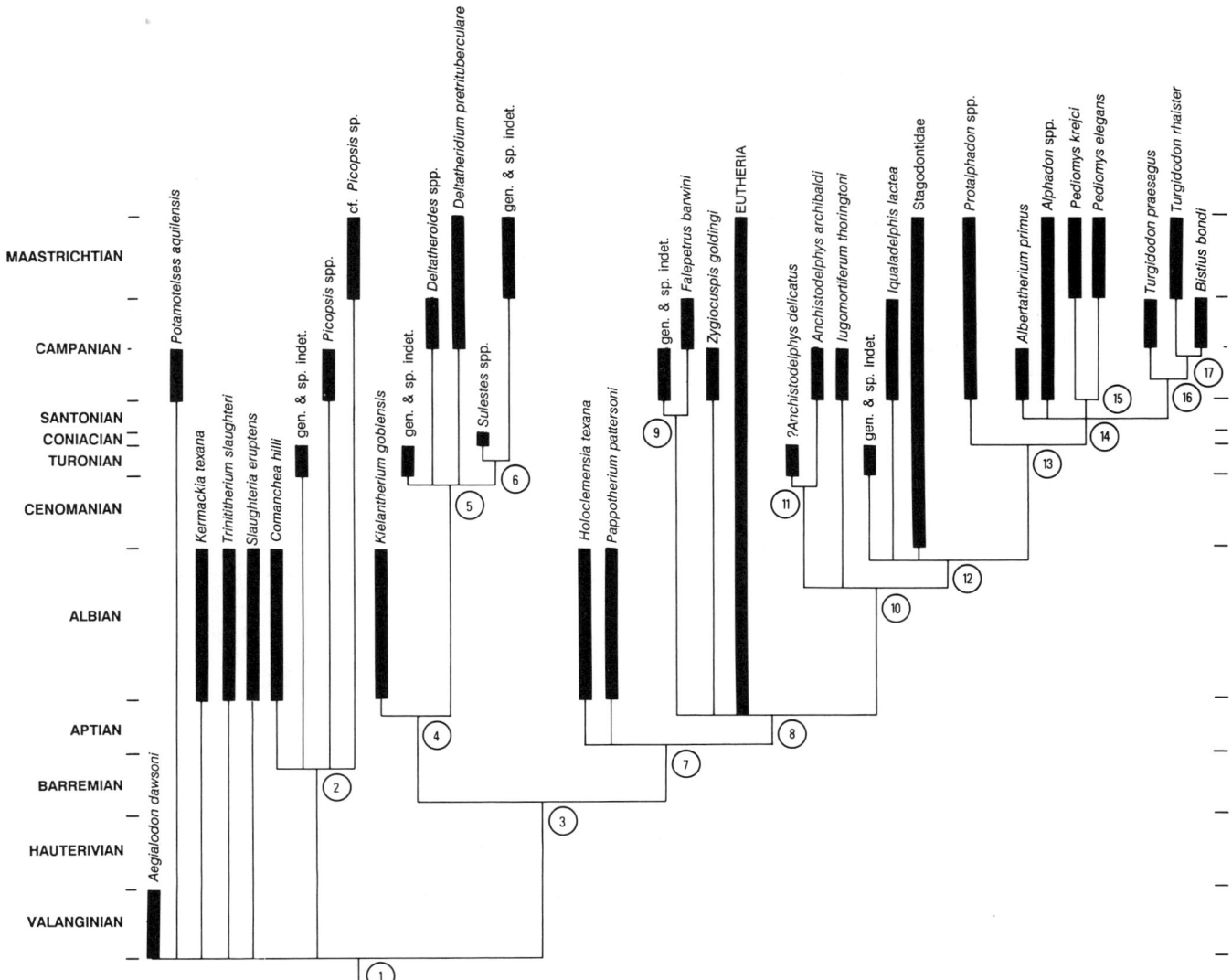

FIGURE 14.3. Temporal distribution and hypothesized relationships among therians of metatherian-eutherian grade, marsupial-like mammals, and selected Cretaceous Marsupialia. *Gallolestes*, tentatively considered a eutherian (Lillegraven, 1976), is excluded, as are the following taxa for which insufficient information was available: *Kumlestes* (Nessov, 1985a), "*Prodelttheridium*" (Kielan-Jaworowska and Nessov, 1990), *Hyotheridium* (Gregory and Simpson, 1926), and *Beleutinus* (Bashanov, 1972). Characters at nodes: (1) (Tribosphenida) upper molars with protocone, lower molars with multicusped, basined talonid; (2) (unnamed taxon) anterior part of stylar shelf reduced, with stylocone reduced or lacking; (3) (unnamed taxon) upper molars with preprotocrista extending labially past the paracone (double rank prevallum/postvallid shearing); (4) (unnamed taxon) postvallum/prevallid shearing emphasized (paraconid enlarged and metaconid reduced on posterior lower molars, at least; postmetacrista strong on upper molars), fourth molars reduced; (5) (Deltatheridiidae) premolars reduced to 3/3; (6) upper molar protocone broadened, conules well developed; (7) (unnamed taxon) distal metacristid lost, cristid obliqua attaches to back of trigonid on lower molars, trigonid cusps form more acute angle; (8) (unnamed taxon) postprotocrista of upper molars extends labially past base of metacone (double rank postvallum/prevallid shearing), conules well-developed, trenchant, and bearing cristae; (9) (unnamed taxon) anterior part of stylar shelf reduced; (10) (Marsupialia) ?premolars reduced to 3/3, upper molar protocone broadened, metacone enlarged; lower molars with labial postcingulid, hypoconulid lingually shifted ("twinned" with entoconid), talonid broadened, and reduced height differential between trigonid and talonid; (11) (*Anchistodelphys*) labial face of hypoconid sharply angular; (12) (unnamed taxon) stylar cusp consistently present in "D" position on upper molars, paraconid lingually placed on lower molars; (13) (unnamed taxon) stylar cusp variably present in "C" position on upper molars, upper molar protocone further broadened, metacone further enlarged, lower molars with talonid further broadened and further reduced height differential between trigonid and talonid; (14) (unnamed taxon) stylar cusp consistently present in "C" position on upper molars; (15) (Pediomyidae) anterior stylar shelf reduced on upper molars, lower molars with cristid obliqua having a labial attachment to trigonid; (16) (unnamed taxon) large size, molars and premolars with inflated cusps and with largely horizontal, apical wear; (17) characters 16 attain greater expression, protoconal cingulae usually present.

the sequence of acquisition of marsupial dental characters is ambiguous (Fig. 14.1E) when characters are simply treated as ordered (Wagner parsimony). However, if the assumption is made that a morphotype for Tribosphenida lacked cusps that can be reliably homologized with marsupial stylar cusps C and D (i.e., ancestral states are defined for these characters), then a sequence of character acquisition emerges (Figs. 14.1F, 14.2, 14.3). The distribution of stylar cusps among both tribosphenic and non-tribosphenic mammals (Clemens and Lillegraven, 1986) suggests that this assumption is well corroborated. *Anchistodelphys* and *Iugomortiferum* species lack or apparently acquired stylar cusps independently of marsupials, but have some other specializations of the group (such as an enlarged metacone, "twinning" of hypoconulid with entoconid, and development of a labial postcingulid) (Cifelli, 1990b, 1990c). Herein, I arbitrarily define the Marsupialia at the trichotomy of these genera with undoubted marsupials (Fig. 14.2), although other definitions are possible. Campanian *Iqualadelphis* (Fox, 1987a; Cifelli, 1990a), an unnamed Turonian genus (Cifelli, 1990b), and Stagodontidae are further derived in having a cusp consistently present in the "D" position on the stylar shelf; the positions of these taxa in the phylogeny corroborates the suggestion (Fox, 1987a, 1987b; Cifelli, 1990a, 1990c) that a "D" cusp was acquired earlier than a "C" cusp. Evidence grouping Stagodontidae specifically with any other taxon within the Marsupialia is weak, and for this reason their branching point is not fully resolved (Fig. 14.2). Stagodontids are of interest in having some specializations seen in Deltatheridiidae (Fig. 14.3). Kielan-Jaworowska and Nessov (1990) have recently advocated placement of deltatheridiid taxa within the Metatheria, as a separate sister taxon (order Deltatheroida) to Marsupialia. The basis for this assignment was a similarity to stagodontids (presumed to be primitive for marsupials; an assumption not supported by the analyses presented herein) in molar morphology and to all marsupials in dental formula and (especially) in having an alisphenoid bulla. A large alisphenoid contribution to an ossified auditory bulla is presumed to be an apomorphy of marsupials (Clemens, 1979); its presence in deltatheridiids depends on interpretation of the condition in an undescribed taxon represented by a skull from Gurlin Cav, Mongolia. Kielan-Jaworowska and Nessov (1990) interpreted this specimen to have an incipient alisphenoid bulla. Although a small posterior projection of the alisphenoid is present on the Gurlin Cav specimen, the significance of this projection and what it represents are not clear. When marsupial and deltatheridiid dental morphotypes are compared, no derived features are found in common, although some advanced deltatheridiids (such as the unnamed Gurlin Cav taxon) are similar to marsupials in features such as an anteroposteriorly expanded protocone. Regarding dental formula, the possession of four molars by both groups is unrevealing, insofar as Tribosphenida primitively may have possessed four molars (Dashzeveg and Kielan-Jaworowska, 1984; Clemens and Lillegraven, 1986). Reduction to three premolariform teeth in each jaw, a characteristic common to Late Cretaceous Deltatheridiidae and all known marsupials, is presumed to be an apomorphy within Tribosphenida (Clemens and Lillegraven, 1986). However, if *Kielantherium* is related to deltatheriids, as evidence considered herein suggests (Fig 14.2C, D), then reduction occurred independently in that group, because *Kielantherium* had four or more premolars yet already possessed presumed synapomorphies of Deltatheridiidae but not Marsupialia. Finally, if deltatheridiids are considered to be a sistergroup to Marsupialia, then reversal of a number of dental characters (Fig. 14.2, nodes 7, 8, 10) must be postulated in order to explain their absence in Deltatheridiidae. Thus, the balance of evidence does not favor a close relationship of deltatheridiids to marsupials.

For remaining North American Cretaceous marsupials, few systematically useful characters are known. Species of *Protalphadon* variably lack or have a small stylar cusp in the C position (Cifelli, 1990a, 1990c); this is reasonably interpreted as an intermediate state between absence and consistent presence (although it was not so coded in the data of Appendix 1). Remaining taxa under consideration (*Alphadon*, *Turgidodon*, *Albertatherium*, *Bistius*, *Pediomys*) have a consistently well-developed stylar cusp C, with the exception of some species of *Pediomys*. As Fox (1987b) has noted, if the absence of a C cusp in species such as *P. krejcii* is primitive and if its presence in other species (such as *P. elegans*) is a synapomorphy with taxa such as *Alphadon*, then polyphyly of the Pediomyidae and of the genus *Pediomys* is implied. However, all pediomyids share at least two synapomorphies: reduction of the anterior stylar shelf and a labial attachment of cristid obliqua to trigonid (Clemens, 1966, 1979). Thus, in the absence of any compelling evidence to the contrary, loss of a C cusp in certain pediomyids better explains the distribution of these characters (Fig. 14.2). Little can be said of relationships among remaining taxa, except that species of *Turgidodon* are united by synapomorphy (Cifelli, 1990a). *Bistius*, described as a tribothere (Clemens and Lillegraven, 1986), is tentatively placed within this clade of large, somewhat specialized marsupials.

Although relationships within Tribosphenida cannot be well-resolved on the basis of evidence in hand, some generalized discussion is possible. One notable aspect of the evolution of tribosphenic mammals is the "bushiness" of the Cretaceous radiations. This appearance partly results from the fact that relationships are not yet well understood, but also (and more importantly) from the fact that this radiation was complex, including a number of other archaic tribosphenidan taxa in addition to an early dichotomy between marsupial and eutherian clades. Consideration of the archaic taxa collectively under the term "therians of metatherian-eutherian grade" obscures the differences between them. One possible group of plesiomorphic taxa is tentatively recognized herein; another, including *Deltatheridium* and allies, is more securely documented. If *Kielantherium* is included in this second group, the clade was distinct since the Albian. Although mainly Asiatic in distribution, the group is represented in North America as early as the Turonian (Cifelli, 1990b); later appearance on the continent (Fox, 1974) may thus be explained by in situ evolution or by a subsequent immigration from Asia. Dental morphology, at least, does not favor a close relationship of deltatheridiids to marsupials. As has been previously recognized (e.g., Fox, 1975; Butler, 1978), at least two of the Trinity therians, *Holoclemensia* and *Pappotherium*, represent a stage in tribosphenic evolution more closely approximating a morphotype for Eutheria and Marsupialia; several other therians of metatherian-eutherian grade can be derived directly from such a morphotype. In terms of practical applicability to the Cretaceous record, no unambiguous synapomorphy of the dentition has yet been recognized for the Marsupialia, but a correlated complex of molar features distinguishes Mesozoic representatives from at least most of their contemporaries. Because these features are present in taxa (such as *Anchistodelphys*) that lack consistently present and readily homologized cusps on the stylar shelf of upper molars, it would appear that these cusps appeared later in marsupial evolution. A northern origin for marsupials has been suggested on the basis of their absence from known Cretaceous faunas of South America (Bonaparte and Kielan-Jaworowska, 1987). In addition to this negative evidence, the presence of a complex marsupial radiation in North America by the Cenomanian (Cifelli and Eaton, 1987; Eaton, 1990), the fact that a broad suite of primitive marsupials or marsupial-like mammals is known from the continent (Cifelli, 1990b, 1990c), and the hypothesized origin of all southern marsupials from an advanced *Alphadon*-like taxon (e.g., Marshall, 1987) together support a northern origin for the group.

Appendix: Character Distribution

Listed here is the character distribution among selected therians of metatherian-eutherian grade, marsupial-like mammals, and Cretaceous marsupials. Characters scored 0 indicate the presumed ancestral condition; missing data are scored ?.

1. Number of lower molars—4 (0) or 3 (1).
2. Number of premolars—4 or more (0) or 3 (1).
3. Upper molar protocone—lacking (0), small (1), somewhat expanded anteroposteriorly (2), or greatly expanded anteroposteriorly (3).
4. Lower molar talonid—small heel (0) or multicusped and basined (1).
5. Lower molar cristid obliqua—incomplete, with distal metacristid present (0) or complete, attaching to back of trigonid (1).
6. Attachment of cristid obliqua—below notch in metacristid (0) or labially placed, at base of protoconid (1).
7. Last molar—subequal in size to penultimate molar (0) or reduced to lost (1).
8. Postvallum/prevallid shearing—modestly developed (0) or strong, with paraconid enlarged, metaconid reduced, and salient postmetacrista (1).
9. Stylar cusp in "C" position on upper molars—absent (0) or consistently present (1).
10. Stylar cusp in "D" position on upper molars—absent (0) or present (1).
11. Stylar shelf labial to paracone—wide (0) or narrowed to absent (1).
12. Trigonid configuration on lower molars—open, with paraconid anteromedian in position (0); more acute, with paraconid more posteriorly located (1); or acute, with paraconid lingually placed (2).
13. Conules—weak or lacking (0); consistently present but lingually placed and without cristae (1); or strong, labially placed, and with winglike cristae (2).
14. Preprotocrista—does not (0) or does (1) extend labially past base of paracone (double rank prevallum/postvallid shearing).
15. Postprotocrista—does not (0) or does (1) extend labially around base of metacone (double rank postvallum/prevallid shearing).
16. Metacone—noticeably smaller than paracone (0), somewhat smaller than paracone (1), or subequal to or larger than paracone (2).
17. Hypoconulid—in posteromedian position (0) or lingually placed and "twinned" with entoconid (1).
18. Labial postcingulid—lacking (0) or present (1).
19. Hypoconid labial face—gently rounded (0) or sharply angular (1).
20. Premolar cusp form—sharp, uninflated (0) or inflated, with apical wear strongly developed (1).
21. Molar cusps—gracile, sharp (0) or robust, inflated (1).
22. Size—small, with maximum molar dimensions less than 3 mm (0), or large (1). Generic abbreviations are given in the legend to Figure 14.1.

Potamotelses	0?1100?000000000000?00
Deltatheridium	0111001100001100000000
Deltatheroides	0111001100001?00000?00
Sulestes	??2100?100001101000?00
Slaughteria	???110?0???1????000000

Trinititherium	???100?0???0????000?00
Kermackia	???100?0???0????000?00
Kielantherium	00?10011???0????000000
Aegialodon	???100?0???0????000?00
Holoclemensia	??1110?011011100000?00
Pappotherium	??1110?001011100000?00
Picopsis	??1100?010100000000?00
Falepetrus	??2????0001?2111????00
Zygiocuspis	??2????0000?2112????01
Eutheria	1011100000012110000000
A. delicatus	??2110000002211111?00
A. archibaldi	??2110000002211111?00
Iugomortiferum	??3110?010012112110?10
new genus	??2110?001022111110?00
Iqualadelphis	??2110000102211110000
Didelphodon	0121100101022112110101
P. krejcii	0131110001122112110000
P. wahweapensis	0?31100001022112110000
Albertatherium	??3110001112211110?00
A. marshi	0131100011022112110000
P. elegans	0131110011022112110000
T. praesagus	0131100011022112110111
T. rhaister	0131100011022112110111

ACKNOWLEDGMENTS. I thank Drs. William A. Clemens, Jason A. Lillegraven, and, especially, Zofia Kielan-Jaworowska for various information and fruitful discussions. This research was supported by NSF grant BSR 8906992.

CORRESPONDENCE ADDRESS. Richard L. Cifelli, Oklahoma Museum of Natural History and Department of Zoology, University of Oklahoma, Norman, OK 73019, USA.

References

Bashanov, V.S. 1972. First Mesozoic Mammalia (*Beleutinus orlovi* Bashanov) from the USSR. *Teriologia, Akad. SSSR, Sib. Otd.* 1:74–80 (in Russian; cited from Kielan-Jaworowska et al., 1979).

Bonaparte, J.F., and Kielan-Jaworowska, Z. 1987. Late Cretaceous dinosaur and mammal faunas of Laurasia and Gondwana. In: *Fourth symposium on Mesozoic terrestrial ecosystems* (Currie, P.J., and Koster, E.H., eds.). Occ. Pap. Tyrrell Mus. Palaeontol. 3:24–29.

Bonaparte, J.F., and Pascual, R. 1987. Los mamíferos (Eotheria, Allotheria y Theria) de la Formación Los Alamitos, Campaniano de Patagonia, Argentina. *IV Congr. Latinoamericano Paleontol., Bolivia* 1:361–378.

Butler, P.M. 1978. A new interpretation of the mammalian teeth of tribosphenic pattern from the Albian of Texas. *Breviora* 446:1–27.

Cifelli, R.L. 1990a. Cretaceous mammals of southern Utah. I. Marsupials from the Kaiparowits Formation (Judithian). *J. Vert. Paleontol.* 10:295–319.

Cifelli, R.L. 1990b. Cretaceous mammals of southern Utah. III. Therian mammals from the Turonian (early Late Cretaceous). *J. Vert. Paleontol.* 10:332–345.

Cifelli, R.L. 1990c. Cretaceous mammals of southern Utah. II. Marsupials and marsupial-like mammals from the Wahweap Formation (early Campanian). *J. Vert. Paleontol.* 10:320–331.

Cifelli, R.L. 1990d. A primitive higher mammal from the Late Cretaceous of southern Utah. *J. Mammal.* 71:343–350.

Cifelli, R.L., and Eaton, J.G. 1987. Marsupial mammal from the earliest Late Cretaceous of western US. *Nature* 325:520–522.

Clemens, W.A. 1966. Fossil mammals of the type Lance Formation, Wyoming. Part. II Marsupialia. *Publ. Geol. Sci., Univ. California* 62:1–122.

Clemens, W.A. 1979. Marsupialia. In: *Mesozoic mammals: The first two-thirds of mammalian history* (Lillegraven, J.A., Kielan-Jaworowska, Z., and Clemens, W.A., eds.). Berkeley: University of California Press, pp. 192–220.

Clemens, W.A., and Lillegraven, J.A. 1986. New Late Cretaceous North American advanced therian mammals that fit neither the marsupial nor eutherian molds. *Contrib. Geol. Univ. Wyoming, Spec. Paper* 3:55–85.

Clemens, W.A., Lillegraven, J.A., Lindsay, E.H., and Simpson, G.G. 1979. Where, when, and what-A survey of known Mesozoic mammal distribution. In: *Mesozoic mammals: The first two-thirds of mammalian history* (Lillegraven J.A., Kielan-Jaworowska, Z., and Clemens, W.A., eds.). Berkeley: University of California Press, pp. 7–58.

Crompton, A.W. 1971. The origin of the tribosphenic molar. In: *Early mammals* (Kermack, D.M., and Kermack, K.A., eds.). *Zool. J. Linn. Soc.* 50(Suppl. 1):65–87.

Dashzeveg, D., and Kielan-Jaworowska, Z. 1984. The lower jaw of an aegialodontid mammal from the Early Cretaceous of Mongolia. *Zool. J. Linn. Soc.* 82:217–227.

Eaton, J.G. 1990. Therian mammals of the Cenomanian (Late Cretaceous) Dakota Formation, southwestern Utah. *J. Vert. Paleontol.* 10:21A.

Fox, R.C. 1971. Marsupial mammals from the early Campanian Milk River Formation, Alberta, Canada. In: *Early mammals* (Kermack, D.M., and Kermack, K.A., eds.). *Zool. J. Linn. Soc.* 50(Suppl. 1):145–164.

Fox, R.C. 1972. A primitive therian mammal from the Upper Cretaceous of Alberta. *Canadian J. Earth Sci.* 9:1479–1494.

Fox, R.C. 1974. *Deltatheroides*-like mammals from the Upper Cretaceous of North America. *Nature* 249:392.

Fox, R.C. 1975. Molar structure and function in the Early Cretaceous mammal *Pappotherium*: Evolutionary implications for Mesozoic Theria. *Canadian J. Earth Sci.* 12:412–442.

Fox, R.C. 1976. Additions to the mammalian local fauna from the upper Milk River Formation (Upper Cretaceous), Alberta. *Canadian J. Earth Sci.* 13:1105–1118.

Fox, R.C. 1980. *Picopsis pattersoni*, n. gen. and sp., an unusual therian from the Upper Cretaceous of Alberta, and the classification of primitive tribosphenic mammals. *Canadian J. Earth Sci.* 17:1489–1498.

Fox, R.C. 1982. Evidence of new lineage of tribosphenic

therians (Mammalia) from the Upper Cretaceous of Alberta, Canada. *Geobios, Mém. Spéc.* 6:169–175.

Fox, R.C. 1987a. An ancestral marsupial and its implications for early marsupial evolution. In: *Fourth symposium on Mesozoic terrestrial ecosystems* (Currie, P.J., and Koster, E.H., eds.). Occ. Pap. Tyrrell Mus. Palaeontol. 3:101–105.

Fox, R.C. 1987b. Palaeontology and the early evolution of marsupials. In: *Possums and opossums: Studies in evolution* (Archer, M., ed.). Sydney: Surrey Beatty & Sons and the Royal Zoological Society of New South Wales, pp. 161–169.

Gregory, W.K., and Simpson, G.G. 1926. Cretaceous mammal skulls from Mongolia. *Amer. Mus. Novitates* 225:1–20.

Jacobs, L.L., Winkler, D.A., and Murry, P.A. 1989. Modern mammal origins: Evolutionary grades in the Early Cretaceous of North America. *Proc. Nat. Acad. Sci.* 86:4992–4995.

Kermack, K.A., Lees, P.M., and Mussett, F. 1965. *Aegialodon dawsoni*, a new trituberculosectorial tooth from the lower Wealden. *Proc. Royal Soc. London* B162:535–554.

Kielan-Jaworowska, Z. 1975. Evolution of the therian mammals in the Late Cretaceous of Asia. Part I. Deltatheridiidae. *Palaeontol. Polonica* 33:103–132.

Kielan-Jaworowska, Z. 1982. Marsupial-placental dichotomy and paleogeography of Cretaceous Theria. In: *Paleontology, essential of historical geology* (Gallitelli, E.M., ed.). Modena: S.T.E.M. Mucci, pp. 367–383.

Kielan-Jaworowska, Z., and Dashzeveg, D. 1989. Eutherian mammals from the Early Cretaceous of Mongolia. *Zool. Scripta* 18:347–355.

Kielan-Jaworowska, Z., Eaton, J.G., and Bown, T.M. 1979. Theria of metatherian-eutherian grade. In: *Mesozoic mammals: The first two-thirds of mammalian history* (Lillegraven, J.A., Kielan-Jaworowska, Z., and Clemens, W.A., eds.). Berkeley: University of California Press, pp. 182–191.

Kielan-Jaworowska, Z., and Nessov, L.A. 1990. On the metatherian nature of the Deltatheroida, a sister group of the Marsupialia. *Lethaia* 23:1–10.

Lillegraven, J.A. 1976. A new genus of therian mammal from the Late Cretaceous "El Gallo Formation," Baja California, Mexico. *J. Paleontol.* 50:437–443.

Marshall, L.G. 1987. Systematics of Itaboraian (middle Paleocene) age "opossum-like" marsupials from the limestone quarry at São José de Itaboraí, Brazil. In: *Possums and opossums: Studies in evolution* (Archer, M., ed.). Sydney: Surrey Beattey & Sons and the Royal Zoological Society of New South Wales, pp. 91–160.

Nessov, L.A. 1985a. New Cretaceous mammals of the Kizylkum Desert. *Vest. Leningradskogo Univ. Geol. Geograf.* 17:8–18 (in Russian).

Nessov, L.A. 1985b. Rare osteichthyans, terrestrial lizards, and mammals from the Cretaceous estuaries and coastal plain zone of the Kizylkum Desert. *Ezheg. Vses. Paleontol. Obshch.* 30:199–215 (in Russian).

Patterson, B. 1956. Early Cretaceous mammals and the evolution of mammalian molar teeth. *Fieldiana: Geol* 13:1–105.

Prothero, D.R. 1981. New Jurassic mammals from Como Bluff, Wyoming, and the interrelationships of nontribosphenic Theria. *Bull. Amer. Mus. Nat. Hist.* 167:277–326.

Reig, O., Kirsch, J.A., and Marshall, L.G. 1987. Systematic relationships of the living and neocenozoic American "opossum-like" marsupials (suborder Didelphimorpha), with comments on the classification of these and of the Cretaceous and Paleogene New World and European metatherians. In: *Possums and opossums: Studies in evolution* (Archer, M., ed.). Sydney: Surrey Beatty & Sons and the Royal Zoological Society of New South Wales, pp. 1–89.

Reshetev, V., and Trofimov, B.A. 1984. Review of the study of fossil mammals from the USSR. In: *Theriology in the USSR* (Solokov, V.E., and Kucheruk, V.V., eds.), pp. 6–29 (In Russian; cited from Kielan-Jaworowska and Nessov, 1990).

Simpson, G.G. 1936. Studies of the earliest mammalian dentitions. *Dental Cosmos* 1936:1–24.

Slaughter, B.H. 1968a. Earliest known marsupials. *Science* 162:254–255.

Slaughter, B.H. 1968b. Earliest known eutherian mammals and the evolution of premolar occlusion. *Texas J. Sci.* 20:3–12.

Slaughter, B.H. 1971. Mid-Cretaceous (Albian) therians of the Butler Farm local fauna, Texas. In: *Early mammals* (Kermack, D.M., and Kermack, K.A., eds.). *Zool. J. Linn. Soc.* 50 (Suppl.):131–143.

Swofford, D. 1989. *PAUP: Phylogenetic Analysis Using Parsimony, version 3.0*. Champaign: Illinois Natural History Survey.

Turnbull, W.D. 1971. The Trinity therians: Their bearing on evolution in marsupials and other therians. In: *Dental morphology and evolution* (Dahlberg, A.A., ed.). Chicago: University of Chicago Press, pp. 151–179.

CHAPTER 15

Metatherian Taxon Phylogeny: Evidence and Interpretation from the Cranioskeletal System

FREDERICK S. SZALAY

Overview

Well-supported ancestral points in time, or nodes of branchings on trees, should be diagnosed by those apomorphies of the group's ancestor that are retained (synapomorphies), or by inference based on their traces. These apomorphic constraints of the last common ancestor are both often discovered through, or should be tested against, causal research. Paraphyletic groups are as diagnosable phylogenetically (from the time of their origin) as holophyletic ones. Although holophyla can be desirable goals for larger clusters of species that share a common complex organizational pattern (Aves, Mammalia, etc), all organisms cannot be accommodated that way in an inclusive classificatory (practical) hierarchy. In addition, the incomplete understanding of evolutionary causality, or the lack of resolution of primitive groups, often makes paraphyla not only natural, but also useful and justifiable monophyletic groupings. The phylogeny and classification of organisms should be tested against a number of taxonomic properties of the groups, which, in turn, are based on paleontological and biological facts, and interpretation of the available evidence.

Archaic metatherians pose a phylogenetic problem in spite of their increasingly rich dental record. The stasis of ancient characters, or of the entire molar crown morphology, necessitates functional-adaptive exploration of teeth and all other available features. In addition to historical and developmental constraints physical properties of diet drive the evolution of teeth, and omnivory strongly buffers selectional forces for change. The adaptive context, therefore, should be considered in the analysis of such issues as, for example, paracone stunting, stasis of traits, convergences, and parallelisms in molar form in archaic ameridelphians. Several recently proposed phylogenies are not concordant with the results of biologically constrained character analyses of teeth, skulls, and selected

Contents

Introduction, 217
Methodological Issues Related to Taxon Phylogeny, 217
Aspects of the Cranioskeletal System, 219
    Dental evidence, 220
    The ear region, 225
    Carpal, tarsal, and other postcranial evidence, 226
Metatherian Taxa and Phylogeny, 227
Classification, 239
Acknowledgments, 240
References, 240

areas of postcranial evidence. Studies of hard anatomy underline the importance of evolutionary morphology, i.e., a perspective and research protocol based on cause and mechanism-oriented character analysis.

The cruropedal evidence in fossil and living metatherians represents a complex homologous area that stretches from the Campanian (Cretaceous) to the Recent. It affords opportunities for character analysis grounded in evolutionary morphology. Traits that occur in diversely adapted living species are the strongest support for the retention of apomorphies of the last common ancestor—synapomorphies of such groups. From a perspective that constructs narrative temporal and causal accounts for the evolution of characters, independent of an assumed taxic framework supplied by a chosen cladistic (i.e., taxic) outgroup, a brief summary of higher taxa and an interim phylogeny of metatherians are presented, monophyletic groups are diagnosed, and this diversity, in addition to the still poorly known but convincingly metatherian order Deltatheroida (see Kielan-Jaworowska and Nessove, 1990), is classifed in three other orders.

Introduction

In another volume (Szalay, 1993), I attempt, in detail, a new synthesis of metatherian (marsupial in a vernacular sense) phylogeny as well as try to show the central role of form-functional, ecological, and paleontologically oriented analyses of selected features in phylogenetics. Here I confine myself to presenting (phylogenetic) diagnoses (in the sense of Szalay et al., 1987, which is similar to that proposed by Rowe, 1987, except for use of both type of monophyletic taxa) of a number of selected, hypothesized monophyletic groups. I emphasize that the analysis of character complexes, molecular evidence included, should be carried out on their own biologically understood merit without the convenient but unwarranted assumptions of parsimonious distribution schemes driven by the use of taxic cladistic outgroups (see Szalay and Bock, 1991). Evolutionary decision making about character polarities should be supported by some causal arguments (historical-narrative explanations; often pejoratively labeled "stories" or "scenarios"). The transformational understanding of one character complex should not be influenced by the polarity decisions relating to others. I consider the recent phylogeny of metatherians proposed by Reig et al. (1987), Marshall (1987), and Marshall et al. (1990), like that of other groups supported primarily by distribution analysis of character matrices, potentially flawed because of their approach to the delineation of characters. Characters are usually selected and atomized, without an attempt to establish them as robust and therefore reliable (probable) taxonomic properties. The cladogenies proposed are consequently not probability based, but reflect a parsimony and taxic outgroup dictated ordering of these characters. An evolutionary causality oriented appraisal of selected dental, cranial, and postcranial features supports the interim phylogeny proposed here, and in Szalay (1993).

The acronyms (in bold face) in the figures and in the text stand for anatomical abbreviations indicated in Table 15.1.

Methodological Issues Related to Taxon Phylogeny

Although the language used in phylogenetic studies is taxon bound, I believe that it is important to maintain the distinction between the understanding of history of *features*, of organisms (species), and the delineation of *taxa* by taxonomists. Delineation of apomorphies for the ancestry of a group does not insure its holophyletic status, nor should it. A lack of such distinction largely characterizes cladistic theory and its subsequent practice. Conflation of these conceptually distinct endeavors is the result of a taxonomic theory–based view of evolution, which has the unfortunate consequence, ideologically driven, of a nearly complete neglect of character analysis in large segments of the systematic community. The Kuhnian paradigm of modern, post-Hennigian (evolutionary) cladistics (conceptually the least clearly enunciated form of panholophylism), with its focus on the mapping of characters onto cladograms and on their weighting, by congruence, as pattern cladistics, has sidestepped the very activities that not only produce robust characters but are necessary for causally assessing factors in these distributions (Bock, 1977a, 1977b, 1981; Szalay, 1977, 1981; Szalay and Bock, 1991; Szalay, this volume, chapter 9). This is certainly not a criticism of character mapping, an exercise that all comparative biologists who reconstruct phylogeny conduct. Yet I believe that taxonomists engaged in phylogenetics should profoundly examine the theoretical validity of some of the procedures of "character" recognition and selection and the taxic, formal, cladistic outgroup-related evaluation of them, particularly the general neglect of the understanding of the evolution of homologies (i.e., the various states of homologous features). As emphasized elsewhere, phylogenetically diagnosable groups are not necessarily holophyletic.

Figure 15.1 is a simple model of evolutionary transformation, through time, of lineages, consisting of organisms. There is a continuity of organisms; note that this is not a cladogram. Cladograms always depict discontinuous "units" (taxa) of phylogeny, and therefore never hypotheses of actual phylogeny. This approach emerged from theory based on taxa, and not the evolution of lineages. If punctuated equilibrium, as an expression of underlying mechanisms related to speciation, were corroborated in living organisms (which it is not), then the model shown here would not be a theoretically acceptable representation. All lines represent continuous populations, i.e., lineages and not taxonomic species, which potentially leave phena as fossils. Splittings (simplified) of the lineage on right (heavy line) which occurred are represented in the figure by numbers 1–5. Thin horizontal arrows (a–d) point to instants in the lineage when detectable (i.e., tested) changes in characters happened. These are the first manifestations of new characters, which are the bases for anything we do in taxonomy-based evolutionary history (phylogeny), unless somebody was there watching and recording the choreography of events and the accompanying changes. The latter case would be a real chronicle. Such real accounts, i.e., chronicles, are in opposition to the concept of *chronicle* redefined by O'Hara, 1988, who advocates the use of cladistic genealogies as being equivalents to a concept of chronicle on which "history" should be based in the pursuit of research in evolutionry biology.

Empirical methods can test only those (monophyletic) taxa that are diagnosed by the apomorphies of the last common ancestor at these points and not at the

TABLE 15.1. Abbreviations for morphology

Bones, joints, and joint facets

ACJ	astragalocuboid joint
ACu	astragalocuboid
AFi	astragalofibular
AN	astragalonavicular
ANJ	astragalonavicular joint
As	astragalus
ATa	astragalotibiale
ATi	astragalotibial
ATia	anterior astragalotibial
ATid	distal astragalotibial
ATil	lateral astragalotibial
ATim	medial astragalotibial
ATip	posterior astragalotibial
Ca	calcaneus
CaA	calcaneoastragalar
CaAd	distal calcaneoastragalar
CaCu	calcaneocuboid
CaCua	auxilliary (australidelphian) calcaneocuboid
CaCud	distal calcaneocuboid
CaCul	lateral calcaneocuboid
CaCum	medial calcaneocuboid
CaCup	proximal calcaneocuboid
CaFi	calcaneofibular
CaMt5l	calcaneal-**Mt5** ligament
CCJ	calcaneocuboid joint
CLAJP	continuous lower ankle joint pattern
CNJ	calcaneonavicular joint
Cu	cuboid
Ec	ectocuneiform
EMt1	entocuneiform-first metatarsal
EMt1J	entocuneiform-**Mt1** joint
EMt1l	lateral **EMt1**
EMt1m	medial **EMt1**
En	entocuneiform
Fe	femur
FFJ	femorofibular joint
Fi	fibula
LAJ	lower ankle joint
Mc	mesocuneiform
Mt	metatarsal
Na	navicular
NaCu	naviculocuboid
NCJ	naviculocuboid joint
Oc	os calcis
Ph	prehallux
SLAJP	separate lower ankle joint pattern
Su	sustentacular
Ta	tibiale
TF	tibiofibular
TFJ	tibiofibular joint, distal
Ti	tibia
TiFi	distal tibiofibular
TMTJ	tarsometatarsal joint
TTJ	transverse tarsal joint
UAJ	upper ankle joint

Topographical bony details, ligaments and tendons, and anatomical directions

ac	astragalar canal
adt	astragalar distal tuber
ampt	astragalar medial plantar tuberosity
at	anterior plantar tubercle
cflf	calcaneofibular ligament facet
cump	cuboidal medial process
gtpl	groove for tendon of peroneus longus

TABLE 15.1. *Continued*

lu	lunula
mc	meniscus
pfp	parafibular process
pp	peroneal process of calcaneus
ppl	process for peroneus longus on **Mt1**
ps	posterior shelf of distal tibia
sa	sulcus astragali
sc	sulcus calcanei
tc	tuber of calcaneus

Abbreviations used both in the text and on the figures for the cruropedal evidence. Specific joints are abbreviated by the combination of the first letters in capitals of the names of those units that contribute to the joint, and the letter *J* for joint. Abbreviations entirely in lower case designate landmarks of specific bones, anatomical directions, or muscles. Abbreviations are listed under two separate headings in order to facilitate retrieval of information; characters abbreviate in this table are specific homology designations rather than only topographical descriptive terms.

furcations which in this model show no character change.

A holophyletic "crown group" could be delimited at nodes 1–5 if one had genuine apomorphies at those nodes. At all other subgroupings below these nodes ("stem groups, "plesia," and other designations), no matter how the cut is made, will result in paraphyletic taxa. Cladistics, which centers on holophyletic classifications (demanding an impossible taxonomic representation of the real world), collapses internodal populations and their evolution and produces and idealized representation of the evolutionary history of the taxa, rather than an account of species through time (i.e., lineages). The fact that taxa, unlike lineages, must be artificially restricted below a "crown group" is ignored (even when the untestable evolutionary species concept is used); the system never grapples with the reality (i.e., the naturalness; as much as artificial delineations such as taxa can be so designated) of paraphyla. The product of the system, isomorphic taxon-cladograms (**taxograms**), is not a real abstraction, but a subtle and fundamental distortion of what monophyletic taxa are capable of summarizing about phylogenetic research.

Notice that on the figure monophyletic groups (such as the one which may be designated as holophyletic at point **a**, and those which are then included as three paraphyletic ones at points **a** to **b**, **b** to **c**, and **c** to **d**) are all phylogenetically diagnosable. All these four monophyletic taxa are natural, testable, taxic segments of the phyletic continuum and its splittings. The ontology of the biological species concept and of lineages (and their approxmiate application through fossils) is available to testing, whereas the phylogenetic species concept, defined by lineage furcation, is not. But in a "real world" (phylogenetic cum evolutionary) framework, the inclusively hierarchic taxonomic representation of evo-

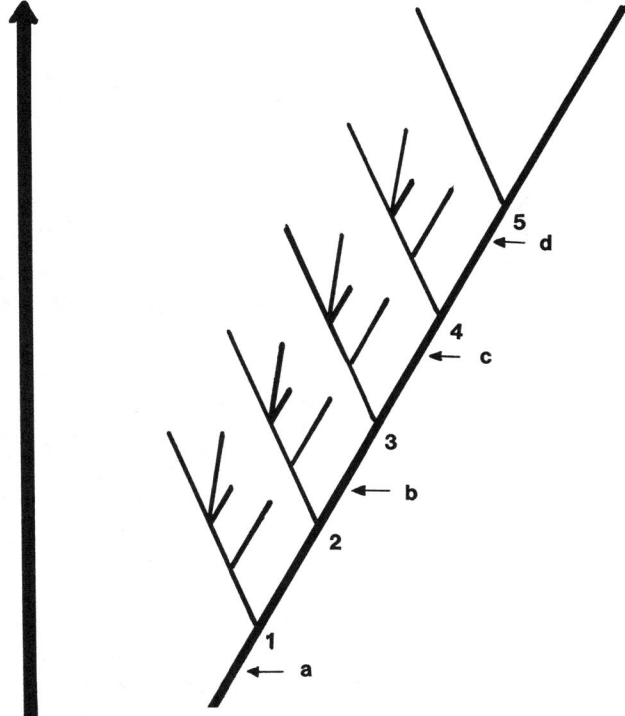

FIGURE 15.1. A model of evolutionary transformation of lineages, but note, not of specific taxa. Delineation of taxa, once phylogenetic relationships have been reconstructed, requires the appropriate data (epistemology), which do not necessarily come into existence at splitting events. Such a model is derived from the understanding and application of evolutionary biology, and not of cladistic taxonomic theory. Vertical arrow is time. See further discussion of this under methodological issues related to taxon phylogeny in text.

lutionary change shown in Fig. 1, below the first diagnosable holophyletic group must be a paraphyletic unit, and so must the others below that, and so on (tortured definitions of formal stem group and plesion concepts notwithstanding). Although it is helpful to know that Life, Aves, and Mammalia are holophyletic, included groups of practical taxonomy within these taxa must be paraphyletically accommodated if they are to be tested and corroborated groups (see Szalay and Bock, 1991). Construction and tests of monophyletic groups and an understanding of their relationships, the aim of phylogenetics, is independent of formally naming or ranking taxa. Because there can not be a *nonarbitrary* theory in classification that is based on evolutionary realities, delineating and ranking monophyletic groups should be carried out with the strictures of testability, the compromises of nomenclatorial tradition, as well as some aim of general usefulness in mind. But there is no known and confidence-generating test in cladistics to determine the specific nature of monophyly of most living species and speciose genera, beyond those formalisms that merely designate endless unanalyzed minutiae as "synapomorphies." The same applies to samples of many suspected lineages, properly named as taxa—there simply is no epistemology available to decide (Szalay and Bock, 1991). Holophyla came from somewhere, and paraphyla, therefore, as shown in Figures 15.1 and 15.15, are unavoidable in any classification that attempts to reflect phyletic events.

To sum up, the monophyletic groups of metatherians summarized below, assuming that the assessments of character transformations are correct, stem from a single common ancestor. The groups are diagnosed, based on the putative and tested apomorphies of that hypothesized first species of the group compared to its ancestry. The fossil record and the causal understanding of complete morphoclines of characters of all sorts, from all aspects of the organisms, supply the only lasting and noncircular foundations of evolutionary history, and allow escape from a perpetual cladistic maze (to use the phrase of Murray et al., 1987). A theoretically coherent phylogenetic method must arise from a complete theory of phylogeny (evolutionary history) that includes chronistics, phyletics, cladistics and phenetics (i.e., the appropriate use of distance measures). Particularly vertical (phyletic), and not only horizontal (cladistic), comparisons of character states (Bock, 1977a) are critical methodological components for deciding on probabilities of character and group relationships. Although algorithms can be very useful techniques to organize data, as widely recognized, the PAUP, Hennig 86, and numerous other algorithms based on parsimony are flawed methods of choice when they are used to generate phylogenies—they are not the theoretically valid methods of taxonomy. Methods must fall out of tested theory. In the particular case of earth history, the theoretical base is an appreciation of the older "rules of correspondence" (of Hutton and others), and also the numerous modern ones which have grown out of plate tectonic theory and its complex foundations. In the case of phylogenetics, it is evolutionary theory (as it accounts for the change and division of lineages) and the multitudes of causal mechanisms that operate within organisms and populations that form the foundations of the method. As they become uncovered, new mechanisms that are responsible for ontogenetic unfolding will undoubtedly strongly influence the probability of inferences in character analysis.

Aspects of the Cranioskeletal System

Rarely is there an adaquate or well-ordered modern comparative account of the musculoskeletal morphology of most living species of mammals. Similarly rare are those dealing with the evolutionary morphology of selected areas of the skeleton. It is not surprising, therefore, that there are relatively few standards of comparisons which would guide the process of integration and interpretation of the fossils. The results are often characters described in an uninterpretive framework of anatomy, akin to static, process-purged, descriptions of land forms. Although such a view of many morphological studies may appear exaggerated to some, it is a fact that the general usage of the term "morphology" refers to the most active sphere of activity within phylogenetics. In fact anatomy (and data derived from it) is, mistakenly, the most often encountered implied synonym of morphological analysis.

One of the critical areas of research for understanding significant aspects of bone, joint, and tooth morphology (the relative absence of which is often lamented in the literature) is the specific study of all known species of marsupials—their behavioral and ecological morphology (a very modern, but little practiced, natural history), particularly positional (postural and locomotor) and feeding behaviors, as well as dietary regimes. Without continued explanation of the evidence of the teeth and bones in the living, with their potential to bridge the time spans of phylogeny, evolutionary understanding is likely to stagnate or merely continue to focus on "consistensies" of taxic schemes. A critical aspect of character analysis is the functional-adaptive analysis of interrelated attributes of organisms, simply because the ecology and naturalistic behavior of the animals is largely causal in the modification of their genotype-based morphology and behavior. Unless such studies are initiated, the near certain wave of approaching faunal extinctions will further deprive natural history of its most powerful explanatory base for phylogeny.

Fossil morphology, of course, cannot be fully appreciated as yet by any of us until the relevant living homologous systems have been subjected to analysis, and until this perspective is transferred by the approach of evolutionary morphology (Bock, 1990) and taxonomy to the fossil evidence.

DENTAL EVIDENCE. When evolutionary change in a character complex is limited, particularly when this relative paucity is spread over enormous periods of time, the problems surrounding evolutionary analysis are much more difficult to surmount than when the change is very rapid, resulting in diversity. This is exactly the problem that faces students of early metatherians (deltatheroidans and ameridelphians) who, relying necessarily in large part on dental evidence, attempt to determine the early ancestral-descendant and branching sequences of metatherian evolution. Because of the early dental record of the Metatheria we have a good understanding of their taxonomic diversity (the extent to which they have diverged, i.e., diversified, based on distance appraisal), not much of their multiplicity (of species and lineages), perhaps a modicum of insight into their feeding diversity, but a very incomplete understanding of their phylogenetic relationships (see also Cifelli, this volume, Chapter 14).

The temptation for mammalian paleontologists to nearly always regard teeth as primary, is based on many sources, and it is understandable given the nature of the fossil record. Teeth are more abundantly represented than skulls or complex bits of postcrania such as parts of the carpus or tarsus, and they appear to evolve more rapidly. Mainly, but not wholly, availability is the reason for greater focus on the dental record, compared to the other potentially rich storehouses of characters in the extant forms. Nevertheless, a tradition from the important practice of stratigraphy, where teeth are great designators of "species-genus" level taxonomy (morphospecies usually being the formally recognized alpha clusters of the fossil record), also carries with it a not always positive bias for assessing phylogenetics. In terms of their form-function, cheek teeth, in general, are well understood. What is still lacking, however, are the ecomorphological studies—in this particular case, of living didelphids, caenolestids, and to lesser extent of sundry australidelphians. The nature of dental differences and similarities of most Cretaceous and Paleogene marsupials and their influence on attempts to understand the early evolution of marsupials are obvious from titles or designations alluded to as "opossum-like" in many contributions. Nevertheless there is confusion from the very difficult past and current array of taxonomic allocations of ameridelphian genera to the Pediomyidae, Peradectidae, Didelphidae, Microbiotheriidae, or even to the Borhyaenidae, by many outstanding taxonomists of marsupials. For example, concepts of *Alphadon* and *Pediomys* (e.g., Fox, 1987; Aplin and Archer, 1987) are still in disarray because of the extreme problems of evaluating similar phenetic clusters. Clearly, the difficulty of separating ancestral from derived patterns has reached a particular zenith in mammal studies with the dentition of the "opossum-like" marsupials. What has emerged from the broad survey of dental and tarsal characters of these taxa is that, while clustering of teeth and grouping of dental attributes are feasible, it is more than likely that a few simple functional and adaptive changes have repeatedly evolved among the amazingly bradytelic molar patterns of early and some living ameridelphians. Figures 15.2–15.4 reveal how narrow the limits of form are within which we characterize selected Cretaceous and Paleogene genera—and the family group taxa based on them (for excellent recent discussions and illustrations of teeth of early therians see Cifelli, 1991a, b, c, this volume, Chapter 14; and Clemens and Lillegraven, 1986). Some of the "extremes" are represented by the dentally derived didelphine didelphids (*Monodelphis*), microbiotherians such as *Dromiciops*, and the perhaps quite primitive ameridelphian dental morph such as *Protalphadon*, or the presumed borhyaenids *Jaskhadelphys* and *Allqokirus* (see

FIGURE 15.2. Comparison of upper and lower molars of selected ameridelphian metatherians to illustrate the retention of metatherian diagnostic attributes and their slight (synapomorphous or parallel) modifications in sundry lineages. Molar topography abbreviations: **entd**, entoconid; **hyd**, hypoconid; **hyld**, hypoconulid; **me**, metacone; **ml**, metaconule; **md**, metaconid; **pa**, paracone; **pad**, paraconid; **pl**, paraconule; **pr**, protocone; **prd**, protoconid.

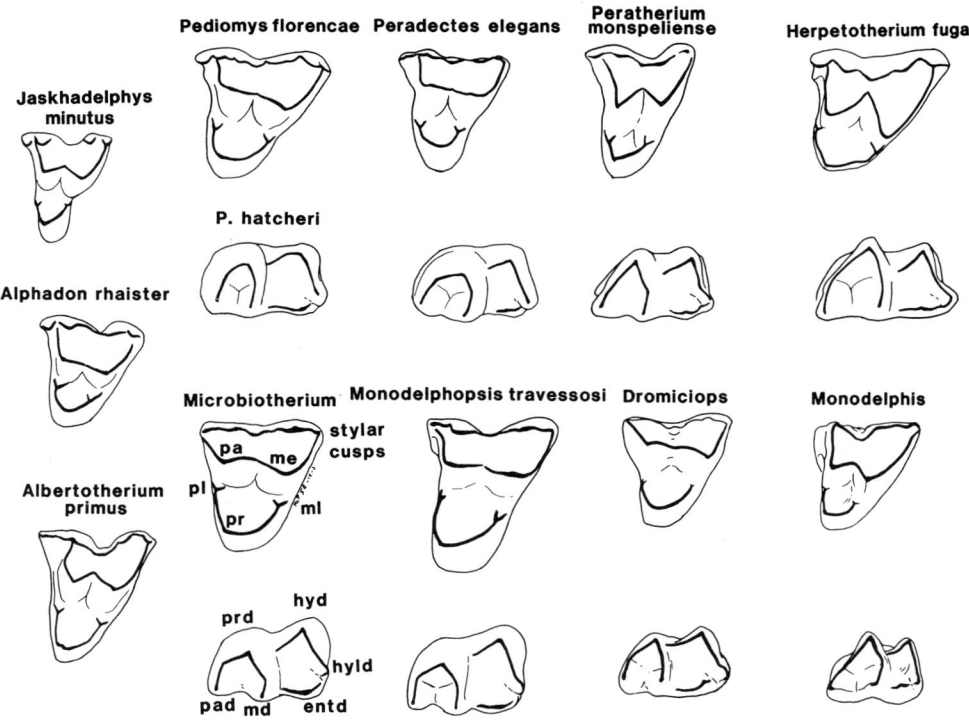

Marshall and de Muizon, 1988, for the latter; see also comment by Kielan-Jaworowska and Nessov, 1990, on the possible, but in my view not likely, deltatheroidan tie of *Jaskhadelphys*).

The potentially confusing dental evidence that has bearing on the early (i.e., Cretaceous and Paleogene) relationships of ameridelphian metatherians was recently studied in detail by Reig et al. (1987), Marshall (1987), and Marshall et al. (1990). While generating a large dental data base, Reig et. al. (1987) have recognized the importance of the Didelphidae, but misinterpreted and misapplied the dental evidence in their analysis of the taxic relationships of the Ameridelphia and Australidelphia. This was largely due, I believe, to the failure to apply the available expertise and judgment derived from development and functional-adaptive understanding to the distributional aspect of the evidence. Most of their characters chosen, as tabulated in their data matrix, were of reasonably contestable significance for the task of ordering intrametatherian relationships. As in many other taxonomic contributions that attempt to build large data matrices without functional-adaptive understanding, the methods and procedures that would have been necessary for character analysis were subsumed and conflated into a mere listing of "characters."

The range of variation that has been used to group dentally relatively unmodified nondeltatheroid marsupial genera with families, and the latter with one or another ordinal group, largely rests on a few simple traits. These are: (a) the conformation of the crest between the two main buccal cusps of the upper teeth, the paracone and metacone; (b) the nature of the stylar cusps (on the buccal margin of the upper teeth); (c) the size and development of the stylar shelf; (d) relative size of the paracone and metacone; (e) the proportions of the upper molars in general; and (f) the alleged emphasis of the metaconule in "later didelphoids." Shearing teeth or bunodonty, the degree to which conules develop, the relative size of the trigonids, and a large number of important minutiae that properly differentiate species and genera in alpha taxonomy may not always be well suited for the analysis of higher taxon relationships when the appropriate fossil (connecting) data are entirely lacking, and in particular, when many of these groups have retained an all but primitive ameridelphian molar crown morphology. My own work with dental evidence in several therian groups suggests an evolutionary lability of such mechanically important features, which may not show traces of their provenance once phyletically altered, and, therefore, of the rampant parallelism that probably occurs in mammals of relatively similar genetic background; this type of evidence is, of course, quite impervious to parsimony-based techniques. This is a causality- and mechanism-based judgment, as is most a priori weighting (but see also Neffs (1986) important but slightly different, allegedly cause- and process-free, views on a priori character weighting).

For example, the buccal crest, including the (mesial) preparacrista, the centrocrista (the crest between the paracone and metacone), and the (distal) postmetacris-

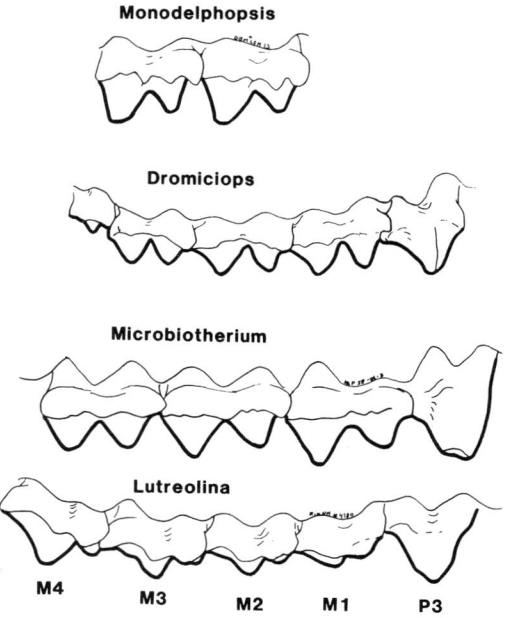

FIGURE 15.3. Relative size of the metacone in relation to the paracone within the same dentitions, and among the taxa of Paleogene itaboraiiform *Monodelphopsis*, the microbiotherians, and the didelphine *Lutreolina*.

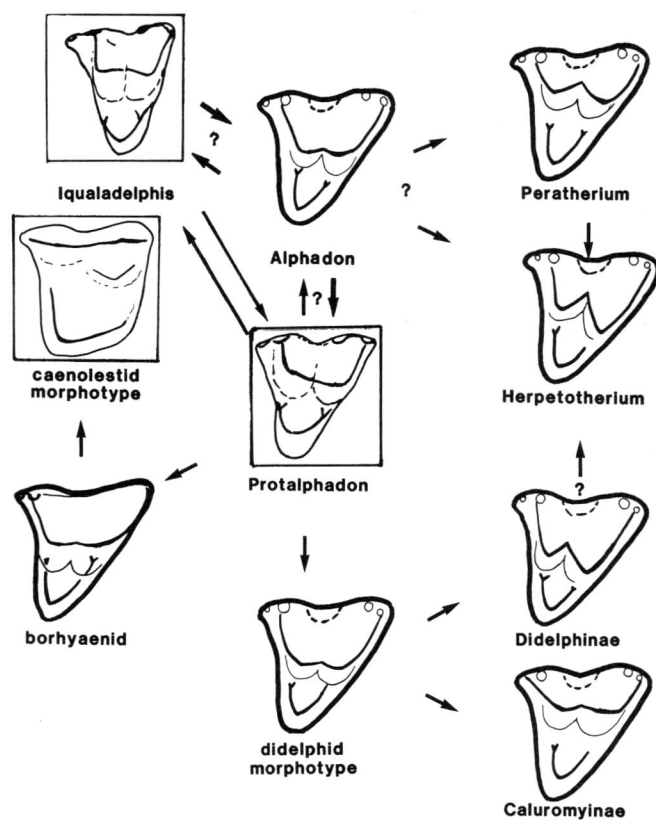

FIGURE 15.4. Suggested, and highly tentative, structural transformations of molar form in the Ameridelphia and Microbiotheria. No taxon phylogeny is implied. The molars in the frames are outlines based on photos of actual specimens (after Cifelli, 1990a), whereas the seven heavily outlined molars have an identical schematic contour in order to allow focus on the characters of the stylar shelf and cusps, and the ectoloph. Note the variation of ectoloph from linear to dilambdodont, and the uncertainties emphasized by the broken lines where stylar cusps **C** and **D** usually occur, an area poorly developed in *Protalphadon* and *Iqualadelphis*.

ta, of didelphids varies between the ***pediomyimorph*** (rectilinear; straight and running essentially apically and basally on the crown, not W-shaped in an occlusal view, but rather U-shaped) and the ***peratheriumorph*** (dilambdodont, in which the centrocrista sweeps buccolingually in addition to apically and basally on the crown, W-shaped in an occlusal view). In the didelphid Caluromyinae (the woolly, black-shouldered, and bushy-tailed opossums), the buccal crest (the ectoloph) is rectilinear, whereas in the other didelphids it is dilambdodont. Furthermore, the relative size of the paracone and metacone is subequal in the caluromyines or more precisely, the relative metacone size seems to covary with the height of the trigonid distal to it. In other words, paracone stunting and metacone "hypertrophy" appear to be partly causally correlated with the enlargement of the mesial shearing wall of the lower trigonid distal to it. If this is the case, then we are observing the consequences of a constantly fluctuating aspect of primitive marsupial dentitions: How much or how little trigonid shear is optimal given each species' fluctuating dietary adaptive treadmill?

In the Caluromyinae (and in the Microbiotheriidae; Fig. 15.3) we see the similar size of the paracone and metacone correlated with a trigonid row that does not increase posteriorly (and therefore probably derived—whether synapomorphously or not is another question). This is a dentition not committed to distal ectoloph shearing to the extent that the homologous tooth row is in other didelphids (Fig. 15.3). So at least in the Didelphidae the ancestral pattern of molars may or may not have been dilambdodont. The reduction of the paracone and the consequent "enlargement" of the metacone seem to vary with the relative importance of bite and shear in the tooth row, given a certain basic adaptation of the cheek dentition. Nevertheless the disparity between the two major buccal upper cusps is ancient and possibly diagnostic of the first nondeltatheroidan metatherian (see also Cifelli, this volume, Chapter 14). In *Microbiotherium*, as in caluromyines, the metacone of the first upper molar occludes against the second strongest trigonid, that of the **M/2**. As predicted, the first upper molars have slightly larger metacones than the teeth behind them (Fig. 15.3). Although these attributes are obviously heritable, sharing of such apomorphies is not a convincing sign of synapomorphies and therefore common descent, because of the high likeli-

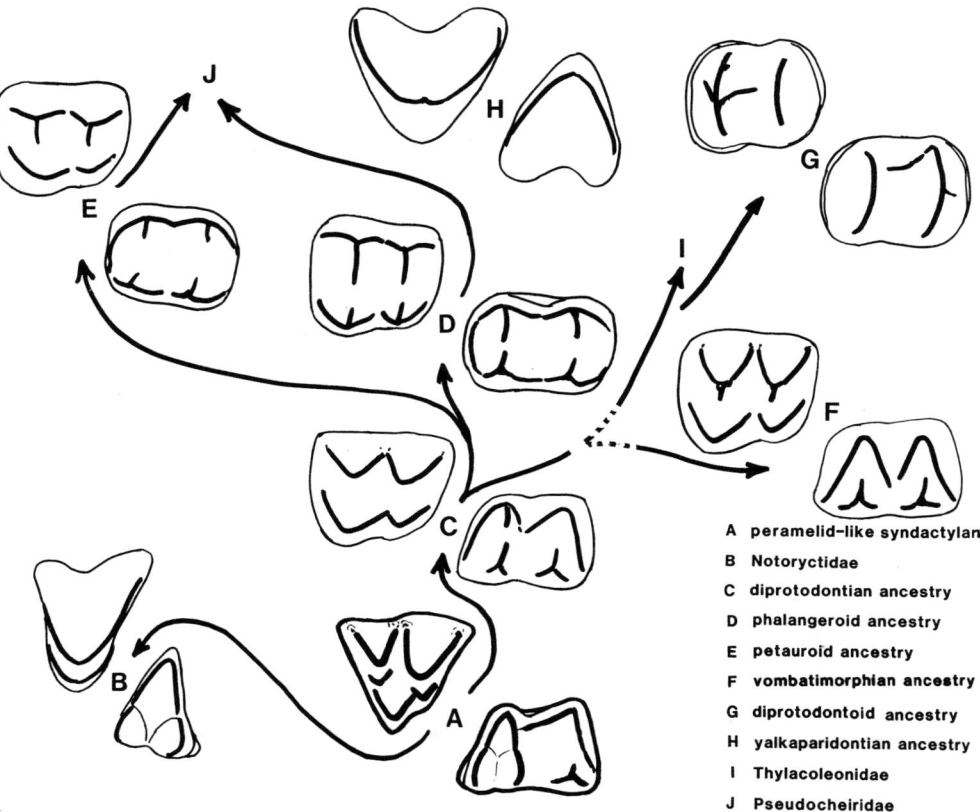

FIGURE 15.5. Highly schematized summary of some hypothesized (and some conjectural) transformations of molar patterns within the Syndactyla. Unsolved problems such as the origins of thylacoleonid (I) and yalkaparidontian (H) or diprotodontoid (G) molar patterns persist. A well-documented phylogeny of molar form and function is still largely unaccomplished, let alone the internal taxon phylogeny of the Diprotodontia.

A peramelid-like syndactylan
B Notoryctidae
C diprotodontian ancestry
D phalangeroid ancestry
E petauroid ancestry
F vombatimorphian ancestry
G diprotodontoid ancestry
H yalkaparidontian ancestry
I Thylacoleonidae
J Pseudocheiridae

hood of parallelism due to the occlusally constrained simplicity of such a progressive change.

What does the causal correlation of paracone reduction, and metacone "enlargement"(?) concerning ectoloph shape mean? The Cretaceous *Albertatherium* and some *Alphadon* have very well developed W-shaped ectolophs, but no significant metacone "enlargement"; the latter is perhaps an expression of relatively minor paracone reduction. *Monodelphopsis* from the Itaborai fissures of Brazil, however, has an archetypical rectilinear ectoloph, yet has an extremely developed disparity in size between the smaller paracone and much larger metacone. This is quite dissimilar to the conformation of the Cretaceous *Pediomys*, with which Marshall (1987) grouped *Monodelphopsis* and microbiotheriids. So it would appear that these two characters can vary independently from each other, perhaps related to nothing else than the degree of paracone reduction (cessation of mitosis in didelphids; Luckett, personal communication). But more important, they are "minor" changes of crown form, not appearing to affect the morphology of teeth very much. Understanding the ecomorphology of such features would help enormously; perhaps they are unaffected by functional demands. But short of such evidence, their presence or absence in various taxa might not represent a character of equal importance to such features that we know manifest themselves even when evolutionary alterations of great magnitude have occurred in a lineage. Yet centrocrista shape as well as fluctuations in paracone/metacone proportions are used far beyond their usefulness in the two most recent attempts to understand early marsupial evolution (Reig et al., 1987; Marshall, 1987). Paracone reduction, because it is a trait that involves progressive simplification of crown morphology, should not be weighted beyond the fact that it appears to be uniquely present at the base of the Ameridelphia. In primitive Deltatheroida that are included in the Metatheria the metacone is smaller than the paracone, as in the tribotherians and early eutherians (see especially Kielan-Jaworowska and Dashzeveg, 1989, and Kielan-Jaworowska and Nessov, 1990). Functional-adaptive analysis is essential in order to counteract the character matrix inflation that is so common due to the atomization of either developmentally or functionally strictly interdependent traits (see also Cifelli, Maier, Zeller, Chapter, 14, 17, and 8, this volume, for similar views).

There are ecologically and evolutionarily sound reasons for unequal evolutionary rates of hard parts, all based on well-tested theory. In contrast to some attributes of the carpus and tarsus, for example, aspects of cheek teeth offer an important set of boundary conditions that should be considered. There is an astronomical number of ways of combining specific food substances in dietary change, in spite of the relatively stringent constraints of the substances needed by most species. Although this offers avenues for diversity and convergence, it is also the basis of very fast evolution-

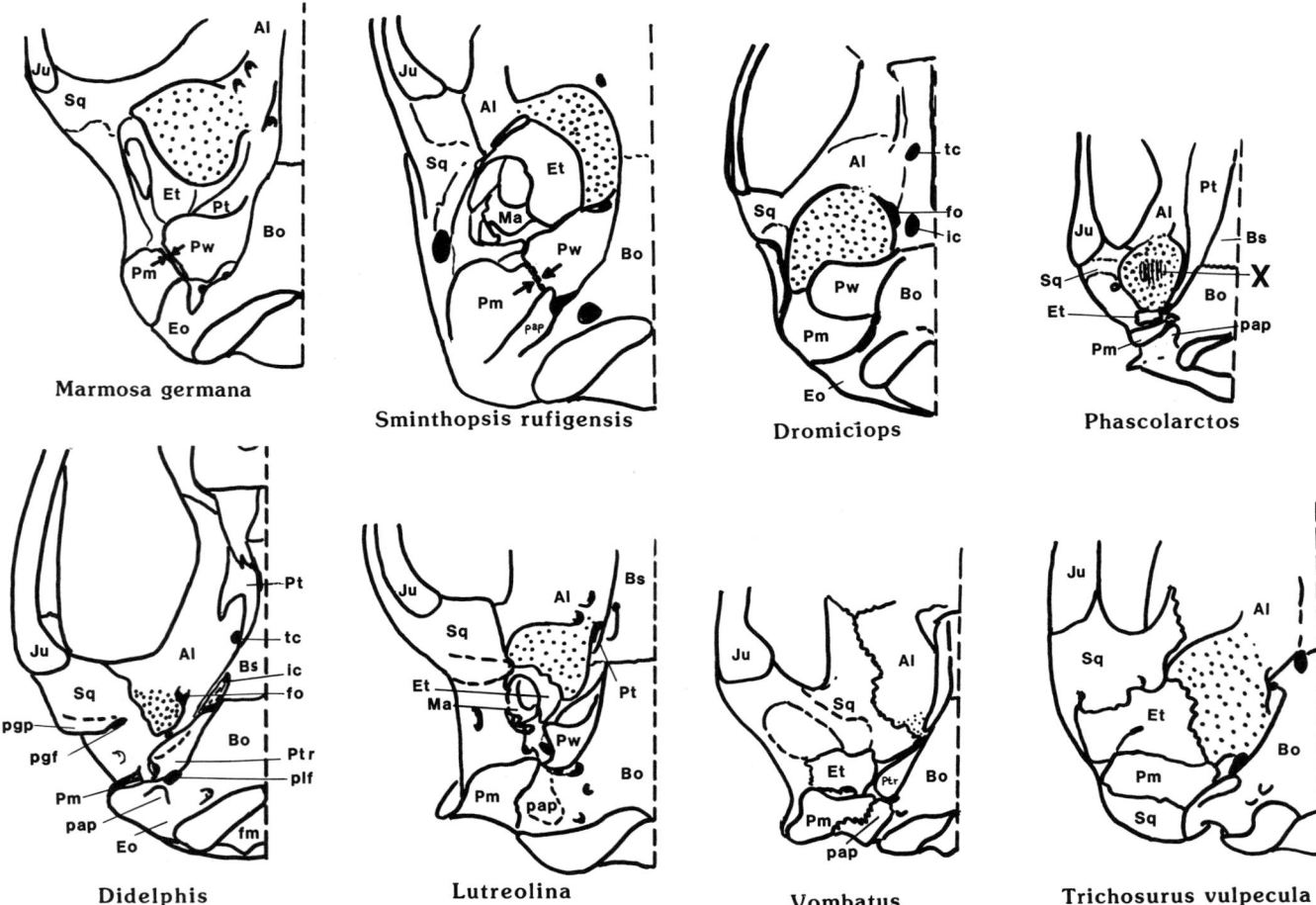

FIGURE 15.6. Selected (both taxa and traits) basicranial patterns in living Metatheria. Stippled areas represent the tympanic wing of the alisphenoid. The following abbreviations are used: **Al**, alisphenoid; **Bo**, basioccipital; **Bs**, basisphenoid; **Eo**, exoccipital; **Et**, ectotympanic; **fo**, foramen ovale; **fm**, foramen magnum; **ic**, entry for promontory branch of internal carotid; **Ju**, jugal; **Ma**, malleus; **pap**, paroccipital process; **pgf**, postglenoid foramen; **pgp**, postglenoid process; **plf**, posterior lacerate foramen; **Pm**, petromastoid; **Pw**, petrosal wing; **Pt**, pterygoid; **Ptr**, petrosal (promontorium); **Sq**, squamosal; **tc**, transverse canal; **X**, ligamentous attachment of hypertrophied adult bulla to mandibular angle (in *Phascolarctos*). *Marmosa*, *Sminthopsis*, *Lutreolina*, and *Trichosurus* were redrawn after Maier (1989).

ary change, which may obliterate the diagnostic apomorphies of ancestral species. It should be seriously considered that: (a) if the complex tarsal adaptations of didelphids are synapomorphies uniting all living genera that share it, and (b) if the non-carnassiform dentition of the caluromyines is a primitive retention from the first didelphids, then (c) the carnassiform molar row of the Didelphinae is independently evolved from other North American Cretaceous and Paleogene taxa that show this character (Fig. 15.4). Clearly, therefore, the shared similarities of the cheek teeth and of the combined molar row in the Caluromyinae, Microbiotheriidae, Peradectinae, and Pediomyinae are probably ancient Cretaceous retentions, and therefore not useful bases for the evaluation of the early branching of the Ameridelphia, the arguments of Reig et. al. (1987), and Marshall (1987) notwithstanding. Something else may be stated about teeth that is strongly suggested by the tarsal evidence discussed in Szalay (1993). The emphasis of the shearing dentition of borhyaenids, an emphasis on the metacone, was probably independent from the stem of the Didelphidae as we must infer from the living didelphids. The tarsal pattern of undoubted borhyaenids does not share any traces of the modified patterns of the protodidelphid (Szalay, 1982, 1993).

At the same time, while the Cretaceous-Paleogene picture is out of focus, the understanding of dental evolution in such metatherian groups as the syndactylans (see Fig. 15.5) during the last hundred years progressed with transformational analysis and an improved fossil record, not through a miracle attributed to cladistic analysis. Dental transformational analysis has been one of the many triumphs of the comparative morphology and paleontological approach. This understanding of transformation of teeth, while still incomplete, has been established through the efforts of Bensley, Gregory, Simpson, Tate, Archer, Clemens, Murray, Tedford, Woodburne, and many others. It is one of the most critical morphological and temporal frameworks that helps to corroborate numerous phyletic origins and nodes in the phylogenetic ordering of taxa. Paleontology is likely to continue to fill in the missing gaps of many osteologically based phena and help resolve problems that seem insurmountable at present.

THE EAR REGION. The taxonomically oriented study of the group-specific differences in basicranial morphology by mammalian paleontologists and morphologists offers a panacea—occasionally. The area, perhaps because of its links to multiple adaptive solutions, and therefore its complexity, appears to be more conservative morphologically and functionally than other cranial complexes or molar teeth. However, *how* basicranial morphology is a compromise adaptation driven by the surrounding mechanical demands is considerably less well understood than the dentition. So the results of basicranial studies are sometimes considered "definitive" when group disputes are in need of resolution, and this can be justified if many of the complex factors affecting an animal's skull and biology are considered. Yet often such statements lack the causal analytical framework that could explain the probable reasons for either similarity or differences of certain patterns (see also Presley's similar views, this volume, Chapter, 3). Like all other character complexes, the ear region may be a good indictor of certain monophyletic clusters of lineages and may show unique specialization, diagnosing a group very well (see Fig. 15.6), but often without indicating any clues about the affinities of that group. Bulla composition in metatherians may even vary drastically within the same population, and vary from population to population in the same species (Norris, 1992). Equally plausible, although not too many cases are known, convergent adaptive modification (bulla inflation, pneumatization, semitubular ectotympanics, carotid entry, etc.) have resulted in confusing assessments that were corrected subsequently. Some of these functional attributes and their adaptive significance are well enough understood and this knowledge is available for systematic application.

The pertinent literature on the evolution of the mammalian middle ear is enormous (see Wible and Hopson, this volume, Chapter 5). The monographic study of Fleischer (1973, 1978), however, is particularly important for understanding some of the causal principles that are likely to channel change in mammalian middle ear evolution, and therefore indirectly help in the recognition of apomorphous ancestral constraints. In fact, until more is known of the probable path of evolutionary transformation of the ear region in marsupials—influenced by the mechanical factors that played an important role in changing cranial morphology, such as the mechanical relationships that integrate aspects of feeding, brain hypertrophy, and hefting of the face on the neurocranium, or the cranium on the vertebral column (along with other critical causal forces)—the cranial similarities by themselves, as in any other character complex, cannot be considered fully reliable tests of taxon phylogenies.

The inflection of the angle in living and fossil marsupials, located in the ear region of the skull, is a well-known diagnostic feature of the last common ancestor of the group, or possibly of the Theria. Its explanation, however, remained elusive. Recently Maier (1987a) has shown in *Monodelphis* that the tip of the angular process, free of muscular insertions, retains contact with a rectangular fenestra on the medioventral floor of the bulla, specifically with a membrane formed of loose connective tissue that covers this fenestra. Thus, we may have the initial and boundary conditions for current jaw mechanics in marsupials constrained by an ancient functional complex—related not to chewing but to hearing. The ear region of the koala (see Fig 15.6) offers a unique opportunity to study the functional-adaptive forces on an obviously highly modified diprotodontian pattern, one which resulted in the hypertrophy of the bulla and as a consequence of its ventral extension produced the (pseudo) reversal of the heritage of the marsupial angle. In fact, the critical developmental connection between the angle and the bulla has remained.

Recent significant contributions to living and Neogene metatherian basicranial development and to taxonomic analysis, which have not conflated the judgments necessary at different levels of character analysis, are those of Maier (1987 a, 1987b, 1989; this volume, Chapter 12) and Murray et al. (1987). Studies by the former author presented invaluable evidence (Fig. 15.6) for developmental dynamics (see above), whereas the contribution of the latter offers a view, and a general model of analysis, of the adaptive and developmental process–guided phylogeny of the syndactylan ear regions (see Fig. 15.7). Wible (1990) has provided excellent and extensive accounts and comparisons within the framework of development dynamics of the rare record of Cretaceous marsupial petrosals.

It is well known that in metatherinans the predominant element involved in an ossified bulla formation is the tympanic wing of the alisphenoid. Whatever the merits of the various views on the beginnings of a bulla, there is obviously great need for caution in interpreting at least the ossified bullar floor evidence (see especially Norris, 1992). The persistence of the alisphenoid wing in various shapes and sizes, coupled with the added floor element produced by the rostral (petrosal wing) or caudal (petromastoid wing) extensions of the petrosal bone, or both (the "tympanic processes of the petrosal"; see Fig. 15.6) must caution our acceptance of

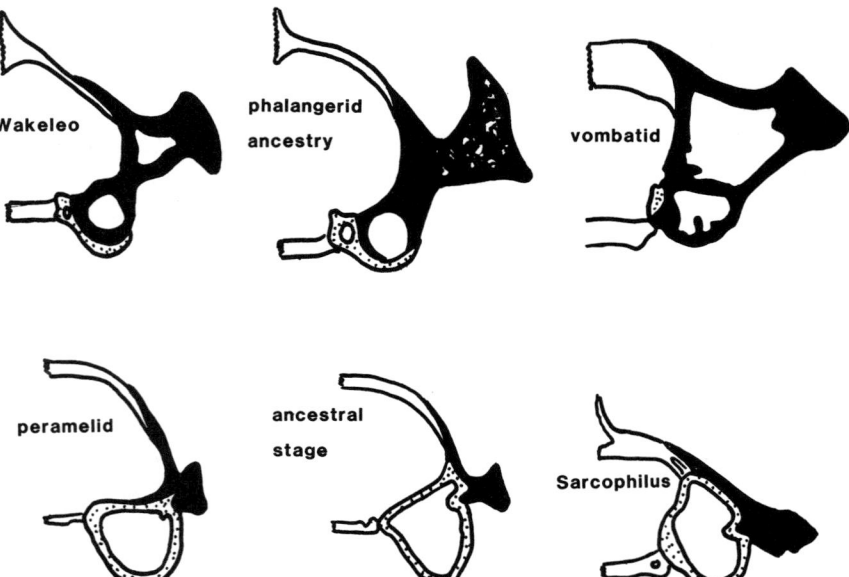

FIGURE 15.7. Schematized cross sections, after Murray et al. (1987), of selected australidelphians to illustrate the pattern of distribution of squamosal (solid black) and alisphenoid (stippled) components in the back end of the skull. Living syndactylans have an intracranial extension of the squamosal, characteristic of the diprotodontians. The latter have a bilaminar bulla construction. Peramelids already show the beginnings of this condition.

any phylogenetic account of metatherians based on the ear region alone. Although there has been no unequivocal evidence for any entotympanics in marsupials (Szalay, 1982, p. 636; Maier 1989) or bullae formed exclusively from the ectotympanic or the petrosal, the most recent analysis of phalanger ear regions by Norris (1992) is revealing of the plasticity of basicranial developmental dynamics. As noted above, he reports not only amazing developmental plasticity in Solomon islands populations but also ossifications that fully qualify as entotympanics.

The phylogenetic study of the cranial attributes of the primitive thaylacoleonid *Wakeleo* described by Murray et al. (1987) presents and adaptive and developmental process–steeped analysis, a causality-oriented methodology that has not been adequately appreciated and therefore practiced. Murray et al. (1987) have opened up the issue of causal ontogenetic and phylogenetic dynamics in the cranial morphology of metatherians, which has been generally outside the approach of cladistic schemes.

A recent assessment of the distribution of some basicranial features of most diprotodontians, with emphasis on the phalangeriforms (present sense), by Springer and Woodburne (1989) has been a useful, cladistic contribution to basicranial diversity. They summarize the presence and clarify the distribution of some characters. I believe, however, that their study is flawed because of the choice of the characters and the manner of their subsequent use of these in polarity analysis. The latter is largely cladistic outgroup-comparison based. The development of a bilaminar condition around the middle ear, due to squamosal invasion and alisphenoid retraction (or the reverse in koalas) requires functional-adaptive analysis as stressed and attempted by Murray et al. (see Fig. 15.7). This peculiarity of the syndactylans may be somehow tied to the pneumatization-related expansion of the squamosal. The repeated use of the same components by diverse lineages can make it nearly impossible to detect, without the appropriate causal analysis, the phylogenetic constraints that may be diagnostic of a particular cluster of lineages. The latter depends entirely on ontogenetic and functional-adaptive information viewed in an evolutionarily transformational context (e.g., in this volume, Maier, Chapter 12; Presley, Chapter 3; Zeller, Chapter 8), and not on merely analyzing the serendipitously recognized and distribution-based selection of (and therefore in some ways *inappropriately* weighted *a priori*) unexplained characters.

CARPAL, TARSAL, AND OTHER POSTCRANIAL EVIDENCE. As a general guideline for using musculoskeletal evidence in phylogenetic analysis, the critical issue is whether some areas or elements will or will not show traces of phyletically antecedent conditions in addition to the group specific traits. Complexity of structure in a functional-adaptive context is the only way to avoid using homoplastic characters. The humerus and femur, both their distal and proximal articulations, have shown themselves to be so often similarly affected by functional demands driven by identical biological roles that their usefulness, unless the whole elbow or knee complex is known, is limited. Demands in the elbow and knee joint on the interacting elements have also produced many detailed convergences, which render the general usefulness of isolated individual elements questionable. Although mammalian skeletal components, and coadapted areas within the skeleton, reflect the mechanics necessary for locomotion and habitual postures to varying degrees, the hand and foot are in particularly close physical contact with the details

FIGURE 15.8. Somewhat schematized patterns of carpal arrangements of the right hand in a number of representative metatherians and the hypothetical stem eutherian. Neither the patterns of diversity (i.e., the range of modifications) of the whole carpus nor the detailed form-function of the individual carpals has been studied in any detail comparable to that of the crus and tarsus (for the latter see Szalay, 1993). **Ce**, centrale; **Cn**, cuneiform (=Triquetrum); **Lu**, lunate; **Mcp**, metacarpal; **Mg**, Magnum (=capitate without the centrale); **Pi**, pisiform; **Pp**, prepollex; **R**, radius; **U**, ulna; **Un**, unciform (=hamate); **Tm**, trapezium; **Tr**, trapezoid.

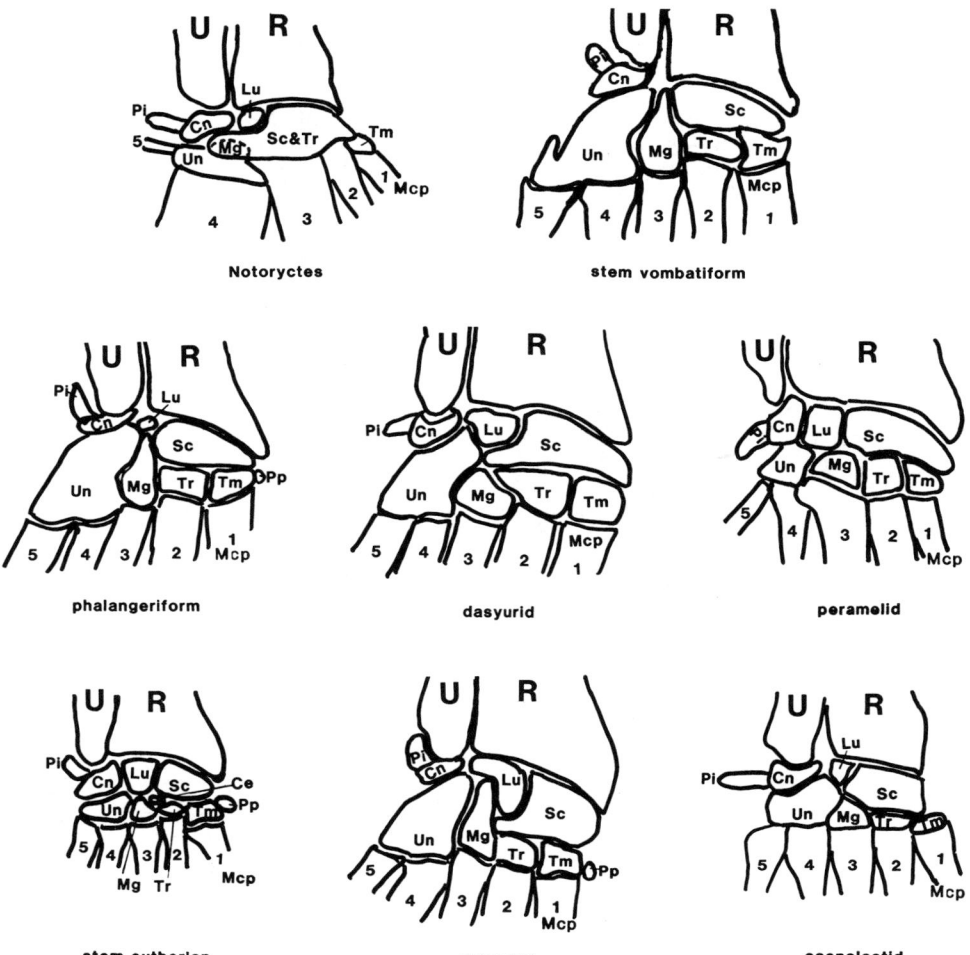

of the substrate. These multiboned units are the immediate arbitrators of many physical forces at the environment-organism interface. They are closely involved in behavioral change, and therefore the selective consequences of these behaviors have the potential to affect them. Considerable adjustments of the hand and foot, however, are brought about by minor movements of the individual elements in relation to one another. Therefore major evolutionary changes in mechanical function of the carpus and tarsus, tracking bioroles, can result from only minor form-function alterations of the individual bony elements. This is one reason why these areas of the skeleton are relatively conservative while they display form-function divergence. The other reason is that there are a limited number of ways of locomoting on either terrestrial or arboreal substrates. The many constraining elements of the carpus and tarsus and the noted small adjustments needed combine to offer regions in which changes from preexisting conditions are well reflected species specific, yet relatively conservative. At the same time, these regions show consistent, ecologically mediated, taxonomic variation at lower categorical levels as well.

In fact, the interpretation of the form-function of the bone as a material with specific load reflecting properties, and that of synovial joints, in a particular adaptive context is often the most critical in the ordering of relevant skeletal information. I have (Szalay, 1993) summarized a list of assumptions that should be kept in mind when ordering skeletal, but particularly diarthrodial, joint evidence. Although studies in behavioral ecology are relevant to some extent, many such excellent endeavors omit form-function considerations of the organisms as these relate to life-history adaptations. Bridges are needed to link behavioral and ecological morphology with historical explanations (Bock, 1990), and these connections will undoubtedly add power to the interpretive perspective focused on the characters. A detailed analysis and interpretation of the tarsal record of metatherians, with some comments on carpal transformations, is found in Szalay (1993). Figs. 15.8–15.14 depict a limited number of selected postcranial traits employed in the various diagnoses below.

Metatherian Taxa and Phylogeny

I have briefly diagnosed a few monophyletic taxa (both holophyletic and paraphyletic ones), some of which

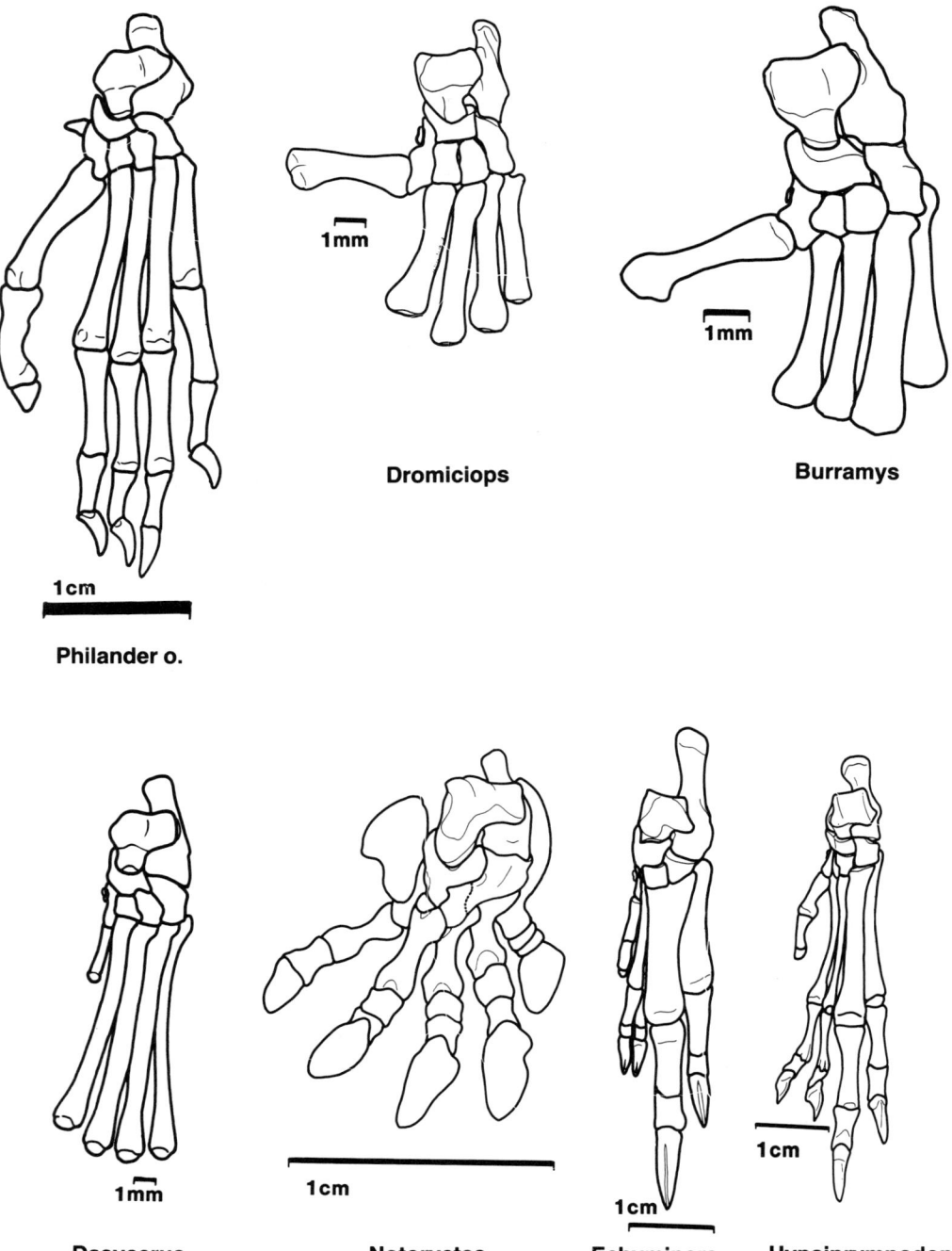

FIGURE 15.9. Representative left foot skeletons of living metatherian groups. In order to keep the relative size of tarsals in perspective, no phalanges are shown for the small *Dromiciops*, *Burramys*, and *Dasycercus*.

Chapter 15. Metatherian Taxon Phylogeny

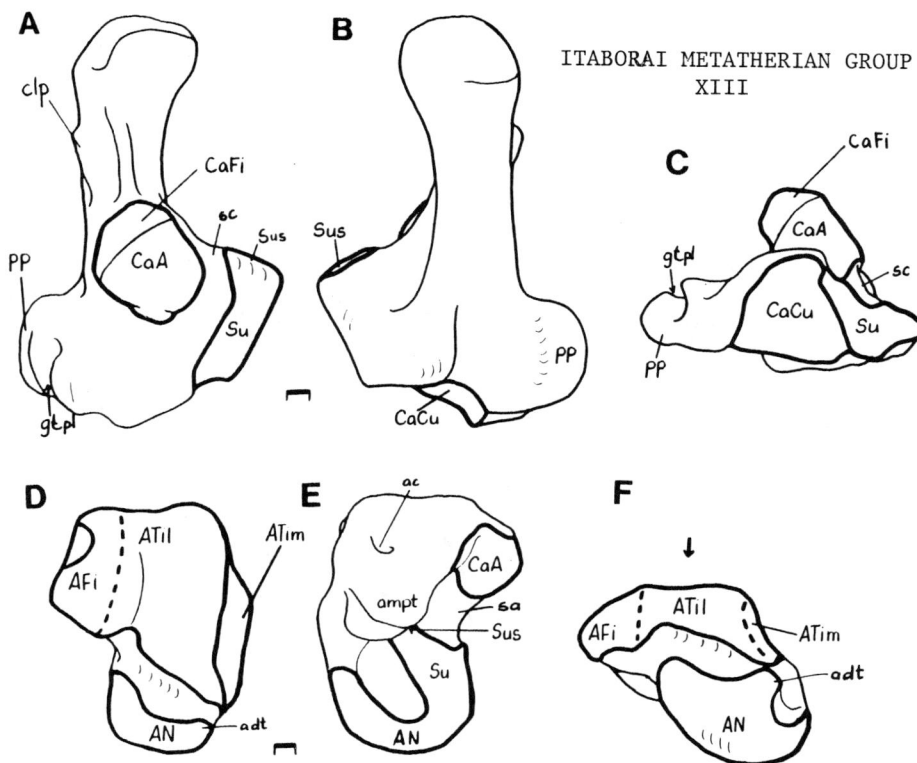

FIGURE 15.10. Right calcaneus and astragalus of Late (or ?Middle) Paleocene South American marsupial. This sample is of the largest sized tarsal phenon from Itaborai. Nonetheless, its morphology is generally closely similar even to the smallest homologous phenon from these fissures, except for the homologous bones of a) two phena of didelphids, b) of a unique tiny metatherian tarsal group, and c) of the polydolopids, all found at Itaborai. From left to right, dorsal, ventral, and distal views. For abbreviations see Table 9.1 Scales represent 1 mm.

have been contentious and problematic, and some of which are new. These few diagnoses (discussed in Szalay, 1993) involve the primary use of osteological (and fossil) evidence, and they list the hypothesized apomorphies of the last common ancestor of a group, supported by an analysis of the members of the group. The increasing fossil record and new evolutionary morphology applied to the evidence, however, will surely change the picture in the future. Due to space limitations here I diagnose only a few of the ancestral conditions designated by numbers in Figure 15.15. These points and nodes are the ones which I consider diagnosable with an acceptable level of confidence. Figure 15.16, the cladogram, is a direct derivation from the most probable tree shown in Figure 15.15.

My practices are a clear "violation" of the rigid and formalistically driven popular "crown group" perspective of the cladistic classificatory paradigm. The latter, I believe, has its roots in the pre-Darwinian Linnean theory of classification, when consideration of time as a factor was nonexistent. I find no other explanation of the fact that most practitioners of cladistics categorically refuse to engage in the practice of phyletics (vertical comparisons) or in the consideration of the chronistic framework of phylogeny. My taxonomic view is dictated as much by the nature of the evidence as by the firm belief that "upward-looking" groups of the post-Darwinian era of taxonomy, employing both (a) Hennig's rule for testing sistergroup relationships with synapomorphies, and (b) the testing of synapomorphies through the methods of functional-adaptive analysis, both of which entail causally ascertained transformation series derived from vertical comparisons, are valid. It is axiomatic that the understanding of the plesiomorphous as well as the apomorphous conditions are of equal significance for constructing phylogenies. Such an approach is in contrast to a view based on the pre-Darwinian heritage of ageless groups perfectly nested in inclusive hierarchies. As noted before, descent of lineages (as opposed to the always arbitrarily delineated *real* taxa) cannot be reflected in such a cladistic system. The fossil samples will continue to yield clusters of phena that will greatly increase understanding of specific character phylogenies but will often remain ambiguous as to their exact taxic position in the continuum and branching of life.

INFRACLASS METATHERIA. The protometatherian, the first species of the phylogenetically and taxonomically formally recognized metatherians (the ancestors of which were very likely marsupials reproductively and developmentally), was probably reproductively, physiologically, cranially, and in some aspects of its molar morphology similar to the putative ancestral therian. This account, therefore, is not a proper diagnosis, as it probably contains characters antecedent to the protometatherian. It is, therefore, a descriptive assessment of this stem metatherian. Furthermore, the documentation recently by Kielan-Jaworowska and Nessov (1990) that the Delatheroida is an early branch of the

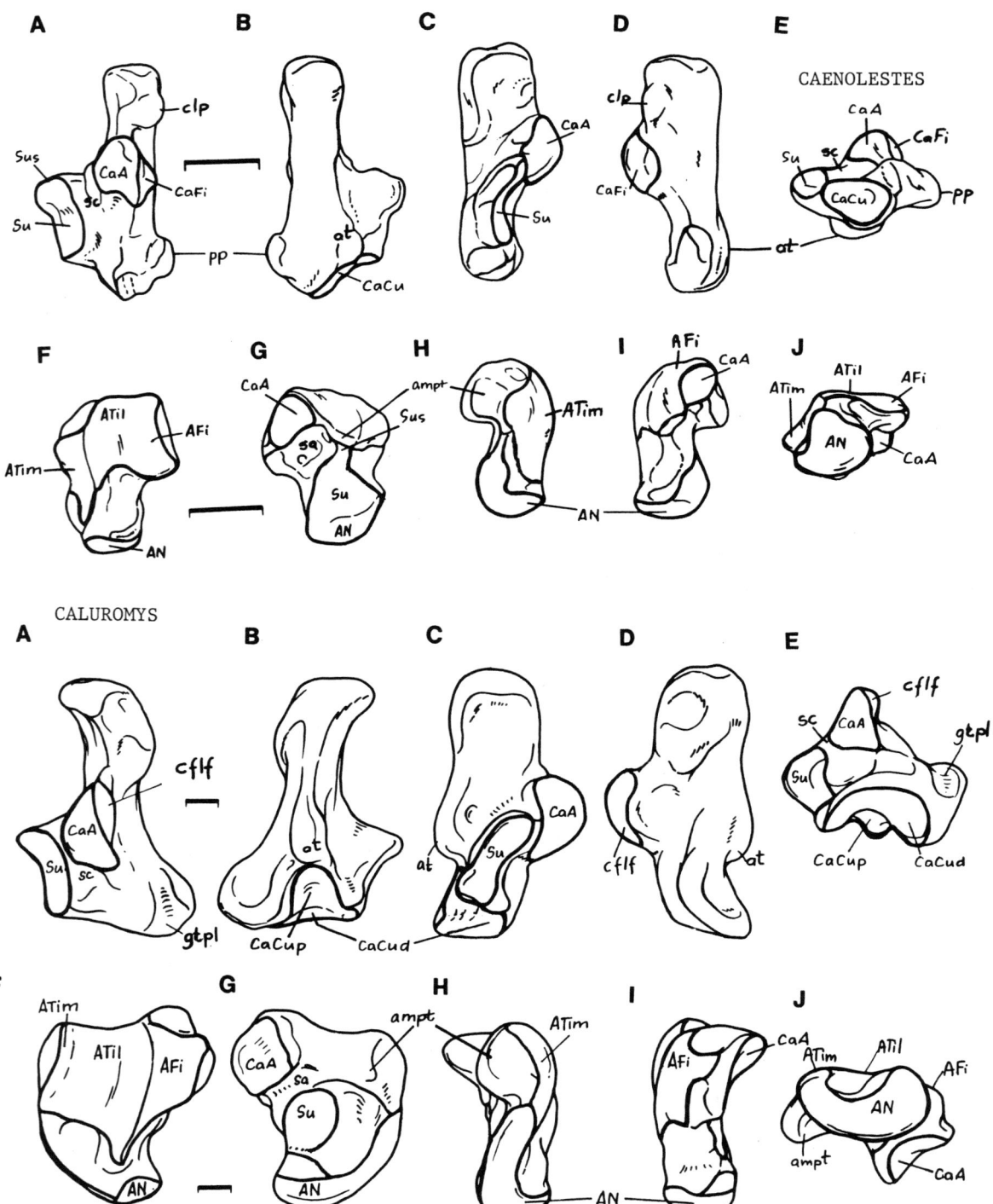

FIGURE 15.11. Comparison of *Caenolestes* sp. (AMNH 62915), left calcanei and astragali, above, and *Caluromys philander* (AMNH 234989), below. From left to right, dorsal, plantar, medial, lateral, and distal views, respectively. Caenolestid tarsals retain more of the detailed similarities to most members of the Itaborai assemblage than didelphids do. The latter, also represented at Itaborai, are apomorphic in several features in comparison to the primitive Itaboraiform phena (see text, and Szalay, 1993). For abbreviations see Table 9.1 Scales represent 1 mm.

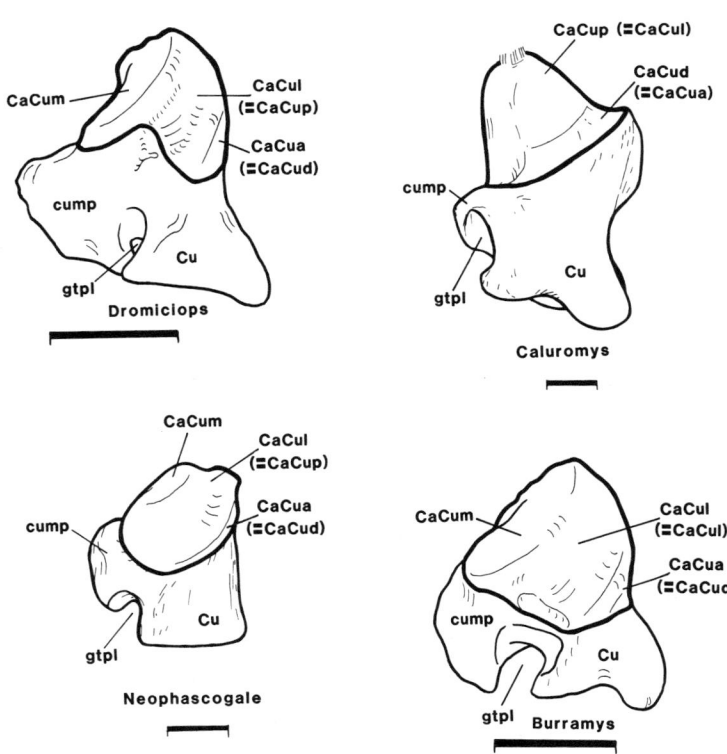

FIGURE 15.12. Comparison of homologous aspects of right cuboids in the didelphid ameridelphians (*Caluromys*), and australidelphians (*Dromiciops*, *Burramys*, and *Neophascogale*). The diagnostic didelphid **CaCud** and **CaCup** facets, apomorphic in respect to the metatherian ancestry, are homologous with the **CaCua** and **CaCup** facets of australidelphians, respectively. Note the tripartite subdivision of the diagnostic australidelphian calacaneocuboid **CaCua**, **CaCul**, and **CaCum** facets, derived from a didelphid one. The **CaCum** facet is a neomorph. Unlike ameridelphians, dasyuromorphians show (betray traces of) the australidelphian facet pattern and the hypertrophy of the cuboidal medial process (**cump**), traits that are apomorphic at the base of the Gondwanadelphia. The tarsal pattern of *Dromiciops* is not an apomorphy shared with stem of the Diprotodontia, but is a retention from the protoaustralidelphian. For abbreviations see Table 9.1. Scales represent 1 mm.

Metatheria significantly extends our perspective on the relationship of the therians. This protometatherian ancestor had: advanced therian reduction of diphyodonty (apomorphic at that point), the therian condition having been more completely retained by most placentals; dental formula **I1,2,3,4,5/4**; **C1**; **Pd1,d2,3**; **M1,2,3,4**; tribosphenic molars with relatively transverse mesiodistally widened and slightly enlarged (compared to the earliest tribosphenic therian) sharp protocone; enlarged talonid and perhaps slightly height-reduced trigonid (compared to eutherian sister lineage); metacone almost subequal to paracone in size; presence of stylar cusps C and D uncertain (because of their ill defined aspect in deltatheroidans, entoconid and hypoconulid twinning on lower molars is unlikely); jugal that extends posteriorly under the zygomatic portion of the squamosal to the anterior margin of the articular surface of the glenoid (probably a premammalian cynodont condition); bony palate fenestrated; unique combination of traits in the basicranial circulation which may be synapomorphies shared with the living metatherians and could be detected in fossils, such as the course of the internal carotid extrabullar medial to the tympanic cavity and the stapedial artery lost in the adults of living marsupials (but presence of adult stapedial artery in metatherian common ancestor is possible); bicrurate stapes retained from therian ancestry, even after loss of stapedial artery in adults; inflected angle of mandible (dentary), probably a retention or a hypertrophied condition from therian ancestry; incomplete, or cartilaginous connection of the lower arch of the atlas; lost or greatly reduced patellar ossification; shoulder-breast complex as in the therian common ancestor; seven cervical vertebrae as in therian ancestry, and probably no more than twenty-six presacral vertebrae, with probably thirteen thoracic vertebrae; scaphoid in carpus which probably incorporated centrale (still widely present in a separate state in eutherians); flabellum of fibula (dorsoposterior process) retained; astragalus with **AFi** facet primitively narrow, as the therian tricontact cruropedal joint, the **UAJ**, is retained; **UAJ** which included the meniscus and lunula of most living metatherians, ligamentously anchored to the calcaneus, fibula, and tibia; relatively weakly developed malleolus on tibia (compared to eutherians), and curved and extended tibial **ATi** articular surface; peroneal process of calcaneus primitively large (as in therian ancestry); astragalar canal possibly more posterior and plantar than in stem eutherians.

While the skull of *Vincelestes*, as presented by Wible and Hopson (this volume, Chapter 5), may be the best current choice of a null group for comparison with therians (see Szalay and Bock, 1991, and Szalay, this volume, Chapter 9) this obviously does not mean that the condition in that genus determines what was or was not diagnostic for the stem therians, marsupials, or eutherians. The apomorphy of a sphenoparietal emissary vein in the protometatherian versus the apomorphy of the cervical emissary vein in the stem eutherian is suggested by Wible (1990). Yet the early developmental patterns of monotremes and living metatherians are identical (Wible, 1990, figs. 3A and 3C), while the adult

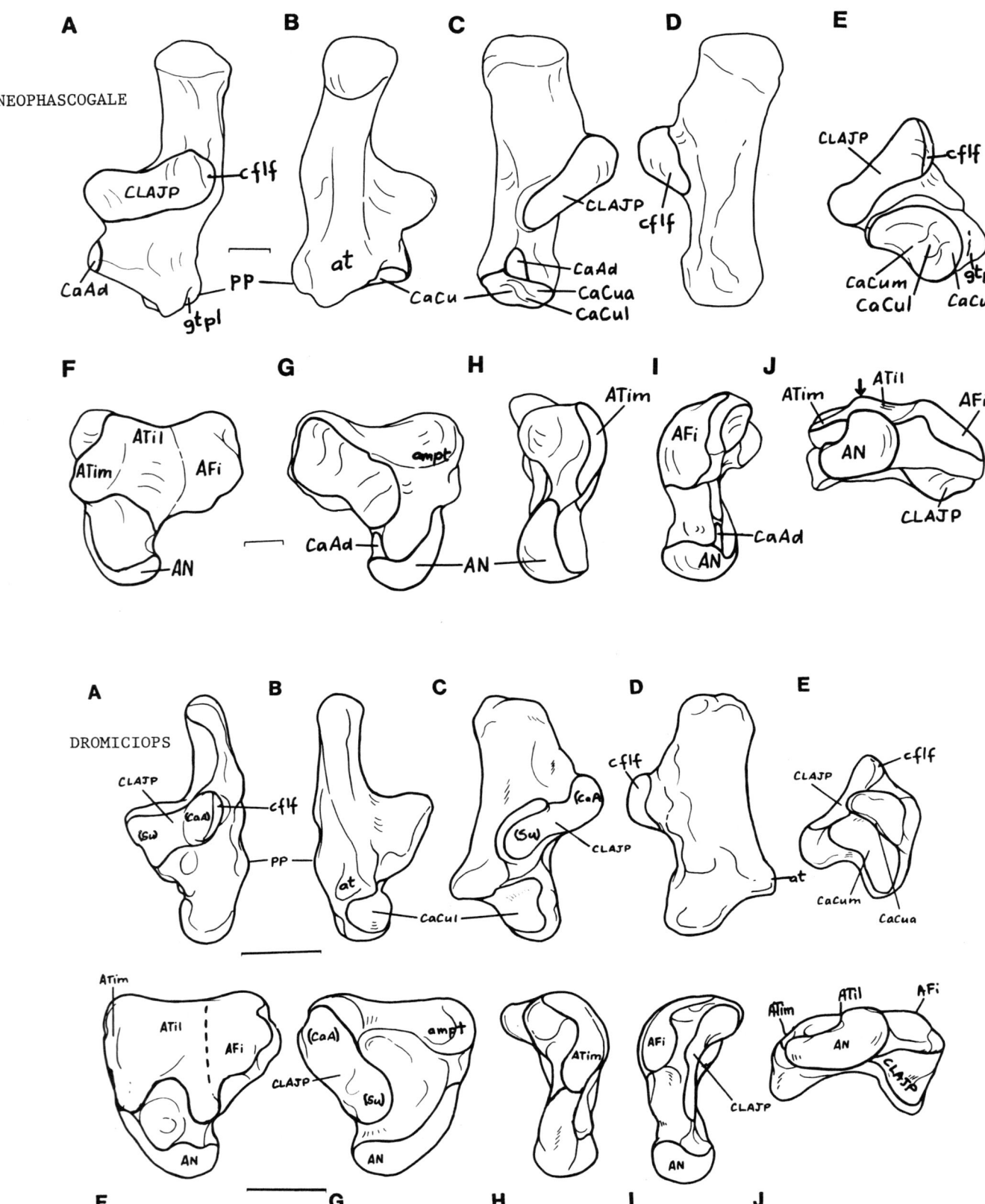

FIGURE 15.13. Comparison of the tarsally primitive australidelphian *Dromiciops australis* (AMNH 97746), below, with an advanced australidelphian pattern of dasyurids such as shown by *Neophascogale lorentzi* (AMNH 101980), above. Left calcanei and astragali. From left to right as in Fig.15.11. For abbreviations see Table 9.1. Scales represent 1 mm.

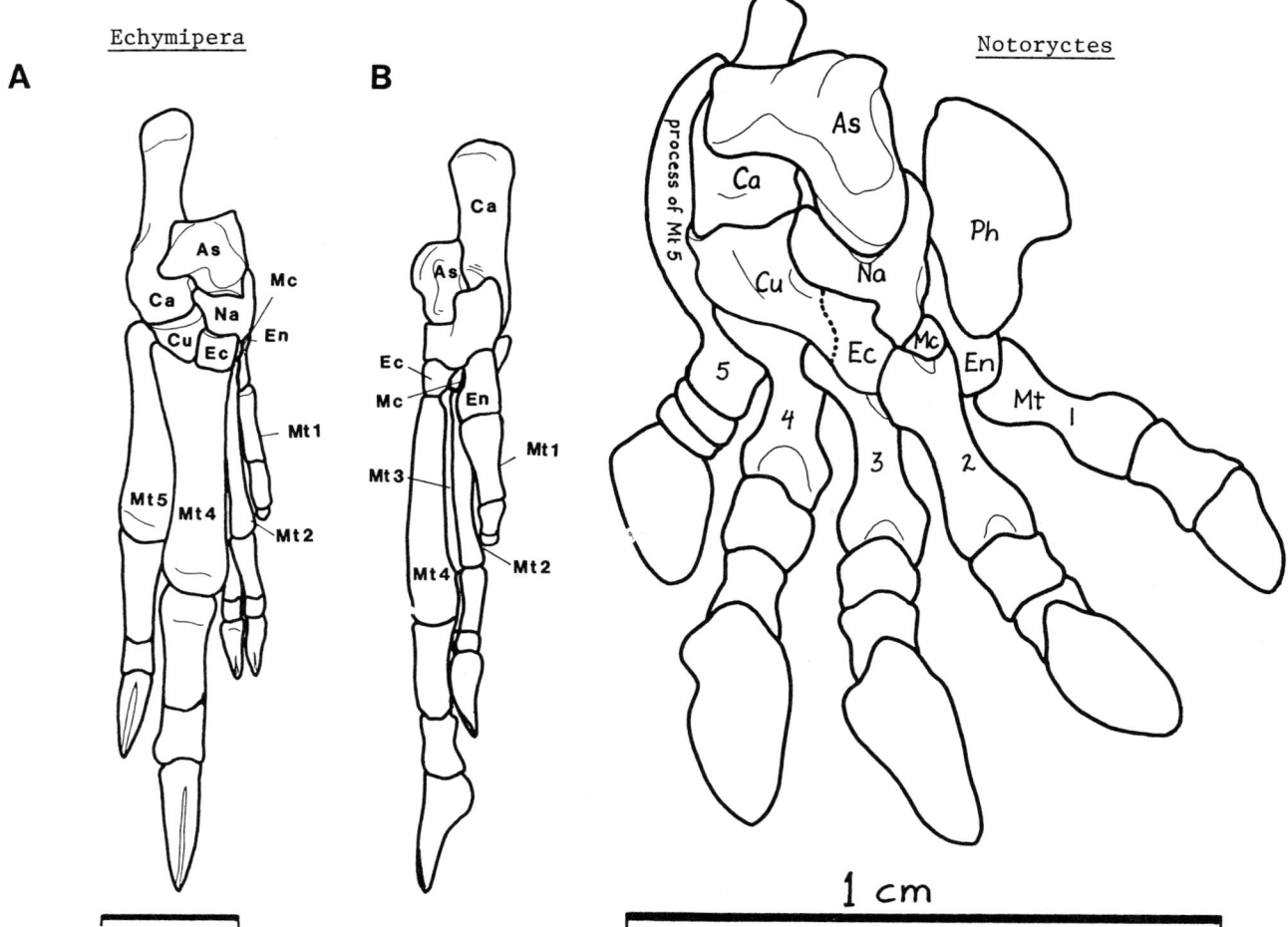

FIGURE 15.14. Comparison of right feet of *Echymipera* (A, B) and *Notoryctes* to show the extreme reduction of the mesocuneiform (**Mc**) in both groups. For abbreviations see Table 9.1. Scales represent 1 cm.

living metatherian condition shows the sphenoparietal vein. The therian morphotypic condition, therefore, may be represented by the ancestral metatherian pattern, and the developmentally "shocked" eutherian one (based on evidence of disturbance in early cranial ontogeny; see references cited by Wible, 1990) may be derived from this "marsupial" pattern. This may have occurred by losing the primitive internal jugular vein (retained in Metatheria) while retaining the lateral branch of the mammalian lateral head vein (the ancestral therian sphenoparietal vein). This speculation is clearly another perspective based on the carefully researched form-pattern analysis of Wible (1990). It is important to note, however, that a perspective provided by vertical comparison of agreed-upon patterns can have interpretive consequences quite distinct from a cladistic, horizontal ordering of information.

COHORT HOLARCTIDELPHIA. This independent, primarily Asiatic, but metatherian branch is known by the Deltatheroida, which includes the families Delatheroidi-dae and Deltatheridiidae, and was recently discussed by Kielan-Jaworowska and Nessov (1990). While the deltatheroidans may represent the primitive metatherian condition in their lack of clearly defined stylar cusps **C** and **D**, they may be diagnosed by a hypertrophied paraconid which is larger than the metaconid and perhaps a secondarily reduced talonid with well separated entoconid and hypoconulid which may have been derived from the relatively wider condition seen in both the earliest eutherians (*Prokennalestes*; see Kielan-Jaworowska and Dashzeveg, 1989) and the earliest ameridelphians in North America. In light of the hypertrophy of trigonid shear, the reduction of the talonid, and therefore its apomorphic condition in contrast to that encountered in the protoameridelphian, is likely as suggested by Kielan-Jaworowska and Nessov (1990). The still uncertain protodeltatheroidan dental formula appears to have been that which characterized the protometatherian.

COHORT AMERIDELPHIA. This cohort includes the heterogeneous assemblages of North and South Amer-

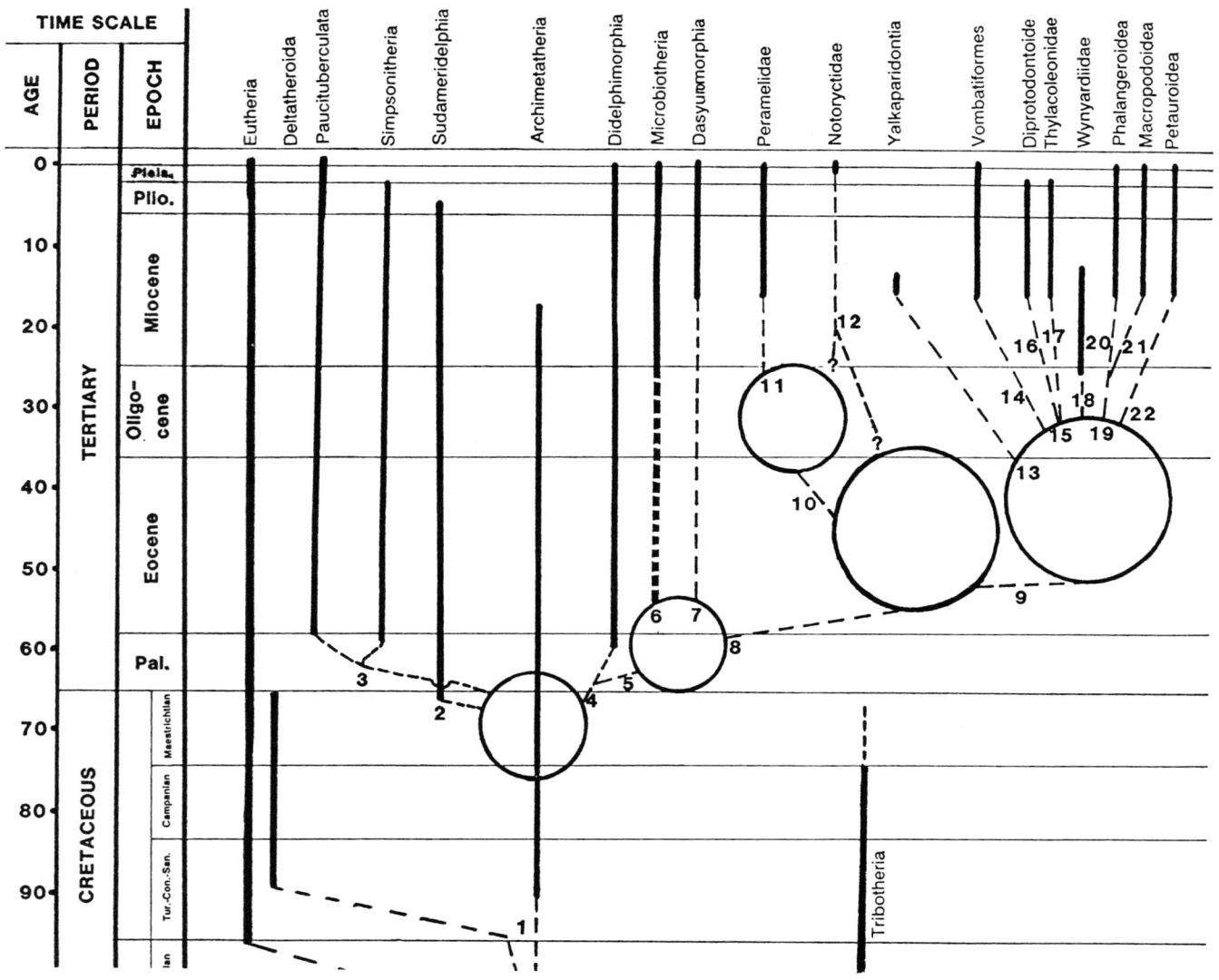

FIGURE 15.15. Phylogenetic hypothesis of metatherian relationships. All named groups are monophyletic (i.e., either paraphyletic or holophyletic). Circles represent the major temporal, phyletic, and/or branching uncertainties. Numbers are the ancestral points, or nodes at branchings, for the taxa that can be diagnosed based on apomorphies at those points. See the selected diagnoses in text. Heavy broken line leading to the Oligocene-Miocene and younger microbiotheriids suggests the uncertainty of recognizing these forms based on the spotty dental record alone. The Deltatheroida, as convincingly argued by Kielan-Jaworowska and Nessov (1990) is the sistergroup of the remaining metatherians, the Ameridelphia and Australidelphia.

ica, which are monophyletic and relatively unmodified archaic metatherians. I classify them in various monophyletic groups in the paraphyletic order Didelphida. These are sundry radiations which at present are nearly impossible to relate dichotomously to one another with any degree of confidence, beyond clustering them in what appear to be well delineated monophyletic family groups. Such families are the Pediomyidae (including Pediomyinae, Alphadontinae, Glasbiinae, and Peradectinae) and Stagodontidae. I include these groups in the paraphyletic suborder Archimetatheria. In addition, the other didelphidan suborders Sudameridelphia, Glirimetatheria (new), and Didelphimorphia accommodate the remaining ameridelphians. The Ameridelphia (and Didelphida and Archimetatheria) can be clearly diagnosed (Szalay, 1993). I choose to retain the Didelphimorphia within this cohort in spite of the fact that an animal that we could recognize as a didelphid only prior to the time of its divergence gave rise to the Australidelphia. The only meaningful way to characterize the first ameridelphian, didelphidan, archimetatherian is to list the traits most probably unique to their postulated last common ancestor. The protoameridelphian had: tribosphenic molars with relatively transverse mesiodistally widened and slightly enlarged (compared to both deltatheroidan eutherian morphotypes) sharp protocone; slightly enlarged talonid and perhaps slightly height-reduced trigo-

Chapter 15. Metatherian Taxon Phylogeny

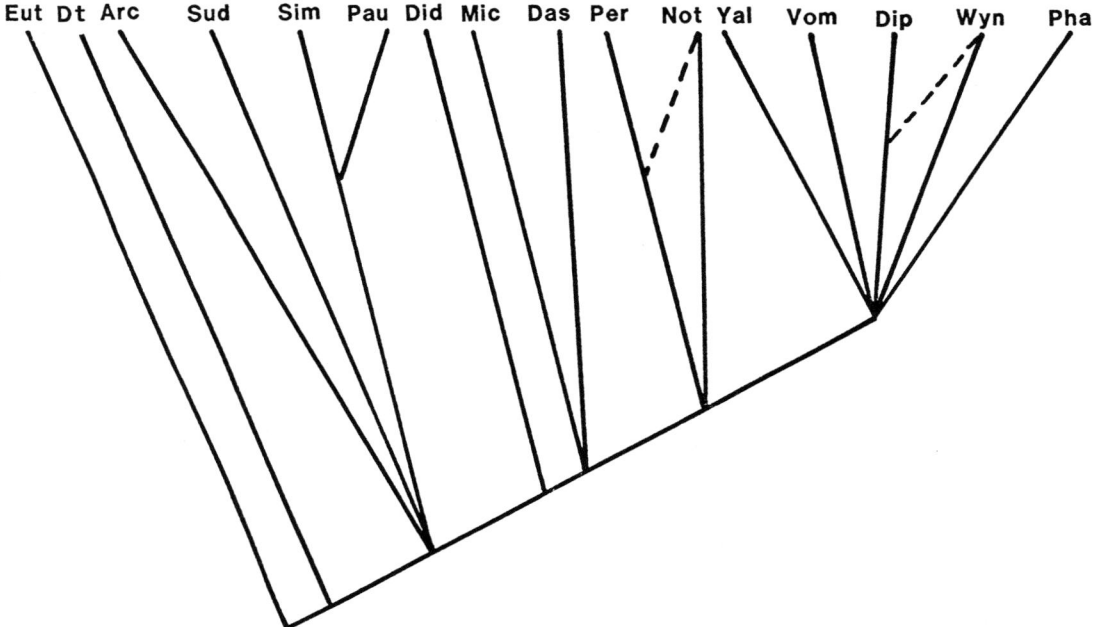

FIGURE 15.16. Stuffenreihe, or cladogram, *derived* from the phylogenetic hypothesis shown above, with considerable loss of the corroborated phylogenetic pattern. Functional-adaptive analysis of homologies, within the context of the fossil record, provided tests for taxonomic properties (e.g., see Fig. 9.1 in Szalay, this volume, Chapter 9) against which the taxon relationships shown in the tree, in Figure 15.15 above were tested, allowing for approximations of ancestral states and groups (hence parphyletic taxa) as sources for the derived groups. Exclusive cladistic practice rejects phyletics, chronistics, and phenetics, and through the use of cladogram driven reseach protocols, rigorously sidesteps those critical areas of phylogenetic inquiry which are a must for probable (believable) character transformation hypotheses. The cladogram, therefore, is *derivative and not foundational* in a phylogenetic (evolutionary) analysis that should be considered theoretically, and consequently methodologically, sound. The methodological use of cladograms in cladistic theory and practice (contrary to their usefulness as a summary device), demanding both process-free "pattern" establishment and its reflection on (and with the circular aid of) the cladogram, has the analytical consequence of being impervious to critical information and (causal) explanatory theory.

nid; metacone almost subequal to paracone in size as in Deltatheroida; presence of all (?) stylar cusps (except, probably, stylar cusp C) in contrast to known Deltatheroida; close twinning of entoconid and hypoconulid on lower molars, again in contrast to known holarctidelphians; cranial and post-cranial characters, as far as known, postulated to be the same as in protometatherian. Yet at the present time we cannot even separate many therian traits from the metatherian ones (see above).

SUBORDER ARCHIMETATHERIA. At present, in spite of the minutely argued points by Reig et al. (1987), Marshall (1987), and Aplin and Archer (1987), there is no evidence that any of the four subfamilies of the archimetatherian Pediomyidae (Alphadontinae, Pediomyinae, Glasbiinae, and Peradectinae) or the Stagodontidae have specific phylogenetic ties (tested by convincing taxonomic properties) to the South American marsupials. Yet Crochet's (1980) important synthesis suggests that a connection through some member of his geographically broad concept of Peradectini (which included *Alphadon*, *Peradectes*, and *Bobbschaefferia*, the latter from Itaborai, among others) may have been likely with the distinct South American radiation. On dental grounds alone the ordering of the increasingly numerous taxa into formal holophyletic groups is not possible with any confidence. The known pedal morphology of non-stagodontid marsupial tarsal remains (Szalay, 1993), as well as the modest range of dental morphology, permits nothing more than the recognition of one family, the Pediomyidae. There are no objectively acceptable taxonomic properties that corroborate special ties (contra Reig et al., 1987; Marshall et al., 1990) of the North American and European taxa of the Cretaceous and Paleogene with the better known later groups. The rapid expansion of dental and skeletal information on archimetatherians (e.g., Marshall and de Muizon, 1988), along with attempts at understanding the evolutionary morphology of dental and skeletal change, will undoubtedly sharpen the picture of their history, which is presently unclear. Good taxonomic properties for diagnostic features will be hard to establish. Suggestions of reduction of cusps (paracone), evolution of dilambdodonty, continuous but difficult to polarize transitions between extremes of size of stylar shelves and hypoconulids, and the degree of separateness of the latter from the entoconid will have to be approached with a causally integrated perspective. Even that may not yield reliable answers.

The pediomyine-alphadontine tarsal material from the Oldman Formation is far more primitive, given the criteria outlined for tarsal transformation, than either the North American Paleogene sample, or some of the most primitive material from Itaborai (e.g., Fig. 15.10). The calcaneal **CaFi** facet is very large and the peroneal process is the least reduced of any marsupial sample known. The meager Tiffany, Wasatch, Bridger, and Tepee Trail tarsal material is morphologically distinct and clusters outside of any other sample. A chronochange in the Eocene is traceable within this group (Szalay, 1993). The reduction of the calcaneal **CaFi** facet is established in the Paleocene, and in the early Eocene a slightly convex **UAJ** seen in the Bitter Creek, and to a lesser degree in the Bridger samples, is transformed into the caenolestid-like (Fig. 15.11, top), highly angulated **ATim-Atil** facets of the upper ankle joint (Szalay, 1993). This is likely to reflect the changes in an increasingly terrestrial lineage in which molar morphology remained conservative. The Pediomyidae, therefore, is either a holophyletic group that gave rise to no others or a paraphyletic one, hiding a cluster that is the ancestor to the sudameridelphian radiation (see Wible, 1990, for the basicranial evidence).

SUBORDER SUDAMERIDELPHIA. The groups included are the infraorders Itaboraiformes (Carolameghiniidae), Polydolopimorphia (Prepidolopidae, Polydolopidae, and Bonapartheriidae), and Sparassodonta (Borhyaenidae and Thylacosmilidae).

The protosudameridelphian had, in addition to its relatively primitive metatherian molar structure and primitive metatherian dental formula: tarsal morphology characterized by a combination of a tricontact **UAJ**, narrow **AFi** facets, somewhat reduced peroneal process, semicircular **CCJ**, very large **ampt**, and a retention of a ribbon-like taper of the **Su** facet proximally (described in detail in Szalay, 1993).

The protopolydolopimorphian had: a combination of enlarged premolars (perhaps originally as in the prepidolopids, and therefore not in a plagiaulacoid mode), reduced trigonid shear and bunodont crown pattern; a combination of drastic reduction of the calcaneal peroneal process, truncated **CCJ**, retention of the primitively elongated calcaneal **Su** facet, as well as the relatively larger calcaneal **CaA** facet compared to the **CaFi** facet.

The protosparassodont, as in a hypothetical hathlyaciynine or *Jaskhadelphys*-like borhyaenid ancestor, had: a (secondarily?) enlarged stylar shelf (due to the demands of postmetacrista-protocristid shear); reduced (but not eliminated) stylar cusps, paracone only slightly smaller than metacone; sizable epitympanic recess in basicranium; tarsal morphology similar to that commonly encountered in primitive itaboraiform groups.

SUBORDER GLIRIMETATHERIA. The infraorders Paucituberculata (Caenolestidae) and Simpsonitheria (Gashterniidae, Groeberiidae, Argyrolagidae, and Patagoniidae) make up this suborder, discussed in Szalay (1993).

The protoglirimetatherian had: incisors reduced to **4/4**, with an initiation of the enlargement of first pair of incisors independently from polydolopimorphians and australidelphian phalangeriforms; essentially the itaboraiiform pedal structure seen in the majority of the Itaborai taxa, and still preserved in the living caenolestids (see Fig. 15.11); molar dentition that was probably characterized by reduced height of trigonids and subsequently extensive horizontal shear, rather than any indication of the early bunodonty seen in the polydolopimorphians (hypocone may be homologous to the metaconule of its ancestry).

The protopaucituberculatan had: emphasized vertical shear between **P3/M1**, although its heritage indicates a loss of the high trigonids, and the upper molars of a probable itaboraiiform ancestry; pedal morphology that was probably primitive sudameridelphian.

The protosimpsonitherian, the stem of a holophyletic infraorder, can be diagnosed as a small, ricochetal marsupial with two functional, greatly elongated metatarsals (3 and 4), highly modified mortise-tenon **UAJ**, fused crus, possibly apomorphic (known only in argyrolagids) deemphasis of the **CaA** facet and the hypertrophy of the **CaFi** facets, together with a "stepped" **CCJ**; short forelimb; skull (where known) with enormous palatal vacuities (probably for cooling), long and pointed rostrum extending well past the enlarged upper teeth, completely ossified alisphenoid bulla, mastoid region of the skull inflated, zygoma deep and downward arching, orbital region expanded and posteriorly isolated from the temporalis musculature; lower jaw deep with small coronoid process; on each side, both above and below, two gliriform, enlarged anterior teeth, probably **I1** and **I2**, and cheek teeth reduced in number (see especially Simpson, 1970; Pascual and Carlini, 1987).

SUBORDER DIDELPHIMORPHIA. The Didelphidae and Sparassocynidae make up this suborder. Undoubtedly, at least two of the Itaborai dental taxa are didelphids, judged by the presence of two didelphid calcaneal phena (Szalay, 1993).

The protodidelphimorphian, dentally probably extremely similar to the morphotypic ameridelphian condition, had: in addition to a full complement of the metatherian dentition, a molar morphology possibly similar to the ones found in caluromyines, namely with a U-shaped rather than the W-shaped ectoloph seen in didelphines, and only a very slight reduction of the paracone compared to the metacone; ear region probably with a small descending wing of the alisphenoid anchoring the fibrous enclosure of the middle ear; tarsal morphology with uniquely derived combination of having an (a) **UAJ** that lost the **CaFi** articulation and extended laterally the **AFi** contact, (b) **LAJ** articulation with an astragalar **Su** facet proximally rounded rather than ribbon-like and tapered, (c) reduced **ampt**, (d) **CCJ** in which the proximally extended new contact breaks up the original modified ovoid articulation into one that has both a **CaCud** facet (the original metatherian **CaCu** facet) and a didelphid **CaCup** facet (the latter condition manifesting itself in a "pivot-like" projection of the cuboid underneath the calcaneus; see Fig. 15.12).

COHORT AUSTRALIDELPHIA. The orders Gondwanadelphia (suborders Microbiotheria and Dasyuromorphia) and Syndactyla (semiorders Peramelina and Diprotodontia, the latter including the suborders Phalangeriformes, Vombatiformes, and Yalkaparidontia) make up this holophyletic cohort.

ORDER GONDWANADELPHIA. The protoaustralidelphian was the first gondwanadelphian, and it was probably South American. It had, in addition to its complete metatherian dental formula: molars probably with stylar cusps as in either caluromyines or stem dasyuromorphians; ear region as in primitive dasyuromorphian, somewhat didelphid-like without a fully ossified bulla, although probably with an alisphenoid wing and a small inflated tympanic wing of the petrosal; complex tarsal modifications in the **EMt1J, UAJ, LAJ,** and **CCJ** as in the living genus *Dromiciops* (see details below).

SUBORDER MICROBIOTHERIA. The South American suborder of the Gondwanadelphia is the Microbiotheria. The protomicrobiotherian, which was derived from the gondwanadelphian ancestry, probably had: a secondarily simplified upper molar pattern; a composite bulla, formed by the tympanic wing of the alisphenoid and the inflated wings of the petrosal (pars petrosa and pars mastoidea); ectotympanic enclosed within the bulla proper; incisors, unlike in didelphids, not peg-like, but mesiodistally expanded, and the lower ones with lingually slightly expanded bases. The latter condition may have been present in primitive Ameridelphia, and therefore not a diagnostic trait of gondwanadelphians or microbiotherians.

The living *Dromiciops* shares with the inferred last common ancestor of the Australasian Syndactyla, in the smallest of the intertaxonally significant details, the inferred attributes of the tarsus. The characters noted are actually present in the phalangeroids and petauroids. These are: (a) the condition of the calcaneal peroneal process reduced to a nubbin; (b) as in didelphids the **ampt** is small; (c) **EMt1J** is saddle-shaped, identical to that seen in the primitive condition of phalangeriforms, and it is (because of its special similarity) derivable from the didelphid condition; (d) in contrast to the didelphids and other ameridelphians the **UAJ** facets on the tibia and astragalus are smoothly continuous as in the putative ancestral syndactylan and diprotodontian (while slightly less angulated than that seen in didelphids, the **AFi** contact, as in didelphids, is expanded); (e) **LAJ** articular facets are confluent resulting in what has been dubbed the continuous lower ankle joint pattern (**CLAJP**), in contrast to the primitive separate lower ankle joint pattern (**SLAJP**) of most synapsids, therians, and non-australidelphian metatherians; more particularly, it is derivable from the condition found in didelphids (the latter apomorphic in the Didelphida in that regard) that have astragalar **Su** facets diagnostically rounded proximally; (f) **CCJ** transformed from the double-faceted dielphid pattern of articulation to the highly modified, triple-faceted, pattern of australidelphians in which the third facet, the **CaCum** one, is a neomorph (see Fig. 15.12).

The complex pattern of the **CCJ** of *Dromiciops* (and that of the protogondwanadelphian), as reflected on both the calcaneus and cuboid, is most clearly derivable from the advanced ameridelphian pattern of didelphids, as the paragraph above suggests. The didelphid **CCJ** can be divided into two slightly angled facets, the distal and proximal ones (Fig. 15.12). In the derivation of the australidelphian pattern a third articular area was added, the **CaCum** facet. The slightly more distal **CaCul** facet is retained approximately in the same proportion as its ancestral homologue, the didelphid **CaCup** facet. The narrow australidelphian **CaCua** facet is the reduced, but persistent, original **CaCud** facet of didelphids, the last remnant of the original therian and metatherian unifaceted, modified ovoid. The latter is the **CCJ** articulation of archimetatherians and most other Ameridelphia.

In stem australidelphians a cuboid process (**cump**) that medially surrounds the large groove for the tendon of the peroneus longus (**gtpl**) develops great medial prominence and proximodistal depth. Both the australidelphian **CaCum** facet as well as the robust **cump** persist in the modified tarsus of the dasyuromorphians. This removes any doubt, and strongly corroborates, that the australidelphian tarsal innovation was *not* like that in dasyuromorphians (contrary to the suggestion of Marshall, 1972), but was similar to the microbiotherian one. The derived nature of the dasyuromorphian tarsal complex in the Australidelphia is discussed below.

SUBORDER DASYUROMORPHIA. This is the Australian suborder of Gondwanadelphia. The protodasyuromorphian had: petrous portion of periotic slihgtly inflated; an incisor formula reduced to **4/3**; relatively primitive ameridelphian dental morphology, perhaps even in the morphology of the incisors, which appear to have been channeled by the conditions noted under Microbiotheria; the talonids of molars reduced relative to the size of the trigonids when compared to the primitive didelphid and microbiotheriid dentitions; carpus with scaphoid retaining the lateral and distal projection; hind foot structure transformed into an elongated and hallux-deemphasized terrestrially adapted one (Fig. 15.7); tarsals modified (Figs. 15.12 and 15.13) under the differing selectional pressures exerted by usage of a primarily non-arboreal substrate, leading to the abandonment of the obligate extreme inversion of the australidelphian ancestral condition; **EMt1J** modified from the diagnostic australidelphian one; **LAJ**, while retaining **CLAJP**, has additional, new contact (**CaAd**); **CCJ** that is secondarily "simplified" (the angled appearance of the australidelphian condition described above is nearly eliminated, but evidence for the ancestral triple-faceting of this joint is retained; see above and Figs. 15.12 and 15.13).

Virtually all previous efforts to understand the origins of dasyuromorphians have centered on dental attributes, with some attention focused on the skull. An excellent review of the literature can be found in Archer (1984). The issue of dental apomorphies of the first dasyurid, probably the stem dasyuromorphian, continues to be beset with as yet insurmountable difficulties for two reasons. The obvious one is the lack of an adequate fossil record of teeth, coupled with the relative uniformity of the molars within the Dasyuridae. The second is the great phenetic proximity of molar structure to that found in some didelphids and other dentally relatively unmodified or convergent ameridelphians. Consequently, those espousing the concept of the Marsupicarnivora have expressed a belief that a relative phenetic uniformity of dental features also mirrors the history of the groups themselves. But if various "marsupicarnivorans" can be shown to be more recently related to other groups not originally included in the concept of the Marsupicarnivora, then that taxon is polyphyletic, and not paraphyletic. Thus, if the sparassodonts prove to be more recently related to the glirimetatherians or the polydolopimorphians, for example, then this would certainly be an argument against the Marsupicarnivora. Dasyuromorphians are more recently related to the Microbiotheria and Syndactyla than to any other group, and these, together as the Australidelphia, are holophyletic. Sarich (personal communication; 1993) pointed out immunological distance measures that corroborate the validity of the Gondwanadelphia.

ORDER SYNDACTYLA. The semiorders Peramelina (Notoryctidae and Peramelidae, the latter including *Macrotis*) and Diprotodontia make up the Synadactyla. The protosyndactylan had: a nearly full complement of the metatherian dentition (last lower incisor lost) and molars with a well developed dilambdodont ectoloph; petromastoid that probably expanded onto occiput; basicranium that was probably not pneumatized, but an internal overlapping by the squamosal process of the roof of the middle ear was probably present; carpus probably primitive didelphimorphian with scaphoid with lateral distal process (probably the incorporated centrale); tarsus virtualy identical to that of the living microbiotherian, but the pes syndactylous.

The dental null group for the dentally most primitive syndactylans (such as we see in dentally primitive peramelids) is the didelphid-dasyurid condition. For the osseous foot structure, however, the null group is the presyndactylan condition, which occurs in the living *Dromiciops*.

SEMIORDER PERAMELINA. There are a number of significant attributes that may corroborate the hypothesis of syndactylan, or a near and possibly more specific peramiloid, affinities of *Notoryctes*. The fourth ray of the hand is hypertrophied, the third somewhat less so. The carpus, while highly modified, has the primitive didelphimorphian distal and lateral process of the scaphoid. The scaphoid appears to have incorporated the trapezoid and dorsally it covers up the magnum, which is nevertheless an independent ossification. The **UAJ** of the marsupial mole is also a highly derived area within the Australidelphia, but derived differently than that seen in the bandicoots. The latter has the "eutherian" mortise-tenon modification and a eutherian-like proximal tibio-fibular contact. The crural mortise of *Notoryctes* (and *Argyrolagus*) is more "marsupial" (or primitive) in having a greater participation by the fibula. In fact this notoryctid condition may represent the one antecedent to the more derived peramelid one in which both the proximal and distal extremities of the fibula are reduced to eutherian-like conformation. Added to this possible transformational association is the peculiar separation of the hypertrophied **Mt4** from **Mt1–3**, and the lateral and plantar, "tucked under", conformation of an externally (but not so much osteologically) reduced fifth ray of the foot. The slight reduction of **Mt5** is compensated by the enormously hypertrophied proximal lateral process of that bone (see Fig. 15.12). This process is a buttress against the peroneal region of the calcaneus; it is functionally (but not morphologically) convergent with the fossorial monotremes, which ligamentously tie their peroneal processes to the lateral proximal section of **Mt5** (see Szalay, this volume, Chapter, 9). The reduced lateral ray of the foot in notoryctids suggests an ancestral constraint in which **Mt5** was also reduced. Unlike the reduced **Mt1** of the protoperamelid, there is every reason to believe that the notoryctid ancestry had a large first pedal ray—a decidedly more primitive australidelphian condition than that displayed by bandicoots (or dasyurids). The skin webbing uniting **Mt1–3**, but not **Mt4**, may represent the medial extension of the original syndactylous association in an ancestor that has opted for terrestrial locomotion emphasizing **Mt4**.

Although notoryctids and peramelids share a rod-like stapes, I consider such a modified condition from an ancestral bicrurate stapes an extremely poor character, usually associated with fossoriality, and would not list it in a diagnosis (see also the views of Rose and Emry, 1993). Similarly, the unquestionably fossorial modifications of the forefoot in the protoperamelids may or may not have been shared in a common ancestry with notoryctids. As I discussed (Szalay, 1993), and as fully noted by Winge (1941), Gregory (1951), and Marshall (1972) for these groups, the syndactyly of ancestry undoubtedly channeled the emphasis onto **Mt4** as opposed to **Mt3**. Given these factors. the unenlarged incisor morphology suggests primitive, nondiprotodont syndactylan ties, but perhaps not specifically with peramelids as early argued by Bensley (1903), but rather with hitherto unknown peramelinans. The great depth of the navicular and contrasting smallness of the mesocuneiform, probably associated with the relatively unloaded syndactylous rays in contrast to the compressively loaded adjacent **Mt4**, is a peramelid-like feature of *Notoryctes* (Fig. 15.14). Yet there is no reason to suspect that **Mt1–3** are not more or less equally loaded while shoveling dirt. This may be an indication of a constraint acquired in an ancestral terrestrial lineage similar to primitive peramelids (but without a chorioallantoic placenta). The great reduction of the epipubic bones is an apomorphic notoryctid condition, and in general the numerous skeletal modifications causally related to a burrowing life are all unique attributes of this group.

The protoperamelid had: chorioallantoic placenta; almost full metatherian complement of teeth (**I1,2,3,4,5/1,2,3**; **C1/1**; **Pd1,d2,3**; **M1,2,3,4**), molars with well-developed *W*-shaped ectoloph; basicranium with relatively large alisphenoid wing and with smaller petrosal wing, and with a slight intracranial overlap of the alisphenoid by the squamosal; highly derived "jointed" neonate shoulder-breast arch in which the bracing of the shoulder and breast region is accomplished through the loading of the humeral head; adult clavicles absent; spinous processes of thoracic vertebrae exceptionally long and inclined posteriorly; spinous processes of lumbar vertebrae exceptionally robust and long and inclined forward; mammillary processes of vertebrae lengthened; lumbar vertebrae with elongated transverse processes; pronation-supination reduced in the forearm, and the ulna distally reduced in importance; hindlimb longer than forelimb; middle three digits well developed on the fossorial manus and lateral ones deemphasized; carpus with large lunula, shallow magnum, and a scaphoid without a distal and lateral process; reduced fibula distally forming the lateral side of the **UAJ** mortise and proximally withdrawn from contact with the femur; fibula articulating with calcaneus; synovial **TFJ** modified into syndesmosis distally; patella larger and more completely ossified than in other living marsupials in which it is usually cartilaginous (with small ossification centers in many taxa); femur with a diagnostic long and deep patellar articulation; first digit of syndactylous pes slightly reduced in size and nongrasping; first digit lined up with the remaining rays; mesocuneiform quite small and possibly related to an enlarged condition of the **Mt4**; **Mt4** more enlarged than the robust **Mt5**; **Mt4** proximally in contact with both the cuboid and the ectocuneiform—the latter shifted its distal articular contact from **Mt3** to **Mt4**, retaining only small and probably nonweight-bearing contact with the third metatarsal.

SEMIORDER DIPROTODONTIA. The suborders Phalangeriformes, Vombatiformes, and Yalkaparidontia are included in this taxon.

The protodiprotodontian had: incisor number probably **I3/3**, with the enlargement of a lower incisor (**I/1** ?) and the less pronounced enlargement of the first upper one; last premolar with shearing emphasis (retained in the dentally more primitive descendants); **M/1** retained a low trigonid, probably because of influence of **P3/3** form-function; **M/2–4** with four cusps emphasized, the protoconids, metaconids, hypoconids, and entoconids, all of which were incipiently and functionally near-bilophodont; upper molars with "hypocone," probably derived from metaconule, also incipiently lophodont but retaining the crescentic (quasi-selenodont) sweep of the preparacone and postmetacone crests; basicranium probably with a relatively small alisphenoid wing but without a completely ossified bulla (bandicoot-like); carpus in which the scaphoid is enlarged but the lunate is present, albeit reduced to a small bone, compared to the more primitive australidelphian condition (which had relatively large lunate); tarsal morphology as described under Australidelphia. Living forms have a superficial thymus gland and share the *fasciculus aberrans*, a connection between the cerebral hemispheres, homoplastic with the eutherian solution of the *corpus callosum*.

Classification

Based on the phylogenetic tree in Figure 15.15, partly on the evidence and interpretations cited above and primarily on information and analysis presented in Szalay (1993), I give a brief outline classification consistent with the tree, and the cladogram in Figure 15.16 which is *derived* from the tree.

It is my conviction that the ordinal classification of Metatheria is highly inflated when compared with the rest of the Mammalia. This is partly driven by the formalism of taxonomy in general, which often equates any node recognition with new ranking. This practice, however, is clearly exacerbated by cladistic ideology that strives for the isomorphy of cladograms and classifications, irrespective of the confidence one has in the level of corroboration of taxa or the nature of differences between branches or segments of lineagaes. As an obviously subjective estimate, living metatherians do not appear to have been more diverse (i. e. divergently modified) and probably considerably less so, reproductively, chromosomally, and morphologically than the generously combined phenetic range of the Carnivora, Primates, and Rodentia. By any measure of multiplicity (i.e., the known number of species) metatherians are nowhere as speciose as the combined range of the eutherian orders noted. In light of the ancient identity of the Deltatheroida from other metatherians I recognize four orders within the Metatheria (for detailed discussions, see Szalay, 1993).

Subclass Theria
  Infraclass Metatheria
    Cohort Holarctidelphia
      Order Deltatheroida
    Cohort Ameridelphia
      Order Didelphida
        Suborder Archimetatheria
        Suborder Sudameridelphia
          Infraorder Itaboraiformes
          Infraorder Polydolopimorphia
          Infraorder Sparassodonta
        Suborder Glirimetatheria
          Infraorder Paucituberculata
          Infraorder Simpsonitheria
        Suborder Didelphimorphia
    Cohort Australidelphia
      Order Gondwanadelphia
        Suborder Microbiotheria
        Suborder Dasyuromorphia
      Order Syndactyla
        Semiorder Peramelina
          Family Notoryctidae
          Family Peramelidae
        Semiorder Diprotodontia
          Suborder Phalangeriformes
          Suborder Vombatiformes
            Semisuborder Vombatimorphia
            Semisuborder Diprotodontiformes
          Suborder Yalkaparidontia

ACKNOWLEDGMENTS. This research was partly supported by PSC CUNY research grants 666141 and 66731, and by a fellowship of the Guggenheim Foundation during the year of 1980. My renewed thanks to Dr. James Warren of Monash University for the hospitality extended to me while working in Australia. I am especially grateful to Dr. Guy G. Musser of the American Museum of Natural History for his patient understanding, Dr. Richard C. Fox of the University of Alberta for his loan of particularly important Cretaceous specimens, and to Dr. Diogenes Compos of the DNPM, Rio de Janeiro, for loan of tarsal material.

I thank Dr. Richard L. Cifelli, who has read the manuscript and made numerous useful suggestions.

CORRESPONDENCE ADDRESS. Frederick S. Szalay, Departments of Anthropology and Ecology and Evolutionary Biology, Hunter College, 695 Park Avenue, New York, NY 10021, USA.

References

Aplin, K.P., and Archer, M. 1987. Recent advances in marsupial systematics with a new syncretic classification. In: *Possums and opossums: Studies in evolution, vol. 1* (Archer, M., ed.). Sydney, Australia: Surrey Beatty & Sons, pp. xv–lxxii.

Archer, M. 1984. Origins and early radiations of marsupials. In: *Vertebrate zoogeography & evolution in Australasia (Animals in Space & Time)* (Archer, M., and Clayton, G., eds.). New South Wales, Australia: Hesperian Press, pp. 585–625.

Bensley, B.A. 1903, On the evolution of the Australian Marsupialia; with remarks on the relationships of the marsupials in general. *Transcr. Linn. Soc. London*, Series 2, 9 (Zoology): 83–217.

Bock, W.J. 1977a. Adaptations and the comparative method. In: *Major patterns in vertebrate evolution* (Hecht, M.K., Goody, P.C., and Hecht, B.M., eds.). NATO ASI series A 14:57–82. New York: Plenum Press.

Bock, W.J. 1977b. Foundations and methods of evolutionary classification. In: *Major patterns in vertebrate evolution* (Hecht, M.K., Goody, P.C., and Hecht, B.M., eds.). NATO ASI series A 14:851–895. New York: Plenum Press.

Bock, W.J. 1981. Functional-adaptive analysis in evolutionary classification. *Amer, Zool.* 21:5–20.

Bock, W.J. 1990. From Biologische Anatomie to Ecomorphology. *Neth. J. Zool.* 40:254–277.

Cifelli, R.L. 1990a. Cretaceous mammals of southern Utah. I. Marsupials from the Kairparowits Formation (Judithian). *J. Vert. Paleo.* 10:295–319.

Cifelli, R.L. 1990b. Cretaceous mammals of southern Utah. II. Marsupials and marsupial-like mammals from the Wahweap Formation (early Campanian). *J. Vert. Paleo.* 10:320–331.

Cifelli, R.L. 1990c. Cretaceous mammals of southern Utah. III. Therian mammals from the Turonian (early late Cretaceous). *J. Vert. Paleo.* 10:332–345.

Clemens, W.A., and Lillegraven, J.A. 1986. New Late Cretaceous North American advanced therian mammals that fit neither the marsupial nor the eutherian molds, Contrib. Geol. Special paper 3:55–85.

Crochet, J.-Y., 1980, L'Occlusion dentaire chez *Peradectes, Amphiperatherium* et *Peratherium*, Marsupiaux du Tertiaire d'Europe. Paleovertebrata, pp. 79–89, Montpellier, Mm. Jubil. R. Lavocat.

Eaton, J.G. 1990. Therian mammals of the Cenomanian (late Cretaceous) Dakota Formation, Southwestern Utah. *J. Vert. Paleo.* 9:21A.

Fleischer, G. 1973. Studien am Skelett des Gehororgans der Saügetier, einschliesslich des Menschen. *Saügetierkundl. Mitteilungen* 21:131–239.

Fleischer, G. 1978. Evolutionary principles of the mammalian middle ear. *Adv. Anat. Embryol. Cell Biol.* 55:1–69.

Fox, R.C. 1987. Palaeontology and the early evolution of marsupials. In: *Possums and opossums: Studies in evolution,* vol. 1 (Archer, M., ed.). Sydney, Australia: Surrey Beatty & Sons and the Royal Zoo. Soc. of N.S. Wales,

Gregory, W.K. 1922. On the "habitus" and "heritage" of *Caenolestes. J. Mammal.* 3:106–114.

Gregory, W.K. 1951. *Evolution emerging: A survey of changing patterns from primeval life to man,* vols. 1 and 2. New York: Macmillan.

Kielan-Jaworowska, Z., and Dashzeveg. D. 1989. Eutherian mammals from the Early Cretaceous of Mongolia, *Zoologica Scripta* 18:347–355.

Kielan-Jaworowska, Z., and Nessov, L.A. 1990. On the metatherian nature of the Deltatheroida, a sister group of the Marsupialia, *Lethaia* 23:1–10.

Maier, W. 1987a. The angular process in *Monodelphis domestica* (Didelphidae: Marsupialia) and its relationships to the middle ear: An ontogenetic and evolutionary study. *Gegenbaurs Morphol. Jahrb.* (Leipzig) 133:123–161.

Maier, W. 1987b. The ontogenetic development of the orbitotemporal region in the skull of *Monodelphis domestica* (Didelphidae: Marsupialia), and the problem of the Mammalian alisphenoid. In: *Morphogenesis of the mammalian skull* (Kuhn, H.-J., and Zeller, U, eds.). Hamburg: Verlag Paul Parey, pp. 71–90.

Maier, W. 1989. Morphologische Untersuchungen am Mittelhor der Marsupialia. *Z. Zool. Syst. Evolut.-forsch.* 27:149–168.

Marshall, L.G. 1972. Evolution of the peramelid tarsus. *Rep. Roy. Soc. Vict.* 85:51–60.

Marshall, L.G. 1987. Systematics of Itaborian (Middle Paleocene) age "opossum-like" marsupials from the Limestone Quarries at San José de Itaborai, Brazil. In: *Possums and opossums: Studies in evolution, vol. 1,* (Archer, M., ed.). Sydney, Australla: Surrey Beatty & Sons, pp. 91–160.

Marshall, L.G., Case, J.A., and Woodburne, M.O. 1990. Phylogenetic relationships of the families of marsupials. In: *Current mammalogy, vol. 2* (Genoways, H.H., ed.). New York: Plenum Press, pp. 433–505.

Marshall, L.G., and de Muizon, C. 1988. The dawn of the Age of Mammals in South America. *National Geographic Research* 4:23–55.

Murray, P., Wells, R., and Plane, M. 1987. The cranium of the Miocene thylacoleonid marsupial *Wakaleo vanderleuri*: Click go the shears—a fresh bite at thylacoleonid systematics. In: *Possums and opossums: Studies in evolution, vol. 2,* (Archer, M., ed.). Sydney, Australla: Surrey Beatty & Sons, pp. 433–466.

Neff, N.A. 1986. A rational basis for a priori character weighting. *Syst. Zool.* 35:110–123.

Norris, C.A. 1992. Changes in the composition of the auditory bulla in southern Solomon Islands populations of the Grey cuscus, *Phalanger orientalis breviceps (Marsupialia, Phalengeridae). Journal of Zoology of London.*

O'Hara, R.J. 1988. Homage to Clio, or toward a historical philosophy for evolutionary biology. *Syst. Zool.* 37:142–155.

Pascual, R., and Carlini, A.A. 1987. A new superfamily in the extensive radiation of South American Paleogene marsupials. In: *Studies in neotropical mammalogy: Essays in honor of Philip Hershkovitz* (Patterson, B.D., and Timm, R. M., eds.). *Fieldiana, Zoology,* n.s.39:

Pascual, R., Carlini, A.A., and de Santis L.J.M. 1986. Dentition and ways of life in Cenozoic South American rodent-like marsupials. Outstanding examples of convergence. In: *Teeth revisited: Proceedings of the Seventh International Symposium on Dental Morphology* (Russell, D.E., Santoro, J.-P., and Sigogneau-Russell, D., eds.). *Mem. Mus. Natn. Hist. Nat., Paris* (series C) 53:217–226.

Reig, O.A., Kirsch, J.A.W., and Marshall, L.G. 1987. Systematic relationships of the living and Cenozoic American "opossum like" marsupials (suborder Didelphimorphia), with comments on the classification of these and of the Cretaceous and Paleogene New World and European metatherians. In: *Possums and opossums: Studies in evolution, vol.1* (Archer, M., ed.). Sydney, Australia: Surrey Beatty and Sons, pp. 1–89.

Rose, K.D. and Emry, R.J. 1993. Relationships of Xenarthra, Pholidota, and fossil edentates: the morphological evidence. In *Mammal Phylogeny: Placentals* (Szalay, F.S., Novacek, M.J. and Mc Kenna, M.C., eds.). New York: Springer-Verlag.

Sarich, V.M. 1993. Mammalian systematics: Twenty-five years among their albumins and transferrins. In *Mammal Phylogeny: Placentals* (Szalay, F.S., Novacek, M.J., and Mckenna, M.C., eds.). New York: Springer-Verlag.

Simpson, G.G. 1970. Addition to knowledge of *Groeberia* (Mammalia, Marsupialia) from the Mid-Cenozoic of Argentina. *Breviora Mus. Comp. Zool, Harvard,* 362:1–17.

Springer, M.S., and Woodburne, M.O. 1989. The distribution of some basicranial characters within the Marsupialia and a phylogeny of the Phalangeriformes. *J. Vert. Paleon.* 9:210–221.

Szalay, F.S. 1977. Phylogenetic relationships and a classification of the eutherian Mammalia. In: *Major patterns in vertebrate evolution* (Hecht, M.K., Goody, P.C., and Hecht, B.M., eds.). NATO ASI series A14:315–374. New York: Plenum Press.

Szalay, F.S. 1981. Functional analysis and the practice of the phylogenetic method as reflected by some mammalian studies. *Amer. Zool.* 21:37–45.

Szalay, F.S. 1982. A new appraisal of marsupial phylogeny and classification. In: *Carnivorous marsupials, vol.2.* (Archer, M., ed.). Sydney, Australia: Royal Zoological Society of New South Wales, pp. 621–640.

Szalay, F.S. 1984. Arboreality: Is it homologous in metatherian and eutherian mammals? In: *Evolutionary biology, vol. 18* (Hecht, M.K., Wallace, B., and Prance, G.T., eds.). New York: Plenum Press, pp. 215–258.

Szalay, F.S. 1985. Rodent and lagomorph morphotype

adaptations, origins, and relationships: Some postcranial attributes analyzed. In: *Evolutionary relationships among rodents—a multidisciplinary analysis* (Luckett, W.P., and Hartenberger, J.L., eds.). NATO ASI series A, *Life Sciences* 92: 83–157. New York: Plenum Press.

Szalay, F.S. 1993. Evolutionary history of the marsupials and an analysis of osteological characters. New York: Cambridge University Press.

Szalay, F.S., and Bock, W.J. 1991. Evolutionary theory and systematics: relationships between process and patterns. *Z. Zool. Syst. Evolutionforsch.*, 29:1–39.

Wible, J.R. 1990. Petrosals of late Cretaceous marsupials from North America, and a cladistic analysis of the petrosal in therian mammals. *J. Vert. Paleo.* 10:183–205.

Winge, H. 1941. *The interrelationships of the mammalian genera, vol. 1, Monotremata, Marsupialia, Insectivora, Chiroptera, Edentata* (written between 1887 and 1918, translated from Danish by E. Deichmann and G.M. Allen, and edited by S. Jensen, R. Sparck, and H. Volsoe). Copenhagen: C.A. Reitzels Forlag.

Topic Index

A
ala temporalis, 23, 100, 172
alicochlear commissure, 21, 23–4, 28, 174
alisphenoid (epipterygoid), 50, 56, 64, 66–8, 97, 138, 141, 172
amniotes, 100
angular process, 140, 174
angulare, 95, 102, 174
appositional bone, 167
aquaeductus cochleae, 56, 95, 100
arteries
    carotid, 21, 23
    promontory, 24
    stapedial, 21, 23, 25, 98, 101
    tympanic region, 23–25, 27, 101
articular(e), 95, 102, 134, 137–8, 140, 167
auditory bulla, 95, 97, 102, 223–225
auditory tube, 66, 99, 135, 223–225

B
basicranium, 24, 26, 95, 137, 148, 223–225
basioccipital, 64, 66, 97, 223–225
basipterygoid articulation, 172
basisphenoid, 64, 66, 97, 100, 223–225
behavior, 14
birth, 13–4, 16
blastocyst, 6–8, 13
bulla tympanica, *see* auditory bulla

C
calcaneus, 108–126, 148, 226–227
calcaneocuboid joint, 112–22, 124
caput mallei, 95, 98, 102
carpus (bones of), 226–227
cavum epiptericum, 28, 50, 67–8, 134, 137, 141, 173
cavum supracochleare, 24, 27–8, 50, 68
centrale, 227
character, *see* phylogenetics, character
chondrocranium, 21–8, 97, 167
chondrogenesis, 22–4
chorda forsatz, 25–6
chorda tympani, 25–6, 98
cochlea, 54, 95, 100, 141
cochlear capsule, 24, 96

coracoid, 135, 140
cranial base, 95, 100
cranial cavity, 96, 98, 100, 104
cranial nerves, 23, 105, 134
crista parotica, 51, 97
crus, 117, 119, 123
cuneiform (triquetrum), 227

D
dentary, trough of, 56, 102, 174
dentition, 96, 130–4, 136–8, 140, 142, 198, 200, 182, 198
    dental development, 182, 185, 194
    dental homologies, 158, 182, 188, 195, 197
    dental lamina, 185, 192
    dental ontogeny, 184, 200
    dentary bone, 95, 101, 135, 137–8, 140
    dilambdodont, 220–23
    enamel, 146, 155, 157–8, 160–3, 187
    serial homology, therians, 184, 198
    stylar cusps, 211, 220–223
    successional lamina, 186, 190–1, 195
    tooth eruption, 148, 187, 193–4, 196
    tooth loss 158, 160–3, 192, 198–9
    tooth replacement, 133, 158, 160–3, 186–7, 194, 196
    tribosphenic, 109, 186, 206, 220–223
    vestigial deciduous, 187, 190, 192, 194, 196
development, 133, 134, 135, 136–7, 139–42
    acceleration, 134, 188, 193, 199
    rates, 184, 193
    retardation, 191, 199
diapause, 13
diarthrodial joints, 111–124
differentiation, 135–6, 186, 189
dimorphism, 1, 12
distal tibiofibular joint (cruropedal joint), 119
ductus cochlearis, 54, 95, 100

E
ear ossicles, 22, 95, 138
ear region, 21–28, 67, 68, 95, 135, 138

ectopterygoid, 66, 68, 97, 179
embryonic, 4–7, 11, 95, 185
endocranium, 95, 167
entotympanics, 26, 97
epipterygoid, 28, 50, 172

F
feeding apparatus, 169
fenestra
    ovalis, 100, 102, 104
    pyriform, 21, 26–8
    rotunda, 100, 104
    vestibular, 64
fetal membranes, 7, 11, 97
fissura metotica, 97, 100
foramen
    condylar, 159–63
    ethmoid, 158, 160–3
    incisive, 158, 160–3
    infraorbital, 64, 70, 158
    for internal carotid artery, 55, 97, 100
    jugulare, 98, 101
    magnum, 98
    masticatorium, 158, 160–63
    perilymphaticum, 95, 97, 100
    postglenoid, 54
    pterygoparoccipital, 54
    stylomastoideum primitivum (N. VII), 26
    supraglenoid, 148
fossa
    fossula fenestrae rotundae, 101
    glenoid, 64, 66, 148
    jugular, 159–63, 224
    post-temporal, 148
    subarcuate, 148
frontal, 64–5, 71, 158, 170
functional-adaptive analysis, 109–111

G
gene, 10–12
genital duct, 4–16
genito-urinary tract, 4–20
genome, 10
goniale/praearticulare, 95, 102
Graafian follicles, 4, 6–7

H
hamate, 227
heterochrony, 184, 199
hyoid, 170

I
incubation patch, 177
incus, 24, 26, 51, 95, 102, 170
infraorbital canal, 66, 70
internal auditory meatus, 53, 67–69

J
jaw apparatus, 95, 102
jaw joint, primary, 24, 96, 98, 102, 169
jugal, 148, 224
junctura basipalatina, 172

L
lacrimal, 64, 66, 68, 158, 170
lactation, 4, 7, 14–17, 177, 195
lamina obturans, 49, 98, 100, 177
leptomeningeal space, 100
locomotion, 125–6
lower ankle joint, 111–124
lower jaw, 89, 95
lunate, 227
luteolytic controls, 7
luteotrophic controls, 7

M
malleus, 25–6, 56, 95, 102, 170
mammary gland, evolution of, 14–17
mastoid (of petrosal), 24
maxillary, 170
Meckel's cartilage, 23, 96, 170
membranous labyrinth, 100
mesenchyme, 22–23
metatarsals, 114
middle ear, 21–29, 69, 95, 174, 223–4
milk composition, 14–17
milk-sucking, 14–17, 177
morphogenesis, 95
Mullerian duct (paramesonephric), 8–9
muscles
   m. stapedius, 26, 99
   m. tensor tympani, 26, 95, 99, 102

N
nasal
   bone, 63–4, 66, 69, 72, 158
   capsule, 175
   cavity, 64
nasopalatine ducts, 176
nerves
   facial, 23, 26, 50, 99, 101
   glossopharyngeal, 23, 101, 104
   trigeminal, 23, 28, 47, 67, 99

O
occipital condyle, 64, 97, 100
occipital region, 68, 97, 100
ontogeny, 95, 182
orbitonasal foramen, 174
orbitosphenoid bone, 64, 66–8
orbitotemporal region, 100
ossification, 25, 97
osteocranium, 25–28, 96
otic capsule, 95
ovary, 4–20
oviduct, 4–20

P
palatal complex, 66
palate, 66
palatine bone, 63, 64, 66, 69, 72, 97, 148, 170
palatine cartilages, 97
papilla lagenae, 101
parietal bone, 64, 66, 158, 170
paroccipital process, 26, 28, 55, 64, 67
perilymphatic duct, 56, 98, 100
perilymphatic foramen, 56, 95, 97, 100
perilymphatic space, 100
periotic, 97
petrosal, 21, 26, 28, 47, 56, 64, 67, 98, 101, 148
phylogenetics, 93, 109–111, 146–154, 217–219
   algorithms, 57, 96, 147, 149–50, 153, 163, 219
   character, 109–111
      coding, 149
      complexes, 111
      development, 225–26
      matrix, 159, 214
      reversal, 110, 152, 160–63
      transformation, 109–11, 217–219
   completeness, 146, 149–52, 159
   consensus tree, 149–52
   convergence (homoplasy), 109–11, 130, 151, 152, 155, 160–3
   functional-adaptive analysis, 109–111
   incompleteness, 146, 149–52
   missing data, 146, 148–52
   monophyletic, 95, 105, 217–219
      holophyletic, 217–219
      paraphyletic, 147, 153–4, 217–219
   null group comparison, 109
   operational taxonomic unit (OTU) 148–50, 157
   parallelism, 104, 109–111
   parsimony, 146–163
   resolution, 146, 149–52
   retention index, 149–50, 52
   taxon, 146–163, 217–219
   taxogram, 218
   taxonomic properties, 109–111
pila antoptica, 53
pisiform, 227
postdentary elements, 56, 96, 102, 174
postcranial morphology, 122–125, 226–27
posterior temporal canal, 24
pouch, 4–17
prehallux, 226–27
premaxilla, 63, 64, 68–9, 72, 95, 101, 158, 160–63
prepollex, 227
presphenoid, 66
prootic canal, 27, 54, 67, 68, 148
prostaglandin, 14
proximal tibiofibular joint, 231

Q
quadrate, 51, 96, 98, 102, 104, 167

R
recessus scalae tympani, 95, 100, 103
Reichert's cartilage, 22, 96
Reichert-Gaupp theory, 96
reproduction, 4–20, 195
resolution, phylogenetic, 146, 149–52

S
scaphoid, 227
secondary sidewall, 105, 172
secondary tympanic membrane, 95, 100
septomaxilla, 63, 69, 72
sex chromosomes, 7
side wall of braincase, 98, 105, 171
skull, 64, 95
sound, 95
sphenoobturate membrane, 171
sphenoid complex, 66
squamosal, 56, 64, 66, 68, 96, 101
squamosal-dentary jaw joint, 57, 95, 99, 101
stapes, 27, 53, 67, 69, 95, 102
sucking, 195
sulcus medialis, 159–63

T
tarsus (bones of), 111–122
taxonomy, *see* phylogenetics, taxon
tegmen tympani, 21, 26–27
testes, 11–13
tibia, 119
tooth, *see* dentition
transverse tarsal joint, 124
trapezium, 227
trapezoid, 227
trigeminal ganglion, 23, 28, 47, 147
triquetrum (cuneiform), 227
trophoblast, 6–7, 11
tympanic (angulare), 95, 102, 175
   cavity, 21–28, 95, 99
   membrane, 22, 95, 97

process, 97
region, 95

U
unciform (hamate), 227
upper ankle joint (cruropedal joint), 123
ureters, 5–10
urethra, 5–10
urogenital sinus, 5–10
uterus, 5–10

V
vagina, 4, 6, 7–11
veins of the tympanic region, 21–28
vomer, 66, 100
vomeronasal (Jacobson's) organ, 176

W
Wolffian ducts, 4–7

X
X-inactivation, 10–12

Z
zygomatic arch, 158, 160–3

Taxon Index

Aegialodon, 206
Albertatherium, 212
Allodon, 153
Allodontidae, 152–3
Allqokirus, 220
Alphadon, 212, 220–223, 235
Alphadontinae, 233
Ameridelphia, 233–4
Amphisbaenia, 100
Anchistodelphys, 211
Anconodon, 153–4, 156
Antechinus, 12, 193, 196
Archimetatheria, 233–4
Arctocyonidae, 119
Arginbaatar, 149–150, 152–3, 155, 157, 159–160, 162
Arginbaataridae, 153
Argyrolagidae, 236
Argyrolagus, 236, 238
Artiodactyla, 184
Asioryctes, 141, 197
Australidelphia, 234, 237

B
Baiotomeus, 150–159
Bienotherium, 212
Bistius, 37
Bobbschaefferia, 235
Boffius, 146–62
Boffiidae, 154
Bolodon, 150–62
Bolodontidae, 152–3, 156, 162
Bonapartheriidae, 236
Borhyaenidae, 236
Buginbaatar, 148–59
Bulganbaatar, 150–161
Burramys, 228, 231

C
Caenolestes, 230
Caenolestidae, 236
Caluromyinae, 221
Caluromys, 230–1
Canis, 196
Captorhinus, 167
Carnivora, 184
Carolameghiniidae, 236

Catopsalis, 53, 150–62
Cavia, 196
Chulsanbaatar, 48, 150–61
Cimexomys, 147–61
Cimolestes, 197
Cimolodon, 150–61
Cimolodonta, 146–8, 150–3, 156, 160, 162
Cimolomys, 150–62
Cimolomyidae, 146–56
Condylarthra, 119
Ctenacodon, 64, 149–62
Cynodont, 26–28, 104, 113, 130, 133, 142
Cynognathus, 104, 133–4, 142

D
Dasycercus, 228
Dasyuridae, 238
Dasyuromorphia, 234, 237
Dasyurus, 190, 192
Deltatheridium, 209
Deltatheroida, 229, 233–4, 240
Deltatheridiidae, 233
Deltatheroididae, 233
Dermoptera, 184
Diademodon, 104, 133–4, 142
Diarthrognathus, 30–42
Didelphida, 233
Didelphidae, 99, 220, 236
Didelphimorphia, 233–4, 236
Didelphinae, 223
Didelphis, 48, 51, 102, 173, 191, 196, 224
Dinnetherium, 30–42, 141
Diprotodontia, 237, 239
Diprotodontiformes, 240
Diprotodontoidea, 224
Djadochtatherium, 150–61
Docodonta, 141
Dromiciops, 221–224, 228, 231–2, 237
Dryolestidae, 103

E
Echidna, see *Tachyglossus*
Echymipera, 228, 233
Ectypodus, 150–60

Elephantulus, 185–6
Eobaatar, 146–62
Eobaataridae, 153
Erinaceus, 186
Essonodon, 150–59
Eucosmodon, 150–61
Eucosmodontidae, 147–154
Eucynodontia, 133
Eupantotheria, 95, 105
Eutheria, 4–20, 60, 70, 122, 141, 148, 185, 197, 209
Exaeretodon, 31, 70, 134, 135, 142, 148

F
Falepetrus, 209

G
Gashterniidae, 236
Glasbiinae, 233, 235
Glirimetatheria, 233, 236
Gobiconodon, 115, 140, 148, 200
Gomphodontidae, 133–4
Gondwanadelphia, 231
Groeberiidae, 236
Groeberia, 236
Guimarotodon, 150–59
Gypsonictops, 197–8

H
Hainina, 150–60
Haldanodon, 58, 68
Haplorhini, 184
Haramiyidae, 141–2, 147–8
Haramyoidea, 147
Helvetiodon, 41
Henckelotherium, 141–2
Herpetotherium, 221–223
Holarctidelphia, 229, 233
Holoclemensia, 209
Hypsiprymnodon, 228
Hyracoidea, 184

I
Insectivora, 99
Iqualadelphis, 211–212, 221–3

Itaboraiformes, 236
Iugomortiferum, 209, 211

J
Jaskhadelphys, 220–223

K
Kamptobaatar, 150–61
Kangaroo, 234
Kayentatherium, 37–8, 137
Kennalestes, 197–8
Kermackia, 207
Kielanodon, 150–59
Kielantherium, 206
Kryptobaatar, 150–61
Kuehneodon, 150–60
Kuehneodontinae, 150–62
Kuehneotherium, 36, 41, 72, 141–2

L
Lacertidae, 105
Lagomorpha, 188
Lambdopsalis, 64–68, 104, 150–62
Liotomus, 146–161
Lutreolena, 223–224

M
Macropodidae, 234
Macropodoidea, 234
Macropus, 190
Macroscelididae, 27, 184
Mammaliaformes, 70, 137–8
Mammaliamorpha, 70, 109, 136, 142
Manda cynodont, 113
Marmosa, 224
Marsupialia, see Metatheria
Marsupicarnivora, 238
Martes, 174
Massetognathus, 30, 141
Megachiroptera, 187
Megazosotron, 30–42
Meniscoessus, 64, 150–62
Mesodma, 150–9
Metatheria, 4–20, 27–28, 60, 70, 122, 148, 189, 206, 211, 216–241
Microbiotheria, 220–224, 237
Microbiotheriidae, 223
Microbiotherium, 221–3
Microchiroptera, 187
Microcosmodon, 150–61
Micropotamogale, 103
Mimetodon, 150–61
Monobaatar 146–59
Monodelphis, 12, 14, 102, 170, 194, 220, 225
Monodelphopsis, 221–223

Monotremata, 5–11, 27–8, 59, 70, 95, 137–41, 148
Morganucodon, 30–42, 48, 51–2, 98, 102, 104, 148, 174
Morganucodontidae, 28, 58, 70, 114–6, 135–40
Multituberculata, 28, 59, 63–74, 95, 104, 119–122, 146–64
Myotis, 186–7
Myrmecobius, 195
Mysticeti, 188

N
Nandinia, 174
Neophascogale, 231–2
Nemegtbaatar, 65, 148–61
Neoliotomus 150–9
Neoplagiaulax, 150–61
Neoplagiaulacidae, 154
Notoryctes, 227–9
Notoryctidae, 224, 234, 238

O
Oligokyphus, 38, 113, 116, 137
Ornithorhynchidae, 115–9, 175
Ornithorhynchus, 9, 48, 51, 53, 95, 115–19

P
Pachygenelus, 30–42, 48, 51
Pantotheria, 95, 103
Pappotheriidae, 209
Pappotherium, 209
Paracimexomys, 147–59
Parectypodus, 150–59
Pariadens, 141, 206
Patagoniidae, 236
Paucituberculata, 234, 236
Paulchoffatia, 65–6, 68, 147–8, 150–62
Paulchoffatiidae, 147–48, 153
Paulchoffatiinae, 153
Pediomyidae, 212, 220, 233, 236
Pediomyinae, 223, 234
Pediomys, 212, 220–223
Pentacosmodon, 150, 154–61
Peradectes, 235
Peradectidae, 220
Peradectinae, 223, 235
Perameles, 190–192, 194
Peramelidae, 234, 238
Peramelina, 237–8
Peratherium, 221–223
Petauroidea, 234
Phalangeroidea, 234
Phalangeriformes, 237, 239
Phascolarctus, 224

Philander, 228
Placentalia, see Eutheria
Plagiaulacoidea, 146, 152–3, 155–7
Plagiaulax, 150–62
Plagiaulacida, 147
Plagiaulacidae, 147–8, 153
Platypus, see *Ornithorhynchus*
Polydolopidae, 236
Polydolopimorphia, 236
Potamotelses, 207, 209
Prepidolopidae, 236
Primates, 98
Probainognathus, 30–32, 40, 134
Proboscidea, 98
Procavia, 186
Prochetodon, 150–61
Prokennalestes, 198, 233
Prosynosuchus, 133
Protalphadon, 220–23
Protungulatum, 120
Psalodon, 150–62
Pseudobolodon, 66, 150–62
Pteropus, 10, 187
Ptilodontidae, 147, 150, 154, 161
Ptilodontoidea, 146–60
Ptilodus, 65–68, 149–61

R
Rhinolophus, 187
Rodentia, 98, 188
Rousettus, 187

S
Sarcophilus, 226
Scandentia, 184
Sciuridae, 184
Simpsonitheria, 234, 236
Sinoconodon, 30–42, 58, 68, 148
Sirenia, 198
Slaughteria, 209
Sloanbaatar, 150–61
Sloanbaataridae, 154
Sminthopsis, 191, 193, 224
Sorex, 175, 186, 196
Sparassocynidae, 236
Sparassodonta, 236
Stagodontidae, 212, 233
Steropodon, 126
Stygimys, 150–61
Sudameridelphia, 233–4, 236
Sulestes, 209
Synapsida, 95
Syndactyla, 237–8

T
Tachyglossus, 5–11, 95, 115–9
Taeniolabidoidea, 146–8, 152–7, 160

Taxon Index

Taeniolabididae, 147–8, 150, 154, 156, 160
Taeniolabis, 67, 150–162
Talpa, 189
Tapirus, 186
Tarsius, 187–189
Theria, 28, 71, 95, 122, 137–141
Theriiformes, 123, 70–1, 140
Theriimorpha, 139–141
Thrinaxodon, 30–2, 37–8, 133
Thylacoleonidae, 222, 234
Thylacosmilidae, 236
Thylogale, 178
Tribosphenida, 198, 206
Tribotheria, 108–9, 234, 206
Trichosurus, 192, 224
Triconodon, 69
Triconodonta, 95, 105, 140–2
Triconodontidae, 58, 115, 140
Trioracodon, 40, 53, 140–2
Tritheledontidae, 30–32, 36, 41–42, 57, 135–8, 142, 148
Tritylodon, 30–42
Tritylodontidae, 30–42, 57, 113–4, 116, 134–8, 148
Tugrigbaatar, 150–61
Tupaia, 98, 100, 102, 186

V
Vincelestes, 48, 59, 69, 105, 141, 148, 231
Vombatiformes, 234, 237, 239–40
Vombatimorphia, 240
Vombatus, 224

W
Wakeleo, 225–6
Wallaby, 9–11
Wynyardiidae, 234

X
Xenarthra, 184

Y
Yalkaparidontia, 222, 234, 237, 239
Yunnanodon, 31, 36
"*Yunnania*," 36

Z
Zygiocuspis, 211

DATE DUE

1-9-97			
MAY 0 1 1999			
MAY 0 8 2000			
JUL 0 6 2004			